Optical Systems Design Detection Essentials

Radiometry, photometry, colorimetry, noise, and measurements

IOP Series in Emerging Technologies in Optics and Photonics

Series Editor

R Barry Johnson a Senior Research Professor at Alabama A&M University, has been involved for over 50 years in lens design, optical systems design, electro-optical systems engineering, and photonics. He has been a faculty member at three academic institutions engaged in optics education and research, employed by a number of companies, and provided consulting services.

Dr Johnson is an IOP Fellow, SPIE Fellow and Life Member, OSA Fellow, and was the 1987 President of SPIE. He serves on the editorial board of Infrared Physics & Technology and Advances in Optical Technologies. Dr Johnson has been awarded many patents, has published numerous papers and several books and book chapters, and was awarded the 2012 OSA/SPIE Joseph W Goodman Book Writing Award for Lens Design Fundamentals, Second Edition. He is a perennial co-chair of the annual SPIE Current Developments in Lens Design and Optical Engineering Conference.

Foreword

Until the 1960s, the field of optics was primarily concentrated in the classical areas of photography, cameras, binoculars, telescopes, spectrometers, colorimeters, radiometers, etc. In the late 1960s, optics began to blossom with the advent of new types of infrared detectors, liquid crystal displays (LCD), light emitting diodes (LED), charge coupled devices (CCD), lasers, holography, fiber optics, new optical materials, advances in optical and mechanical fabrication, new optical design programs, and many more technologies. With the development of the LED, LCD, CCD and other electo-optical devices, the term 'photonics' came into vogue in the 1980s to describe the science of using light in development of new technologies and the performance of a myriad of applications. Today, optics and photonics are truly pervasive throughout society and new technologies are continuing to emerge. The objective of this series is to provide students, researchers, and those who enjoy self-teaching with a wide-ranging collection of books that each focus on a relevant topic in technologies and application of optics and photonics. These books will provide knowledge to prepare the reader to be better able to participate in these exciting areas now and in the future. The title of this series is Emerging Technologies in Optics and Photonics where 'emerging' is taken to mean 'coming into existence,' 'coming into maturity,' and 'coming into prominence.' IOP Publishing and I hope that you find this Series of significant value to you and your career.

Optical Systems Design Detection Essentials

Radiometry, photometry, colorimetry, noise, and measurements

Robert M Bunch

Department of Physics and Optical Engineering, Rose-Hulman Institute of Technology, Terre Haute, IN, USA

IOP Publishing, Bristol, UK

ISBN   978-0-7503-2252-2 (ebook)
ISBN   978-0-7503-2250-8 (print)
ISBN   978-0-7503-2253-9 (myPrint)
ISBN   978-0-7503-2251-5 (mobi)

DOI   10.1088/978-0-7503-2252-2

Version: 20210601

IOP ebooks

British Library Cataloguing-in-Publication Data: A catalogue record for this book is available from the British Library.

Published by IOP Publishing, wholly owned by The Institute of Physics, London

IOP Publishing, Temple Circus, Temple Way, Bristol, BS1 6HG, UK

US Office: IOP Publishing, Inc., 190 North Independence Mall West, Suite 601, Philadelphia, PA 19106, USA

This book is dedicated to my wife Anne Lawson Bunch for her constant encouragement and support.

Contents

3   Wave optics and light propagation   3-1

Preface

We encounter optical systems everywhere. Typical examples are imaging systems like telescopes, microscopes, and cameras. In a camera, light from a source is reflected by an object, the camera lens collects this light placing the image at a detector plane, and the detector provides an image for the user. Sometimes this is called the image model. But even optical systems that are not based on imaging follow the same approach. In a fiber optic communication system, the source is a modulated laser or LED carrying the encoded time signal, the system contains connectors, fiber, splices, and multiplexers that can degrade the signal, and an optical detector in the receiver decodes the signal.

This textbook covers essential optical science topics but in the context of defining an optical system specification or requirement. Chapter 1 begins with an introduction to systems and a review of some mathematical functions and operations important to optical systems. The optical science of rays, waves, and photons is required to describe optical radiation inputs and propagation within systems. These concepts are discussed in chapters 2–4. Optical system parameters such as f-number and field of view are defined providing a measurement of system requirements. The interaction of electromagnetic waves at surfaces and within material media is described along with models used to obtain the optical properties of materials. The concept of the photon is introduced in chapter 4 and used to describe the operation of thermal sources and photonic devices such as light emitting diodes and lasers. Chapter 5 builds on the systems approach to electromagnetic wave propagation using this formalism to understand diffraction phenomena. Diffraction within an optical system is quantified using a Fourier optics approach to obtain system functions such as the modulation transfer function which predicts the performance limitations of systems. Components of a system modify the amount of radiation transferred as well as the spectral content of the radiation. The topic of radiometry is discussed in chapter 6 to define the fundamental quantities used to specify the energy characteristics of sources and processes of detecting electromagnetic radiation. When a human observer is involved in the system photometry and color are important concepts that describe the outputs from these systems and are also described. Since systems transfer information, this information must be detected. Chapter 7 covers the basic properties of detection along with a description of detection mechanisms and specific detectors for various spectral regions. In addition, noise mechanisms are described and used to define figures of merit, like signal-to-noise ratio, that are used as a system specification. Finally, chapter 8 discusses some common layouts for forming illuminating beams and the interaction of these beams with scanners and modulators. Application examples are included in most chapters to illustrate one or more practical concepts of a system, device, or experimental measurement. Each chapter contains end-of-chapter problems.

Acknowledgements

As I began my career, I was fortunate to have two role models that introduced me to optics, Dr Jeffrey Davis, and Dr Rajpal Sirohi. I owe a debt of gratitude to Dr Jerry Wagner and Professor Paul Mason who taught a young professor what it meant to be a good teacher. To my friends and colleagues, Dr Sergio Granieri, Dr Charles Joenathan, and Dr Paul Leisher, I thank you for our many thoughtful discussions and a shared passion for teaching optical engineering. Last but not least, I thank the many students that I have taught over the years. Seeing that spark of understanding on your faces in the classroom, laboratory, during office hours, and talks in the hallway always confirmed my decision to become a teacher.

Author biography

Robert M Bunch

Robert M Bunch is a Professor Emeritus of Physics and Optical Engineering at Rose–Hulman Institute of Technology. He joined the faculty at Rose–Hulman in 1983 and was one of the founding faculty members of the BS and MS Optical Engineering degree programs where he developed and taught undergraduate courses, graduate courses, and laboratories. From 2000 until his retirement in 2020 he also served as an Innovation Fellow at Rose–Hulman Ventures working on design and development projects with industrial clients in the areas of optics-based products, sensors, fiber optics, optical instruments, and lighting. He continues to consult with industry and is co-inventor on two patents. In 2000, he received the Rose–Hulman Board of Trustees Outstanding Scholar Award.

Optical Systems Design Detection Essentials

Radiometry, photometry, colorimetry, noise, and measurements

Robert M Bunch

Chapter 1

Introduction to optical systems

1.1 What is a system?

All engineers and scientists must understand the behavior of systems. Systems are a collection of components, physical parts, or logical steps that when linked together result in a desired output that cannot be achieved by one component alone.

A high-level approach to understanding systems is to think of the system as a 'black-box' object that is immersed in some environment in which the system interacts. This approach is commonly called a top-down view. In many instances, we are interested in how an input from the environment to the system is accepted, processed, and modified by the system, which then produces an output back to the environment. Figure 1.1 shows the simple model diagram used to represent this concept.

Blanchard and Fabrycky [1] list the elements that comprise a system as (1) components, (2) attributes, and (3) relationships. The components consist of the input, output, and process that occur within the system. Attributes characterize the desired properties of the system resulting in meeting system requirements. Relationships are the connecting links or interfaces between the components and system attributes.

Of course, this top-down view hides much of the complexity of the system. Every component and system can be broken down into other components or subsystems forming a series of hierarchical levels. A system at one level can be a component at another higher level.

1.2 Technical systems design

Hubka and Eder coined the term technical system to describe the development of products and processes made by humans [2]. The work of engineers is to design and construct a technical product, system, component, or process that meets a stakeholder need based on a set of requirements. In this context, stakeholders are individuals or groups that can affect or be affected by the product or process.

doi:10.1088/978-0-7503-2252-2ch1

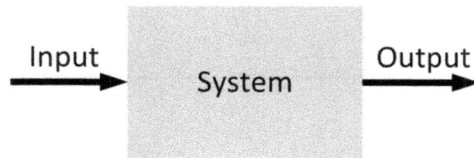

Figure 1.1. Block diagram representation of a general system.

In addition, in order to meet these requirements, the designers must also consider various constraints. It is important to note that the engineering design process is not random but is a logical sequence of steps with the result of arriving at the best solution to a specific problem. A technical system design is a success when it brings value to stakeholders.

There are numerous methodologies and approaches used to design and develop products from a systems perspective. Meeting stakeholder needs implies that a product development process considers the entire lifetime of the product, life-cycle engineering [1]. The product life cycle is divided into an acquisition phase and a utilization phase. Conceptual design, preliminary design, and detail design and development are part of the acquisition phase. The National Society of Professional Engineers in 'Engineering Stages of New Product Development' [3] identified six stages (1) Conceptual, (2) Technical Feasibility, (3) Development, (4) Commercial validation and production preparation, (5) Full scale production, and (6) Product support. The engineering activities performed in each stage result in information that is used in the next stage to continue development. Another approach is from Ulrich and Eppinger [4] who also divide the product development process into six phases (1) Planning, (2) Concept development, (3) System-level design, (4) Detail design, (5) Testing and refinement, and (6) Production ramp-up. The names of each stage or phase are not as important as the activities performed at each step of the process. The engineering design process and the results of a design influence every stage or phase.

Rarely are the stages of product development completed in a strictly step-by-step manner where the result of one stage is provided as information to the next stage. This simple approach is often called serial or sequential engineering. Instead, concurrent engineering, simultaneous engineering, or parallel engineering is most often used where a cross-functional multidisciplinary team considers all stakeholder/ customer needs. Concurrent engineering has been shown to decrease design and development time thus reduce costs of product development [5].

Any design begins with a set of stakeholder/customer needs that the system must meet. The role of the design engineer is to translate these needs into a list of metrics with a corresponding physical value that specifies what the system must do. It is then up to the designer to determine how the system addresses the needs using these metrics [4]. The list of metrics and corresponding values are commonly called product specifications and are the items listed in the specification sheets of most products.

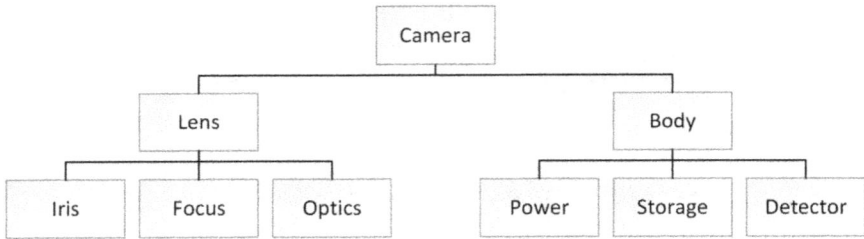

Figure 1.2. Hierarchical tree diagram of a digital camera.

Decomposition is one of the most common techniques used in generating and selecting concepts during the design of technical systems [6, 7]. It is a top-down systems technique. The designer defines the input, output, and behavior of the system and identifies top-level functions/attributes. This process continues at successively lower levels until all subsystems/components have been identified. Since optical systems are comprised of passive and active optical components, such as sources of light, lenses, mirrors, and detectors, they can often be well described using a functional decomposition technique. This technique also illustrates how the optical elements relate to other aspects of the entire system.

Two different approaches are used in system decomposition to begin a design, design in the physical domain or design in the functional domain. Both approaches provide the designer with a roadmap for developing a product. Design in the physical domain often begins with a physical decomposition. The physical decomposition separates various physical subsystems into smaller hierarchical elements. This approach can help a designer organize components and the interactions between components. Functional decomposition separates the system into a set of functional subsystems that follow an input state to a desired output state.

A hierarchical tree diagram is often used to represent physical domain decomposition. An example of a digital camera is shown in figure 1.2. In this case, the lens and the body are considered separate components of the system and different components are allocated as shown.

As an example of functional decomposition, figure 1.3 shows one possible representation of a high-level design diagram for a camera. The system requirement is to collect a light field from the input object scene and provide an output of the image in some form. Obviously, there are many ways to accomplish the desired task and that is the freedom given to the designer based on the specified requirements. This diagram specifies an iris to control the amount of light and some type of focus mechanism to control the image quality as subsystems in the 'modify light' function. Also, for clarity no connections are made between any function and the power source.

Inputs in traditional functional decomposition are information, energy, and material. These inputs are often referred to as the 'flows' [4] because they connect to lower-level subsystems. The advantage of functional decomposition is that it allows for organization and allocation of subsystem components into different physical arrangements. As seen in this example, the designer might choose to have

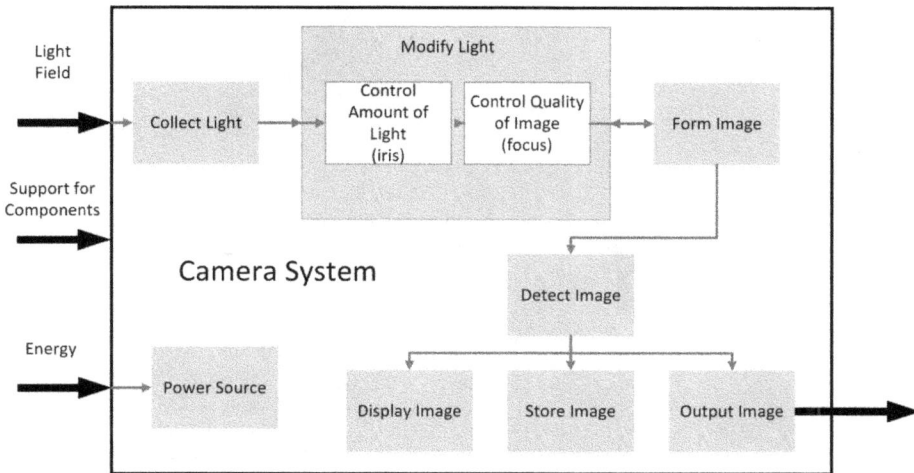

Figure 1.3. Functional decomposition of a camera, sample representation.

the lens allocated as a separate component or the lens could be integrated into the body of the camera.

1.3 Optical systems design

First a note about the use of the term light in optical systems. Historically, the term light has been used to describe only the radiation in the visible portion of the electromagnetic spectrum. However, many optical systems operate in a wide range of the spectrum from the ultraviolet to the far-infrared. The term light has become a common description used to represent the energy flow of electromagnetic radiation or photons in optical science, optical engineering, and photonics. While we know that the usage of the term 'light' in these spectral regions is not exactly correct, it is common terminology.

Even though systems engineering is often thought to be useful only for very complicated systems it is also helpful for simpler systems. All designs should consider issues such as stakeholder/customer needs, costs, and interactions between parts of the entire system. In particular, optical system designs often fit into a larger system as a subsystem component. Following system engineering approaches in designing optical systems will assist in seamlessly incorporating the optical system into a larger system.

The simple block diagram of a general system in figure 1.1 can be used to model many optical systems [8, 9]. A simple single lens imaging system is a good example. The input to the system is the object, the lens along with the object and image distances is the system, and the output is the image. This concept is illustrated in figure 1.4 for a visible imaging system taking a three-dimensional scene and forming an image in the image plane. Note that three-dimensional inputs are transformed by the system to display an output in two dimensions. The light falling on the image plane is the result of the illumination and reflectance of the object at every point.

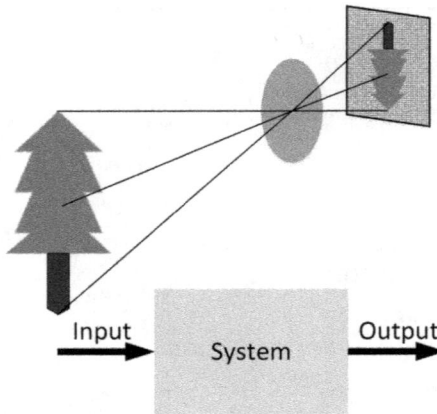

Figure 1.4. A simple single lens image formation of an object along with its systems model.

In imaging systems this is sometimes called the image formation model and used extensively in image processing applications [10]. The image formed in the output plane is an analog light field until it is transformed into electronic signals as in a digital display representation.

Another model of a general optical system is to assign the input as a source of light, the system as the surfaces and medium in which the light propagates, and the output as the sensor collecting the light [11]. This would be the case for a light projection system or a non-imaging optics illumination system. As we shall see, taking a systems view of optics rather than simply discussing various optical phenomena provides insight into the design of any system employing optical devices and components.

An alternative way to think of the input to a general optical system is as a source of information. This input can then take on several different physical forms. It could simply be a traditional constant light field like the imaging system example above. However, the input could also be an optical source that is modulated in time such as a laser diode in a fiber optic communication system. The input may be a pulsed source from a laser. These inputs would be characterized in a temporal domain. Another physical approach considers the input object plane as an electromagnetic field function in space. This is similar to a common eye chart used in testing visual acuity. This type of input object would be a spatial domain function. The system models and mathematics discussed above can be used to parameterize any of these sources of input information. Regardless of the type of input, the systems approach provides a framework for analyzing the performance of the system.

The systems view of optical design described here is not just lens design. Hobbs [12] states this well, 'Optical design (in the restricted sense used here) is concerned with sticking lenses and other parts together to get the desired result'. In fact, the role of optics in most contemporary systems is said to be an enabling technology [13]. This concept is well described in the book *Harnessing Light: Optical Science and Engineering for the 21st Century*, 'Although optics is pervasive in modern life, its role is that of a technological enabler: It is essential, but typically plays a supporting role in a larger system'. The authors go on to state, 'The remarkable breadth of optics' enabling role is both an indicator of the field's importance and a source of challenges.' The challenges are created because so many disciplines need to make use of elements of optical science and engineering.

All optical engineers and designers are faced with a need to understand not just the optical operation of the system, but aspects of the entire system including electrical/electronic interfaces, mechanical considerations, thermal issues, and software [11–14]. Thus, successful optical systems or optical subsystems must be designed as part of an established engineering design process [6, 13, 15].

Based on the discussion above designing optical systems requires knowledge about both the functional aspects of optics (science of light and light propagation) as well as the physical aspects (choice of specific passive and active components). Both are essential to a successful optical design. We build optical systems in the physical domain, but we understand their detailed operation in the functional domain.

Generally, the basic high-level requirements for an optical system are to collect a sufficient amount of radiant energy as the input from a finite scene and provide an output signal in some form to a user. The optical design engineer must decompose the system requirements into subsystems and allocate physical components to meet system requirements. Alternatives for components can be investigated. Each component or building block comes with its own set of specifications and limitations that must be factored into the overall design. This allows tradeoffs to be made between components that lead to conflicting requirements.

In the traditional study of optical science, light has been classified according to its properties and interaction with materials, what we described above as the functional approach. This classification leads to several models or pictures of light based on experimental observations and the size of the interaction of light in materials. At the smallest level is the particle picture of light in the form of a photon or bundle of energy. The energy arises at the atomic level when an electron makes a transition from a higher quantized energy state to a lower state releasing energy equivalent to the photon energy. The next picture of light is that of a wave. This picture follows directly from electromagnetic theory described by Maxwell's equations where the wave is composed of an oscillating electric and magnetic field. Some optical phenomena such as interference and diffraction can be treated using simple scalar fields instead of vector fields with interactions on the order of the wavelength of light. The simplest theory used to model light propagation is the ray. The ray theory follows from the laws of geometrical optics and is valid when sizes of interactions are much larger than the wavelength of light. In practice, one uses the picture of light and light propagation that best suits the size and interaction involved in the phenomena of interest.

This text takes a systems level approach to any product/process design that incorporates optics, optical components, and photonic devices as building blocks. Basic optics principles are discussed with emphasis placed on defining optical system-level parameters and functions that are necessary in characterizing the system to meet requirements.

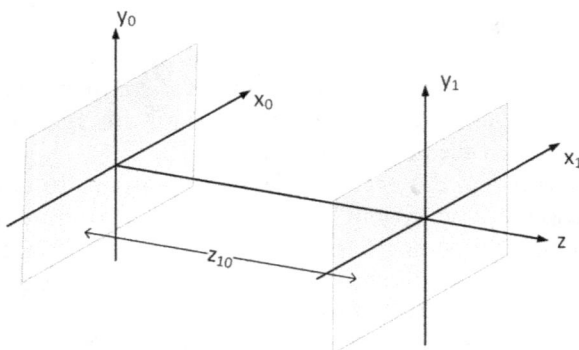

Figure 1.5. Plane-to-plane geometry and notation used for optical system designs.

1.4 Modeling input and output system functions

To apply a systems approach requires that we model physical input and output quantities using mathematical functions. There are numerous special functions and mathematical methods that can be used to arrive at an appropriate functional representation for system inputs, especially for optical systems [8, 10, 16]. Here we review models for representing both continuous functions and discrete/sampled functions along with notation used in describing these functions.

1.4.1 Coordinate geometry and notation

Optical systems transform a light field in one plane to the field in another plane. To describe this process, we need a measure of both position in a plane and the amount of light at that position.

Keeping track of spaces and domains requires a defined set of notation. If only a single dimension is needed then we will use a single coordinate, generally either x or y. If only a single two-dimensional plane is being used then we use general (x,y) coordinates. When multiple planes are necessary then the geometry and notation used is shown in figure 1.5 where the input plane has coordinates (x_0,y_0) and the next successive plane has coordinates (x_1,y_1). The z-axis is a common line of symmetry in most optical systems called the optical axis. The z-axis connects vertices of each optical surface that coincides with the origin of each successive plane. The distance between the two planes is defined as z_{10}, a positive distance from Plane 0 to Plane 1.

The physical quantity in a particular plane is represented by a function of the appropriate coordinate. For example, if Plane 0 was an input plane to an optical system, the input function in Plane 0 would be specified as $f(x_0,y_0)$. In this case the function might represent the value of the electric field signal at points in the plane or the amount of light at a point.

1.4.2 Light measurements

In applying a systems approach to designing optical systems we follow the flow of energy from plane-to-plane in a three-dimensional coordinate. In geometrical optics, we follow rays from the input plane through the system to an output plane. With wave optics, a light field in the input plane propagates through the system and a modified field is observed in the output plane. In all the chapters that follow we will use this geometry to follow the flow of energy whether it be a ray, wave, or photon.

Recently a major revision was made in defining SI units related to a set of seven defining fundamental constants of nature and technical constants [17]. This change has impacted the way measurement of light is specified. All other units and dimensions may be derived from these fixed values. These defining values and units are provided in table 1.1. By definition, there is no uncertainty associated with any of these values.

Along with these fundamental constants are the SI base units. See table 1.2. Methods are provided in obtaining the typical base unit values from the fundamental constants [17].

Physical measurements of light related to a flow of electromagnetic radiation, radiant energy, are adequately addressed using base units. However, a specific set of definitions, terms, and nomenclature have been developed over the years related to physical quantities needed in measuring light energy [17–19]. Measuring radiant energy and power has evolved into a field of its own called radiometry [14, 18–24].

The addition of the unit of the candela and the definition of the luminous efficacy constant has slightly altered the way in which physical radiometric measures are related to luminous measurements. Luminous measurements are based on the intensity of a source as perceived by a human observer, the so-called standard observer. When radiant power, which is a physical measure from a source in Watts, is perceived by a human observer, we now have a precise way of computing the perceived visual flux measured in units of lumens through the luminous efficacy. All visual response quantities fall under the subject of photometry.

Further details about the subjects of radiometry, photometry, and detection of light will be described in later chapters. However, it is necessary to introduce some of these terms in order to discuss the quantities related to functional descriptions in optical systems. Table 1.3 lists selected radiometric quantities, typical symbols, and units [18–24], where t is time, A is the surface area receiving radiant flux, ω is the solid angle in the given direction at a point on the surface of a source, and θ is the

Table 1.1. Values and units of the seven defining SI constants.

Hyperfine transition frequency of Cs	$\Delta\nu_{Cs}$	9192 631 770	Hz
Speed of light in vacuum	c	299 792 458	m s^{-1}
Planck constant	h	$6.626\ 070\ 15 \times 10^{-34}$	J s
Elementary charge	e	$1.602\ 176\ 634 \times 10^{-19}$	C
Boltzmann constant	k	$1.380\ 649 \times 10^{-23}$	J K^{-1}
Avogadro constant	N_A	$6.022\ 140\ 76 \times 10^{23}$	mol^{-1}
Luminous efficacy	K_{cd}	683	Lm W^{-1}

Table 1.2. Base units in the SI system.

Time	second	s
Length	meter	m
Mass	kilogram	kg
Electric current	ampere	A
Thermodynamic temperature	kelvin	K
Amount of substance	mole	mol
Luminous intensity	candela	cd

Table 1.3. Selected radiometric quantities, symbols, and typical units of measure.

Quantity	Symbol	Unit (MKS)
Radiant energy	Q_e	Joule
Radiant energy density	w_e	Joule m^{-3}
Radiant flux (radiant power)	$\Phi_e = \frac{dQ_e}{dt}$	Watt
Irradiance (radiant flux areal density)	$E_e = \frac{d\Phi_e}{dA}$	Watt m^{-2}
Radiant intensity	$I_e = \frac{d\Phi_e}{d\omega}$	Watt sr^{-1}

angle between the surface normal and the direction of the solid angle. The subscripts e explicitly show that the quantities are related to energy units.

Radiant flux/power and radiant intensity are terms related to the source. Irradiance describes the amount of radiant flux falling on a surface or detector.

1.4.3 Fourier series and periodic functions

If a complicated input signal is periodic in time or space a Fourier series representation is often used to characterize the function [12, 16]. The assumption is that a general continuous but periodic function of x in space (spatial period of length L or angular frequency $2\pi/L$) can be represented as a superposition of many specific frequency components. A function is considered periodic if $f(x) = f(x + L)$. Using a cosine and sinusoidal basis functions, the Fourier series representation is defined as,

$$f(x) = \frac{a_0}{2} + \sum_{m=1}^{\infty} a_m \cos\frac{2\pi}{L}mx + \sum_{m=1}^{\infty} b_m \sin\frac{2\pi}{L}mx$$

where the a_m and b_m coefficients are determined by applying Fourier analysis.

As an example, suppose we are given the simple periodic function in figure 1.6(a). One can see from analysis or inspection that the function is,

$$f(x) = \frac{4}{2} + (1)\cos\left(\frac{2\pi}{L}x\right)$$

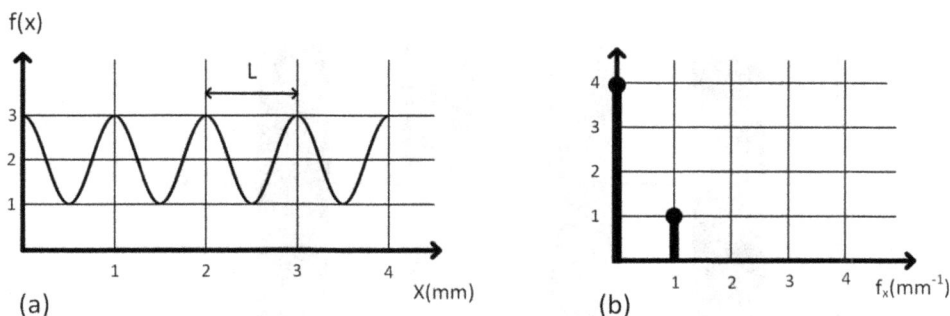

Figure 1.6. Periodic function with period $L = 1$ mm. (a) The input function and (b) the Fourier series coefficients describing the input function in the spatial frequency domain.

where $a_0 = 4$, $a_1 = 1$, $L = 1$ mm, and all other coefficients of the general Fourier series are zero. This same function can also be represented in another way just using the Fourier series coefficients as shown in figure 1.6(b). In this graph the two coefficients a_0, and a_1 are plotted at points on an axis called the spatial frequency, where $f_x = m/L$. In other words, the single values of each coefficient are shown at the appropriate spatial frequency.

As another example, figure 1.7(a) shows an (x,y) coordinate image of a series of black bars where the width of each bar is equal to the separation between bars. In this case the separation is 1.0 mm. Taking a slice across the image along the x-direction yields a signal where 0 represents the black region and 1 represents the white region. This $f(x)$ signal is shown in figure 1.7(b) and is a typical square wave function. Since this signal is periodic in the space coordinate x, $f(x)$ can also be represented mathematically as a Fourier series function. The resulting Fourier series function contains many terms and spatial frequency components. The first spatial frequency (or dominant spatial frequency) for this function is $f_x = 1.0$ line/mm.

To find the Fourier series coefficients, we make use of the orthogonality of the basis functions. Multiplying the general function $f(x)$ by $\cos(n\omega x)$ and integrating over any spatial period we have,

$$\int_{x_0}^{x_0+L} f(x)\cos(n\omega x)dx = \int_{x_0}^{x_0+L} f(x)\cos(2\pi f_x x)dx$$

where n is a general integer that could be different from the mth coefficient, $\omega = 2\pi/L$, $f_x = n/L$, and x_0 is a constant distance. The result of this integration leads to an equation for the a_n, coefficients, as

$$a_n = \frac{2}{L}\int_{x_0}^{x_0+L} f(x)\cos(n\omega x)dx$$

Following a similar process but multiplying by $\sin(n\omega x)$ and integrating over any spatial period leads to an equation for the b_n coefficients as,

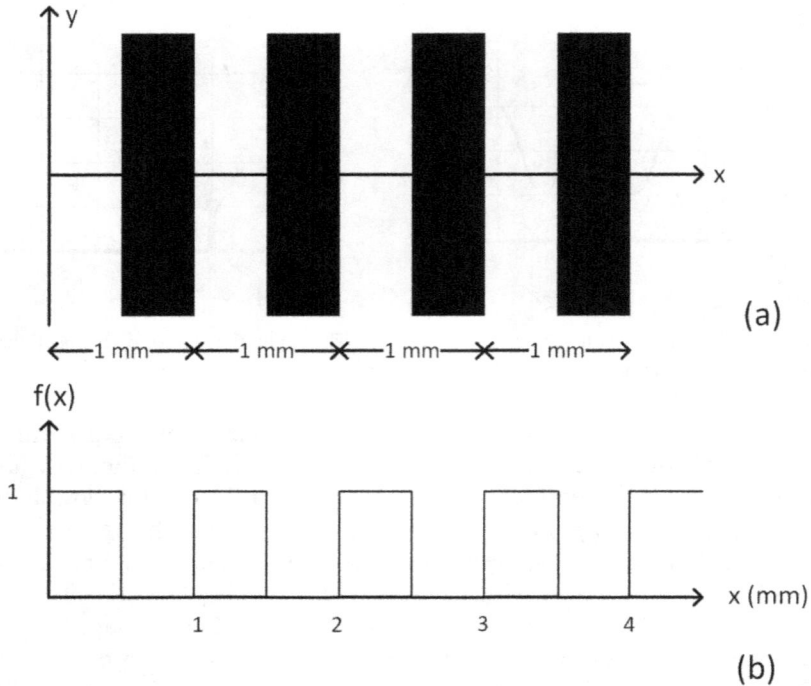

Figure 1.7. (a) Periodic function of black bars and (b) square wave function observed from the values along a row in the x-direction of the same image.

$$b_n = \frac{2}{L} \int_{x_0}^{x_0+L} f(x)\sin(n\omega x)dx$$

The a_0 constant term is obtained by simply integrating over the spatial period to obtain,

$$a_0 = \frac{2}{L} \int_{x_0}^{x_0+L} f(x)dx$$

Therefore, if the function is known over just one period, the coefficients of the general Fourier series can be computed from these integral relations.

Applying these results to the function in figure 1.7, we take $x_0 = 0$ and $L = 1$ mm we find the coefficients to be,

$$a_0 = 1; a_n = 0 \text{ for all } n; b_n = \begin{cases} \frac{2}{n\pi} & n \text{ odd} \\ 0 & n \text{ even} \end{cases}$$

Thus, the resulting function is,

$$f(x) = \frac{1}{2} + \sum_{m=1}^{\infty} \frac{2}{(2m-1)\pi} \sin\left[\frac{2\pi}{L}(2m-1)x\right]$$

Wherein index m of the series is now any positive integer.

An alternate form of the Fourier series expansion using complex exponentials is sometimes useful. Writing the cosine and sine terms using the Euler identity, it is straightforward to show that,

$$f(x) = \sum_{m=-\infty}^{\infty} c_m e^{j\frac{2\pi}{L}mx}$$

where $j = -1$ and c_m are a new set of coefficients related to the a_m and b_m coefficients. Note that the summation of the series index m is from $-\infty$ to $+\infty$ including $m = 0$. Any c_n coefficient can then be found using,

$$c_n = \frac{1}{L} \int_{x_0}^{x_0+L} f(x)e^{-jn\omega x}dx$$

1.4.4 Discrete input and output functions

Many optical systems, especially imaging systems, use array detectors in the image plane as a means of extracting the output function. Input functions can also be specified in a similar manner. In the image plane we obtain a sampled function across a two-dimensional grid or set of pixels. A pixel is not a particular physical entity, but a sample of the continuous image light field (as a two-dimensional function) specified by a coordinate location and a value proportional to the light collected. The size and separation of a pixel will be different depending on the particular device. Mapping the continuous image function into a discrete set of coordinate locations is called sampling. The process of mapping the continuous value of the light field collected at a pixel location into a discrete value is called quantization. Thus, the output function of a system containing an array detector is a sampled and quantized representation of the continuous image formed by the optical system.

Figure 1.8 illustrates the principles of sampling and quantization of a continuous function. For simplicity, the (x,y) coordinate space is sampled into a rectangular array of pixels with M rows and N columns. Typically, at each (M,N) location the function is quantized into a set of discrete values. The quantized values are known as gray-levels when a monochromatic light field is being sampled. If the quantization is a discrete value, L, it is often written as a power of 2 levels. So, for k-levels, the number of quantized gray-levels is,

$$L = 2^k$$

For example, a quantized 8-bit monochromatic image has $L = 2^8 = 256$ levels. By convention, the lowest level, $L_{min} = 0$, is considered black and the highest level,

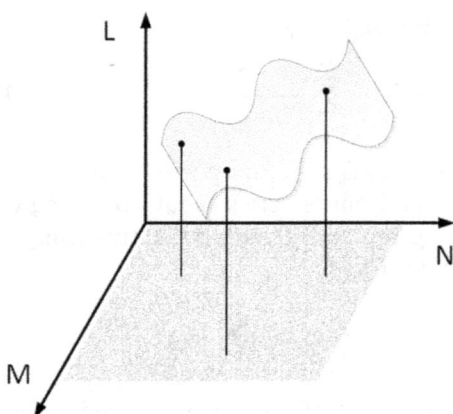

Figure 1.8. Sampling and quantization of an image plane function. Three different sample points are shown with the gray-levels at those locations.

$L_{\mathrm{max}} = 255$, is considered white. Three specific sampled points of the given light field function are shown in figure 1.8 along with their resulting quantized values.

The sampled quantized function of the image is generally represented by an $M \times N$ matrix of integer values. The matrix equation representing this function is,

$$f(M, N) = \begin{bmatrix} f_{0,0} & \cdots & f_{0, N-1} \\ \vdots & \ddots & \vdots \\ f_{M-1,0} & \cdots & f_{M-1, N-1} \end{bmatrix}$$

Typical, notation used for image representation uses an index beginning with zero [10]. Effects of sampling and quantization of electronic/digital images depend on the characteristics of the detector. This will be discussed in a later chapter. For color images, we use three sets of gray-levels for the red, green, and blue combinations that provide a value in color space. More will be discussed about color, color spaces, and color coordinates in a later chapter.

As another example of a function representation in two dimensions, take the photograph of a brick wall, shown in figure 1.9. We need to represent the brick wall object as a mathematical function. In this case, we need a two-dimensional function in both x and y coordinates with the value of the function associated with the amount of light collected at each (x,y) coordinate. As we discussed, this image was collected with the detector array of a camera, so it is a sampled image.

To simplify the functional representation this image was converted to an 8-bit quantized grayscale image as shown in figure 1.10. In addition, two one-dimensional slices were made across the image, one slice in the x-direction along a row and one in the y-direction along a column. The signal function associated with each of these slices is shown in figure 1.11 where the gray-level at a particular pixel location is plotted as a function of the coordinate along a particular slice.

Notice that the grayscale values have a definite periodicity in space dictated by the mortar between bricks. Along the x-direction there are three large peaks at pixel

Figure 1.9. Image of a brick wall.

Figure 1.10. 8-bit grayscale image of the brick wall shown in figure 1.9. The blue lines in the image highlight the positions of two slices taken across the image.

locations 160, 520, and 880. Thus, the separation between each peak is a value of 360 pixels representing the distance in pixels of the long length of a brick. Along the y-direction there are four peaks at pixel locations 60, 180, 300, and 420. This time, the separation between peaks is 120 pixels and as expected a shorter distance associated with the smaller brick width. This gives a 3:1 ratio of length to width. On the original object the length of a brick is 200 mm and the width is 6.7 mm in the same 3:1 ratio. For this specific camera, the size of each pixel is given as 5.0 μm. So, on the image recorded by the camera the 360 pixel separation distance between bricks along the x-axis is 1.8 mm. Similarly, the 120 pixel separation distance between bricks along the y-axis is 0.60 mm.

Since both of the signals in figure 1.11 are periodic, a Fourier series representation could be used as a model to describe each function. Along the x-axis the spatial period to use is 1.8 mm and along the y-axis the spatial period to use is 0.6 mm. The first order or dominate spatial frequency in the image for the x-axis signal and the y-axis signal is then,

$$f_x = \frac{1}{1.8 \text{ mm}} = 0.56 \frac{\text{lines}}{\text{mm}} \text{ and } f_y = \frac{1}{0.6 \text{ mm}} = 1.7 \frac{\text{lines}}{\text{mm}}$$

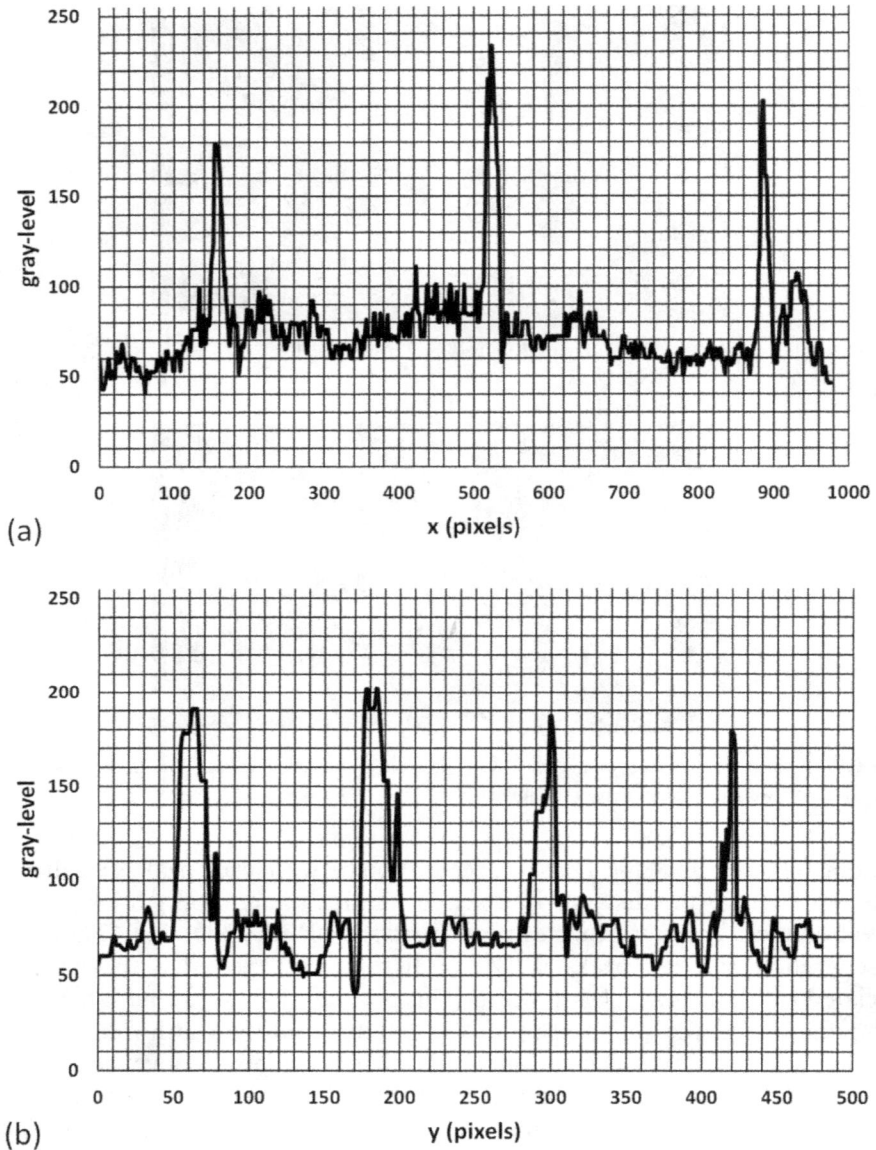

Figure 1.11. Plots of the gray-level values across the brick wall image (a) along a row in the *x*-direction and (b) along a column in the *y*-direction.

Therefore, any two-dimensional image is comprised of overlapping one-dimensional signals. The one-dimensional Fourier series representation of signals also applies to a two-dimensional function where each row or column represents information about an input or output from a system.

The examples above all show why specific targets were historically developed to test optical system resolution. The standard MIL-STD-150A Photographic Lenses defined a simple three-bar resolving power test target element for use in determining the minimum resolving power of lenses [25]. Numerous other tests were also developed to test a variety of specifications for optical design [26]. Figure 1.12(a) shows an image of a USAF 1951 chart based on this standard. The same principle is used in the familiar optometrist's test charts to characterize visual acuity, as shown in figures 1.12(b) and (c). The 'Tumbling-E' chart is essentially a three-bar variation.

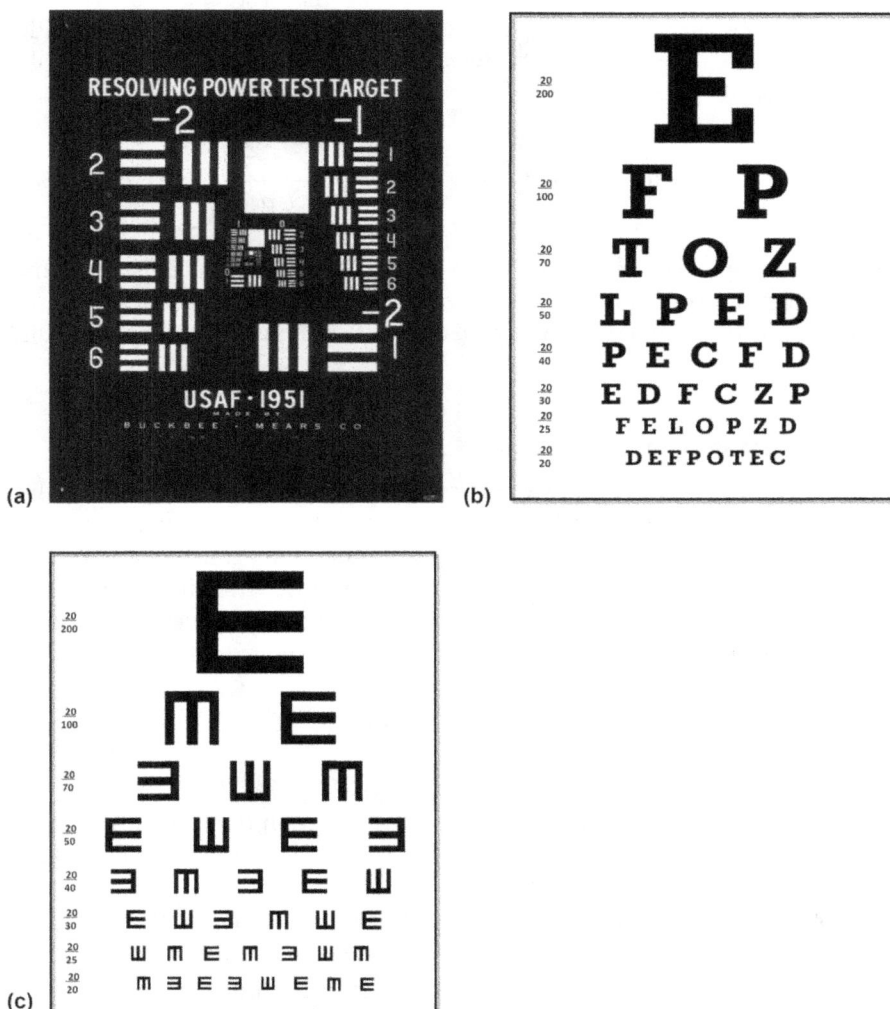

Figure 1.12. Resolution test charts. (a) USAF 1951 test target showing groups of three-bar elements (the original uploader was Alemily at English Wikipedia, CC BY-SA 2.5 <https://creativecommons.org/licenses/by-sa/2.5>, via Wikimedia Commons), (b) common visual acuity eye chart, and (c) 'Tumbling-E' eye test chart.

1.4.5 Special functions and their combinations

To represent physical quantities like incident optical fields or intensity distributions, a number of special functions have been defined [8, 9, 16]. Table 1.4 lists a few of these functions in one dimension along with a plot of each function. These are also included in the appendix along with some definitions of selected two-dimensional functions. As an example, the rectangle function is a shorthand way of writing the field function of a slit. In two dimensions, the function would represent a rectangular bar, so that three two-dimensional rectangle functions can be used as the transmittance function of each three-bar element in the test chart of figure 1.12. The transmittance function for a plane wave incident on three slits of width, a, along the x-axis each separated by the distance, a, would be written as a sum of three rectangle functions,

$$\text{rect}\left(\frac{x}{a}\right) + \text{rect}\left(\frac{x - \frac{3}{2}a}{a}\right) + \text{rect}\left(\frac{x + \frac{3}{2}a}{a}\right)$$

As we will see, the delta function is a representation for a point source and the Gaussian function represents the profile of a laser beam. These functions or combinations of these functions will be used often in describing the input function for a system.

1.4.6 Fourier transform and operators

The power of using these functional representations of physical quantities is that the function can be described in the time/space domain or equivalently in the temporal/spatial frequency domain. We saw that periodic functions can be represented as a Fourier series. If the input function is aperiodic, a Fourier transform representation can be used to obtain a function in the temporal/spatial frequency domain. The Fourier transform of a space domain function $f(x)$ is defined as,

$$F(f_x) = \int_{-\infty}^{+\infty} f(x)e^{-j2\pi f_x x}dx$$

where $F(f_x)$ is the spatial frequency domain representation function. (We use $j \equiv -1$ as the complex number.) Note the similarity between this equation and the exponential Fourier series expansion.

The Fourier transform of a general function can be found directly by integration over the input coordinate variable. As an example, the Fourier transform of the rectangle function of width, a, $f(x) = \text{rect}\left(\frac{x}{a}\right)$ would be computed as,

$$F(f_x) = \int_{-\infty}^{+\infty} \text{rect}\left(\frac{x}{a}\right)e^{-j2\pi f_x x}dx$$

Table 1.4. Special functions, definitions, and graphical representations.

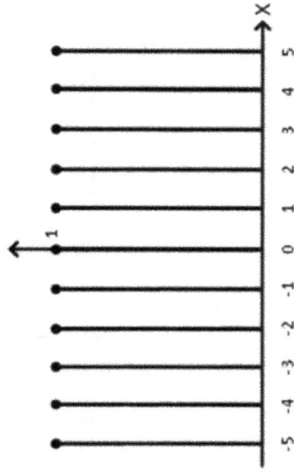

Delta function	$\delta(x - x_0) = \int\limits_{-\infty}^{+\infty} e^{+j2\pi u(x-x_0)} du$	
Comb function	$\mathrm{comb}(x) = \sum\limits_{n=-\infty}^{+\infty} \delta(x - n) = \sum\limits_{n=-\infty}^{+\infty} e^{j2\pi nx}$	
Step function	$\mathrm{step}(x) = \begin{cases} 1, & x > 1 \\ \frac{1}{2}, & x = 0 \\ 0, & x < 0 \end{cases}$	

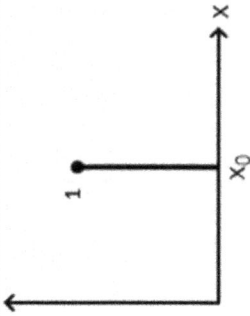

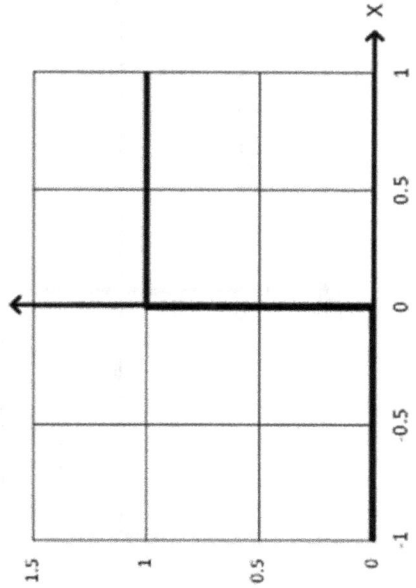

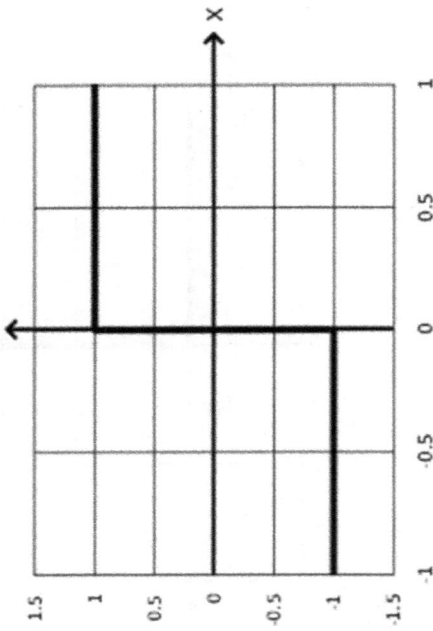

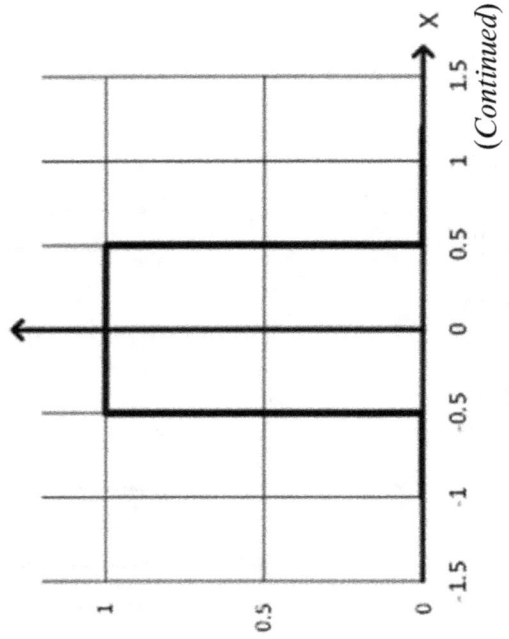

(Continued)

Sign function

$$\text{sgn}(x) = \begin{cases} 1, & x > 1 \\ 0, & x = 0 \\ -1, & x < 0 \end{cases}$$

Rectangle

$$\text{rect}(x) = \begin{cases} 1, & |x| < \frac{1}{2} \\ \frac{1}{2}, & |x| < \frac{1}{2} \\ 0, & |x| < \frac{1}{2} \end{cases}$$

Table 1.4. (*Continued*)

Triangle	$\mathrm{tri}(x) = \begin{cases} 1 - |x|, & \& \ |x| < 1 \\ 0, & \& \ |x| \geqslant 0 \end{cases}$

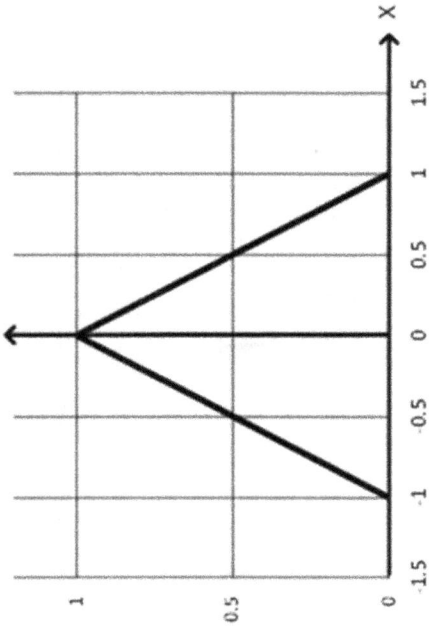

Sinc	$\mathrm{sinc}(x) = \dfrac{\sin(\pi x)}{\pi x}$

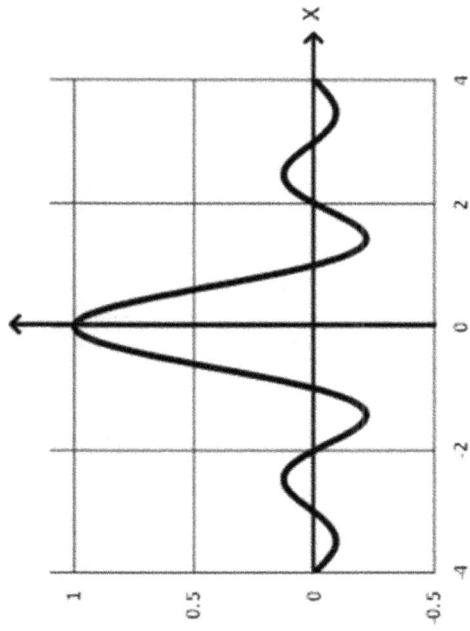

1-20

Sinc-squared

$$\text{sinc}^2(x) = \frac{\sin^2(\pi x)}{(\pi x)^2}$$

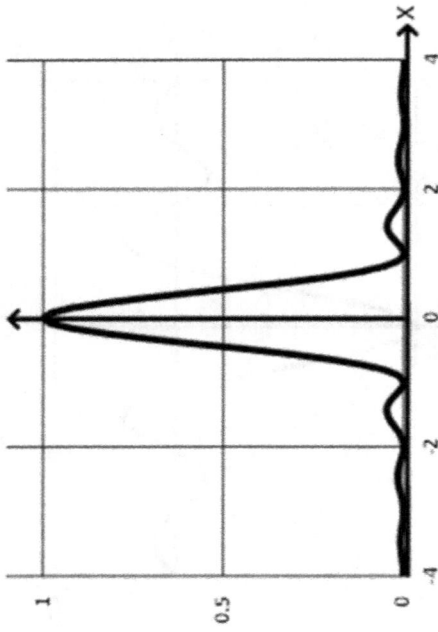

Gaussian

$$\text{Gaus}(x) = \exp(-\pi x^2)$$

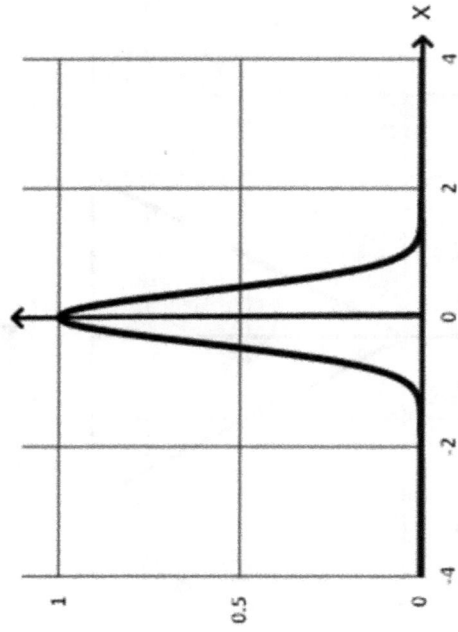

And from the definition of the rectangle function,

$$F(f_x) = \int_{-\frac{a}{2}}^{+\frac{a}{2}} e^{-j2\pi f_x x} dx$$

since the values of the rectangle function are zero for all $x < -a/2$ and for $x > +a/2$. After integration, the resulting function in the spatial frequency domain is,

$$F(f_x) = |a| \frac{\sin(\pi a f_x)}{\pi a f_x} = |a| \operatorname{sinc}(a f_x)$$

The Fourier transform of several functions including all of the special functions of table 1.4 are tabulated in the appendix. These results along with the general properties of the Fourier transform can be applied to assist in calculations that describe the effects that a system imposes on the input function to the system.

In the final analysis, the resulting mathematical output response function must then be interpreted as a physical phenomenon. Obtaining the output response from an input stimulus requires a set of rules or mathematical operations. A common method used to mathematically describe this process uses operator formalism [8, 9]. Given an input function represented by $f(x)$, the resulting output function $g(x)$ is obtained by application of the system operator $\mathcal{H}$ as,

$$\mathcal{H}\{f(x)\} = g(x)$$

Take a single point source, delta function, as the input to a general linear system. The point source input function or impulse is then defined at a general point w so that,

$$f(x) = \delta(x - w)$$

and the magnitude of the function is one. Using the operator above, the output of a system in response to the input results in,

$$g(x) = \mathcal{H}\{f(x)\} = \mathcal{H}\{\delta(x - w)\} \equiv h(x, w)$$

Thus, a new function $h(x,w)$ is defined as the impulse response function or system function and is the response of the linear system to a unit impulse input at any general point where $x = w$. Therefore, several methods are available for writing input functions in a system model. We will review these concepts again later and use them extensively in other system contexts with numerous examples.

Exercises and problems

1. A portion of a USAF test chart contains a series of lines and is used as the input to an optical imaging system. If a certain set of dark lines in a group are each 3 mm wide and separated by 3 mm of white space, what is the dominant spatial frequency of these lines?

2. A co-sinusoidal amplitude grating is specified by the function,

$$f(x, y) = 2\cos(20\pi x)$$

where x and y are in units of mm. What is the spatial frequency of the grating?

3. The image of an optical transmission grating is shown on the diagram (figure P1.3) as observed on a 2-D array detector. The data is plotted as relative gray-level from 0 (black/dark) to 1 (white or bright).

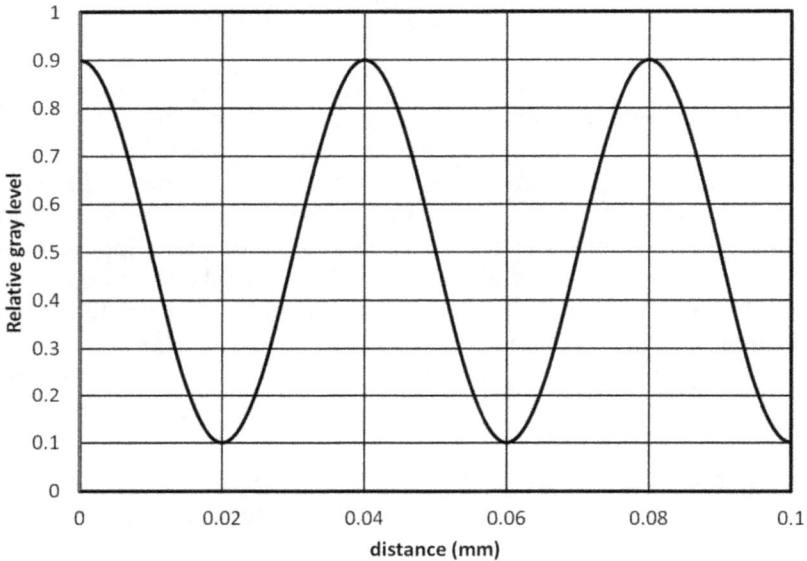

Figure P1.3. Gray level scan across a detector array from an image of a transmission grating.

(a) What is the spatial frequency of this grating?

(b) Specify the function that you would use to describe this grating. Take the distance coordinate to be a variable x. Assume that it is periodic and continues over all space.

4. Find the values of the five Fourier series coefficients specified below for the function of time shown in figure P1.4.

$a_0 = $ _____

$a_1 = $ _____

$b_1 = $ _____

$a_2 = $ _____

$b_2 = $ _____

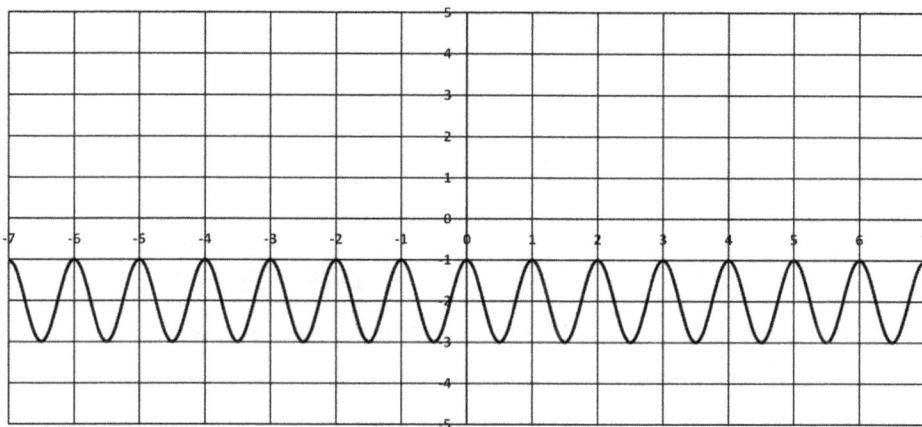

Figure P1.4. Data taken from an oscilloscope with settings of 1 millivolt per division on the voltage axis and 1 microsecond per division on the time axis.

5. The graph (figure P1.5) for a section of the spatial variation in the electric field for a wavefront propagating along the z-axis is shown below. The units of x are in microns and the field has a maximum amplitude of $E_0 = 5.0$ N/C. This periodic function can be represented by a trigonometric Fourier series expansion.

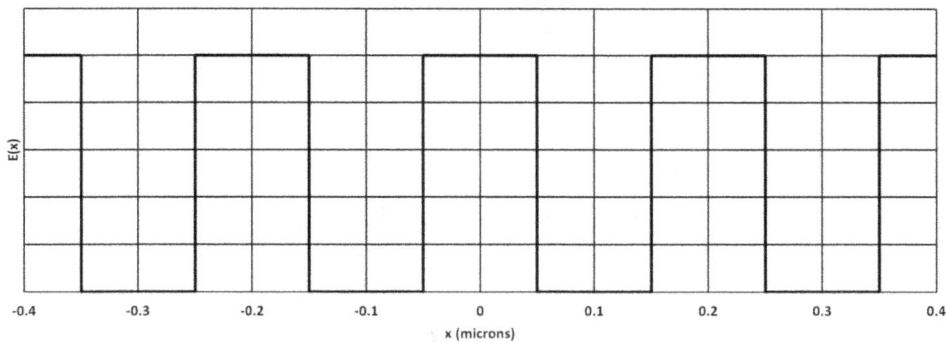

Figure P1.5. Electric field as a function of position perpendicular to the direction of propagation of an electromagnetic wave.

    (a) What is the longest spatial period associated with the Fourier series expansion of this waveform?

    (b) Compute all of the coefficients in the Fourier series expansion for this wave.

6. Compute the exponential Fourier series coefficients (c_n values) of the function and write out the complete series.

$$f(x) = \mathrm{comb}\left(\frac{x}{L}\right)$$

7. Sketch a graph of these functions:

$$f(x) = 2 \, \text{rect}\left(\frac{x+1}{2}\right)$$

$$g(w) = 3 \, \text{tri}(w + 2)$$

8. Calculate the Fourier transforms of the following functions by direct integration. The variables 'a' and 'A' are positive parameters.

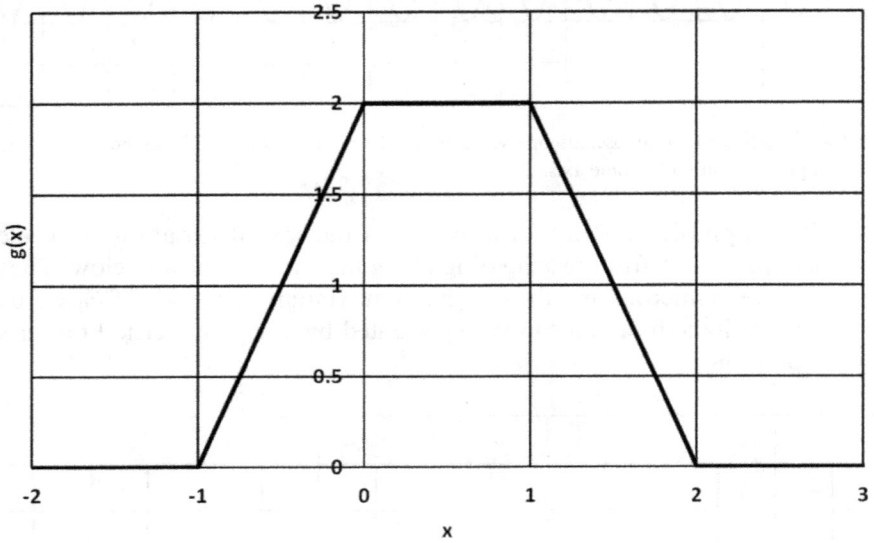

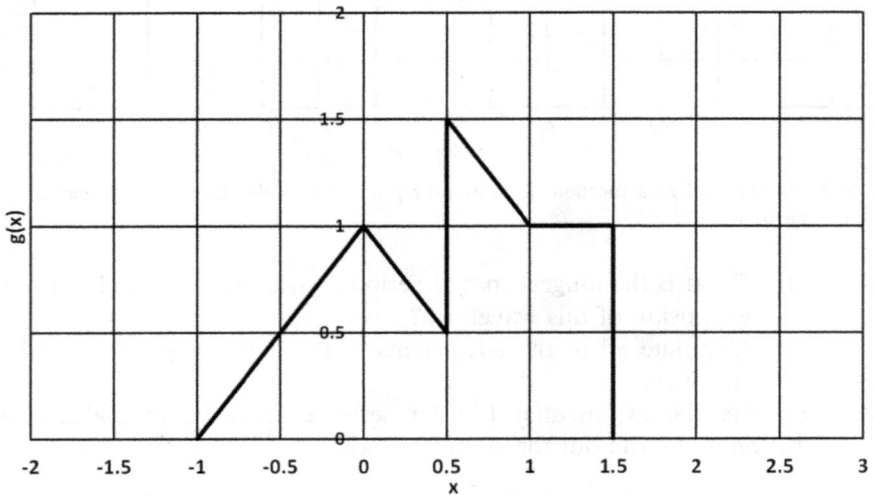

Figure P1.10. Spatial amplitude functions in normalized position units.

(a) $f(x) = \exp(-a \mid x \mid)$
(b) $f(x) = A\delta(x - a)$
(c) $f(x) = A\cos(2\pi a x)$
9. Given the following functions:

$$f_1(x) = \text{rect}(x + 1) + \text{rect}(x - 1)$$

$$f_2(x) = \text{tri}(x + 2) + \text{tri}(x - 2)$$

Sketch each function and compute the Fourier transform of the function. (Hint: the Fourier transform of a sum is the sum of the Fourier transforms.)
10. Find the Fourier transform function $G(f_x)$ for the functions $g(x)$ shown in figure P1.10. (Hint: represent these functions in terms of special functions and use the Fourier transform tables in the appendix.)

References

[1] Blanchard B S and Fabrycky W J 1998 *Systems Engineering and Analysis* (London: Prentice Hall)
[2] Hubka V and Eder W E 1988 *Theory of Technical Systems* (Berlin: Springer)
[3] LeFevre E W 1990 *Engineering Stages of New Product Development, NSPE Publication #3018* (Alexandria, VA: National Society of Professional Engineers)
[4] Ulrich K T and Eppinger S D 2012 *Product Design and Development* 5th edn (New York: McGraw-Hill)
[5] Loch C and Terwiesch C 2000 Product development and concurrent engineering *Encyclopedia of Production and Manufacturing Management* ed P M Swamidass (Boston, MA: Springer)
[6] Dieter G E and Linda C S 2009 *Engineering Design* (Boston, MA: McGraw-Hill Higher Education)
[7] Pahl G and Beitz W 1996 *Engineering Design: A Systematic Approach* (Berlin: Springer)
[8] Gaskill J D 1978 *Linear Systems, Fourier Transforms and Optics* (New York: Wiley)
[9] Goodman J W 2017 *Introduction to Fourier Optics* 4th edn (New York: W H Freeman)
[10] Gonzalez R C and Woods R E 2018 *Digital Image Processing* 4th edn (New York: Pearson/Prentice Hall)
[11] Kasunic K 2011 *Optical Systems Engineering* (New York: McGraw-Hill Education)
[12] Hobbs P C D 2000 *Building Electro-Optical Systems: Making It all Work* (New York: Wiley)
[13] National Research Council 1998 *Harnessing Light: Optical Science and Engineering for the 21st Century* (Washington, DC: The National Academies Press)
[14] Smart A E 1994 Folk wisdom in optical design *Appl. Opt.* **33** 8130–32
[15] Willers C J 2013 *Electro-optical System Analysis and Design: A Radiometry Perspective* (Bellingham, WA: SPIE Press)
[16] Tyo J S and Alenin A S 2015 *Field Guide to Linear Systems in Optics* (Bellingham, WA: SPIE Press)
[17] National Institute of Standards and Technology Special Publication 330 2019 *Natl. Inst. Stand. Technol. Spec. Publ. 330* pp 122 (this publication is available free of charge from: https://doi.org/10.6028/NIST.SP.330-2019)

[18] IES Nomenclature Committee 1986 *American National Standard Nomenclature and Definitions for Illuminating Engineering, ANSI/IES RP-16-1986* (New York: Illuminating Engineering Society of North America)

[19] Murray J J, Nicodemus F E and Wunderman I 1971 Proposed supplement to the SI nomenclature for radiometry and photometry *Appl. Opt.* **10** 1465–68

[20] Hecht E 2002 *Optics* 4th edn (Reading, MA: Addison-Wesley)

[21] McCluney W R 1994 *Introduction to Radiometry and Photometry* (Boston, MA: Artech House)

[22] Boyd R W 1983 *Radiometry and the Detection of Optical Radiation* (New York: Wiley)

[23] Pedrotti F L, Pedrotti L S and Pedrotti L M 2007 *Introduction to Optics* 3rd edn (Englewood Cliffs, NJ: Prentice Hall)

[24] Grant B G 2011 *Field Guide to Radiometry* (Bellingham, WA: SPIE Press)

[25] MIL-STD-150A 1959 *Military Standard, Photographic Lenses* (Washington, DC: U S Govt. Printing Office)

[26] MIL-HDBK-141 1962 *Military Standardization Handbook, Optical Design* (Washington, DC: U S Govt. Printing Office)

IOP Publishing

Optical Systems Design Detection Essentials
Radiometry, photometry, colorimetry, noise, and measurements
Robert M Bunch

Chapter 2

Ray optics and optical system parameters

2.1 Introduction

Using the ray model of light propagation is the most common approach to start an optical system design. A number of system parameters can be elicited from the design's requirements by only making a few assumptions. The ray model is especially useful in imaging systems where effects of interference and diffraction can be neglected. Analysis of these types of systems falls under the realm of geometrical optics. This chapter reviews concepts of geometrical optics and ray propagation then uses these concepts to define optical system parameters. Many optical systems operate within the visible portion of the electromagnetic spectrum where the radiation is referred to as light. Even though optical systems can operate outside the visible portion of the electromagnetic spectrum, the term light is used here generically as for propagation of electromagnetic radiation.

2.2 Rays

The initial assumption in modeling light propagation in optical systems is to consider the object of interest to be comprised of numerous point sources each one radiating light in all directions. In this case, since we are neglecting interference and diffraction effects, every single point source is independent from all other sources. The wavefront from each point source spreads out with the same velocity in all directions from the source (assuming a homogeneous and isotropic medium). A medium is homogeneous if its physical and optical properties are the same at any one point within its volume. The medium is isotropic when it has similar properties in all directions. A wavefront surface is defined by a set of points where the wave has a constant phase, or in other words, has traveled a distance from the original point source in a fixed amount of time. We define a ray as the path along which light travels and energy flows, thus rays are directed line segments perpendicular to a wavefront surface. Figure 2.1 illustrates the rays and wavefronts in a cross-section of space near a point source within a homogeneous isotropic medium. As can be seen,

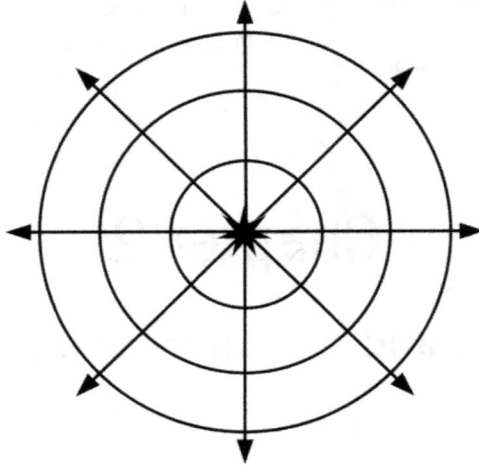

Figure 2.1. Rays and wavefronts near a point source within a homogeneous isotropic medium.

light rays travel in a straight line in a homogeneous isotropic media (rectilinear propagation) and all wavefronts emanating from a point source are spherical [1].

2.3 Refractive index and optical path length

Electromagnetic theory tells us that waves travel at a fixed speed in a vacuum symbolized as c, or the speed of light in vacuum. For light waves propagating in a material medium the wave speed is less than c due to the electromagnetic permittivity and permeability of the medium. This results in a macroscopic parameter, n, the refractive index, defined as the ratio of c, the speed of light in vacuum to v, the speed of light in the medium.

$$n = \frac{c}{v}$$

For a homogeneous isotropic medium the refractive index is a constant value, however it is a function of the wavelength of the light. As an example, figure 2.2 shows the refractive index as a function of wavelength for three different common glasses, NBK7, NSF5, and fused silica (quartz) over a wavelength range through the visible portion of the spectrum and into the near infrared. The basic shape of each curve is the same, that is, the refractive index decreases as the wavelength increases. This phenomenon is called normal dispersion and is characterized by the slope of the $n(\lambda)$ curve, $dn/d\lambda$ at each wavelength [2].

One method used to relate the amount of dispersion of a given material is to use empirical models for the functional form of the dispersion curve. One of the simplest models is the two parameter Cauchy equation [4],

$$n(\lambda) = A + \frac{B}{\lambda^2}$$

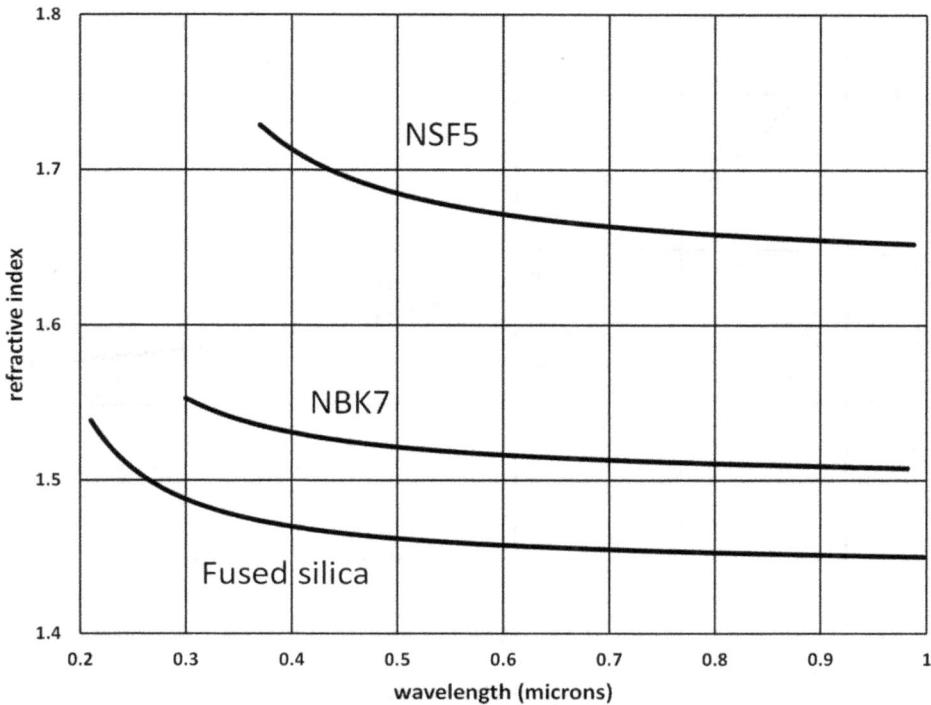

Figure 2.2. Refractive index of selected glasses as a function of wavelength [3].

where A and B are values to the determined from a fit to refractive index data as a function of wavelength. Note that A is unitless and B has units of square wavelength. A more detailed form used for wider wavelength ranges is the Sellmeier equation [4],

$$n^2(\lambda) - 1 = \sum_{i=1} \frac{A_i\lambda^2}{\lambda^2 - \lambda_i^2}$$

where A_i are fit coefficients and the λ_i values are wavelengths at which there occurs high absorption or resonances. We will discuss more about this issue in a later chapter. There are many other empirical dispersion formulas in use by various manufacturers of glass products [4].

A quick way to identify and classify glasses according to their dispersive behavior is through the Abbe number parameter [2, 4]. The symbol used for this parameter is either V or v. There are actually several alternative definitions for the Abbe number depending on the refractive index value at a given reference wavelength. For example, a D subscript is used when the refractive index at the sodium D spectral line is used. A subscript d indicates that the Fraunhofer d helium line at 587.6 nm was used for the refractive index reference. For this case, the Abbe number is defined as,

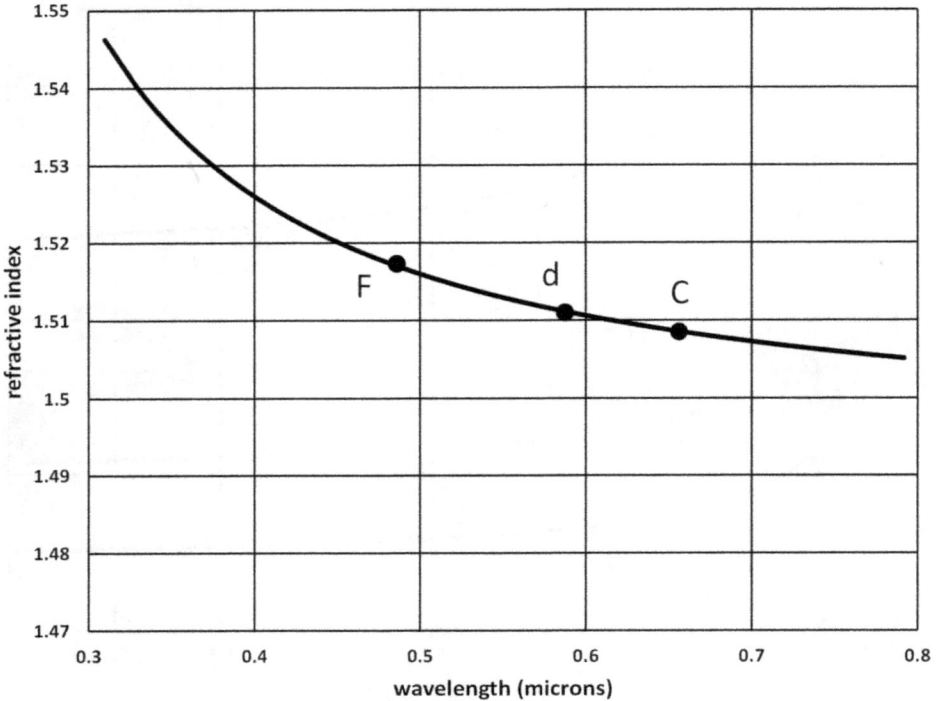

Figure 2.3. Normal dispersion curve indicating the d, F, and C wavelength positions of n_F, n_C, and n_d.

$$V_d = \frac{n_d - 1}{n_F - n_C}$$

where n_F and n_C are the refractive indices at the Hydrogen F-line (486.1 nm) and the Hydrogen C-line (656.3 nm), respectively. The lower the value of the Abbe number the higher the dispersion. The applicability of this definition can be understood by examining a characteristic normal dispersion curve as in figure 2.3. The numerator is a scaling factor representing the difference between the reference wavelength index to vacuum. A large difference between the refractive indices in the denominator indicates a high slope or higher dispersion. This difference is called the principal dispersion [5].

For comparing different types of glasses in terms of their dispersion characteristics an Abbe diagram is used. This graph plots the refractive index value at the given reference wavelength (n_d in this case) versus the value of the Abbe number, as seen in figure 2.4. In this plot, the labeled sections refer to the Schott Glass letter codes for material compositions [5].

In an inhomogeneous material, the refractive index is a function of position, such as in birefringent materials. In addition, the refractive index may need to be treated mathematically as a complex number, for example in the case of light interaction with conductive materials like metals. More details will be provided in the following chapters.

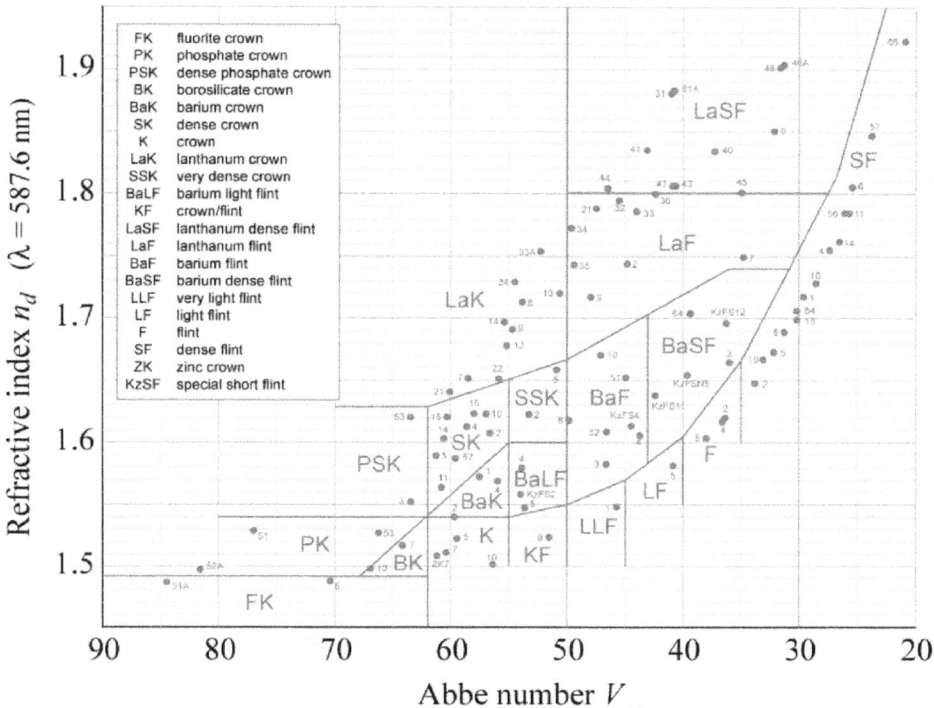

Figure 2.4. Abbe diagram for various glass types (Bob Mellish, CC BY-SA 3.0 <http://creativecommons.org/licenses/by-sa/3.0/>, via Wikimedia Commons).

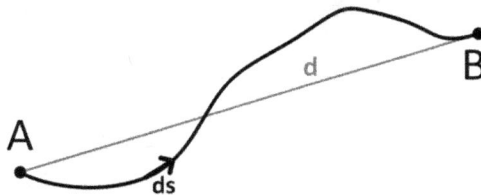

Figure 2.5. General path for a wave between two points A and B in a medium where d is the linear distance between the two points.

If the refractive index changes with position or there are multiple media present, then we must take this into account when examining how a ray will propagate through this material or system. Fermat's principle, or the principle of least time, says that the light passing between two points will traverse the path in least time [1]. Take two points in space A and B as shown in figure 2.5. The time it takes for light to travel from point A to point B in a medium of refractive index, $n(s)$, along any path ds is given by,

$$t_{AB} = \int_A^B \frac{n(s)}{c} ds$$

If $n(s) = n$ is a constant, or a homogeneous isotropic medium, then the time for light to travel from A to B is,

$$t_{AB} = \frac{n}{c}d$$

where the path length is simply the linear geometric distance, d, between point A and point B.

Another related term often used in these situations is the optical path length (OPL). OPL is a distance that a wave travels proportional to the time t_{AB} [2, 6]. In a homogeneous isotropic medium the OPL is simply the product of the constant refractive index and the geometric distance, or,

$$OPL = \int_A^B \frac{n}{c}ds = nd$$

We often compare an OPL to a length measured in some other medium. If the refractive index is a reference medium is n_0, then the OPL in the reference medium is $n_0\, d$. The difference between these two OPLs is called the optical path difference OPD.

$$OPD = (n - n_0)d$$

2.4 The law of refraction and the law of reflection

Using the OPL concepts, let us follow the path of a ray across the plane surface boundary between two homogeneous media as shown in figure 2.6. We wish to find the relationship that describes the path of the wave as it travels from point A in medium n to point B in medium n'. There are of course many possible paths so we

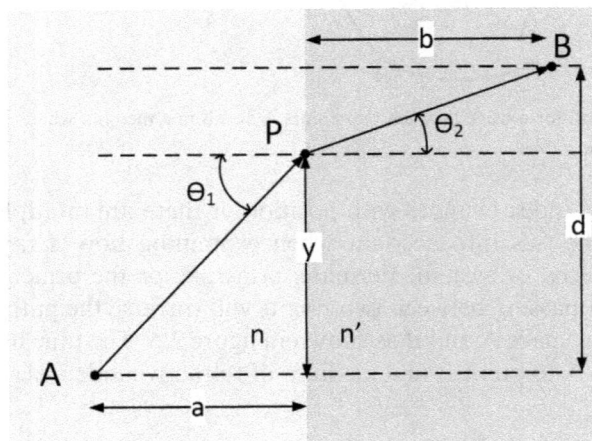

Figure 2.6. Applying Fermat's principle to obtain the ray paths at a boundary between two media. For this diagram $n' > n$.

must use Fermat's principle to find the path. A general point P is chosen at a height y along the boundary surface. The total travel time is then,

$$t_{AB} = t_{AP} + t_{PB}$$

Using the parameters of figure 2.6 along with some basic geometry, the total time of propagation can be written as a function of the unknown variable y as,

$$t_{AB} = \frac{\sqrt{a^2 + y^2}}{\left(\frac{c}{n}\right)} + \frac{\sqrt{b^2 + (d - y)^2}}{\left(\frac{c}{n'}\right)}$$

Applying Fermat's principle we minimize t_{AB} for all distances y by taking the derivative dt_{AB}/dy and setting the result equal to zero. This results in the relationship for Snell's Law, or the law of refraction.

$$n \sin \theta = n' \sin \theta'$$

Thus, a ray of light incident on the boundary between two materials will refract or change its angular direction at the boundary. The ray AP is said to be the incident ray and the ray PB is the refracted ray. All angles are measured with respect to the surface normal.

We can perform a similar calculation for the case where point B is now above the surface in the n medium but with the incident ray still approaching point P at angle θ with respect to the boundary. Applying Snell's law again tells us that since $n' = n$ then $\theta' = \theta$ which is the law of reflection. This is often stated, as the angle of incidence is equal to the angle of reflection.

Another interesting reflection phenomenon occurs when the incident ray originates in a medium with a greater refractive index than the refractive medium, $n > n'$. As the angle θ increases in magnitude, the angle $\theta' > \theta$. So, at some finite angle of incidence $\theta' = 90°$. Since the $\sin(90°) = 1$ we define the angle $\theta_c = \theta$ as the critical angle, where,

$$\sin \theta_c = \frac{n'}{n}$$

When $\theta > \theta_c$ we experimentally observe that the ray reflects back into the n medium. This phenomenon is called total internal reflection and as before, the angle of incidence is equal to the angle of reflection.

2.5 Optical surfaces, curvature, and sign convention

In arriving at Snell's law, we used a simple planar surface for the boundary separating two optical media. However, the result is the same for any surface shape. Optical components are designed, manufactured, and sold with a wide variety of surface shapes [7]. The designer picks a particular shape of lens, mirror, prism, window, etc. in order to meet a specific system requirement [8]. Here we discuss some of the geometrical features and characteristics of common planar and spherical surfaces along with a convention of signs for geometrical dimensions used in analyzing a specific design.

As discussed in chapter 1, we use a standard plane-to-plane approach in following energy flow through an optical system. At each plane resides a surface whose vertex is at the origin of that plane. We need a way to describe the surface shape with respect to this coordinate plane as well as a set of rules (sign convention) in assigning positive and negative values to all the distances and angles that we use in ray propagation.

The sign convention used here is:

- All rays travel from left-to-right in a uniform medium. This means that the rays travel in the positive z-axis direction sequentially from plane-to-plane in an optical system.
- Positive curvatures are obtained when the radius of curvature is to the right of the vertex. Negative curvatures occur when the radius of curvature is to the left of the vertex. A planar surface has a curvature of zero and a radius of curvature of infinity.
- Distances from the vertex to the right are positive (positive z-axis direction) and distances from the vertex to the left are negative (negative z-axis direction).
- Heights in the positive y-axis direction (above the z-axis) are positive and heights in the negative y-axis direction (below the z-axis) are negative.
- Angles with a positive slope with respect to the $+z$-axis (optical axis) direction are positive. Angles with a negative slope with respect to the $+z$-axis (optical axis) direction are negative.
- Spherical surfaces are formed by a set of points equidistant from a center of curvature.

   The distance from the center of curvature to any point on the surface is the radius of curvature. Drawing a line segment from the center of curvature to a point on the circle identifies a vertex point. When a center of curvature is to the right of a vertex the surface is convex. When the center of curvature is to the left of a vertex the surface is concave.

Figure 2.7 shows a spherical surface with refractive index n' within an external medium of refractive index n. The point C is at the radius of curvature of the surface and the point V is the surface vertex.

According to our sign convention, the radius of curvature R is a positive value when R is measured from V to C, the positive z-axis direction. A convenient parameter used to characterize spherical surfaces is curvature. Curvature is simply defined as the inverse of the radius of curvature.

$$C \equiv \frac{1}{R}$$

Also shown in figure 2.7 is the surface sag parameter, s, at a height y on the surface. Since the height y to point P forms a right angle on the optical axis, the surface sag can be computed using the Pythagorean Theorem.

$$R^2 = (R - s)^2 + y^2$$

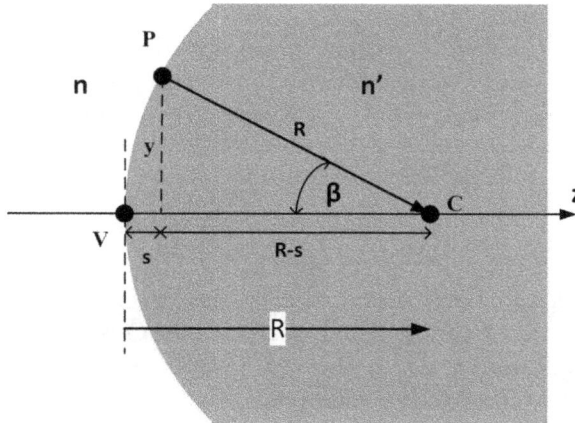

Figure 2.7. A spherical surface of index of refraction n' in a medium of refractive n with positive curvature. The parameter s is the surface sag at height y.

Solving for s, as a function of the height y, we obtain,

$$s(y) = R - \sqrt{R^2 - y^2}$$

Surface sag can be measured directly using a spherometer instrument with known value of y allowing the radius of curvature to be computed.

$$R = \frac{s^2 + y^2}{2s}$$

In many cases, the distance s is small compared to y. A convenient approximation for surface sag when the radius of curvature is known is,

$$s \cong \frac{y^2}{2R}$$

often referred to as the parabolic approximation.

Note that surface sag is small when either y is small, or R is large compared to y. When this occurs the height, y approaches the arc length of the angle β subtended at point C. Since

$$\sin \beta \equiv \frac{y}{R} \approx \beta$$

This approximation is in the form of the small angle approximation. We shall see that this is called the paraxial approximation for rays near the optical axis of the system.

2.6 Image formation

As mentioned earlier, the ray approach is especially useful in analyzing imaging systems. By simply applying the laws of geometrical optics (rectilinear propagation, refraction, and reflection), we can predict the location and size of the image of an object. The process is called ray tracing. In ray tracing, we follow the path of a few rays from any point on the object and find the point at which these rays cross. This point identifies the location of the object point on the image. The formation of images by planar surfaces, spherical reflecting surfaces, spherical refracting surfaces, and lenses using ray tracing is described in this section.

2.6.1 Image formation by spherical refracting surfaces

Figure 2.8 shows the geometry of a ray traced from an on-axis object point source to its corresponding image location. The object at point O is in front (to the left) of a spherical refracting surface. As in the last section, we use a convex surface of refractive index n' surrounded by a medium of refractive index n. Begin the ray trace from an on-axis object point O is located a distance $-t$ from the vertex V of the surface. We also identify point C as the radius of curvature to the right of the vertex. We then trace a ray from O to a general point P somewhere on the convex surface. At point P, the incident ray refracts according to Snell's law. The image point is then located at the point I where the refracted ray crosses the optical axis and defines the distance from the vertex to the image point as t'. This layout is shown in figure 2.8.

Now, we apply geometry to the triangles formed from the traced rays. We define the angle at point O between the optical axis and point P as α. Similarly, we define the angle at point C between the optical axis and point P as β and the angle at point I between the optical axis and point P as γ. Note that the dotted line between the radius of curvature is in a direction normal to the surface at point P and the distance from point C to Point P is the radius R. For simplicity, we will use the small angle approximation for these angles taking point P at height y above the optical axis.

$$\alpha \cong \frac{y}{-t} \beta \cong \frac{y}{R} \gamma \cong \frac{y}{t'}$$

Since the sum of the angles of any triangle must equal 180° we find that $\gamma + \theta' = \beta$ and $\alpha + \beta = \theta$. Snell's law gives us a relationship between the angles θ and θ'. Again, to simplify the calculation we assume that the angles are small or $\theta \approx \sin \theta$ (paraxial

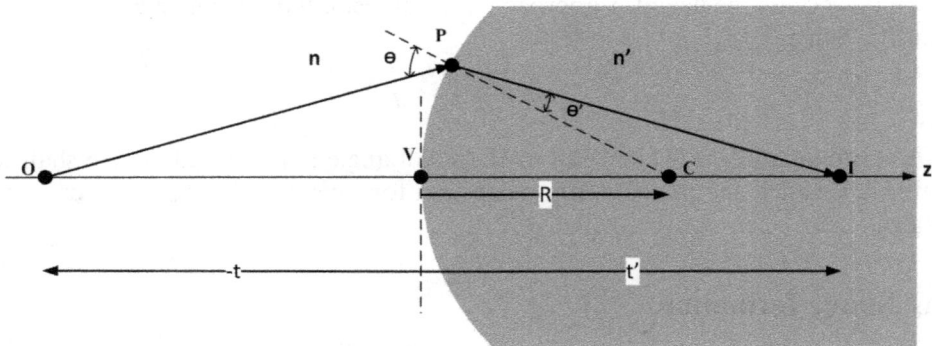

Figure 2.8. Optical layout for imaging a point source by a spherical refracting surface.

approximation), so that the angle relations can be made linear. Snell's law can then be reduced to, $n\,\theta = n'\,\theta'$. Solving these three equations simultaneously and using definitions of the angles α, β, and γ we obtain,

$$\frac{n'}{t'} - \frac{n}{t} = \frac{n' - n}{R}$$

which is the paraxial equation for imaging by a single refractive surface. The term on the right-hand side of this equation has special significance; it is the surface power of a single refractive surface. It is often convenient to use this parameter to simplify and linearize many of the equations related to imaging. Surface power, φ, is defined as,

$$\phi = \left(\frac{n' - n}{R}\right) = (n' - n)C$$

The unit of measure of surface power is the diopter or m^{-1}.

Suppose that the object point is far away from the vertex so that t approaches infinity. This means that a ray arriving at the vertex plane at point P is parallel to the optical axis. This ray crosses the optical axis at a particular point $t' = f'_R$ such that,

$$\frac{n'}{t'} = \frac{n' - n}{R} = \frac{n'}{f'_R}$$

The distance f'_R is the rear focal length of the surface. Likewise, we take an incident ray at an angle such that the refracted ray at point P is parallel to the optical axis, or $t' = \infty$. Then,

$$-\frac{n}{t} = \frac{n' - n}{R} = -\frac{n}{f_F}$$

The distance t, in from of the vertex is defined as the front focal distance, f_F.

The effective focal length of the spherical refracting surface is then defined as a value independent of medium so that,

$$f \equiv f_E = \frac{f'_R}{n'} = -\frac{f_F}{n} \equiv \frac{1}{\phi}$$

We see that the ratio of the front and rear focal lengths is the ratio of the front medium to the rear medium.

2.6.2 Image formation by thin lenses

Single lenses are optical elements formed from a specific material with two refracting surfaces. The surfaces need not necessarily be spherical. Special lenses called aspheres are designed specifically to improve performance of a lens system, such as minimize aberrations in an imaging system or alter the profile of the light emitted from a diode laser. Other lens types such as cylindrical lenses are also convenient for applications where light is required along a one-dimensional line. However, since we are interested here in general system properties and not details of a design, we will restrict the discussion to spherical surfaces.

Figure 2.9 shows a lens of refractive index n' immersed in a medium of refractive index n. The radius of curvature of the first surface is R_1 and the radius of curvature of the second surface is R_2. Even though the figure shows the lens with a finite thickness, for now, we will assume that the thickness is small compared to the image and object distances.

Using the result from the previous section, we apply the relation for single surface imaging successively for each of the two surfaces. The image distance for surface 1, t_1', becomes the object distance for surface 2. Since the object is to the right of the surface, our sign convention gives the new object distance for surface 2 to be $t_2 = t_1'$. Since we are assuming that the thickness of the lens is small ($t_{lens} \approx 0$) we take t as the object distance and t' as the image distance with respect to the center of the lens to obtain the lensmaker equation for a thin lens.

$$\frac{1}{t'} - \frac{1}{t} = \left(\frac{n' - n}{n}\right)\left[\frac{1}{R_1} - \frac{1}{R_2}\right]$$

Most often the refractive index of the external medium, especially for a thin lens system is $n = 1$. In this case, the equation reduces to the Gaussian form of the thin lens imaging equation for a thin lens of focal length f in air.

$$\frac{1}{t'} - \frac{1}{t} = \frac{1}{f}$$

As we shall see, this equation applies to spherical mirrors as well as thin lenses.

Specific values chosen for the radii of curvature describe the basic character of most single lenses, or singlets. There is also a nomenclature used in naming these different configurations based on lens shape, convex, concave, or planar. Figure 2.10 shows the cross-section of both positive and negative focal length lenses along with the symbol used for optical system layouts incorporating thin lenses.

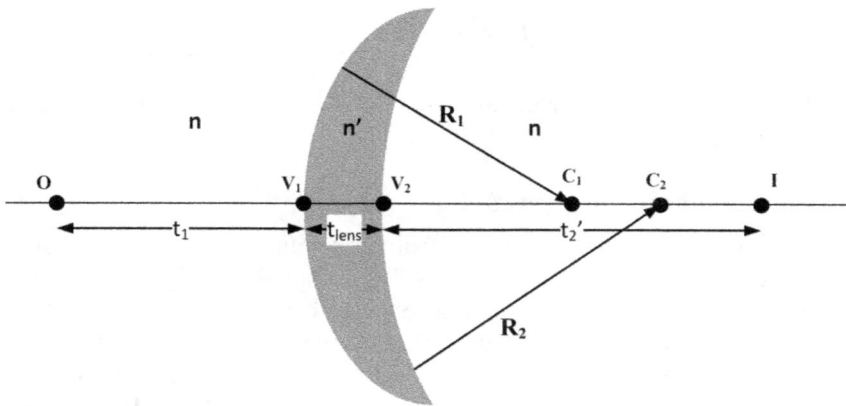

Figure 2.9. Geometry of a lens of refractive index n showing two different radii of curvature.

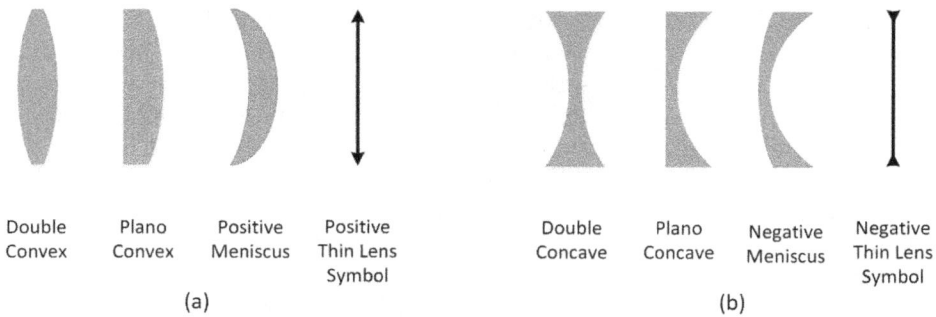

Figure 2.10. Shapes common types of spherical singlets along with the thin lens symbols used in thin lens layouts. (a) Positive focal length or converging lenses and (b) negative focal length or diverging lenses.

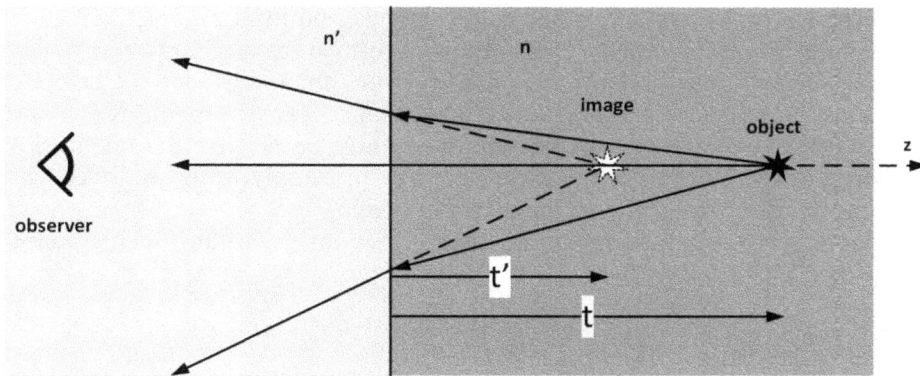

Figure 2.11. Apparent depth of an object embedded in a medium of higher refractive index than the external medium of the observer.

2.6.3 Image formation by planar surfaces

We analyzed the rays incident on a plane surface to arrive at Snell's law. Using the results of section 2.6.1, a plane surface between a medium of refractive n and n' is found by setting the radius of curvature at infinity, or curvature to zero. Thus,

$$\frac{n'}{t'} - \frac{n}{t} = 0$$

As an example, suppose we are viewing an object embedded in medium n, from a medium of refractive index n', where $n > n'$. Figure 2.11 shows the geometry of this imaging case. Three real rays of light are traced from the object and refract at the surface according to Snell's law. As viewed by the observer in medium n', the rays appear to be coming from a different point which is the image of the original object.

Using our equation above, we see that the image distance is,

$$t' = \frac{n'}{n}t$$

since $|n| > |n'|$, then $t' < t$ as shown from the ray trace. Both t and t' are positive values according to our sign convention. This phenomenon is called apparent depth.

Now suppose that the plane surface is a reflecting surface such as a plane mirror. Take an object point source located in front (to the left) of a plane mirror, reflecting surface. The space in which the object is located is called object space. This is the space of medium n in our planar imaging equation. Light rays emitted by the source are directed toward the mirror and reflect according to the law of reflection. See figure 2.12 showing the point source and three rays that are traced toward the mirror and reflect. However, none of the real reflected rays actually cross, so where is the object located? In this case, each of the reflected rays are extended (dotted lines) behind the mirror to follow the law of rectilinear propagation. The extended rays, or virtual rays, cross at a point behind the mirror surface. This point identifies the location of the image. Since the image location is found using virtual rays, this type of image is said to be a virtual image. The image is an apparent source for the reflected rays.

Now we can apply geometry to the triangles formed from the traced rays. Using similar triangles shows that the magnitude of the image distance $|t'|$ behind the mirror must be equal to the magnitude of the distance from the mirror to the object, $|t|$. However, to follow our sign convention, t' must be positive for t defined as a negative distance. We can account for this sign by simply using the rule that on reflection the image space media has a negative sign, or $n' = -n$.

Thus, from the equation above, we see that the image distance of the reflected image is,

$$t' = \frac{n'}{n}t = \frac{-n}{n}t = -t$$

so that when the object distance t is negative the image distance t' is a positive value.

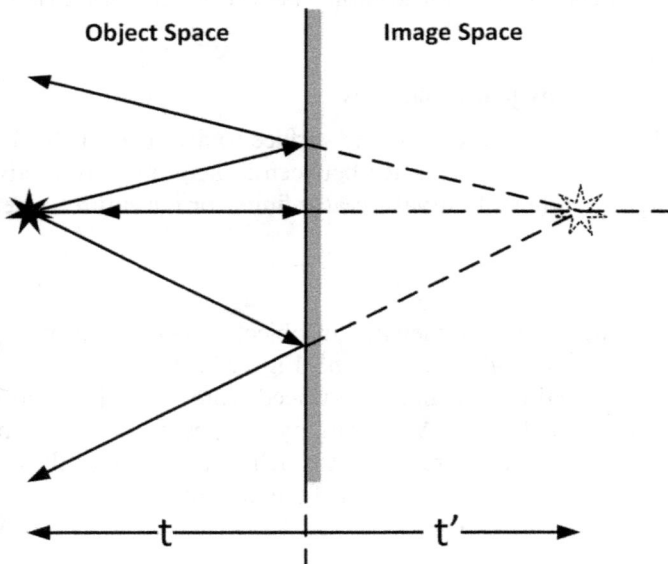

Figure 2.12. A point source located in front of a mirror with three rays traced to locate the image of the object.

Figure 2.13 shows an extended arrow object of height, h. Rays could be traced from any point on the object to find the image location. For convenience, we use just two points, the bottom of the object (on the optical axis) and the top tip of the arrow object. As before the bottom of the object locates the image position at the distance t'. Two rays are shown traced from the top tip of the object. Ray 1 is parallel to the optical axis and reflects back on itself. Ray 2 is traced from the top tip toward the point where the optical axis intersects the mirror and reflects according to the law of reflection. Ray 1 and Ray 2 are then extended back as virtual rays and they cross at the location of the image of the top tip of the arrow, forming a virtual image of the extended arrow object.

Using similar triangles, and the fact that $t' = -t$ we find $h' = h$ or the image height is equal to the object height. The magnification, m, of an image is defined as the ratio of the image height to the object height.

$$m = \frac{h'}{h}$$

So, the magnification of the image in a plane mirror is one. Extending this definition further we can also say that in general the magnification is given by,

$$m = \frac{n}{n'} \frac{t'}{t}$$

2.6.4 Image formation by spherical reflecting surfaces

Take an on-axis object point O a distance t from the vertex V of a concave mirror, as shown in figure 2.14. The center of curvature is identified as point C and the resulting image is at point I. The line of symmetry connecting points O, V, and C is defined as the optical axis of the system. The image point is located by tracing a ray from O to a

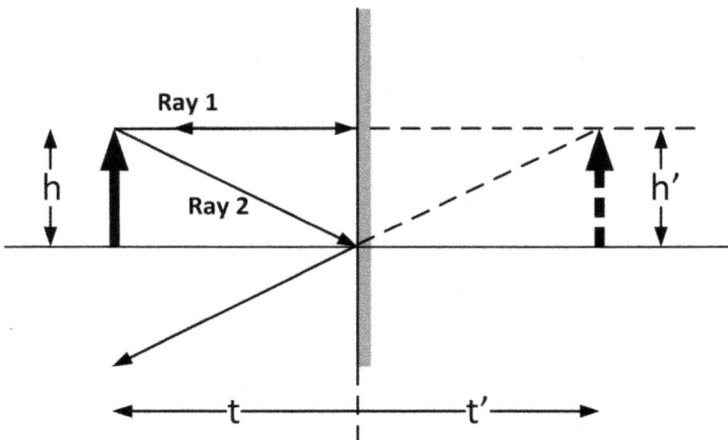

Figure 2.13. An extended object and the virtual image formed behind a plane mirror.

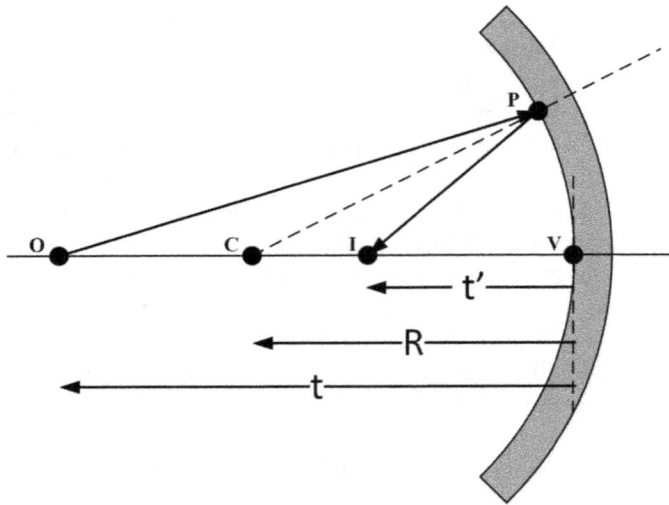

Figure 2.14. Image location of an object in front of a concave spherical mirror.

point P on the surface of the mirror. This ray reflects at P to I so that the angle of incidence is equal to the angle of reflection and the reflected ray crosses the optical axis a distance s' from the vertex. Note that the distance from point C to point P is the radius of curvature R. This layout is similar to that of figure 2.8 for a single refracting surface

As with planar reflecting surfaces we can simply apply the paraxial imaging equation for single surfaces where $n' = -n$, the reflecting surface condition. This means that a concave mirror, as shown in figure 2.14, having a negative radius of curvature has a positive optical power.

$$\phi = \frac{-2}{R}$$

When the object distance is far from the mirror vertex the incident ray is essentially parallel to the optical axis. The reflected ray then crosses the optical axis at a distance,

$$t' = \frac{R}{2} = f$$

where f is the focal length of the mirror. Likewise, any ray crossing the focal point will be reflected back parallel to the optical axis at the height of intersection of the ray with the mirror surface. These observations form the basis for rules in tracing rays with mirrors.

2.7 Analysis methods in paraxial optics

There are several common methods used to analyze a simple optical system based on the basic physics of ray propagation and image formation. Analysis of a system can only proceed once a sufficient number of components are specified and information is known about the components. First, if any two of the parameters in the imaging equation (s, s', or f) are known then we can easily compute the other unknown. Second, we can also graphically trace rays from object points to locate corresponding points in the image and determine the location and form of the image. In addition, there are two more analysis methods based on paraxial ray tracing principles that are commonly used, YNU ray tracing and matrix methods as summarized below. All of these methods are complementary and will be used as needed.

2.7.1 YNU ray tracing

The YNU ray tracing method is a geometric approach that follows a ray sequentially from surface to surface through a system. Again, we are using the general notation of figure 1.5 in following a ray from plane 0 to plane 1 in a system. For simplicity, we will constrain our rays to be in the y–z-plane only using a two-dimensional layout of the systems.

Two parameters specify the ray at any plane in a system, (1) the ray height, y, and (2) the angle of inclination, u. Since these are optical rays, we must also account for the refractive index of the medium, n. Thus, the name YNU ray tracing.

Figure 2.15 shows the geometry of a ray traced from plane 0 to plane 1 separated by a distance t (taken as positive) in a uniform medium of refractive index n. The ray height in plane 0 is denoted by y_0 and it is inclined at an angle u_0. This ray intersects plane 1 with height y_1 and inclination angle u_1. Since the ray is in a uniform medium $u_1 = u_0$. From geometry, we find $y_1 = y_0 + t \tan(u_0)$. Since we are restricting our calculations to paraxial rays then $\tan(u) \approx u$, so,

$$y_1 = y_0 + t u_0$$
$$u_1 = u_0$$

These equations provide a means of computing the new ray height and angle for translation from one plane to another in a uniform medium.

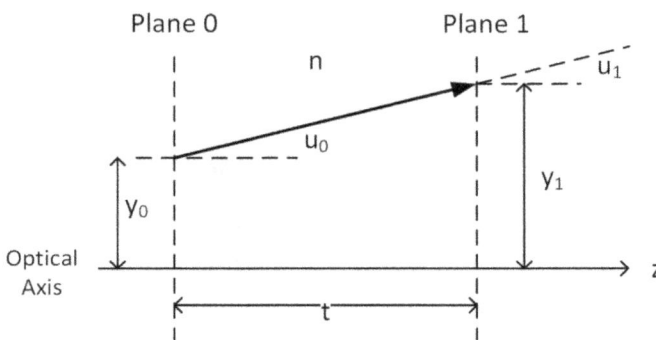

Figure 2.15. Translation of a ray from plane 0 to plane 1 in a uniform medium of refractive index n.

Using a similar process, we will analyze how a ray refracts at a point on the boundary between two refractive media using this notation. We take the boundary to be a section of a sphere with radius of curvature R and again we are only considering paraxial rays. The geometrical layout is given in figure 2.16. Note that since we are using small angles the angle $\beta \approx y/R$ and $y' = y$.

From Snell's law for paraxial rays we have,

$$n\theta = n'\theta'$$

or

$$n(u + \beta) = n'(u' + \beta)$$

Solving for u_2 in terms of u_1 and substituting the angle β from above we obtain another system of equations for refraction at a spherical boundary of radius R separating two media.

$$y' = y$$
$$u' = \left(\frac{n - n'}{n'}\right)\frac{1}{R}y + \frac{n}{n'}u$$

Therefore, the system of equations for refraction can also be written as,

$$y' = y$$
$$u' = -\frac{\phi}{n'}y + \frac{n}{n'}u$$

In order to analyze a paraxial optical system comprised of many surfaces separated by a distance between each surface we simply apply these equations for each corresponding translation and refraction.

Systematic methods for paraxial ray tracing are commonly employed such as ray trace tables [9–11] or worksheets [6, 11]. The designer specifies the physical and optical parameters at each surface (radius of curvature, separation between planes and refractive index) then computes the ray height and angle at each successive surface for a series of rays beginning in the input plane.

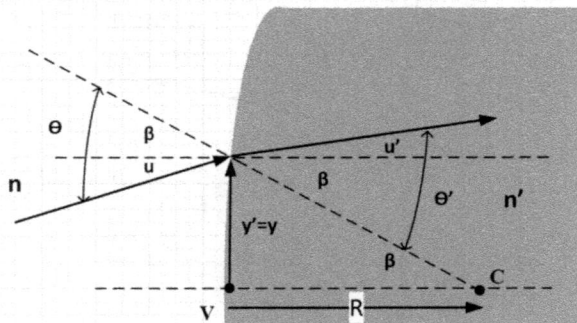

Figure 2.16. A ray intersecting a boundary of radius R separating two media.

2.7.2 Matrix methods in paraxial optics

Another common analysis method to trace paraxial rays through an optical system is to use matrices [2, 10–13]. Since we are following both a ray height and angle of inclination through successive planes, we form a vector of these two parameters at any ith surface in a system. The specific form of this vector is not standard. Many references use different vector forms, and a user must be careful and apply the correct matrices for that vector form. Here we define the ray vector as,

$$\begin{pmatrix} y_i \\ u_i \end{pmatrix}$$

Using this definition, the systems of equations for ray translation and ray refraction can be specified by 2×2 matrices. For translation from plane 0 to plane 1 the matrix equation is,

$$\begin{pmatrix} y_1 \\ u_1 \end{pmatrix} = \begin{pmatrix} 1 & t \\ 0 & 1 \end{pmatrix} \begin{pmatrix} y_0 \\ u_0 \end{pmatrix}$$

For refraction at a spherical surface between media n and n' with surface power φ the matrix equation is,

$$\begin{pmatrix} y' \\ u' \end{pmatrix} = \begin{pmatrix} 1 & 0 \\ -\dfrac{\phi}{n'} & \dfrac{n}{n'} \end{pmatrix} \begin{pmatrix} y \\ u \end{pmatrix}$$

Let us apply these concepts to the lens geometry of figure 2.9. The matrix equation connecting rays entering vertex plane 1, through the lens and leaving vertex plane 2 results in the product of three matrices, refraction at plane 1, translation to plane 2 and refraction at plane 2. Note that the order of matrix multiplication is important since we begin at plane 1 and end at plane 2.

$$\begin{pmatrix} y_2 \\ u_2 \end{pmatrix} = \begin{pmatrix} 1 & 0 \\ -\dfrac{n-n'}{nR_2} & \dfrac{n'}{n} \end{pmatrix} \begin{pmatrix} 1 & t_{\text{lens}} \\ 0 & 1 \end{pmatrix} \begin{pmatrix} 1 & 0 \\ -\dfrac{n'-n}{n'R_1} & \dfrac{n}{n'} \end{pmatrix} \begin{pmatrix} y_1 \\ u_1 \end{pmatrix}$$

From the properties of matrices, we know that the product of any number of 2×2 matrices leads to a single 2×2 matrix. So, our matrix equation reduces to,

$$\begin{pmatrix} y_2 \\ u_2 \end{pmatrix} = M_{\text{lens}} \begin{pmatrix} y_1 \\ u_1 \end{pmatrix}$$

As before we let the lens thickness be small making the translation matrix through the lens thickness approach the unity matrix. This results in a thin lens system matrix,

$$M_{\text{thin lens}} = \begin{pmatrix} 1 & 0 \\ -\dfrac{n-n'}{nR_2} & \dfrac{n'}{n} \end{pmatrix} \begin{pmatrix} 1 & 0 \\ -\dfrac{n'-n}{n'R_1} & \dfrac{n}{n'} \end{pmatrix}$$

Multiplying these matrices leads to a single matrix for a thin lens in a medium of refractive index n,

$$M_{\text{thin lens}} = \begin{pmatrix} 1 & 0 \\ -\frac{\phi}{n} & 1 \end{pmatrix} = \begin{pmatrix} 1 & 0 \\ -\frac{1}{f_n} & 1 \end{pmatrix}$$

where f_n is the focal length in the medium of refractive index n.

A summary of the various system matrices found in this section are provided in table 2.1.

Suppose we want to map rays leaving a plane 0 and arriving at plane 4 with a different surface at each ith plane. Using the notation of table 2.1, the matrix equation for this mapping would simply be,

$$\begin{pmatrix} y_4 \\ u_4 \end{pmatrix} = T_{43}\mathcal{R}_3 T_{32}\mathcal{R}_2 T_{21}\mathcal{R}_1 T_{10}\begin{pmatrix} y_0 \\ u_0 \end{pmatrix}$$

No matter how many surfaces are in a system, the ray transfer results in one system matrix that describes the ray transfer.

2.8 First-order optical system design

The small angle assumptions made in the previous sections mean that the form and shape of the physical lenses and mirrors will lead to errors in the ray paths and ray intercepts for non-paraxial rays. These types of errors are called aberrations and occur for both monochromatic light (light of a single wavelength) and light that is a mixture of wavelengths (chromatic). However, even with the errors, we can still define a number of system parameters and properties necessary to begin an optical design. Although the consideration of aberrations are critical for a detailed design

Table 2.1. Matrices used to trace a ray of height y and angle u defined with the vector $\begin{pmatrix} y_i \\ u_i \end{pmatrix}$.

Translation From surface i to j over distance t_{ji}	T_{ji}	$\begin{pmatrix} 1 & t_{ji} \\ 0 & 1 \end{pmatrix}$
Refraction At ith boundary between n and n' media with a planar surface	$\mathcal{R}_i$	$\begin{pmatrix} 1 & 0 \\ 0 & \frac{n}{n'} \end{pmatrix}$
Refraction At ith boundary between n and n' media with radius of curvature R_i or surface power φ_i	$\mathcal{R}_i$	$\begin{pmatrix} 1 & 0 \\ -\frac{n'-n}{n'R_i} & \frac{n}{n'} \end{pmatrix} = \begin{pmatrix} 1 & 0 \\ -\frac{\phi_i}{n'} & \frac{n}{n'} \end{pmatrix}$
Mirror Radius of curvature R	M_m	$\begin{pmatrix} 1 & 0 \\ -\frac{2}{R} & -1 \end{pmatrix}$
Thin lens in air Focal length f or power φ	M_{thin}	$\begin{pmatrix} 1 & 0 \\ -\frac{1}{f} & 1 \end{pmatrix} = \begin{pmatrix} 1 & 0 \\ -\phi & 1 \end{pmatrix}$

they are not considered here and are described in detail in other references [6, 8–11]. Higher order design can then be applied as necessary at a later stage of the total design process.

From chapter 1 we saw that a single lens imaging system is a good example of a general system concept where the input to the system is the object, the lens along with the object and image distances is the system, and the output is the image. Figure 2.17 shows the layout of a thin lens imaging system based on first-order optics along with the corresponding systems model diagram.

2.8.1 Thin lens imaging systems

In figure 2.17, the system includes the combination of object space, thin lens and image space. The system provides the map for rays emitted from points in the object plane to a corresponding point in the image plane. As in the previous sections, we could simply apply the Gaussian imaging equation, do a graphical ray trace, or apply the YNU method in order to find the image location given any object distance and focal length.

Using the YNU ray tracing equations, we trace two different rays from the top tip of the object of height $y_0 = h$ in order to obtain relationships between system parameters. For the first ray we translate from plane 0 to plane 1 at angle $u_0 = 0$, or parallel to the optical axis. This means that,

$$y_1 = y_0 = h$$

In plane 1 the lens refracts the ray toward plane 2, so that,

$$u_1 = -\phi h$$

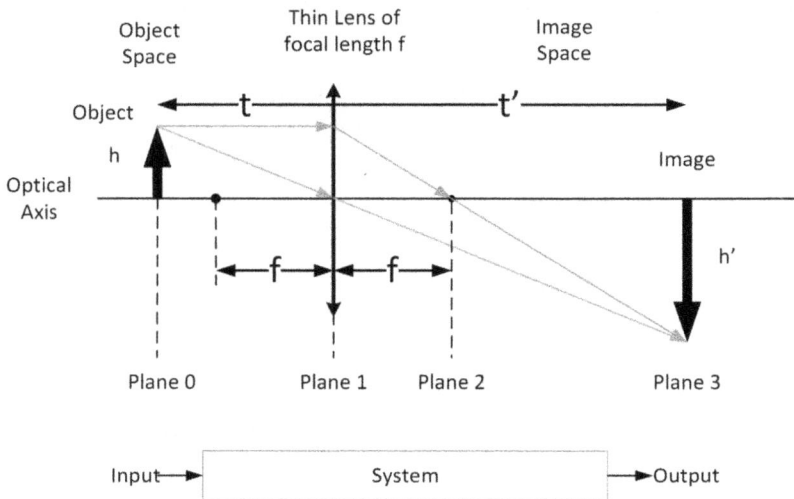

Figure 2.17. First-order optics layout of a thin lens imaging system and corresponding systems level diagram. Two rays are shown leaving plane 0 at the top tip of the object of height h. The corresponding image is in plane 3 and has height h'.

To locate the point at which this ray crosses the optical axis, we set or $y_2 = 0$, and translate an unknown distance z along the axis to find this location.

$$y_2 = 0 = h + zu_1 = h + z(-\phi h)$$

which gives the result that $z = f$. This ray crosses the optical axis at the focal point in image space. If we continue this ray to plane 3 it would intersect the top tip of the image.

Now, let us follow another ray from the object tip in plane 0. This ray leaves plane 0 and intersects the lens at a point where $y_1 = 0$. Again, $y_0 = h$, but for this ray we take the ray angle to be $u_0 = y_0/t = h/t$. (Remember t is negative by our sign convention meaning the u_0 is a negative angle.) Now, since $y_1 = 0$ for this ray, $u_1 = u_0$. This ray is said to be the un-deviated ray since the lens has no refractive effect as it passes through the center of the thin lens. Continuing to plane 3, the image plane, is simply a matter of applying a translation through the distance t', where $u_1 = y_3/t' = h'/t'$. Thus, this ray trace provides us with the relation for the magnification of the image in terms of the image and object distance.

$$m = \frac{h'}{h} = \frac{t'}{t}$$

Again, because t is a negative value in figure 2.17 the image is inverted and $h' = y_3$ is negative.

The matrix method used above is another way to take a systems view of this simple imaging system. To find the image location using matrices we follow a general ray from the object plane, translate to the thin lens, apply the thin lens refraction, and then translate to the image plane. Assume that the image and object are in the same medium. Taking plane 0 to be in object plane, plane 1 to be the lens plane, and plane 2 to be the image plane we map rays from plane 0 to plane 2 with the following matrix equation,

$$\begin{pmatrix} y_2 \\ u_2 \end{pmatrix} = \begin{pmatrix} 1 & t' \\ 0 & 1 \end{pmatrix}\begin{pmatrix} 1 & 0 \\ -\frac{1}{f} & 1 \end{pmatrix}\begin{pmatrix} 1 & -t \\ 0 & 1 \end{pmatrix}\begin{pmatrix} y_0 \\ u_0 \end{pmatrix}$$

Again, reducing the product of the three matrices to a single matrix gives,

$$\begin{pmatrix} y_2 \\ u_2 \end{pmatrix} = M_{\text{sys}}\begin{pmatrix} y_0 \\ u_0 \end{pmatrix}$$

where M_{sys} is the thin lens object-image system matrix. This matrix reduces to,

$$M_{\text{sys}} = \begin{pmatrix} 1 - \frac{t'}{f} & -t + t' - \frac{tt'}{f} \\ -\frac{1}{f} & 1 + \frac{t}{f} \end{pmatrix} \equiv \begin{pmatrix} M_{11} & M_{12} \\ M_{21} & M_{22} \end{pmatrix}$$

At first glance, this matrix may look a bit complicated but notice that if the matrix element $M_{12} = 0$ then we are back to the Gaussian form of the imaging equation. For any system matrix describing imaging an object, no matter how complicated, we can find the image location by setting the matrix element M_{12} to be zero. This is called the imaging condition for paraxial optics system matrices.

Another relationship or property of general systems matrices can also be observed if we take the determinant of the matrix.

$$\det M_{\text{sys}} = M_{11}M_{22} - M_{12}M_{21}$$

Since in this case $M_{12} = 0$ then the determinant is,

$$\det M_{\text{sys}} = \left(1 - \frac{t'}{f}\right)\left(1 + \frac{t}{f}\right) = 1$$

Remember that earlier we assumed that the object and image were in the same medium. If this is not the case then the determinant results in,

$$\det M_{\text{sys}} = \frac{n_{\text{input}}}{n_{\text{output}}}$$

where n_{input} and n_{output} are the refractive index of the corresponding input and output regions of the system.

If a lens system is comprised of a combination of several thin lenses, we can also use the paraxial matrix methods to reduce the lens system matrix to one effective M_{lens} matrix. Consider two thin lenses in air, one with focal length f_1 and the other with focal length f_2 and separated by a distance t, as shown in figure 2.18. We can find a single matrix for this lens system simply by multiplying matrices, or,

$$M_{2\text{ lenses}} = \begin{pmatrix} 1 & 0 \\ -\phi_2 & 1 \end{pmatrix}\begin{pmatrix} 1 & t \\ 0 & 1 \end{pmatrix}\begin{pmatrix} 1 & 0 \\ -\phi_1 & 1 \end{pmatrix}$$

In analogy with the above system matrix the M_{21} matrix element $-\phi = -1/f$, the system power. Multiplying these matrices results in,

$$\phi = \phi_1 + \phi_2 - \phi_1\phi_2 t$$

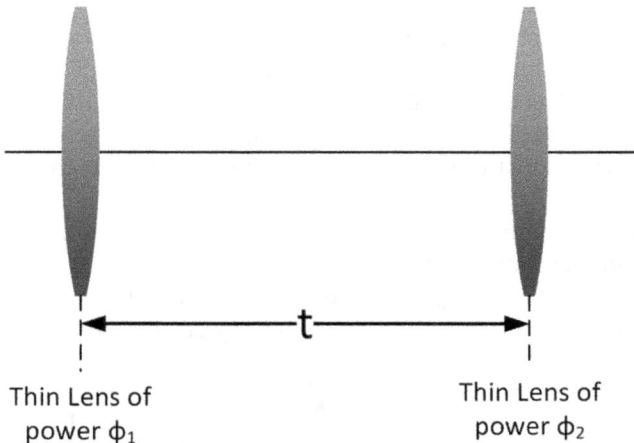

Thin Lens of
power ϕ_1

Thin Lens of
power ϕ_2

Figure 2.18. Two thin lenses of powers ϕ_1 and ϕ_2 separated by a distance t.

which is the effective optical power of the two-lens system in air. This illustrates the process known as Gaussian reduction where multiple surfaces or elements can be combined to form a single effective element. From a systems perspective, Gaussian reduction of an optical system is related to a general systems decomposition.

2.8.2 Thick lenses and cardinal points

The results of section 2.6 showed that mapping input rays through any number of surfaces to an output plane is equivalent to having a system mapping for one equivalent surface. To illustrate this concept, we again return to the lens geometry of figure 2.9. However, now we assume that the lens thickness is not small compared to other physical dimensions of the lens. This is called a thick lens.

Matrix methods are a convenient way of analyzing this situation. In fact, we began the thin lens matrix discussion by writing down the ray mapping for rays from vertex plane V_1 to vertex plane V_2 as,

$$\begin{pmatrix} y_2 \\ u_2 \end{pmatrix} = \begin{pmatrix} 1 & 0 \\ -\frac{\phi_2}{n} & \frac{n'}{n} \end{pmatrix} \begin{pmatrix} 1 & t_l \\ 0 & 1 \end{pmatrix} \begin{pmatrix} 1 & 0 \\ -\frac{\phi_1}{n'} & \frac{n}{n'} \end{pmatrix} \begin{pmatrix} y_1 \\ u_1 \end{pmatrix}$$

Now, we no longer allow the lens thickness, t_l, to be small. To simplify this calculation, we will also assume that the lens is in air so that $n = 1$ and $n' = n_l$, where n_l is the refractive index of the lens material. The result of this matrix multiplication, however complicated, again results in one matrix that describes the thick lens system.

$$M_{\text{thick}} = \begin{pmatrix} 1 - \frac{\phi_1 t_l}{n_l} & \frac{t_l}{n_l} \\ -\left(\phi_1 + \phi_2 - \phi_1 \phi_2 \frac{t_l}{n_l} \right) & 1 - \frac{\phi_2 t_l}{n_l} \end{pmatrix} \equiv \begin{pmatrix} M_{11} & M_{12} \\ M_{21} & M_{22} \end{pmatrix}$$

Notice that matrix element M_{21} is the negative of the effective optical power (reciprocal of the effective focal length) of the thick lens. This result is like the case for two thin lenses in air, the only difference is that the separation between two lenses, t, has been replaced by the effective optical thickness of the lens, $t = t_l/n_l$. As a check, the thick lens matrix reduces to the thin lens matrix when $t_l = 0$. The validity of this system matrix can also be verified by taking the determinant and showing that it is unity.

As before, we now apply the thick lens to an imaging situation using the thick lens ray matrix. We define a distance s' as the distance to the image plane of vertex of surface 2 and define s as the object distance measured from the vertex of surface 1. Then, M_{IO}, the matrix mapping rays from the object plane to image plane is,

$$M_{\text{IO}} = \begin{pmatrix} 1 & s' \\ 0 & 1 \end{pmatrix} \begin{pmatrix} 1 - \frac{\phi_1 t_l}{n_l} & \frac{t_l}{n_l} \\ -\phi & 1 - \frac{\phi_2 t_l}{n_l} \end{pmatrix} \begin{pmatrix} 1 & -s \\ 0 & 1 \end{pmatrix}$$

Since this is an imaging system, we know that the resulting matrix element M_{12} must be zero to follow the imaging condition. After performing the matrix multiplication, we obtain,

$$M_{IO} = \begin{pmatrix} 1 - \frac{\phi_1 t_l}{n_l} - s'\phi & 0 \\ -\phi & 1 - \frac{\phi_2 t_l}{n_l} + s\phi \end{pmatrix}$$

Comparing this matrix to the thin lens imaging system matrix there are obvious similarities. First, the M_{21} matrix element is the negative of total effective system power in both cases. Second, the M_{11} and M_{22} matrix elements in both the thin lens and the thick lens imaging matrices involve an object distance and an image distance, respectively. Therefore, by equating the matrix elements from each, we can then solve for the effective thin lens image and object distances in terms of the thick lens parameters. This result is,

$$t = s - \frac{\phi_2 t_l}{\phi n_l} \quad \text{and} \quad t' = s' + \frac{\phi_1 t_l}{\phi n_l}$$

To understand the meaning of these new effective distances, see figure 2.19.

The extra terms in each of the Gaussian object and distance equations are related to a shift away from the vertex planes to new locations identified by points P and P' as the thickness of the lens increases. The location of these points forms two new planes perpendicular to the optical axis called the principal planes. The distances to these locations with respect to the lens vertices are defined as,

$$d = \frac{\phi_2 t_l}{\phi n_l} \quad \text{and} \quad d' = -\frac{\phi_1 t_l}{\phi n_l}$$

Since the principal planes relate to the locations of the Gaussian image and object distances, they are planes in which rays effectively refract in following a graphical ray trace. A ray parallel to the optical axis from object space will refract at the P'

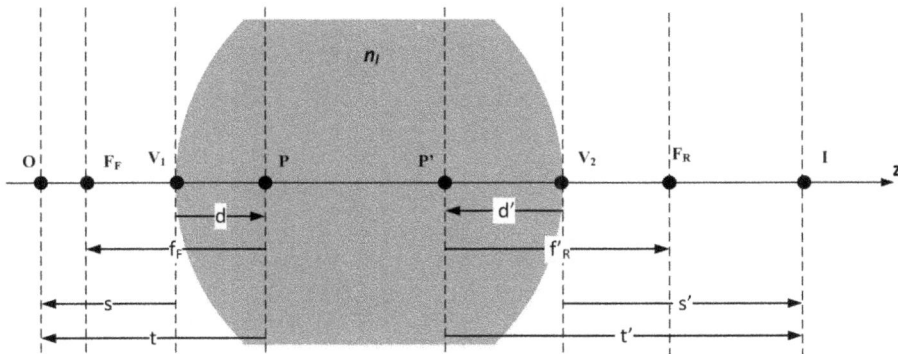

Figure 2.19. A thick lens in air showing vertex distances and effective image and object distances. Also shown are the location of principal planes including the front and rear focal planes.

plane toward the rear focal point F_R, a distance f'_R from the P' plane. A ray at some angle going through the front focal plane, F_F, a distance f_F in front of the P plane, will refract at the P plane and travel into image space parallel to the optical axis.

The points identified in this section and shown in figure 2.19 are referred to as the cardinal points of an optical system. These are the front and rear focal points (F_R and F_F) and the principal plane points (P and P'). In addition to these points there are two other cardinal points called nodal points. In a lens immersed in a uniform medium, the principal points and the nodal points coincide. If the media on either side of the lens is different, then the points P and P' shift away from the nodal points.

As discussed before, Gaussian reduction tells us that all optical systems can be reduced to a single effective element. Thus, all optical systems, no matter how many single components, have cardinal points that describe the ray propagation through the system.

In practice, physical measurements can only be made with respect to vertex planes and not the principal planes. Knowing the location of the front and back focal points with respect to the lens vertices is a straightforward measurement. Therefore, a common terminology is to define a BFD, back focal distance, and FFD, front focal distance. Since focal point distances are measured with respect to the principal planes, we can compute BFD and FFD as,

$$\text{FFD} = f_F + \frac{\phi_2 t_l}{\phi n_l} = f_F + d \ \text{ and } \ \text{BFD} = f'_R - \frac{\phi_1 t_l}{\phi n_l} = f'_R + d'$$

In designing a system, we would typically not know the complete prescription of the lens needed to meet requirements. However, we might be able to determine the matrix elements directly from a mapping of rays from an input to an output. If matrix elements are known, then locations of all cardinal points can be computed. Examining the thick lens matrix, we see that element $M_{12} = t_l/n_l$, which is the reduced thickness of the lens. Also, matrix element $M_{22} = -\phi$, or effective lens power. The other two matrix elements can be written in terms of values that we calculated earlier. For example,

$$M_{11} = 1 - \frac{\phi_1 t_l}{n_l} = 1 + \phi d' = 1 - M_{21} d'$$

So,

$$d' = \frac{1 - M_{11}}{M_{21}}$$

In a similar manner,

$$d = \frac{M_{22} - 1}{M_{21}}$$

And,

$$\text{FFD} = \frac{M_{22}}{M_{21}} \ \text{ and } \ \text{BFD} = -\frac{M_{11}}{M_{21}}$$

If the complete ray matrix of a lens system is known, all cardinal points may be located, and the lens specified.

A two-lens optical system is commonly used in experimental layouts and prototypes where one lens is not available or cannot achieve the desired specifications. See figure 2.18. This lens combination is an effective thick lens. We found earlier that the effective power of a system comprised of two lenses of power ϕ_1 and ϕ_2 is,

$$\phi = \phi_1 + \phi_2 - \phi_1\phi_2 t$$

In terms of the focal length of each of the lenses, f_1 and f_2, the effective focal length of the system is,

$$\text{EFL} = \frac{f_1 f_2}{f_1 + f_2 - t}$$

By using either YNU ray tracing or ray matrices one finds that the back focal distance for this system is,

$$\text{BFD} = \frac{f_2(f_1 - t)}{f_1 + f_2 - t}$$

2.8.3 Stops, apertures, and pupils

Another factor to consider in first order optical designs is the amount of light transmitted. Tracing just a few rays through a system locates image positions and cardinal points of a system. However, to determine the flux transfer through a system we must trace a large number of rays, each carrying a fraction of the light flux. This light flux is controlled by the size of physical stops in the system. Stops are any object, located at a plane in the system, which limits rays emitted from an on-axis object point. The specific stop whose size limits the angular cone of rays from an on-axis object point is called the aperture stop of the system.

A simple single lens imaging system can be used to illustrate the effect of stops. Figure 2.20 shows a ray launched from the on-axis object point and intersecting the lens plane at the edge of the lens of diameter D. This ray is defined as a marginal ray since the lens will not collect any rays launched at angles larger than the marginal ray angle. The smallest marginal ray angle identifies the aperture stop of the system. The aperture stop is a physical stop, lens rim or other separate stop. In this figure, the lens is the aperture stop. Also shown in this figure is the largest image height h' that can be observed due to the finite size of the detector in the image plane. The ray traced from the top tip of this largest object to the edge of the detector is the chief ray. The chief ray goes through the center of the aperture stop. The angle made by the chief ray in both image space and object space is the half-angle field of view. This angle represents the largest angular sized object that can be observed by the system. The finite detector size is also considered a stop, called the field stop. The field stop limits the angle of chief rays.

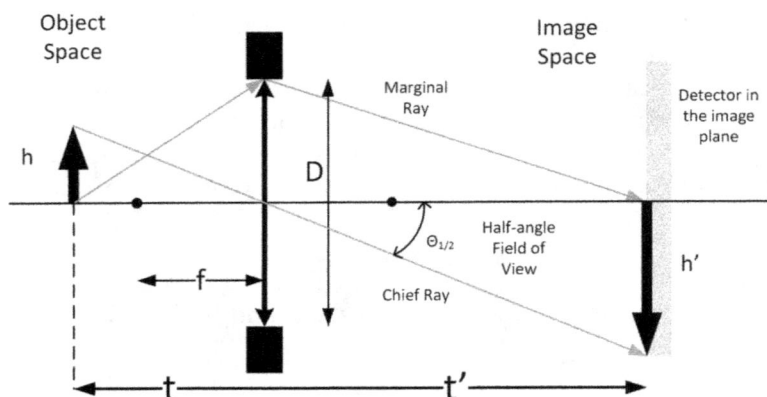

Figure 2.20. Single thin lens imaging system with lens of focal length f and diameter D. A marginal ray identifies the lens rim as the aperture stop. The chief ray specifies the half-angle field of viewing the lens.

Using the system dimensions in figure 2.20 we can develop some relationships for the half-angle field of view.

$$\tan \theta_{\frac{1}{2}} = \left| \frac{h}{t} \right| = \left| \frac{h'}{t'} \right| \equiv \text{HFOV}$$

Sometimes the field of view (FOV) or full field of view (FFOV) is used. If the detector has rectangular geometry then one must specify whether a horizontal, vertical, or diagonal measure is used for the detector size. When imaging distant objects the image is formed in the back focal plane of the system. In these cases, the field of view is,

$$\tan \theta_{\frac{1}{2}} = \left| \frac{h'}{f} \right|$$

A rule-of-thumb often used for distant objects is when the object is 10–20 times the focal length then the object can be considered at infinity and the image is formed in the focal plane [6].

In a general system, the aperture stop may not be the first element of the system, as is the case in figure 2.20. To analyze this situation, one first identifies the aperture stop from the given object position. If there are optical components in front of the aperture stop then these components influence the actual cone of rays emitted from the on-axis object point. If one traces a ray from the center of the aperture stop back through the components in front of the aperture stop, the position where this ray crosses the optical axis in object space is the image plane of the aperture stop. This process identifies the entrance pupil location. The entrance pupil is defined as the aperture stop or image of the aperture stop in object space if there are optical components in front of the aperture stop.

Likewise, if optical components are behind the aperture stop, they influence the cone of rays that fall on the image plane. Again, tracing a ray from the center of the aperture stop through the components behind the aperture stop, the position where this ray crosses the optical axis in image space is the image plane of the aperture stop. This process identifies the exit pupil location. The exit pupil is defined as the aperture stop or image of the aperture stop in image space if there are optical components behind the aperture stop.

Figure 2.21(a) shows a single lens system collecting light from a distant point object. In this case the aperture stop, the entrance pupil, and the exit pupil is the lens of diameter D. The incident light is illustrated as a bundle of rays parallel to the optical axis. The effect of the lens creates a cone of light that is focused to a point in the image plane, in this case the rear focal plane. The angle of the cone with respect to the lens is indicated as θ_0 in the figure.

Since light enters an optical system from object space and the aperture stop controls the amount of light throughput, the amount of light collected by an optical system is controlled by the size of the entrance pupil. A bundle of light rays from a distant object, traveling parallel to the optical axis, is collected across the entire entrance pupil. This bundle is transformed by the system into a cone of light focused on the image plane. We define a system parameter, $f/\#$ as,

$$f/\# = \frac{f}{D_{EP}}$$

where f is the effective focal length of the system and D_{EP} is the diameter of the entrance pupil.

Figure 2.21(b) shows a single lens imaging a distant circular object whose image size is d. In this case, the aperture stop in front of the lens is much smaller than the lens diameter and limits the amount of light found in the image. As the diameter of the entrance pupil increases the amount of light in the image plane increases, figure 2.21(c). Note that only on-axis ray bundles are shown in the figure. All ray bundles collected by the lens are limited by the entrance pupil diameter. As the entrance pupil diameter increases the amount of light collected by the system increases and the image formed in the focal plane will have more light (be brighter). The amount of light per unit area in the image is proportional to the amount of light per unit area of the entrance pupil. This implies that the amount of light per unit area in an image is inversely proportional to the square of the $f/\#$. The smaller the numerical value of the $f/\#$ the more light will be collected by system.

Another system parameter that is often specified for optical systems is the numerical aperture (NA). From figure 2.21(a), the numerical aperture is defined as,

$$NA \equiv n_0 \sin(\theta_0)$$

where n_0 is the refractive index of the image space medium. For a thin lens in air the numerical aperture can also be written as,

$$NA = \sin(\theta_0) \approx \tan(\theta_0) = \frac{\left(\frac{D}{2}\right)}{f} = \frac{1}{2f/\#}$$

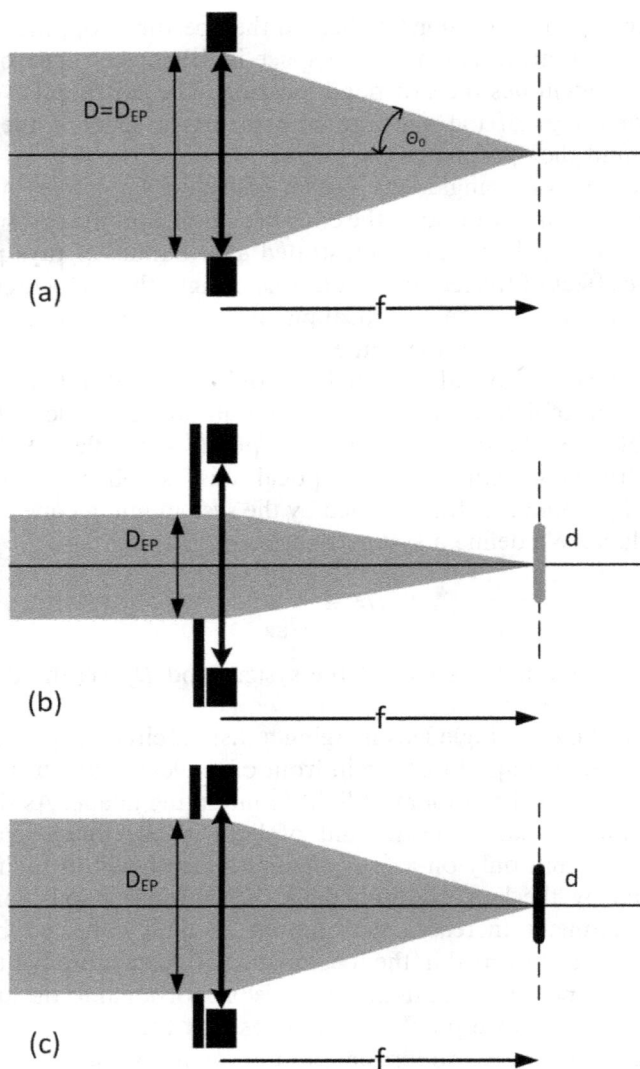

Figure 2.21. Single thin lens with lens of focal length f and diameter D. (a) A bundle of light from a distant object. (b) Lens with a small aperture stop just in front of the lens forming an image of a circular distant object. (c) Lens with a larger aperture stop in just in front of the lens forming an image of a circular distant object. (Only on-axis ray bundles are shown.)

2.8.4 Depth of field and depth of focus

Another effect on the system created by stops is a practical tolerance associated with image formation. The longitudinal distance in image space over which an image appears to be acceptably in focus is called the depth of focus. The conjugate distance in object space, or distance that an object may be moved for the same acceptable focus, is called the depth of field. The question to answer is what is acceptable. As we will see later there are other effects of diffraction as well as system aberrations that can also cause the image to be blurry. But in practice a model for an acceptable blur can be used in first order system designs [6].

Take the case of simple point source imaging. The paraxial image plane is considered the plane where the observed image size is the smallest or appears to have the smallest blur spot size. Figure 2.22(a) illustrates the concept and indicates the distance along the optical axis of the depth of focus. The amount of acceptable blur diameter, B', could be identified as a system specification. As one can see from geometry, a smaller diameter lens with a smaller exit pupil increases the depth of focus, or

$$\text{DOF} \approx \pm \frac{B'}{2\text{NA}}$$

where NA is the numerical aperture of the system. Note that since DOF is similar to an error condition the plus-minus sign is applied to indicate the distance in either direction from the paraxial image plane location. A typical rule-of-thumb is that an acceptable blur diameter to be on the order of 30 μm or $f/1000$ whichever is larger.

In image space there are two different distances from the original object position near the lens and far from the lens that create the same blur size. A ray trace is shown in figure 2.22(b). Within the two defined distances, L_{near} and L_{far} is the depth of field [6].

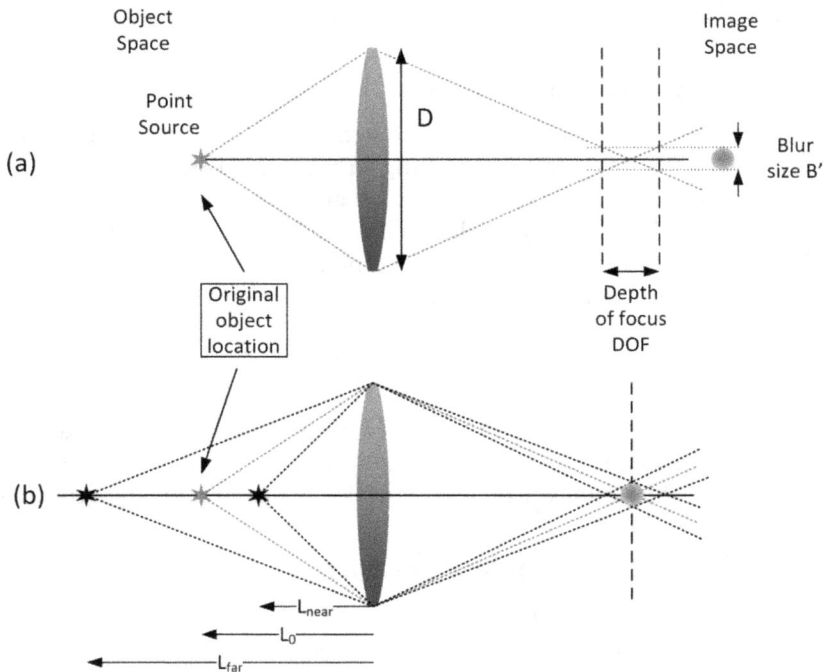

Figure 2.22. Ray diagrams for (a) depth of focus and (b) depth of field.

As with the depth of focus, applying geometry to ray tracing triangles and applying the imaging condition yields [6],

$$L_{near} \approx \frac{L_0 f D}{f D - L_0 B'} = \frac{L_0}{1 - \frac{L_0 B'}{f D}}$$

$$L_{far} \approx \frac{L_0 f D}{f D + L_0 B'} = = \frac{L_0}{1 + \frac{L_0 B'}{f D}}$$

From these relations we can see that as the lens diameter increases both far and near distances approach the nominal object position L_0, thus showing that the depth of field decreases with increasing pupil diameter.

However, one may want to have a large range of the depth of field as is required with fixed focus cameras. When the value of L_{far} is set to infinity we find that,

$$L_{near} \approx \frac{-f D}{2 B'} \equiv \frac{L_H}{2}$$

where L_H is called the hyperfocal distance. The hyperfocal distance is then,

$$L_H = \frac{-f D}{B'}$$

Using the rule-of-thumb that the value of $B' = f/1000$, a 50 mm focal length lens at $f/8$ would result in a hyperfocal distance of 6.25 m making $L_{near} = 3.125$ m. Such a system would then be in an acceptable focus from 3.125 m out to infinity.

2.8.5 Detector array limitations on systems

As discussed in chapter 1, many optical systems, especially imaging systems, use array detectors in the image plane resulting in a sampled output function at each pixel location. As we saw earlier, the field of view of an optical system is limited by the field stop. In a detector array, each specific pixel location essentially becomes its own field stop. In this case, the value of the pixel size, y_{pixel}, divided by the focal length, f, is defined as the instantaneous field of view (IFOV).

$$IFOV = \frac{y_{pixel}}{f}$$

IFOV is always a full-angle field of view. The result of sampling and quantization limit the ultimate resolution that can be achieved by any optical system. The pixel size may also provide a limit on the blur size desired as previously discussed.

2.9 Issues in assembling optical systems

First-order designs are done, not only as a starting point of a more complicated lens system but often as the starting point for a prototype optical system development. By definition, a first-order design assumes that the components in the design introduce no aberrations. However, some of the effects due to spherical aberration and chromatic aberration can be minimized by making appropriate choices of specific types of lenses. Small errors in placing any optical component into a beam can cause large errors in the expected image positions both on and away from the optical axis. In addition, the dimensional tolerances of lenses and mechanical mounts can also introduce errors in ray angles and locations. While this section is not meant to be a complete list, understanding at least some of these issues is necessary in assembling an optical system.

2.9.1 Picking off-the-shelf lenses for designs

Many times, often for cost saving, both the system design and system construction are constrained to have only off-the-shelf components. Even if the system is designed using the paraxial approximation and thin lens relationships, making the correct choice of components and placement of components in a system layout can help minimize some of the effects of aberrations. With so many lenses and lens types available, the first question to answer is which type of lens is best for my specific application.

Figure 2.10 shows cross-sections of common lens types. Each type has advantages and disadvantages depending on the application. Below are some considerations in picking lenses for a design using off-the-shelf components.

Positive focal length lenses, bi-convex and plano-convex are used in imaging applications and to focus collimated beams and collimate point sources. Bi-convex lenses are best used in imaging applications where the object is on one side of the lens and the image is formed on the opposite side. Plano-convex lenses take incident collimated light and focus to a point or collect incident light from a point source and form a collimated beam. To minimize spherical aberration when using plano-convex lenses, orient the curved surface of the lens to focus an incident collimated beam, and orient the planar surface toward the point source in forming a collimated beam.

Positive and negative focal length meniscus lenses are generally designed to minimize spherical aberration, and when used in conjunction with a stop can minimize other aberrations. In a system design combining other lenses, using a positive meniscus lens will generally decrease the focal length and the $f/\#$ of the system. In a system design combining other lenses, using a negative meniscus lens will generally increase the focal length and the $f/\#$ of the system. As with plano-convex lenses in collimated beams, to minimize spherical aberration, the convex side of the lens should be oriented toward the source.

Negative focal length lenses, bi-concave and plano-concave are both used to diverge collimated beams. When used in a system design with multiple lenses, they can help offset the effects of spherical aberration caused by other positive focal length lenses. To minimize spherical aberration, orient the curved surface of the lens toward the source.

There are several specific lens designs worth mentioning that are also off-the-shelf components. One design is called a 'best form' lens where the radius of curvature for each surface is chosen to minimize both spherical aberration and coma. This may be a good choice when one lens is needed in a prototype development. Another specific design is the achromatic doublet, a combination of a positive and negative lens that work together to minimize chromatic aberration. The design is based on using lenses made from two different materials with different dispersion curves, refractive index as a function of wavelength. Each lens will have a different focal length at two different wavelengths, but the combined effective focal length is the same at the two wavelengths. There are also aspheric lenses, where the surfaces are not spherical, designed for specific applications. These applications include laser diode collimation, fiber optic coupling, and spatial filtering.

2.9.2 Beam deviations by plates and prisms

Even simple component geometries and elemental shapes can cause beam deviations and image position errors. As a starting point, we simply examine ray propagation through a parallel plate using YNU ray tracing. Figure 2.23 shows the geometry of a plate with refractive index n and thickness d immersed in a medium of refractive index n_a. The incident ray encounters the plate at point P_1, refracts and propagates to point P_2, then again refracts into the external medium. The rays drawn are for the case when $n > n_a$.

Take the initial ray with height $y_0 = 0$ and angle u_0. After refraction, the new ray height and angle are,

$$y_1 = y_0 = 0$$
$$nu_1 = n_a u_0$$

Translating to point P_2, the new ray height and angle are,

$$y_2 = u_1 d = \frac{n_a}{n} u_0 d$$

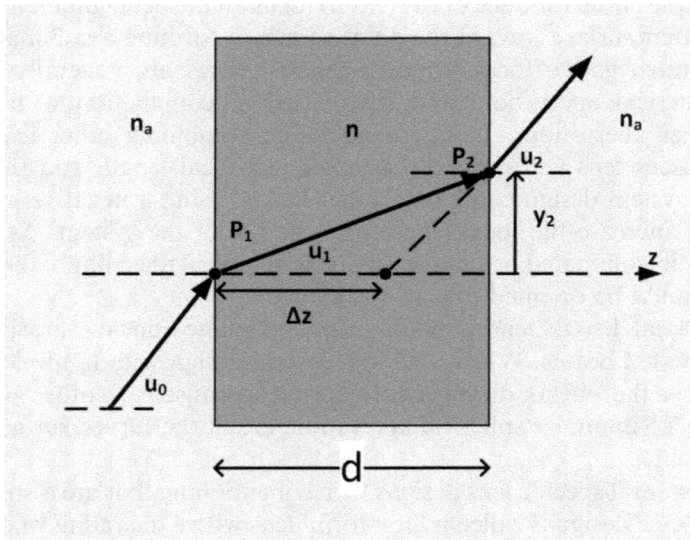

Figure 2.23. Ray propagation through a parallel plate.

And refracting again at point P_2,

$$u_2 = \frac{n}{n_a} u_1 = u_0$$

So, the exiting ray angle is equal to the incident ray angle, but the ray exits the plate off axis at height y_2. We can also trace the exit ray back to a location where it appears to originate at a plane in front of the exit face where the ray crosses the axis at $y = 0$, shown as a dotted line in figure 2.23. The distance in front of the exit plane is, $-d(n_a/n)$ resulting in a shift of,

$$\Delta z = d\left(\frac{n - n_a}{n}\right) = \frac{\text{OPD}}{n}$$

This same relationship also predicts the z-axis shift of and image point when a plate is placed in a converging beam [9, 14]. A typical value for the refractive index of a glass material is 1.5. Using this value in the equation above, we find that the image shift by a parallel plate of glass in air is $\Delta z \approx d/3$. Figure 2.24(a) illustrates the shift of the image point along the z-axis when a parallel plate is placed into the converging beam. Also shown in figure 2.24(b) is the beam shift of an incident ray by a rotated parallel plate. If the parallel plate is rotated in a converging beam, the image point shifts along the z-axis and along the y-axis

Another elemental shaped plate that causes a ray to deviate is a thin prism wedge. Figure 2.25 shows the cross-section of a wedge of angle α and refractive index n with an incident ray parallel to the base in a medium of refractive index n_a.

The incident ray is transmitted through the first surface along the z-axis and refracts at the second surface. Applying Snell's law at this boundary, we obtain,

$$n \sin \alpha = n_a \sin(\alpha - \delta)$$

remembering that δ is a negative angle. If the wedge angle, α, and the deflection angle, δ are both small, then,

$$\sin \alpha \approx \alpha; \cos \alpha \approx 1; \sin \delta \approx \delta; \cos \delta \approx 1$$

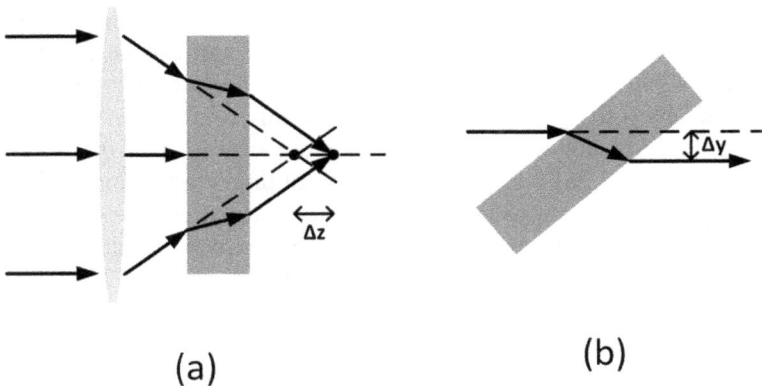

(a)

(b)

Figure 2.24. Rotated parallel plate in a parallel beam and in a converging beam.

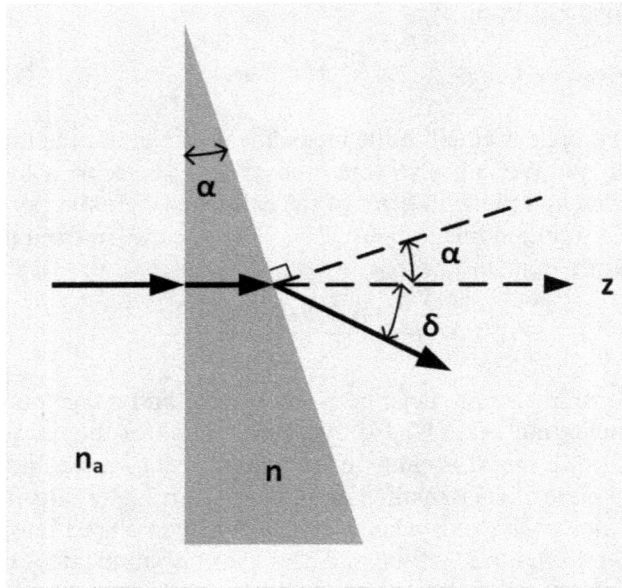

Figure 2.25. Ray deviation by a thin wedge prism.

Expanding Snell's law and substituting these approximations into the relation results in,

$$\delta \approx -\left(\frac{n - n_a}{n_a}\right)\alpha$$

This is fine for monochromatic light but if the incident ray contains multiple wavelengths then we must account for the effects of dispersion [6].

Larger prisms, like the right-angle prism component shown in figure 2.26(a), are used in optical systems to create large ray angle deviations. Since the ray is reflected by total internal reflection, at least for ray angles near the axis, this component choice has the advantage that there are lower losses than using a mirror. The disadvantage of using a prism like this in a system design can be the weight of the glass component. The reflected ray changes the optical axis and creates a coordinate break. However, by folding the component about any total internal reflection plane a ray tracing analysis can be performed as if the optical axis of the system was maintained. See figure 2.26(b) for an example of folding a right-angle prism. The end result is that the folded prism placed in an optical beam alters incoming rays exactly as a parallel plate. This diagram of the folded prism is called a tunnel diagram [6, 14]. The overall effect, like a parallel plate, is a reduced path length. The tunnel diagram is especially helpful in finding the clear aperture for any type of prism and prism orientation used in a system design.

Using a similar approach, we next investigate rays incident on plane mirrors and the effect of the reflected rays when the mirror is tilted. Figure 2.27(a) shows a single

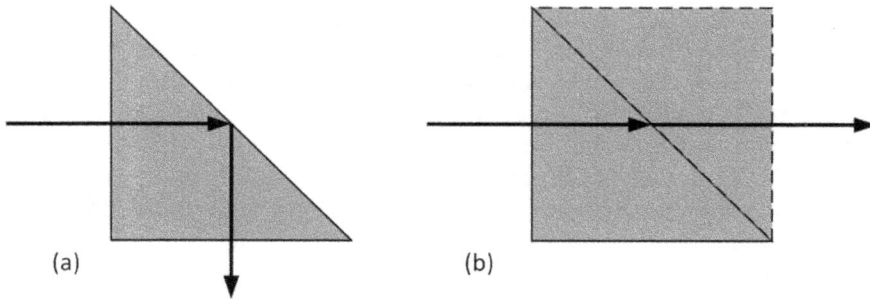

Figure 2.26. (a) Ray deviation by a right-angle prism and (b) tunnel diagram created by a fold about the total internal reflection surface. The effect on the central axis ray is to create an effective parallel plate.

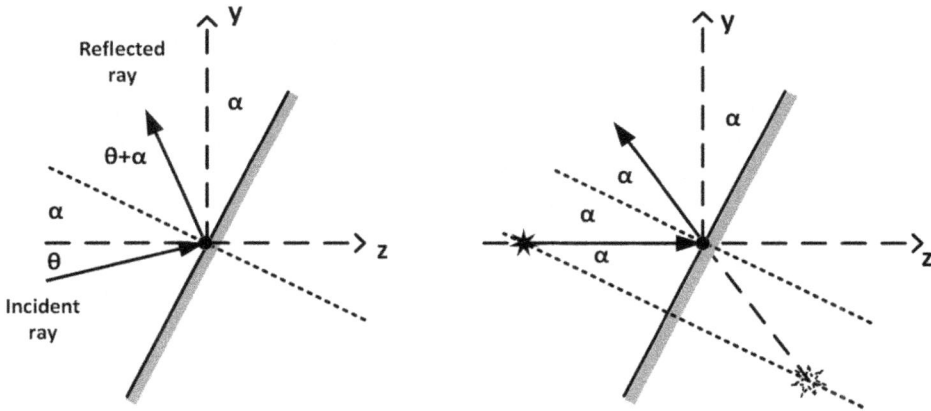

Figure 2.27. Effect of rays and image locations due to mirror tilt.

ray incident on a plane mirror at the origin of the (y,z) coordinate with angle θ. If the mirror was not tilted, then the ray would reflect at angle θ with respect to the z-axis. However, when the mirror surface is tilted by an angle α with respect to the y-axis the ray deviates to a higher angle. As the ray diagram shows, the reflected ray is now at angle $\theta + \alpha$. This means that tilting a plane mirror by angle α produces a deflection of 2α with respect to the ray direction with no tilt.

Figure 2.27(b) shows the formation of the virtual image of an object in front of a tilted plane mirror. In this case the rays deviate by angle 2α as before, but the image rotates only by angle α with respect to the z-axis.

Placing prisms or mirrors in an imaging system will alter the orientation of the final image produced. This will depend on the number of reflections that occur [6].

2.9.3 Component placement errors

In taking an optical design from a theoretical layout to construction either in a laboratory setup or in a prototype build, the challenge is to precisely place the physical component on a defined optical axis at a specific location along the axis. As described previously, the line joining centers of curvature defines the optical axis of a lens or mirror. However, the mechanical axis of the mount holding the component may be different than the optical axis [14] or could change due any number of environmental conditions such as temperature and vibration [15].

Figure 2.28 illustrates the types of potential errors that can occur in placing components in relation to a given optical axis defined as the z-axis. Although a single lens is shown in the diagram, the concept would be the same for any other component such as a mirror, window, or prism. Aligning an optical component's optical axis with the mechanical axis of a mount is called centering. Figure 2.28(a) shows a centering error of distance Δy_L away from the optical axis. A component may also be tilted within a mount producing an angular deviation $\Delta \theta_L$, figure 2.28(b). In mounting multiple lenses together in a system, the separation distance between elements must also be held to an acceptable tolerance defined by the designer [15]. This would lead to an error in spacing of components by a distance Δz_L, from the desired location as shown in figure 2.28(c). Any one of these effects could potentially change the first-order design predictions for beam and image locations. It is important to remember that since we are talking about errors, all these linear and angular displacements are assumed to be small deviations. With these issues in mind, it is useful to analyze the effects of beam shifts due to mechanical motion of components [14, 16].

Lateral errors (or decenters) as shown in figure 2.28(a) will deflect an incident parallel beam through an angle. Since all parallel rays must go through the focal point the deflection angle δ will be,

$$\delta = \frac{\Delta y_L}{f}$$

This is also true for lateral motion of mirrors.

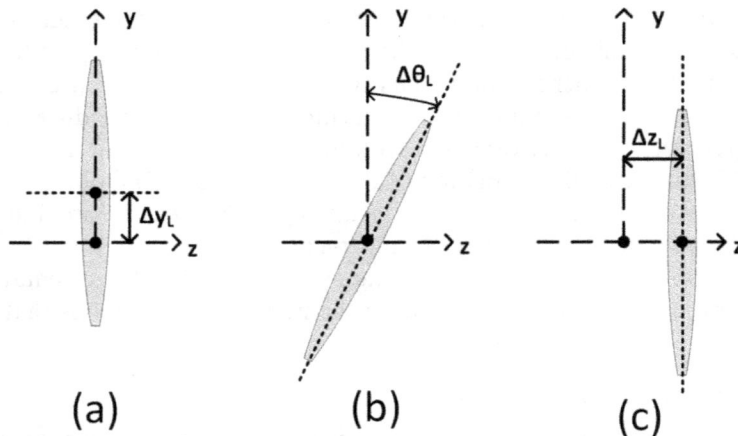

Figure 2.28. Potential linear and angular displacement error effects in placing a lens at a particular coordinate.

Errors in image positions will also arise from lateral errors. The image location of an on-axis object point for a decentered lens can be found using calculations or graphical ray tracing. The result is that the image is displaced from the optical axis by an amount, Δy_I

$$\Delta y_I = (1 - m)\Delta y_L$$

where m is the image magnification [14, 16]. For an object at infinity, $\Delta y_I = \Delta y_L$.

Now, consider rotating optical elements with power, like lenses and mirrors. If the rotation is about a nodal point, as shown with the thin lens in figure 2.28(b), there is no effect on the predicted image location. An incident ray along the optical axis is the un-deviated ray in a geometric ray tracing. Tilting the thin lens will have no effect on this ray. Tilting a mirror with power about a nodal point deviates the ray as if the mirror were a plane mirror.

For a thin lens imaging system, we know that the image location t' is can be found from the object location and the focal length.

$$\frac{1}{t'} - \frac{1}{t} = \frac{1}{f}$$

For an object position error of dt, the error in the image position, dt', can be found by taking the appropriate derivatives as,

$$dt' = dt\ m^2$$

where m is the magnification, $m = t'/t$. Using this result, if a lens is shifted along the axis by a distance Δz_L, as shown in figure 2.28(c), then the error in the image location, Δz_I, is

$$\Delta z_I = (1 - m^2)\Delta z_L$$

The relationships above are particularly useful when developing an optical system in a laboratory or when developing a prototype. Observing and measuring a ray intercept error or image location error will lead back to the root cause of the error in optical element placement. They also provide a designer with some initial guide to tolerances that may arise in constructing a system.

Of course, tilting a thick lens about any other point besides a nodal point will cause a ray to deviate [17]. Again, because we are only interested in potential errors in ray intercepts and image locations all angles are assumed to be small. We know from ridged body rotation that a rotation about one point is geometrically equivalent the rotation about another point plus a translation. Figure 2.29(a) shows a thick lens rotated about point C (a distance r in front of the principal plane P) by an angle α with respect to the z-axis. This total motion is equivalent to a translation of the lens by height Δy_L, as shown in figure 2.29(b) and rotation about the nodal point of plane P by angle α, figure 2.29(c). Also shown in figures 2.29(b) and (c) are the effects of incident rays due to translation and tilt. Thus, the effect of a ray incident on a thick lens rotated about any point on axis is a combination of the ray deviations from a translation and a tilt.

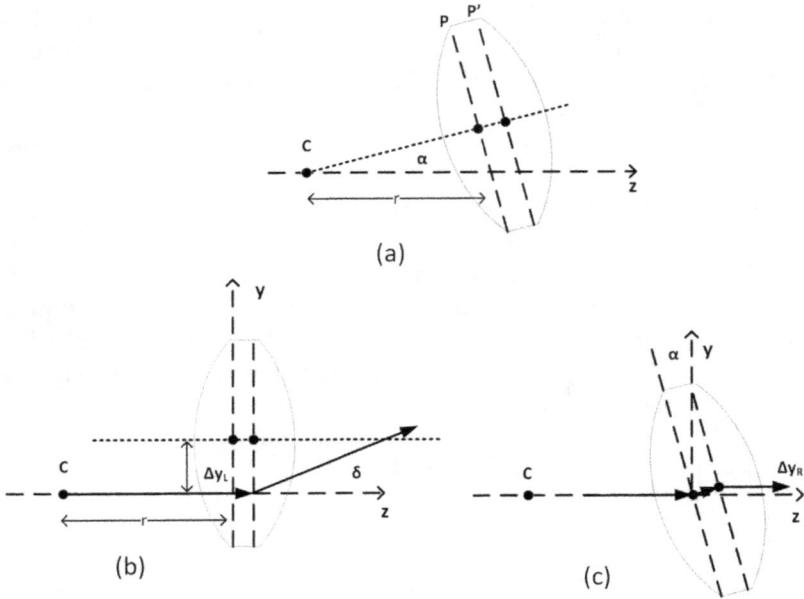

Figure 2.29. (a) Rotation of a thick lens about a point C on the optical axis by angle α, (b) lateral translation of the lens by Δy_L, and (c) rotation about a nodal point.

From the geometry and the optics for figure 2.29(b), we find,

$$\Delta y_L \cong \alpha r \ \text{ and } \ \delta \cong \frac{\Delta y_L}{f}$$

where f is the focal length of the lens. For the rotation by angle α in figure 2.29(c), the shift in ray height due to rotation about a nodal point is,

$$\Delta y_R \cong \alpha d_{PP'}$$

where the distance $d_{PP'}$ is the distance between principal planes. Therefore, the total ray intercept error Δy in the image plane a distance t' from the P' plane is,

$$\Delta y \cong \alpha \left(\frac{r}{f} \delta t' + d_{PP'} \right)$$

A general a set of relations has also been developed to analyze the effects of the motion of individual elements in more complicated optical systems. These relations are especially useful in understanding issues of the stability of a system and sensitivity of the system to vibration [17].

2.9.4 Mounting errors

Holding an optical component in place within a system requires some type of mount. There are numerous manufacturers and distributers of off-the-shelf individual mounts and mounting systems [18, 19].

A common practice used in lens mounting is to place the lens within the inner diameter of a tube. These tubes can be custom designs or commercial prototype tube products [18]. Tube geometry mounts rely on the lens to be centered in the tube with the rim in close proximity to the inner diameter of the tube. The assumption is that the mechanical axis of the mount aligns with the optical axis of the element. The optical axis is defined as the line that connects the two centers of curvature. Ideally, the optical axis and the mechanical axis would be identical since the center of curvature of each lens surface would be located on the mechanical axis. However, this is not always the case and the optical axis of the lens may be decentered and no longer align with the mechanical axis of the mount [15]. An exaggerated drawing of this is shown in figure 2.30.

As shown in the figure, a fabrication error could create a difference in the thickness of the edges of the lens. The edge thickness difference, ETD, is,

$$\text{ETD} = t_{\max} - t_{\min}$$

Creating a ray deviation similar to the addition of a thin prism wedge. The wedge angle is then simply,

$$\alpha = \frac{\text{ETD}}{D}$$

where D is the diameter of the lens. This wedge angle will deviate an incident ray away from of the mechanical axis of the system complicating system alignment.

Spherical lenses mounted in tubes or circular mounts are typically held in place by retaining rings that apply axial force on the lens. These forces can introduce stress in the lens material depending on the amount of the restraining force or preload [15].

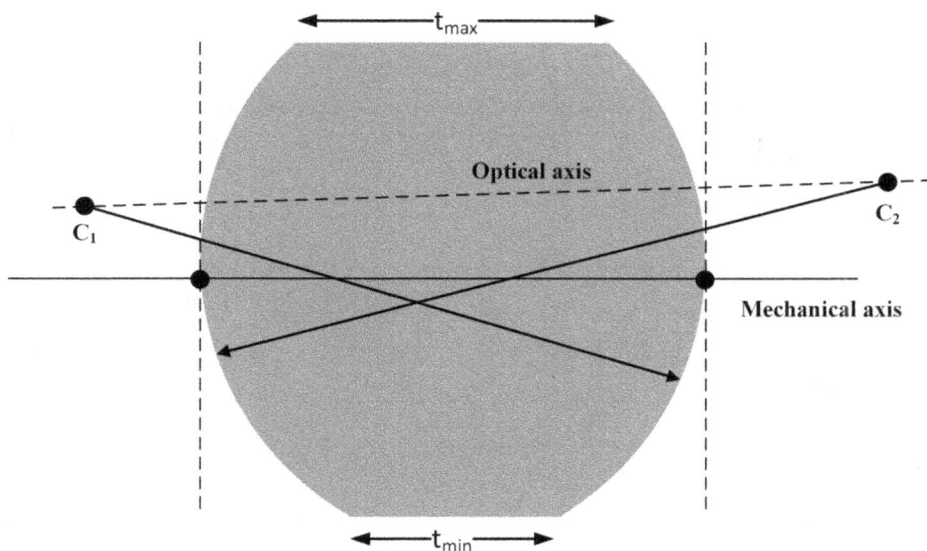

Figure 2.30. Difference between the mechanical axis and the optical axis of a lens.

2.10 Application: optical fiber numerical aperture

Optical fibers have quickly become ubiquitous in communication systems and networks. While the concept of communicating with light waves is not new the ability to guide light from point-to-point with little loss and at high data rates is unprecedented [20, 21].

The physical structure of an optical fiber is a concentric cylindrical geometry with an inner core and an outer cladding. The core and the cladding can be made from many transparent materials, but most communication fibers are made from glass. Figure 2.31 shows a schematic of a fiber. The inner core has a radius, a, and is composed of a low loss transparent material of index n_1. When the core is one only material this is called a step-index fiber. Other core structures are designed and manufactured to optimize light guiding properties. The cladding surrounds the core and has a refractive index n_2 which must be less than n_1. The buffer is a protective coating that is applied to the inner fiber structure [22].

A parameter called the numerical aperture (NA) is one of the specifications provided by optical fiber manufacturers for any fiber. The NA controls the light collecting ability of the fiber. We can use ray optic principles to find a relation for the numerical aperture. From a cross-section view of a fiber core and cladding we will examine rays incident on the core from an external medium at various angles, figure 2.32. The plane of figure 2.32 is called a meridional plane and can be any plane containing the central axis of the fiber. We will also assume for simplicity that

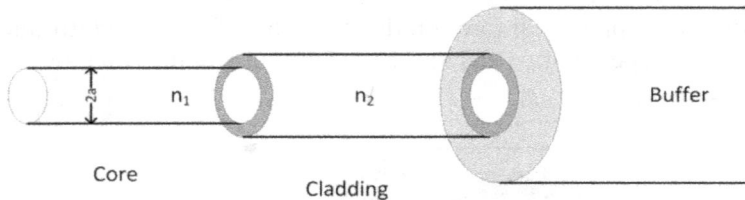

Figure 2.31. Optical fiber structure with an inner core of radius, a, and refractive index n_1 surrounded by a cladding material of n_2 and all surrounded by a protective buffer coating.

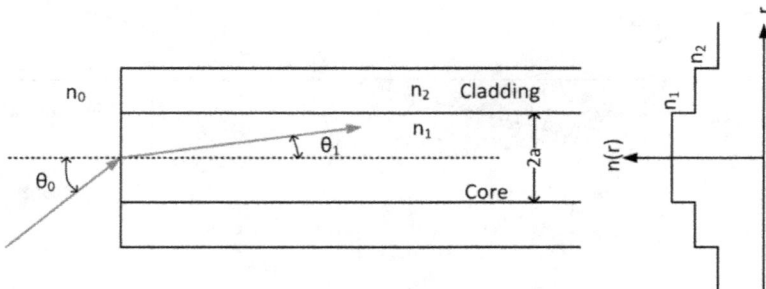

Figure 2.32. Cross-section of a step-index optical fiber in one plane called the meridional plane. A plot of the refractive index as a function of radius is shown corresponding to the core and cladding. The external medium has refractive index n_0. A general ray at angle θ_0 to the axis is refracted into angle θ_1 into the core.

the rays entering the fiber are on the axis from external medium of refractive index n_0.

For a ray incident at angle $\theta_0 = 0$ the ray will simply propagate down the axis with $\theta_1 = 0$. If we trace the ray for a large incident angle it will refract into the core, encounter the core-cladding boundary, refract into the cladding and eventually escape the fiber. However, there is an incident angle $\theta_0 = \theta_{max}$ such that the ray incident on the core-cladding boundary is the critical angle for total internal reflection. This situation is shown in figure 2.33.

Applying Snell's law at the incident face of the fiber, and using the geometry of the triangle formed by the refracted ray, we find,

$$n_0 \sin \theta_{max} = n_1 \sin \theta_1 = n_1 \cos \theta_C$$

The quantity $n_0 \sin \theta_{max} \equiv NA$, the numerical that we discussed earlier related to the $f/\#$ optical system parameter.

The total internal reflection condition at the core-cladding boundary gives,

$$n_1 \sin \theta_C = n_2$$

From this relation and geometry, we also know that,

$$\cos \theta_C = \sqrt{1 - \sin^2(\theta_C)} = \sqrt{1 - \left(\frac{n_2}{n_1}\right)^2}$$

Resulting in the relationship for the numerical aperture of an optical fiber,

$$n_0 \sin \theta_{max} = \sqrt{n_1^2 - n_2^2} = NA$$

All rays with angles less than the value θ_{max} will be guided within the core of the fiber due to total internal reflection. This forms a cone of incident rays often called the numerical aperture acceptance cone.

As with any optical system, the light collecting ability of an optical fiber is governed by numerical aperture. Specifically, for the fiber geometry, the numerical aperture is determined by the refractive index of the core and cladding. For an

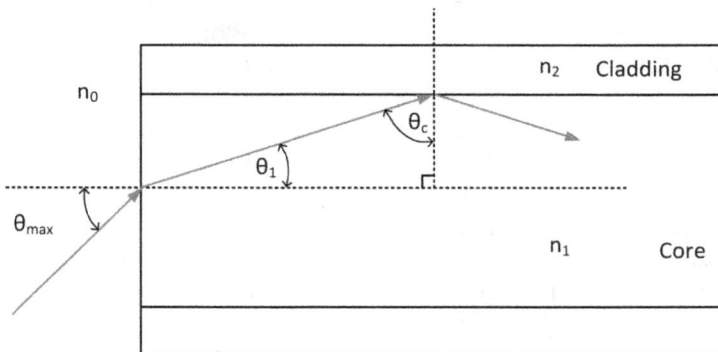

Figure 2.33. Ray geometry for determining numerical aperture of a step-index optical fiber.

optical fiber to collect a maximum amount of light we desire that the numerical aperture be as large as possible. From our simple analysis here, that would tell us that the core refractive index must be made large compared to the cladding refractive index. However, other considerations lead to a condition that requires the different between the core and cladding refractive indices be small limiting the value of the NA [22]. Values of numerical aperture for various types of optical fibers range from 0.15 to 0.5.

Exercises and problems

1. Find the $f/\#$ of a single lens that had a 50 mm focal length and an aperture of 25 mm.
2. Why are equally spaced stop positions ($f/\#$) on a camera listed as 1.44, 2, 2.8, 4, 5.6, 8, 11, 16? Explain how the amount of light collected by the camera varies with each f-stop.
3. An object 2.0 cm high is positioned 5.0 cm to the left of a positive thin lens with focal length of 10 cm. Completely describe the characteristics of the resulting image.
4. Compute the focal length in air of a thin bi-convex lens ($n = 1.5$) having radii of +20 cm and +40 cm. Locate and describe the image of an object placed 40 cm in front of the lens.
5. Design an eye for a robot using a concave spherical mirror such that the image of an object 1.0 m tall and 10 m away fills its 1.0 cm × 1.0 cm photosensitive detector (which is moveable for focusing purposes). Where should this detector be located with respect to the mirror? What should be the focal length of the mirror?
6. Two thin lenses with 5.0 cm diameters have effective focal lengths of 10.0 cm and −50.0 cm, but are cemented together to form a thin lens doublet. A field stop of 5.0 cm (diameter) is placed at the back focal point of the doublet to image distant objects.
   (a) What is the ($f/\#$) for the doublet system?
   (b) What is the full field of view?
7. In paraxial optics, Snell's law is written as $n\theta = n'\theta'$ using the small angle approximation. Take $n = 1$ and $n' = 1.5$. Plot the approximate and exact results for θ' versus θ in the range of $\theta = 0$ to 90°. Over what angular range of θ is the paraxial approximation good to 0.1%, 1%, and 10%?
8. A lens is used to image a 1.0 cm high object placed 20.0 cm in front of the lens. This lens is a double convex lens (the magnitudes of the radii are equal on both surfaces) with a focal length of +10.0 cm and refractive index of 1.60. (a) Find the radii of curvature of this lens. (b) A second thin lens with a focal length of +4.0 cm is placed a distance 25.0 cm behind the first lens. Where is the final image located with respect to the second lens? (c) What is the height of the final image?
9. A concave mirror forms the image of an object on a screen so that the image is twice as large as the object. When both object and screen are moved the

image on the screen is now three times the size of the object. If the screen is moved 75 cm in the process, how far was the object moved and what is the focal length of the mirror?

10. A diverging thin lens and a concave mirror have equal magnitude focal lengths. An object is located 15 cm to the left of the diverging lens, and the mirror is located 30 cm behind the lens. Find the position of the final image formed by the system.

11. The left end of a long glass rod of index 1.555 is ground and polished to a convex spherical surface of radius 2.50 cm. (a) The beam from a laser pointer (620 nm wavelength) is incident from the left, in the air, along a line parallel to the axis of the cylinder at a height of 0.5 mm above the axis. At what position, measured with respect to the vertex of the spherical surface, does the laser beam cross the axis? (b) An object of height 0.5 mm located in the air and on the axis at a distance of 9.0 cm to the left of the vertex. Find the image distance.

12. An object measures 2.0 cm high above the axis of an optical system consisting of a 2.0 cm diameter aperture stop and a thin convex lens of 5.0 cm focal length and 5.0 cm aperture. The object is 10.0 cm in front of the lens and the stop is 2.0 cm in front of the lens. Determine the position and size of the image, entrance pupil, and exit pupil.

13. A thin lens with a 50.0 cm effective focal length and $f/\#$ equal to 12.5 is used to image an object 20.0 cm in front of it. A screen with a 2.0 cm diameter hole is placed 3.0 cm to the left (in front) of the lens. What are the locations of both the entrance and exit pupils with respect to the lens?

14. An object is positioned 60 cm in front of a thin lens (L1) of focal length +20 cm. A second thin lens (L2) is placed 12.0 cm behind the first lens. An image of the object is observed on a screen 20 cm behind L2. (a) What is the focal length of lens L2? (b) Find the magnification of the system.

15. The matrix below was computed to describe a thick glass lens in air.

$$\begin{bmatrix} 0.66667 & 2.0000 \text{ cm} \\ -0.166667 \text{ cm}^{-1} & 1.0000 \end{bmatrix}$$

(a) Find the effective focal length of the lens.
(b) Determine the location of the principal planes with respect to each vertex.
(c) An object is placed 15 cm to the left of the left lens vertex. Find the position of the image in relation to the right vertex.

16. The dimensions and refractive index of an optical element is unknown. Using experimental techniques, you find that the distance from the front focal plane to the left vertex is 6.0 cm, the distance from the right vertex to the back focal plane is 4.0 cm, and the focal length is +6.0 cm. Find the system matrix for this element in air.

17. A positive thin lens of focal length +10.0 cm is placed in front of a negative thin lens whose focal length is −10.0 cm. The system matrix for this combination lens is:

$$\begin{pmatrix} \dfrac{1}{2} & 5 \text{ cm} \\ \dfrac{-1}{20} \text{ cm}^{-1} & \dfrac{3}{2} \end{pmatrix}$$

    (a) What is the separation between the two lenses?
    (b) An object is placed 40 cm to the left of the positive lens. Where is the image located with respect to the negative lens?

18. An object of height +2.0 cm is located in front of an optical imaging system 25.0 cm from the left vertex. An image of the object was found 10.0 cm behind the right vertex with a height of −5.0 cm. A ray leaving the top tip of the object at an angle of +0.025 rad goes through the top tip of the image at an angle of −0.050 rad. The object and image are in air. Find the values of the image-object matrix which connects rays from the object plane to the image plane.

19. Analysis of a symmetric double convex lens (refractive index of 1.600) in air resulted in the following system matrix.

$$\begin{bmatrix} 0.9400 & 1.0000 \text{ cm} \\ -0.1164 \text{ cm}^{-1} & 0.9400 \end{bmatrix}$$

Compute the vertex-to-vertex thickness of the lens and the magnitude of the radius of curvature of the lens surfaces.

20. A Schmidt-type prism is shown in figure P2.20. The apex angle is 30° and the base angles are 75°. The refractive index is 1.5.

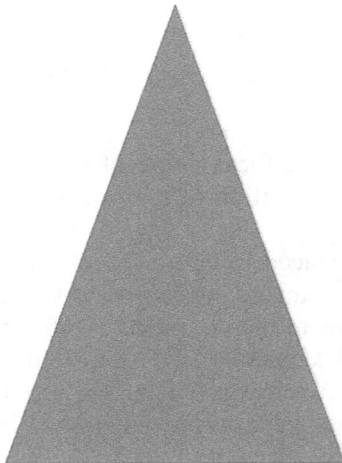

Figure P2.20. Schmidt prism with a 30° apex angle.

(a) The base is usually silvered to form a mirror surface. Why do you think this might be necessary?

(b) Draw a tunnel diagram for this prism. For two incident rays from the left, parallel to the prism base, sketch their propagation through the system.

(c) Comment on the clear aperture available. (You do <u>not</u> have to do a calculation of the clear aperture.)

(d) For an incident right-handed axis, determine the orientation and handedness of the final image. Comment on how this prism might be used in an optical system.

21. You need to view an object a large distance away from a single lens system in front of a 5.00 mm square detector array with 800×800 pixels. The lens system requirement is that the full field of view be $10.0°$. What focal length lens is required?

22. Draw a tunnel diagram for the 45-degree deflecting prism in the space to the right of the figure P2.22. The ray path is shown as the dotted-arrow line segment.

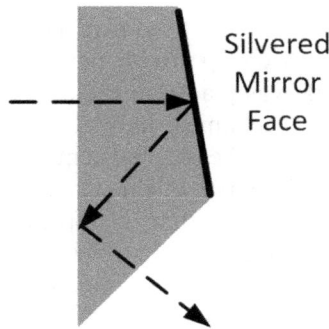

Figure P2.22. Deflecting prism cross-section.

23. (a) A light ray is incident on a plane mirror at an angle θ to the normal. If the normal vector of the mirror is rotated by angle α (in the plane of the incident and reflected ray), what is the angle of the reflected ray with respect to the incident ray?

(b) A light ray is incident on a concave mirror (focal length +25.0 mm) along the optical axis. If the mirror is translated by 2.0 mm perpendicular to the optical axis, what is the angle of the reflected ray with respect to the incident ray?

(c) Repeat part (b) for a convex mirror.

(d) What do these results tell you about tolerances of mirrors in optical systems?

24. Two positive thin lenses are separated by 5.0 cm. Lens1 has a diameter of 60 mm and a focal length of 90 mm. Lens2 has a diameter of 40 mm and a focal length of 30 mm. A circular iris with a 10 mm diameter is placed between the two lenses 2.0 cm from Lens2. (a) Locate the aperture stop for

an axial point 12 cm in front of Lens1. (b) Compute the location and size of both the entrance and exit pupils. (c) What system parameters change if Lens2 is shifted by 3.0 mm along the axis away from the iris?

25. An $f/8$ aplanatic lens of diameter 25.00 mm images a distant object.
    (a) Where is the image location with respect to the lens position? (b) A right angle NBK7 prism (base length 29.35 mm) is placed behind the lens to fold the beam path. What is the location of the new paraxial image position with respect to the lens?

26. A lens of focal length 100.0 mm and diameter 25.0 mm was picked to image an object 150 mm in front of the lens. However, when testing occurred it was found that the image was formed 2.0 mm further away from the computed image location. What was the error in the placement of the lens?

27. A 25 mm diameter bi-convex lens $f/10$, refractive index 1.5 is mounted into a lens prototype tube using the mechanical axis for centering. After mounting it was found that a beam of light sent through the tube's center deviated the beam by 1.0 mrad. (a) Compute the effective wedge angle of the lens. (b) Assuming a perfect lens barrel, how much decenter would be required to align its optical axis to the barrel axis?

28. A laser bean was aligned directly down the mechanical axis of a 2′ inner diameter lens mount. A cheap double convex BK7 ($n = 1.517$) lens with a +100 mm focal length was placed in the mount so that its mechanical axis was aligned with the mechanical axis of the mount. The beam was now found to be 0.20 mm from the original central spot in the focal plane. What is the edge thickness difference of this lens?

29. You are given the task of designing a simple imaging system to view a steel disk of diameter 50.00 mm moving down an assembly line (figure P2.29).

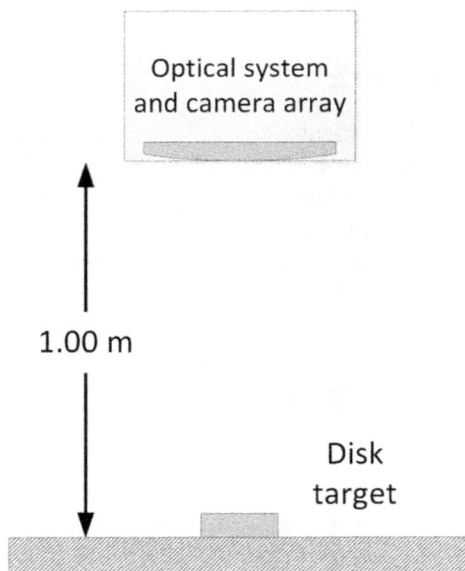

Figure P2.29. Imaging system layout for viewing a disk on an assembly line.

2-48

To keep it simple you decide to use a single lens imaging system, however, the closest distance you can place a lens to image the disk is 1.00 m above the assembly line. The array detector in your camera is a square array of 5.00 mm on a side with 800 × 800 pixels.

(a) Compute the required magnification.

(b) What focal length lens do you require so that the image of a single steel disk fits within the camera array area?

(c) In constructing the test system, the center of the lens was offset from the axis of the camera by 2.00 mm. Compute the change in the image position observed by the camera.

30. A 25 mm focal length <u>thin</u> lens is used in a beam collection system as shown in figure P2.30. The detector is a circular detector of diameter 5.0 mm. (a) By how much can the mirror rotate and still collect light from the incident collimated beam?

(b) The deviation in the focal plane due to errors in mirror rotation was determined to be 30 μm. For a lens decenter error of 40 μm, what is the total error for the focused beam position?

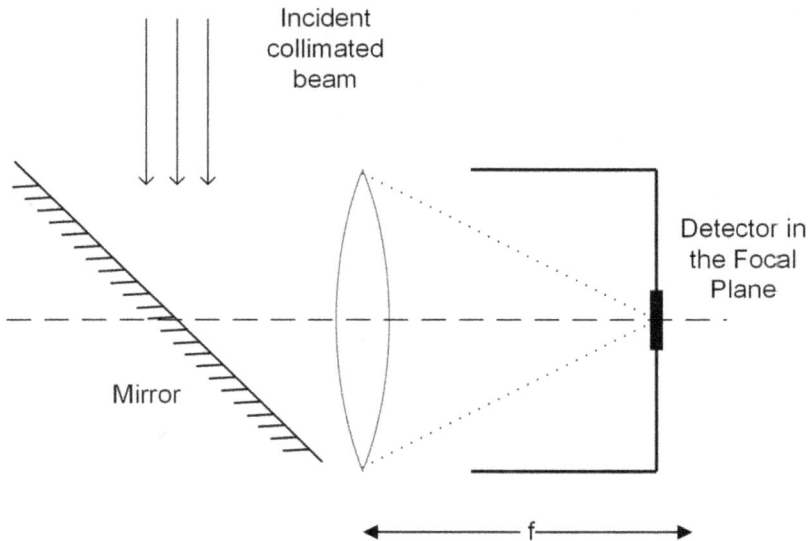

Figure P2.30. Layout of a beam collection system.

31. The beam from a laser is expanded and collimated and incident on a *f*/10 lens which focuses the beam to a spot. A plane parallel plate with a thickness $t = 5.0$ mm, index $= 1.5$ is placed in the converging cone of light normal to the optical axis (see figure P2.31(a)) (a). By how much does the image move axially and laterally?

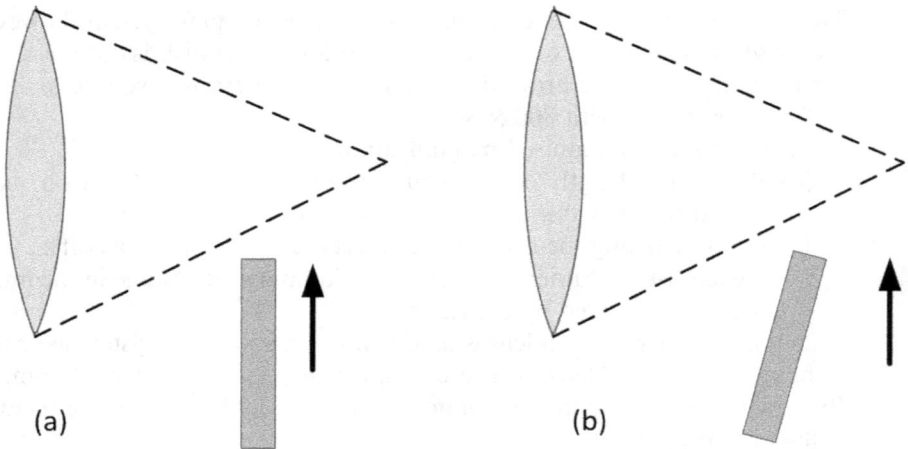

Figure P2.31. (a) Parallel plate inserted into a converging beam perpendicular to the axis and (b) Tilted parallel plate inserted into a converging beam.

(b) By how much does the image move axially and laterally when the plate is rotated by 15° and inserted into the beam? (figure P2.31(b))

32. A high quality NBK7 bi-convex lens with a specified focal length of 250 mm and assumed thin was mounted at the end of a 1.0 inch long lens tube. A collimated and expanded laser beam was directed through the center of the tube showing that the optical axis and mechanical axis were in alignment. When the tube is rotated by an angle of 2° around point A as shown on the figure P2.32, what happens to the focus spot?

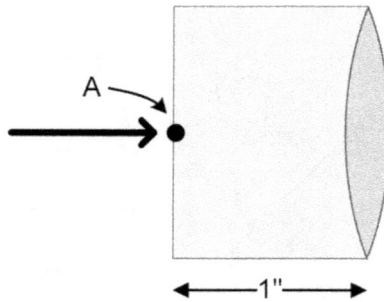

Figure P2.32. Lens mounted in a tube.

33. Your boss wants you to measure the focal length of a thick objective lens. He tells you that there is an old focal collimator in the metrology lab, but he does not know how to use it. A schematic of the focal collimator is shown in figure P2.33.

(a) Given the geometry of the focal collimator, determine how the device measures the focal length of the lens under test, i.e., give a formula for f in terms of D, D', and f_0.

(b) Suppose that you find the calibration data for the device, which has the values of D and f_0 listed as: $D = 25.4$ mm $\pm$ 1% and $f_0 = 300.00$ mm $\pm 0.02\%$

Placing the objective lens in the device you measure the image size to be $D' = 8.46$ mm ± 0.05 mm. What is the focal length of the lens under test (with uncertainty)?

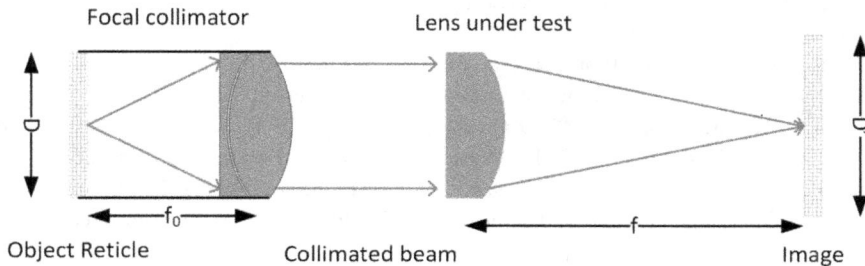

Figure P2.33. Schematic of the focal collimator system.

References

[1] Goodman D S 2009 General principles of geometric optics *Handbook of Optics, Third Edition Volume I: Geometrical and Physical Optics, Polarized Light, Components and Instruments* ed M Bass, E W Van Stryland, D R Williams and W L Wolfe (New York: McGraw-Hill) ch 1

[2] Hecht E 2002 *Optics* 4th edn (Reading, MA: Addison-Wesley)

[3] Polyanskiy M N *Refractive Index Database* https://refractiveindex.info (Accessed 13 November 2020)

[4] Tropf W J, Thomas M E and Harris T J 1995 Properties of crystals and glasses *Handbook of Optics, Second Edition Volume II: Devices, Measurements, and Properties* ed M Bass, E W Van Stryland, D R Williams and W L Wolfe (New York: McGraw-Hill) ch 33

[5] SCHOTT Technical Information TIE-29 *Refractive Index and Dispersion* (https://us.schott.com/)

[6] Greivenkamp J E 2004 *Field Guide to Geometrical Optics* (Bellingham, WA: SPIE Press) (doi: 10.1117/3.547461)

[7] Karov H H 1993 *Fabrication Methods for Precision Optics* (New York: Wiley)

[8] Kingslake R, Johnson R and Barry R 2010 *Lens Design Fundamentals* 2nd edn (Bellingham, WA: SPIE Press)

[9] O'Shea D C 1985 *Elements of Modern Optical Design* (New York: Wiley)

[10] Ditteon R 1998 *Modern Geometrical Optics* (New York: Wiley)

[11] Pedrotti F L, Pedrotti L S and Pedrotti L M 2007 *Introduction to Optics* 3rd edn (Englewood Cliffs, NJ: Prentice Hall)

[12] Gerrard A and Burch J M 1975 *Introduction to Matrix Methods in Optics* (New York: Dover Publications)

[13] Kloos G 2007 *Matrix Methods for Optical Layout* (Bellingham, WA: SPIE Press) (doi: 10.1117/3.737850)

[14] Schwertz K and Burge J 2012 *Field Guide to Optomechanical Design and Analysis* (Bellingham, WA: SPIE)

[15] Yoder P R 2008 *Mounting Optics in Optical Instruments* 2nd edn (Bellingham, WA: SPIE Press) (doi: 10.1117/3.785236)

[16] Schwertz K and Burge J H 2010 Relating axial motion of optical elements to focal shift *Proc. SPIE* **7793** 779306

[17] Burge J H 2006 An easy way to relate optical element motion to system pointing stability *Proc. SPIE* **6288** 628801

[18] Thorlabs Inc. 2020 www.thorlabs.com (Accessed 8 June 2020)

[19] Edmund Optics Inc. 2020 www.edmundoptics.com (Accessed 8 June 2020)

[20] National Research Council 1998 *Harnessing Light: Optical Science and Engineering for the 21st Century* (Washington, DC: The National Academies Press) (doi: 10.17226/5954)

[21] Hecht J 2006 *Understanding Fiber Optics* 5th edn (Englewood Cliffs, NJ: Prentice Hall)

[22] Keiser G 2011 *Optical Fiber Communications* 4th edn (New York: McGraw-Hill)

IOP Publishing

Optical Systems Design Detection Essentials
Radiometry, photometry, colorimetry, noise, and measurements
Robert M Bunch

Chapter 3

Wave optics and light propagation

3.1 Introduction

When light propagates through an optical system and interacts with elements of the system on the order of the wavelength, we observe deviations that cannot be explained using ray optics. Phenomena such as diffraction, interference, and polarization are explained using wave optics [1–5]. From a systems perspective, wave optics also provides a means of predicting system parameters such as resolution. This chapter reviews some basic optical science topics such as coherence, interactions of light waves with surfaces, interference, and polarization. Also included is an electromagnetic treatment of light wave propagation in vacuum and in material media and relates this to the optical properties of materials.

3.2 Light as a wave

The treatment of light as a wave follows directly from electromagnetic theory described by Maxwell's equations [1, 4]. Maxwell assembled the relations that describe electric and magnetic phenomena in free space and in matter; Gauss' law, Faraday's law, Ampere's law, and Gauss' law for magnetism and showed that the solutions for the electromagnetic fields were all consistent. Maxwell also introduced the concept of a displacement current component into Ampere's law for compatibility with charge conservation [4]. Since Maxwell's equations are a set of linear differential equations, if two or more of the fields are solutions then the sum of the fields will also be a solution according to the principle of superposition.

Defining the electric field as $\vec{E}$, the magnetic induction as $\vec{B}$, the electric displacement as $\vec{D}$, and the magnetic field as $\vec{H}$, a general form for Maxwell's equations including interactions of the fields in material matter are,

$$\vec{\nabla} \cdot \vec{D} = \rho_f$$

$$\vec{\nabla} \times \vec{E} = -\frac{\partial \vec{B}}{\partial t}$$

$$\vec{\nabla} \cdot \vec{B} = 0$$

$$\vec{\nabla} \times \vec{H} = \vec{J}_f + \frac{\partial \vec{D}}{\partial t}$$

where ρ_f is the density of free charges, and $\vec{J}_f$ is the free charge current density [4]. Inherent in these equations are the relations between fields if they are in a material medium or,

$$\vec{D} = \varepsilon_0 \vec{E} + \vec{P} \equiv \varepsilon \vec{E} \text{ and } \vec{H} = \frac{\vec{E}}{\mu_0} - \vec{M} \equiv \frac{\vec{B}}{\mu}$$

with ε the permittivity of the medium, and μ the permeability of the medium, Two new fields have been introduced because of the medium; $\vec{P}$, the electric polarization or dipole moment per unit volume in the material, and $\vec{M}$, the magnetization or magnetic moment per unit volume in the material. Both the electric polarization and magnetization provide a bridge between the macroscopic fields of Maxwell's equations and the microscopic view of these fields within the material.

In a linear medium the electric polarization is proportional to the electric field. The constants of proportionality that describe material characteristics can be expressed in different ways depending on the context of the discussion, but the two relationships we will use most often are,

$$\vec{P} = \varepsilon_0 \chi_e \vec{E} \text{ and } \vec{P} = \mathcal{N} \alpha \vec{E}$$

where χ_e is the electric susceptibility, $\mathcal{N}$ is the number density of electric dipoles, and α is the polarizability of the medium. Using these relations, the permittivity of the medium is,

$$\varepsilon = \varepsilon_0(1 + \chi_e) = \varepsilon_0\left(1 + \frac{\mathcal{N}\alpha}{\varepsilon_0}\right)$$

We will specifically discuss more about the electric polarization and its importance in understanding the optical properties of materials later.

For the majority of optical systems, and for electromagnetic wave propagation in materials, we are only interested in the value of the fields at a point at a particular time. This means that we can treat each component of a vector field separately as scalars. In addition, we do not usually have a situation where free charges must be considered. With these assumptions, the macroscopic Maxwell's equations (using MKS units) can be written for the electric field, $\vec{E}$, and magnetic field, $\vec{H}$, as,

$$\vec{\nabla} \cdot \vec{E} = 0$$

$$\vec{\nabla} \times \vec{E} = -\mu \frac{\partial \vec{H}}{\partial t}$$

$$\vec{\nabla} \cdot \vec{H} = 0$$

$$\vec{\nabla} \times \vec{H} = \sigma \vec{E} + \varepsilon \frac{\partial \vec{E}}{\partial t}$$

where ε is the permittivity of the medium, μ is the permeability of the medium, and σ is the conductivity of the medium.

For the special case of free space, $\sigma = 0$, $\varepsilon = \varepsilon_0$, and $\mu = \mu_0$. Taking the curl of both sides of Faraday's law and using a vector identity, we obtain.

$$\vec{\nabla} \times \vec{\nabla} \times \vec{E} = \vec{\nabla}(\vec{\nabla} \cdot \vec{E}) - \nabla^2 \vec{E} = -\mu_0 \frac{\partial \vec{\nabla} \times \vec{H}}{\partial t}$$

Now substituting into this equation from the equations above results in,

$$\nabla^2 \vec{E} = \mu_0 \varepsilon_0 \frac{\partial^2 \vec{E}}{\partial t^2} = \frac{1}{c^2} \frac{\partial^2 \vec{E}}{\partial t^2}$$

which is the general form of the wave equation. Thus, the electric field vector in free space is a harmonic wave propagating with speed c, the speed is the speed of light in free space. A similar process can be used to show that the magnetic field is also a harmonic wave traveling with speed c in free space.

3.2.1 Harmonic waves, plane waves, spherical waves

Consider the electromagnetic fields in a three-dimensional (x, y, z) coordinate system. If we write the electric and magnetic field vectors as components we obtain,

$$\vec{E} = E_x \hat{x} + E_y \hat{y} + E_z \hat{z}$$

$$\vec{H} = H_x \hat{x} + H_y \hat{y} + H_z \hat{z}$$

where unit vectors along each coordinate are $(\hat{x}, \hat{y}, \hat{z})$. The wave equation for each field vector can be reduced to a single one-dimensional scalar wave equation for each component along a particular coordinate. For example, the y-component of the electric field would be given by,

$$\frac{\partial^2 E_y}{\partial y^2} = \frac{1}{v^2} \frac{\partial^2 E_y}{\partial t^2}$$

Thus, all field components could be found by solving one general wave equation. Note that each field component is a function of both position and time, $E_y(y,t)$. Using the separation of variables technique, we write, $E_y(y,t) = Y(y)T(t)$. This leads to two differential equations,

$$\frac{\partial^2 Y}{\partial y^2} + k^2 Y = 0 \text{ and } \frac{\partial^2 T}{\partial t^2} + \omega^2 T = 0$$

where $k^2 = \frac{\omega^2}{v^2}$ links the two differential equations. The parameter k is called the wave propagation constant, and ω is the angular frequency, such that,

$$k \equiv \frac{2\pi}{\lambda} \quad \text{and} \quad \omega \equiv 2\pi\upsilon = \frac{2\pi}{T}$$

where λ is the wavelength, υ is the wave frequency, and T the period of the wave oscillation.

The solutions for each equation are of the same form of a complex exponential function. The general solution can be simplified for wave propagation if we assume that ω is always positive. The phase velocity of the wave is then,

$$v = \frac{\omega}{k} = \frac{c}{n} = \lambda\upsilon$$

The sign of k dictates the direction of propagation, either positive or negative. Combining the space and time dependence solutions together results in a general form for the y-component of the electric field as simply,

$$E_y(y, t) = E_{y0}e^{j(ky - \omega t + \varphi)}$$

where E_{y0} is the real amplitude of the field. In this relation we use the notation $j \equiv \sqrt{-1}$ for the complex number. The part of the argument of the exponential $ky - \omega t + \varphi$ is called the phase of the wave. This means that as y and t change together as the wave propagates so that the total phase remains constant, the value of the field remains constant. Positions of constant phase are what we referred to earlier as wavefronts and these wavefronts are perpendicular to the direction of propagation and are separated in space by a wavelength. The value φ is the phase constant of the wave when $y = 0$ and $t = 0$. Even though we are using a complex representation for the wave solution, the actual value of the physical field must of course be real which means that we take the real part for the final wave solution.

Figure 3.1(a) shows a plot of a section of the real part of a harmonic wave at a point in space as it oscillates in time with frequency ν_0. The phase shift of this wave is

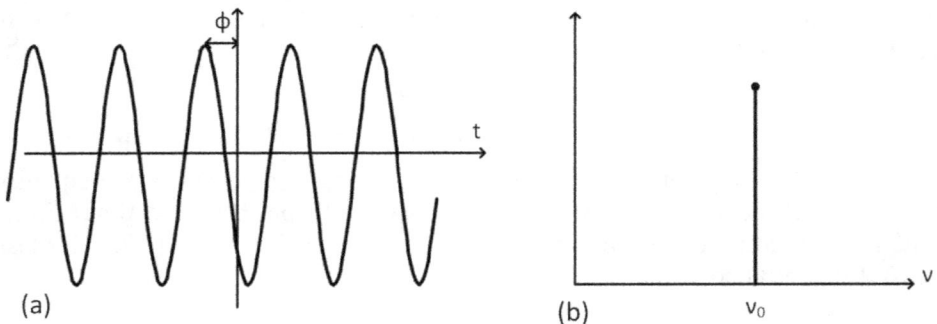

Figure 3.1. (a) Temporal oscillation of the real part of a harmonic wave and (b) frequency representation of this wave at frequency ν_0.

also shown. In theory this wave extends for all time and space. Another equivalent representation of this wave is in the temporal frequency domain, figure 3.1(b) [1]. Since there is only one specific frequency the representation of the function in this domain is an impulse function that can be mathematically described using a Dirac delta function at the frequency ν_0.

Extending this solution for the fields into three dimensions, the one-dimensional propagation constant becomes a vector with components along each coordinate axis.

$$\vec{k} = k_x\hat{x} + k_y\hat{y} + k_z\hat{z}$$

Also, defining a general position vector in this coordinate system, we have,

$$\vec{r} = x\hat{x} + y\hat{y} + z\hat{z}$$

We can easily show that the solutions for the electric field and magnetic field vectors are,

$$\vec{E} = \vec{E_0}e^{j(\vec{k}\cdot\vec{r}-\omega t)} \text{ and } \vec{H} = \vec{H_0}e^{j(\vec{k}\cdot\vec{r}-\omega t)}$$

where the individual components of the amplitude vectors must be related through Maxwell's equations. For these equations, the field amplitudes are vectors whose components are in general complex to account for any phase shifts when $r = 0$ and $t = 0$. Again, when the total phase term is a constant, a series of planes perpendicular to the direction of propagation are formed as surfaces of constant phase. This type of wave is called a plane wave whose wavefronts are separated by a wavelength. Figure 3.2 shows a plane wave propagating in a three-dimensional coordinate system along a general $\vec{k}$-direction.

The concept of a k-vector ties the wave picture of light propagation with the ray approach. In fact, $\vec{k}$ is in the same direction as a ray vector, perpendicular to the wavefront.

For the simple case of a wave propagating along the +z-direction, $\vec{k} = k\hat{z}$, first note that,

$$\vec{\nabla} \times \vec{E} = ik(-E_y\hat{x} + E_x\hat{y}) = jk(\hat{z} \times \vec{E}) \text{ and } -\mu\frac{\partial\vec{H}}{\partial t} = +j\mu\omega\vec{H}$$

Substituting these relations into Maxwell's equations results in,

$$k(\hat{z} \times \vec{E}) = \omega\mu\vec{H}$$

showing that both wave functions $\vec{E}$ and $\vec{H}$ are oriented perpendicular to the direction of propagation and to each other. Figure 3.3 illustrates this case and includes the electric field in the x-direction and the magnetic field in the y-direction with wave vector $\vec{k}$ for two successive wavefronts in the plane wave. This illustration also assumes that the fields remain in a single plane as they oscillate which is the special case of a linearly polarized wave [2, 3]. While this is computed for a special

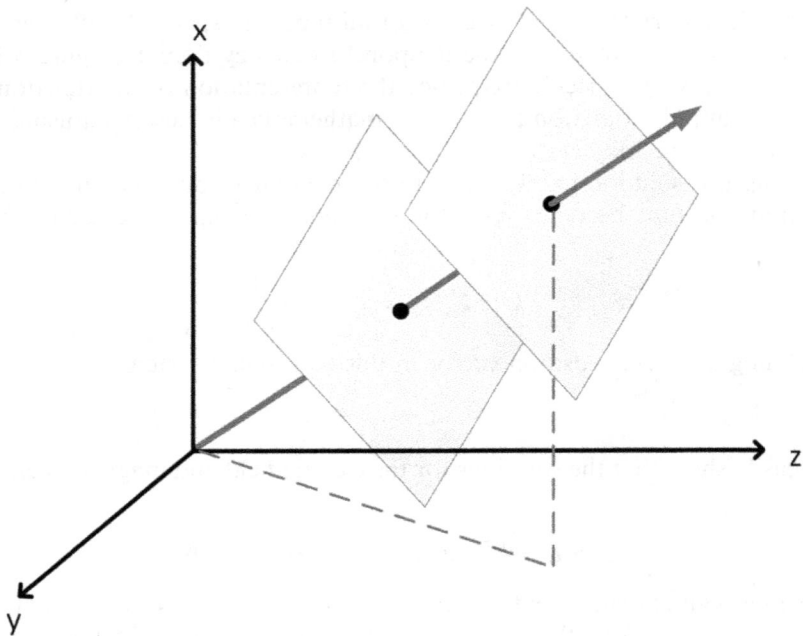

Figure 3.2. A plane wave propagating along a general $\vec{k}$-direction showing surfaces of constant phase or wavefronts. The k-vector is normal to each wavefront surface.

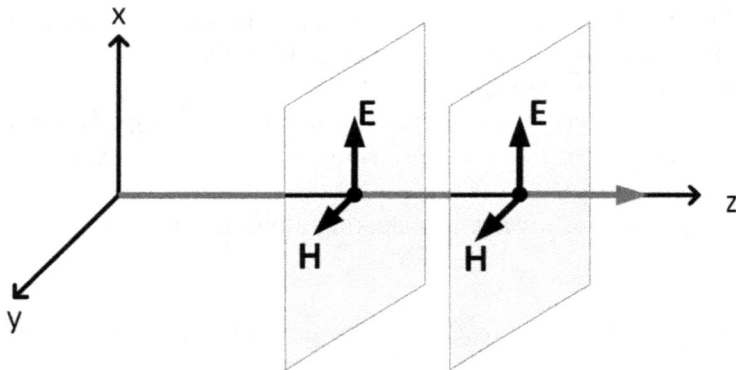

Figure 3.3. A plane wave propagating in the z-direction with the electric field in the x-direction and the magnetic field in the y-direction at two wavefronts.

case, it is a general result for all electromagnetic waves that the fields and propagation vector form a mutually perpendicular set of vectors.

The relationship between the fields and a general $\vec{k}$-direction can be obtained by following a similar procedure. In general,

$$(\vec{k} \times \vec{E}) = \omega\mu\vec{H}$$

Plane harmonic waves are only one example of a solution to the wave equation. Another particular and important solution to the wave equation is a spherical wave that arises from the wave equation solution in spherical coordinates [2]. The physical significance of a spherical wave solution is that it models the waves emitted by a single point source. Because of symmetry the magnitude of the field is only a function of the radial distance from the point source providing a spherical surface of constant phase wavefronts. Refer to figure 2.1 in chapter 2 where this concept was briefly discussed. Let the function, u, represent the magnitude of the electromagnetic field function. Then a spherical wave takes the form,

$$u = \frac{A}{r}e^{j(kr-\omega t)}$$

where r is the radial distance from the point source, k is the propagation constant, and ω is the angular frequency. The amplitude of a spherical wave is the combination value A/r where A is a constant, sometimes called the source strength [2]. Thus, the amplitude of a spherical wave decreases the further away one moves from the point source and the power per unit area decreases as the inverse square of the distance. Also, for large values of r, a small area of the wavefront surface of a spherical wave appears as an approximate plane wave.

As we have seen, solutions of the wave equation are numerous. We have also seen that a complex representation of the fields is a natural way to express these functions. A typical approach to the field solution representations is to separate the space and time dependence of the fields by assuming that the field has a complex amplitude with a temporal harmonic component, or,

$$u(\vec{r}, t) \equiv u(\vec{r})e^{-j\omega t}$$

Substituting this relation back into the wave equation leads to another form for the spatial wavefunction,

$$\nabla^2 u(\vec{r}) + k^2 u(\vec{r}) = 0$$

which is called the Helmholtz equation. This is useful as a starting point when only the spatial function of the complex wave is needed. Another useful wavefunction representation is a paraxial wave (as with paraxial rays in ray optics). In this case the complex field can be further divided into an amplitude and phase propagation solution in the direction of propagation (z-axis in this case). The wavefunction is then,

$$u(\vec{r}) = A(\vec{r})e^{jkz}$$

where $A(\vec{r})$ is another complex amplitude function. Of course, this must also satisfy the Helmholtz equation.

3.2.2 Light flux and the Poynting vector

We know from electrostatics and magnetostatics that the time average total energy density, w_e, contained within a volume of free space in which there exists both electric field with magnitude, E, and a magnetic field with magnitude, H is [4]

$$w_e = \frac{1}{2}\varepsilon_0 E^2 + \frac{1}{2}\mu_0 H^2$$

Examine a value, $\vec{S}$, called the Poynting vector [4, 5] which is defined as,

$$\vec{S} \equiv \vec{E} \times \vec{H}$$

For the specific situation shown in figure 3.3, the electric field magnitude is the component in the x-direction and the magnetic field magnitude is the component in the y-direction with the wave propagating in the z-direction. From the definition of the Poynting vector, we have,

$$S_z = E_x H_y$$

By examining the energy contained within a volume of space in which the wave is propagating, it is straightforward to show that this quantity represents the flow of energy in the electromagnetic wave per unit area per unit time. When the flow of energy is measured with an optical detector, the detector records a time average of the fields so that,

$$S_{\text{ave}} = \frac{1}{2}E_0 H_0 = \frac{1}{2}\varepsilon_0 c E_0^2$$

where E_0 and H_0 are the amplitudes of the electric and magnetic fields, respectively. This quantity is the power per unit area of the propagating wave or irradiance (MKS units of Watts m^{-2}). So, the magnitude of the time average Poynting vector magnitude is the irradiance of the light falling on a surface area and is proportional and the square of the electric field magnitude. (Note: for historical reasons most introductory physics texts continue to define this quantity as intensity. The modern and standard definition of intensity will be discussed in more detail later in the context of radiometry.) [6].

3.2.3 Wave superposition and interference

Wave equation solutions obey the superposition principle. Harmonic waves are periodic and as we saw in chapter 1, a periodic function can be represented by a Fourier series of the addition of wave harmonics. Let us take two electric field functions propagating in space and in time. For simplicity, assume that the two waves have the same wavelength and frequency but overlap in space and have different phase constants, ϵ_1 and ϵ_2. The scalar fields for these waves are defined as,

$$E_1 = E_{01} \cos(kr_1 - \omega t + \epsilon_1)$$
$$E_2 = E_{02} \cos(kr_2 - \omega t + \epsilon_2)$$

Now, define two new variables $\phi_1 \equiv kr_1 + \epsilon_1$ and $\phi_2 \equiv kr_2 + \epsilon_2$, and add the two fields using the superposition principle. The resultant electric field of the propagating wave is,

$$E = E_{01} \cos(\phi_1 - \omega t) + E_{02} \cos(\phi_2 - \omega t)$$

After some algebraic manipulation and using trigonometric identities, we find,

$$E = E_0 \cos(\alpha - \omega t)$$

where E_0 and α are a new amplitude and phase of the combined wave. These values are,

$$\tan(\alpha) = \frac{E_{01} \sin(\phi_1) + E_{02} \sin(\phi_2)}{E_{01} \cos(\phi_1) + E_{02} \cos(\phi_2)}$$

And,

$$E_0^2 = E_{01}^2 + E_{02}^2 + 2E_{01}E_{02} \cos(\phi_1 - \phi_2)$$

The resultant wave has a new amplitude and phase constant. But, from the previous section, we found that the square of the electric field is the irradiance or time average Poynting vector. Using this definition, we find,

$$S_0 = S_{01} + S_{02} + 2\sqrt{S_{01}} \sqrt{S_{02}} \cos(\phi_1 - \phi_2)$$

This relation is the interference condition for the superposition of two waves of the same frequency. (Note: many references use $I = S_{\text{ave}}$ as intensity or irradiance.)

The value of the phase difference between the two waves is,

$$\Delta\phi = \phi_1 - \phi_2 = k(r_1 - r_2) + (\epsilon_1 - \epsilon_2) = \frac{2\pi}{\lambda}\text{OPD} + \Delta\epsilon$$

The phase difference between the two waves can occur if there is any optical path difference (OPD) or initial phase difference between the two wave sources.

Now, examine some conditions of the third term, the interference term. First take the phase constant difference to be zero, $\Delta\epsilon = 0$. When, $r_1 - r_2 = \frac{\lambda}{2}$ then $\Delta\phi = \pi$ and $\cos(\Delta\phi) = -1$. This is the condition of destructive interference since the value of the irradiance at this point on the wave is minimum.

$$S_{0\,\text{min}} = S_{01} + S_{02} - 2\sqrt{S_{01}} \sqrt{S_{02}}$$

when $r_1 - r_2 = \lambda$, then $\Delta\phi = 2\pi$ and $\cos(\Delta\phi) = 1$. The irradiance is a maximum and we have constructive interference.

$$S_{0\,\text{max}} = S_{01} + S_{02} + 2\sqrt{S_{01}} \sqrt{S_{02}}$$

Interference conditions can be summarized by: destructive interference results when, $\Delta\phi = (2m + 1)\pi$, and constructive interference results when, $\Delta\phi = 2m\pi$, where m is an integer greater than or equal to zero.

The classic experiment that illustrates interference between the waves from two sources is Young's double slit [2, 3, 5]. The two sources are produced by placing two slits, separated by a distance, $2a$, behind a single source. This is shown in figure 3.4.

The light distribution falling on a screen a distance z away is the interference pattern and originates because each slit is acting as a point source. At a general point on the screen at height y from the axis, interference between the waves from each slit is observed. Two path lengths, r_1 and r_2, are shown and the difference in the path is,

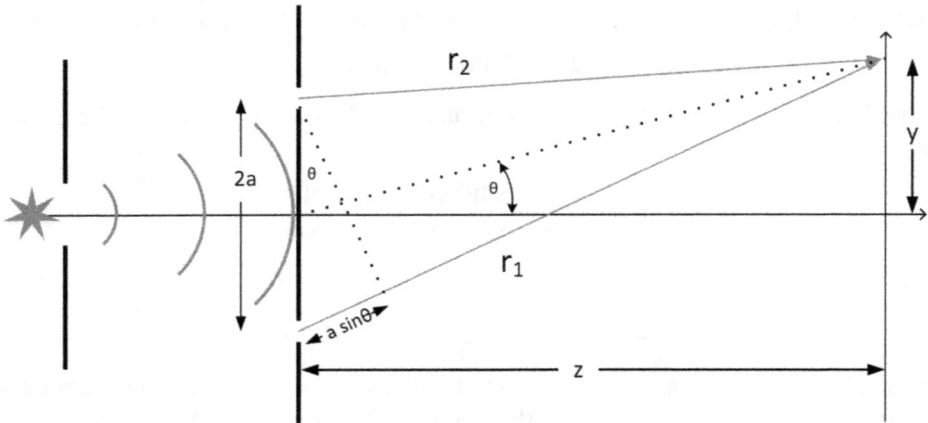

Figure 3.4. Layout of Young's double slit experiment.

$OPD = a \sin \theta$. For small angles, we have, $\sin \theta \cong \tan \theta = \frac{y}{z}$. Using the interference relation from above we find the irradiance distribution as a function of the y-position to be,

$$E_e(y) = 4E_e(0)\cos^2\left[+\pi\frac{2a}{\lambda z}y\right]$$

where the irradiance at $y = 0$ is $E_e(0)$. This is the irradiance value at the center of the output plane. It is also the value of the irradiance that would be observed if either one of the sources were covered and we were just measuring the irradiance in the output plane from a single source.

Figure 3.5 shows a plot of the normalized irradiance across the output plane located 1.0 meters away from the two, point sources with a wavelength of 0.50 μm separated by a distance, $2a = 50$ μm. In this case each source has the same output flux. The maxima and minima correspond to positions of constructive and destructive interference in the addition of the waves from each source.

When we view the interference of two wavefronts experimentally, what is observed are called fringes, bands of bright and dark regions corresponding to the maximum and maximum interference at a point. A parameter knows as visibility gives a measure of the contrast of the observed fringes. The visibility is defined as,

$$V \equiv \frac{S_{0\,max} - S_{0\,min}}{S_{0\,max} + S_{0\,min}}$$

As in this example, when the irradiance from each source is the same the visibility is unity. If the amount of light from each source is not the same then the visibility is a value,

$$V = \frac{2\sqrt{S_{01}}\sqrt{S_{02}}}{2(S_{01} + S_{02})}$$

Less than the maximum visibility, $V = 1$.

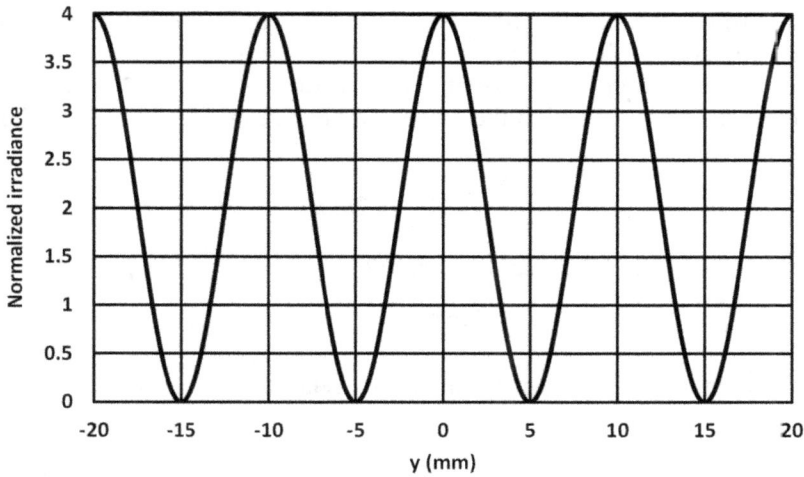

Figure 3.5. Normalized irradiance across the output plane a distance 1.0 m away from two, point sources with a wavelength of 0.50 μm separated by a distance of $2a = 50$ μm.

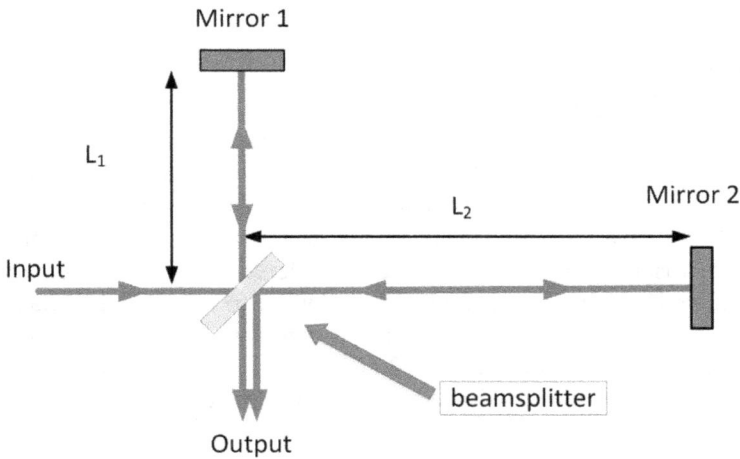

Figure 3.6. Michelson interferometer.

An important instrument whose operation is based on interference is the Michelson interferometer [2, 3]. This interferometer is known as an amplitude splitting interferometer since the incident wave is split into two beams by a beamsplitter and then recombined to observe the interference pattern. Figure 3.6 shows a simple Michelson interferometer arrangement where the light of wavelength λ is incident from the left onto a beamsplitter separating the wave into two equal amplitude parts. One wave propagates a distance L_1 to mirror M_1, is reflected and sent back through the beamsplitter. The other wave propagates a distance L_2 to mirror M_2, is reflected and sent back through the beamsplitter. The two beams then overlap, and interference fringes are observed at the output.

For constructive interference (bright fringes) the optical path traveled by beam 1 is $2L_1$ from the beamsplitter to the mirror and back to the beamsplitter. The phase condition is then,

$$\frac{2\pi}{\lambda}(2L_1) = 2m_1\pi$$

where m_1 is an integer. Likewise, for beam 2,

$$\frac{2\pi}{\lambda}(2L_2) = 2m_2\pi$$

Adding these equations together results in the relation,

$$\Delta L = m\frac{\lambda}{2}$$

where ΔL is the difference in path and $m = m_2 - m_1$, an integer. This tells us that each fringe observed in the output interference pattern is the result of a half-wavelength difference in optical path. See the references for the numerous applications of this interferometer [2, 3].

3.3 Coherence

The coherence of wave emanating from a source is related to the ability of the wave source to maintain a constant phase as it propagates. Coherence is ultimately associated with the characteristics of the source. Figure 3.7 shows a general wavefront propagating in space. Our ability to predict the phase of the wave and thus its coherence properties is done by measuring the phase at two points. However, there are two orientations that must be considered. From a system perspective, consider that we have two phase detectors and want to monitor the phase difference at any two points. Figure 3.7(a) shows the longitudinal orientation which is along the wave propagation direction. Figure 3.7(b) shows the orientation where the phase is measured perpendicular to the wave propagation direction, or the transverse orientation.

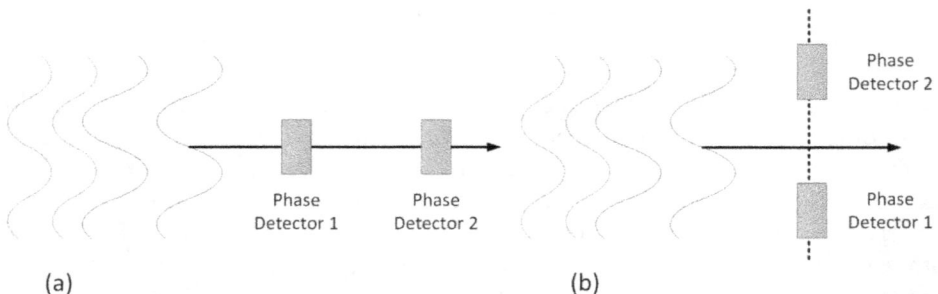

(a)   (b)

Figure 3.7. Propagating wavefront and phase measurement/detection for (a) longitudinal and (b) transverse directions.

Phase is measured using interferometry. Our phase measurement concept can be carried out using a Michelson interferometer for the test in figure 3.7(a) and a Young's interferometer for the test in figure 3.7(b). For the longitudinal orientation, the difference in path between Phase Detector 1 and Phase Detector 2 is just the difference associated with translating one of the mirrors in one arm of a Michelson interferometer. In this situation we are observing the phase difference related to the time of propagation between two points in a wavefront or the amount of temporal coherence. As long as we observe some interference (fringe contrast) the source will be coherent over that length. In the transverse orientation, the separation difference of Phase Detector 1 and Phase Detector 2 in the transverse direction is like the separation between two slits. If the fringe visibility is greater than zero, the source will be coherent. This type of coherence is spatial coherence.

3.3.1 Temporal coherence

As described earlier, a harmonic wave propagates in space and oscillates in time. The temporal oscillation was briefly discussed earlier in this chapter for an ideal oscillation resulting in a single frequency or monochromatic light. However, sources of light are not ideal and can only be considered harmonic over a finite period of time. This time is the coherence time of the source. The concept is illustrated in figure 3.8 where two different time frames are shown for a particular wave oscillation between fluctuations of the source. The wave is coherent within a time frame Δt [7]. This implies that the frequency domain representation is no longer a monochromatic single frequency but a spread of frequency components over a frequency range $\Delta \nu$ [1]. In this case, Δt is of the order of $1/\Delta \nu$.

As this wave propagates in space at speed c, we can then define a length Δl over which the wave segment is coherent. The longitudinal coherence length of the source is then,

$$\Delta l = c\Delta t \sim \frac{c}{\Delta \nu}$$

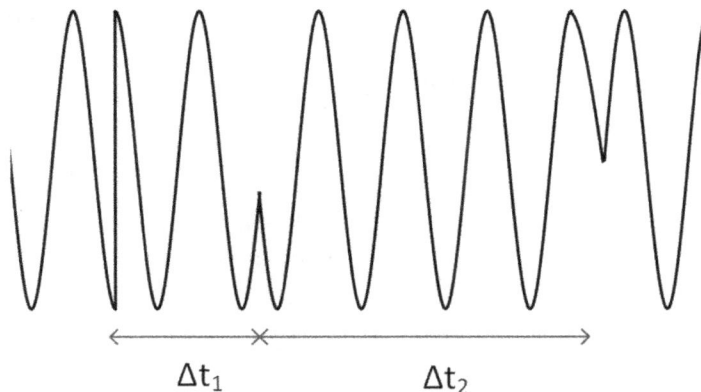

Figure 3.8. Harmonic wave with temporal sections indicating time frames over which the wave is coherent.

Thus, the wider the frequency bandwidth of the source the smaller the coherence length of the source. This relation shows why a laser source, with a very small spectral bandwidth, has a longer coherence length than a source such as a light emitting diode. Temporal coherence arises from the finite frequency bandwidth of the light source.

3.3.2 Spatial coherence

This maintenance of a constant phase difference not only occurs in time as the wave propagates but also occurs in space between two different points along a wavefront from a source. This is referred to a spatial coherence [7]. As discussed previously, measurements of spatial coherence can be performed using a system like a Young's double slit. Earlier we found the irradiance associated with this interferometer was,

$$S_0 = S_{01} + S_{02} + 2\sqrt{S_{01}}\sqrt{S_{02}}\cos(\phi_1 - \phi_2)$$

Using a more detailed analysis of the phase term, the cosine factor is a special case of a general function called the complex degree of coherence, γ_{12} [7]. In the case above, the cosine function is the real part of the complex degree of coherence. When $\gamma_{12} = 1$ there is no phase difference, and the source is coherent. When $\gamma_{12} = 0$ the entire interference term is zero and the source is completely incoherent (irradiance values simply add). Any value between zero and one tells us that the source is only partially coherence. This means that an ideal point source is completely coherent. The value of the degree of coherence is just the visibility of the source when the two beams creating an interference pattern are equal.

In designing and analyzing optical systems some assumptions must always be made related to the sources used and their degree of coherence. If spatial phase is important then we will use spatial functions of the complex electromagnetic field amplitude and assume complete spatial coherence. If there is no phase importance, then we will use spatial functions that are related to the light flux and proportional to the absolute square of the field amplitude (incoherent light) [8].

In summary, any physical source is not completely monochromatic but has a finite spectral width and a point source will always have some finite spatial extent [1]. All sources are only partially coherent and thus have a degree of coherence [7].

3.4 Light waves in material media

Using the same procedure as before for the solution to Maxwell's equations in a vacuum but now assuming that the waves are in material medium results in similar differential equations for the electric and magnetic fields. Take the case of an isotropic non-conductive dielectric media, with ε the permittivity of the medium, and μ the permeability of the medium. Being non-conductive implies that the conductivity, $\sigma = 0$. This results in,

$$\nabla^2\vec{E} = \mu\varepsilon\frac{\partial^2\vec{E}}{\partial t^2} = \frac{1}{v^2}\frac{\partial^2\vec{E}}{\partial t^2} \text{ and } \nabla^2\vec{H} = \mu\varepsilon\frac{\partial^2\vec{H}}{\partial t^2} = \frac{1}{v^2}\frac{\partial^2\vec{H}}{\partial t^2}$$

where v is the speed of propagation of the electromagnetic wave in the medium. Taking the refractive index of the medium to be n (again assuming $\sigma = 0$) the speed of propagation of the wave is the phase velocity in the medium, or,

$$v = \frac{1}{\sqrt{\mu\varepsilon}} \equiv \frac{c}{n}$$

Thus, the refractive index of a linear isotropic medium is,

$$n = \sqrt{\frac{\varepsilon\mu}{\varepsilon_0\mu_0}}$$

3.4.1 Waves in conducting media, absorption

Earlier we found the solutions for electromagnetic waves propagating in free space and in dielectric materials. Now we assume that the wave is propagating in a medium where the conductivity is not zero, however, we still assume that there are no free charges. Remember the material constants that describe the medium (ε, μ, σ) are all real values. As before, we take the curl of both sides of Faraday's law and using a vector identity, we obtain

$$\vec{\nabla} \times \vec{\nabla} \times \vec{E} = \vec{\nabla}(\vec{\nabla}\cdot\vec{E}) - \nabla^2\vec{E} = -\mu\frac{\partial\vec{\nabla} \times \vec{H}}{\partial t}$$

Now substituting into this equation from the equations above results in,

$$\nabla^2\vec{E} - \mu\sigma\frac{\partial\vec{E}}{\partial t} - \mu\varepsilon\frac{\partial^2\vec{E}}{\partial t^2} = 0$$

Assume that we know that the time dependence of the field is of the same form as waves in a dielectric, or,

$$\vec{E} = \vec{E_0}e^{-j\omega t}$$

Then our general equation for the electric field in the medium reduces to,

$$\nabla^2\vec{E} = -(\omega^2\mu\varepsilon + j\omega\mu\sigma)\vec{E}$$

Thus, we now find in general that the propagation constant k is complex,

$$k^2 = \omega^2\mu\varepsilon + j\omega\mu\sigma$$

And the square of the phase velocity becomes,

$$v^2 = \left(\frac{c}{N}\right)^2 = \frac{1}{\mu\varepsilon + j\frac{\mu\sigma}{\omega}}$$

again, a complex number, implying that the refractive index of the medium is also complex. The symbol N is now used for the refractive index indicating that it is in general complex and has a real and imaginary part. The refractive index also includes an explicit frequency dependence not relying on the inherent frequency dependences of the other material constants to create a dispersion relation. We define n_R and n_I as the real and imaginary parts of the complex refractive index [3], and

$$N^2 = \mu\varepsilon c^2 + j\frac{\mu\sigma c^2}{\omega} \equiv (n_R + jn_I)^2$$

with the propagation constant k of,

$$k = \frac{\omega}{c}(n_R + jn_I) = \frac{\omega}{c}N = \frac{2\pi N}{\lambda}$$

Many references us different notation for the real and imaginary parts [2, 3, 5]. However, the point of this calculation is to describe the character of the field as it propagates. Remember that the general form of the electric field propagating along the z-axis direction is,

$$\vec{E} = \vec{E_0}e^{j(kz-\omega t)} = \vec{E_0}e^{-j\omega t}e^{+j\frac{\omega}{c}zn_R}e^{-z\frac{\omega}{c}n_I}$$

The term containing the imaginary refractive index is an exponential damping term on the amplitude of the field. The other terms are the same oscillatory terms we saw for a wave propagating in a dielectric and in free space. Therefore, as an electromagnetic wave propagates through a material with non-zero conductivity, the amplitude dampens and thus the amount of light decreases. This phenomenon is macroscopic absorption.

Macroscopic absorption is measured by examining the reduction in light flux through a medium. Since the absolute square of the electric field is proportional to the irradiance, the irradiance of a wavefront after propagating through distance z is then,

$$E_e(z) = E_e(0)e^{-\alpha z}$$

where the coefficient α is the macroscopic absorption coefficient and $E_e(0)$ is the incident irradiance at $z = 0$.

3.4.2 Dispersion theory-modeling the optical properties of materials

From basic physics we know that electric dipoles align along electric field lines. This fact is used as the basis for modeling the effect of a macroscopic field applied to a microscopic region within a solid or gas. The assumption is that the outer electrons of an atom are displaced creating an electric dipole at the location of the atom. If the field is oscillating in time, as an electromagnetic wave, then the effect of the electron displacement will also oscillate in relation to the average field experienced creating a dipole moment. This dipole moment is at the microscopic level; however, the applied field is macroscopic. These fields are different because neighboring dipoles alter the microscopic field. There are three different electric field values of interest in applying this model, the macroscopic field, the microscopic field, and what we will call the local field which accounts for the effects of the microscopic fields averaged over a volume of material. This local field is proportional to the electric polarization.

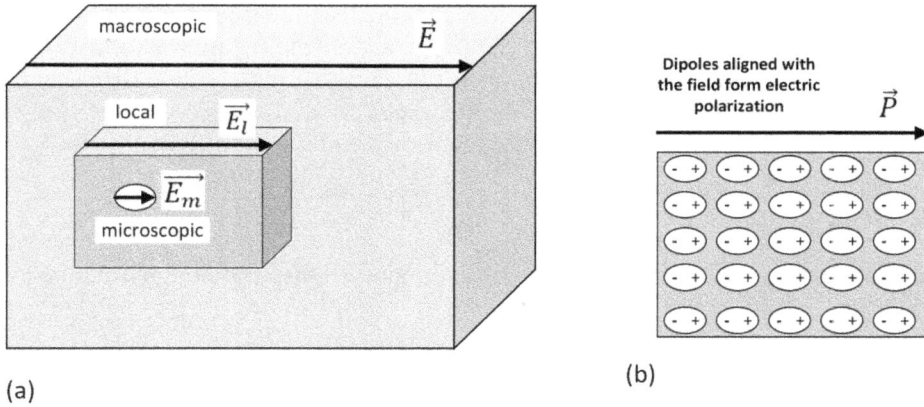

Figure 3.9. (a) Relation between macroscopic, local, and microscopic fields, and (b) region with dipole moments aligned to the applied field creating an electric polarization.

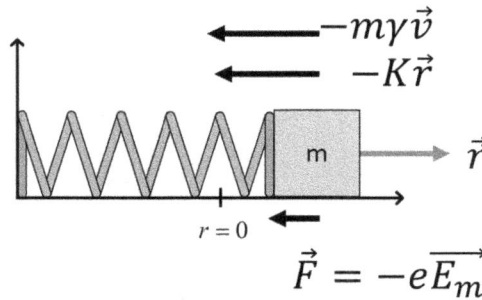

Figure 3.10. Forced and damped oscillator dynamic model for a microscopic region of a material.

Figure 3.9 illustrates the concept of these fields and the relationship between their regions, and also shows the electric polarization due to the collective dipole moment in a volume of the material.

The microscopic model is based on mechanical dynamics of the forces acting on an electron, mass m and charge $-e$, surrounding the atom. This model is often referred to as the Lorentz model or the Lorentz–Drude model [9]. The applied field causes a displacement from equilibrium specified by the displacement vector, $\vec{r}$. The force at each microscopic site is the Lorentz force due to the microscopic field, $\overrightarrow{E_m}$ at that site. (Note: the $\vec{v} \times \vec{B}$ term of the Lorentz force is assumed to be small.) The electron is taken as being held in place by an effective spring that follows Hooke's law with spring constant K. In addition, a spring damping factor, γ, is also included. In this model, damping is assumed to be proportional to the velocity of the particle, where $\vec{v} = \frac{d\vec{r}}{dt}$. Figure 3.10 shows a schematic representation of the model concept.

The equation of motion can be written using Newton's laws as,

$$m\frac{d^2\vec{r}}{dt^2} = -e\overrightarrow{E_m} - m\gamma\frac{d\vec{r}}{dt} - K\vec{r}$$

As a first case, let us assume that the field is a static field and does not change in time. We expect an electric polarization to be setup within the medium as illustrated in figure 3.9(b) with a constant dipole moment at each site with a fixed displacement. Since there is no time dependence the acceleration and velocity terms are zero resulting in an equation for the displacement of the static dipoles of,

$$\vec{r} = -\frac{e}{K}\overline{E_m}$$

The electric polarization is then found by averaging all displacement within the local region, or,

$$\vec{P} = -\mathcal{N}\,e\langle\vec{r}\rangle$$

where the bracket operation represents an average of the displacement with the local region surrounding several dipoles and $\mathcal{N}$ is the number density of electric dipoles. Substituting the displacement vector solution from the model, we find,

$$\vec{P} = -\mathcal{N}\,e\left[-\frac{e}{K}\langle\overline{E_m}\rangle\right]$$

But the average of all microscopic fields is the local field, $\overline{E_l}$. Since the electric polarization is $\vec{P} = \mathcal{N}\alpha\overline{E_l}$ we find that the polarizability of the medium in a static field is a constant, or,

$$\alpha = \frac{e^2}{K}$$

For the second case of most importance, take the field to be a harmonic wave, such as a light wave propagating through the material with frequency ω and speed v. The field is of the form used in our earlier definitions as,

$$\overline{E_m} = \overline{E_0}e^{-j\omega t}$$

This field will create a driving force that oscillates in time and from our model we assume that the displacement will also oscillate in time about an equilibrium position, giving a solution for the displacement vector of,

$$\vec{r} = \vec{r}_0 e^{-j\omega t}$$

Taking the appropriate derivatives of the displacement to obtain the acceleration and velocity of the electron and substituting into the general equation of motion gives,

$$-m\omega^2\vec{r} = -e\overline{E_m} + j\omega m\gamma\vec{r} - K\vec{r}$$

Solving this equation for the displacement vector results in,

$$\vec{r} = \frac{-e\overline{E_m}}{-m\omega^2 - j\omega m\gamma + K}$$

Using the same averaging procedure as before over the local region we obtain the electric polarization,

$$\vec{P} = \frac{\mathcal{N}e^2}{-m\omega^2 - j\omega m\gamma + K}\vec{E_l}$$

when $\omega = 0$, this result reduces to our first case of a static field. Defining a resonance frequency of a typical oscillating spring as, $\omega_0 = \sqrt{\frac{K}{m}}$, the complex polarizability of the medium becomes

$$\alpha = \frac{e^2/m}{\omega_0^2 - \omega^2 - j\omega\gamma}$$

From the polarizability of the medium we can compute the dielectric constant and thus the refractive index of the medium. For non-magnetic materials, $\mu = \mu_0$, so an equation for the complex refractive index is,

$$N^2 = 1 + \frac{\mathcal{N}e^2/\varepsilon_0 m}{\omega_0^2 - \omega^2 - j\omega\gamma}$$

where again we have used the notation that an upper-case N represents a general complex value of the refractive index. This model predicts that in a general optical media there will be absorption, like a conducting media discussed above. In the general case absorption is related to the damping factor of the model and occurs in the frequency range near the spring resonance frequency where the imaginary term dominates the denominator.

A commonly used parameter that appears in our equations is the plasma frequency defined as,

$$\omega_p = \sqrt{\frac{\mathcal{N}e^2}{\varepsilon_0 m}}$$

This represents the oscillation frequency of a charged particle that is displaced from the other particles and with respect to the fixed background of opposite charges [5]. With this definition, the relation for the square of the refractive index can be written as,

$$N^2 = (n_R + jn_I)^2 = 1 + \chi_R + j\chi_I = 1 + \frac{\omega_p^2}{\omega_0^2 - \omega^2 - j\omega\gamma}$$

where χ_R and χ_I are defined as the real and imaginary parts of the susceptibility. Although this equation is straightforward, an analytic solution for the real and imaginary parts of the refractive index is algebraically intensive. For computation, the best method to use is find the susceptibility functions and use them to compute the refractive index values. From above, the complex susceptibility is,

$$\chi_R + j\chi_I = \frac{\omega_p^2(\omega_0^2 - \omega^2)}{(\omega_0^2 - \omega^2)^2 + \omega^2\gamma^2} + j\frac{\omega_p^2\omega\gamma}{(\omega_0^2 - \omega^2)^2 + \omega^2\gamma^2}$$

(a)

(b)

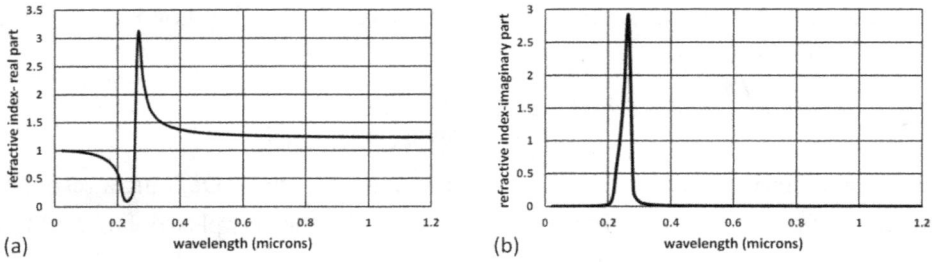

Figure 3.11. Graphs of the (a) real and (b) imaginary parts of the refractive index as a function of wavelength from dispersion theory. The parameters used for these graphs were: $\omega_p = 0.5 \times 10^{16}$ rad s^{-1}, $\omega_0 = 0.7 \times 10^{16}$ rad s^{-1}, and $\gamma = 2.0 \times 10^{14}$ rad s^{-1}.

Then,

$$n_R^2 - n_I^2 = 1 + \chi_R \ \text{ and } \ 2n_R n_I = \chi_I$$

And these equations can be solved simultaneously to obtain n_R and n_I. Figure 3.11 shows plots for the real and imaginary parts for a material corresponding to the model as a function of wavelength. The parameters used $\omega_p = 0.5 \times 10^{16}$ rad s^{-1}, $\omega_0 = 0.7 \times 10^{16}$ rad s^{-1}, and $\gamma = 2.0 \times 10^{14}$ rad s^{-1}. The large peak observed in the imaginary part indicates the region of absorption. For these variables, this is in the ultraviolet region of the spectrum. The real part of the refractive index also shows a resonant behavior about the resonant frequency. In this region the index value increases with increasing wavelength, unlike common transparent materials, thus, this type of dispersion is called anomalous dispersion.

If we consider only the range over the visible part of the spectrum well away from the resonant frequency, the real part of the refractive index appears more familiar. This spectral region is shown in figure 3.12 in a plot of the real part of the refractive index. This curve exhibits what we defined as normal dispersion in an earlier chapter. The imaginary part of the refractive index in the visible for this set of parameters is small at all wavelengths.

3.4.3 Optical properties of gases

A gas can be considered a low-density dielectric since the molecules are separated at relatively large distances. Because of this separation, the local field is approximately the applied field. Further, the polarizability will be a small contribution. Under these assumptions, we can again apply the dispersion theory model.

Returning to the complex refractive index relation in terms of the polarizability, we have,

$$N^2 = (n_R + jn_I)^2 = 1 + \frac{\mathcal{N}}{\varepsilon_0}\alpha = 1 + \frac{\omega_p^2}{\omega_0^2 - \omega^2 - j\omega\gamma}$$

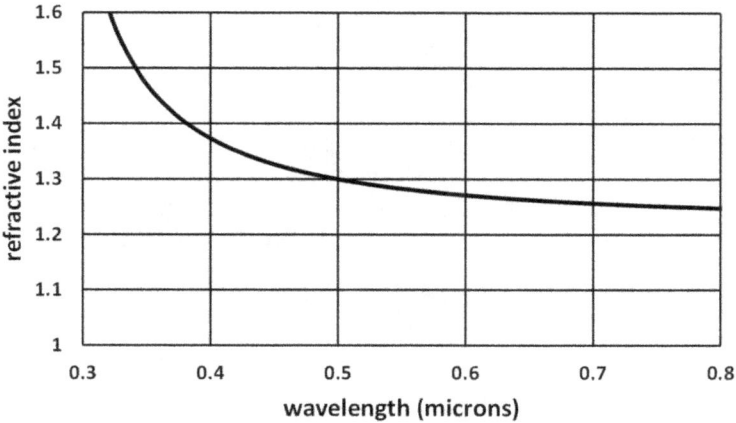

Figure 3.12. A graph of the real part of the refractive index from the model calculations of figure 3.11.

Define a dummy variable x so that,

$$x \equiv \frac{\omega_p^2}{\omega_0^2 - \omega^2 - j\omega\gamma}$$

with this definition the refractive index may be written as,

$$N = (n_R + jn_I) = \sqrt{1 + x}$$

But since α is small x is also small and we can expand the square root term using a series expansion only keeping the first term. This results in,

$$N \cong 1 + \frac{1}{2}x = 1 + \frac{1}{2}\frac{\omega_p^2}{\omega_0^2 - \omega^2 - j\omega\gamma}$$

Extracting the real and imaginary parts directly give relations for the real and imaginary parts of the refractive index directly

$$n_R = 1 + \frac{\omega_p^2(\omega_0^2 - \omega^2)}{2[(\omega_0^2 - \omega^2)^2 + \omega^2\gamma^2]} \quad \text{and} \quad n_I = \frac{\omega_p^2\omega\gamma}{2[(\omega_0^2 - \omega^2)^2 + \omega^2\gamma^2]}$$

This not surprising since we expect the refractive index of a gas to be close to 1 (like air) except in the region near a resonance. Resonant frequencies are where we experimentally observe absorption of energy into the atom or molecule making up the gas and promoting an electron from a lower to an upper energy level.

3.4.4 Optical properties of metals

A metal is characterized by a conduction band where electrons can easily travel allowing current to flow. Our dispersion theory models the dynamics of an electron in an applied electric field. If we apply this model to a metal, the oscillating particles become a cloud of nearly free electrons (plasma) moving through a positively charged background. Since the electrons are not fixed to a particular point there is no effective spring needed in modeling a metallic material. However, damping can still be applied and interpreted as the rate at which collisions occur. Macroscopically, collisions of charged carriers manifest itself as resistance [5].

It is instructive to first consider a collisionless plasma where there are no damping effects. From these results we can extract some general characteristics of electromagnetic radiation incident on metals. For the case of no damping, and no Hooke's law force, the force equation of the dispersion model is simply,

$$m\frac{d^2\vec{r}}{dt^2} = -e\overrightarrow{E_m}$$

The solution can be found by integrating this equation twice over time. Using the definitions from above, the displacement vector is,

$$\vec{r} = +\frac{e}{m\omega^2}\overrightarrow{E_m}$$

And the polarizability is,

$$\alpha = -\frac{e^2}{m\omega^2}$$

Resulting in a relation for the complex refractive index of,

$$N^2 = 1 - \frac{\omega_p^2}{\omega^2}$$

From this relation we note that when $\omega > \omega_p$ then N^2 is positive and N is real. But if $\omega < \omega_p$ then N^2 is negative and N is purely imaginary. Thus, for frequencies less than the plasma frequency a propagating electromagnetic wave cannot be supported, and the wave will be reflected. An electromagnetic wave will be supported for frequencies greater than the plasma frequency defining the frequency for the onset of transmittance within a metal.

When damping (or resistance) is included in the model, the dynamic model equation for the displacement becomes,

$$m\frac{d^2\vec{r}}{dt^2} = -e\overrightarrow{E_m} - m\gamma\frac{d\vec{r}}{dt}$$

Following the same steps used above in obtaining the complex refractive index of the medium, we find,

$$N^2 = 1 - \frac{\omega_p^2}{\omega^2 + j\omega\gamma}$$

The real and imaginary parts of the dielectric constant, ε_R and ε_I, are then,

$$\frac{\varepsilon_R}{\varepsilon_0} + j\frac{\varepsilon_I}{\varepsilon_0} = 1 - \frac{\omega_p^2}{\omega^2 + \gamma^2} + j\frac{\omega_p^2\gamma}{\omega(\omega^2 + \gamma^2)}$$

These equations are commonly called the Drude model of the dielectric constants [9]. As with the collisionless plasma model we can determine the condition for the onset of transparency for a resistive metal, that is the condition when $\varepsilon_R > 0$. This implies that,

$$\omega^2 > \omega_p^2 + \gamma^2 \equiv \omega_c^2$$

where ω_c is a critical frequency for the material to become transparent.

As an illustration, model predictions for the real and imaginary refractive index of a metal with a plasma frequency of 1.35×10^{16} rad s^{-1} and a damping factor of 4.1×10^{13} rad s^{-1} is shown in figure 3.13 as a function of the angular frequency. These values were taken as a representative average of some reported constants for gold [10]. As the frequency increases, the real part of the refractive index asymptotically approaches a value of one. Figure 3.13(b) is the same curves but plotted on a different scale to highlight the location of the plasma frequency which occurs where the two curves cross.

This same model was used as a comparison to experimental data [10]. Figure 3.14 shows the real and imaginary parts of the refractive index of gold plotted along with the model prediction, this time as a function of wavelength. This is not intended to be a fit to the data but only a comparison. The general form of the model does follow the data, especially away from the plasma frequency as expected. However, near the plasma frequency these is a discrepancy where the assumptions of the simple Drude model begins to break down.

3.4.5 Optical properties of dense dielectrics

As a material becomes even more dense, the electric polarization of the medium itself affects the local field [4]. It can be shown that the electric susceptibility must be modified to account for this contribution becoming,

$$\chi_e = \frac{N\alpha/\varepsilon_0}{1 - N\alpha/3\varepsilon_0}$$

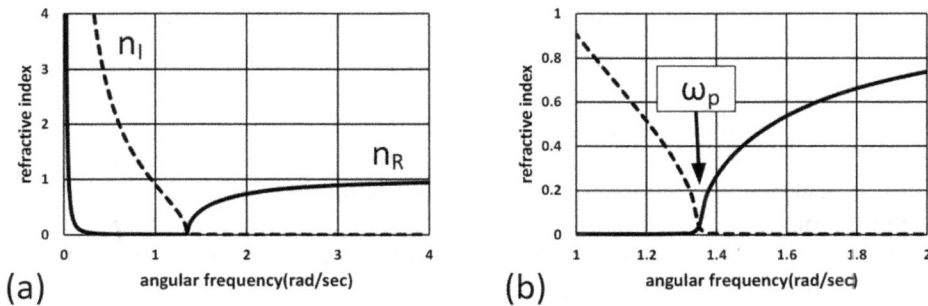

Figure 3.13. (a) The real (solid curve) and imaginary (dashed curve) refractive index as a function of angular frequency. Model parameters are: $\omega_p = 1.35 \times 10^{16}$ rad s^{-1} and $\gamma = 4.1 \times 10^{13}$ rad s^{-1}. (b) The same curves plotted on a different scale to highlight the location of the plasma frequency which occurs where the two curves cross.

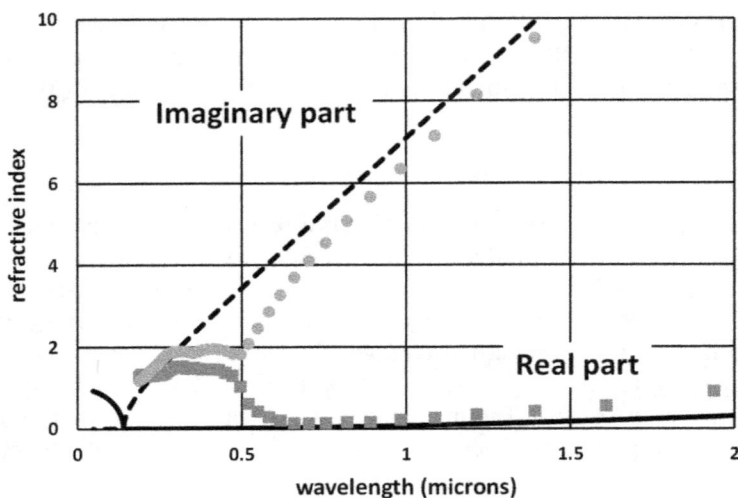

Figure 3.14. Measured values of the real and imaginary refractive index of gold plotted along with the dispersion theory model for comparison.

where α is the same polarizability that is obtained from dispersion theory [5]. The dielectric constant can now be calculated from the definition to be,

$$\varepsilon = \varepsilon_0(1 + \chi_e) = \varepsilon_0\left(1 + \frac{\mathcal{N}\alpha/\varepsilon_0}{1 - \mathcal{N}\alpha/3\varepsilon_0}\right)$$

where from before,

$$\mathcal{N}\alpha/\varepsilon_0 = \frac{\omega_p^2}{\omega_0^2 - \omega^2 - j\omega\gamma}$$

Substituting this into the equation for the dielectric constant results in,

$$N^2 = 1 + \frac{\omega_p^2}{\left(\omega_0^2 - \frac{1}{3}\omega_p^2\right) - \omega^2 - j\omega\gamma}$$

a similar form as the general model except that there is an additional term in the denominator whose presence produces a new effective resonant frequency.

The model we used only considered one oscillating electron species in the interaction of the electromagnetic wave. In more complex materials and alloys containing many different species there can in fact be multiple oscillators that are mixed together with different resonances. For these situations, the complete polarizability of the medium becomes a sum over all the polarizabilities of each separate oscillator. Further, each oscillator will have an oscillator strength associated with it. The result is an extension of the general dispersion model to,

$$N^2 = 1 + \omega_p^2 \sum_i \frac{f_i}{\omega_{i0}^2 - \omega^2 - j\omega\gamma_i}$$

where f_i is the oscillator strength, ω_{i0} is the resonant frequency, and γ_i is the damping constant for each ith species [5, 9].

As an example, figure 3.15 shows experimental data for the refractive index of a common clear soda lime glass [11, 12]. The data shown is a compilation of the real and imaginary parts of the refractive index over a wavelength range from the near ultraviolet to the far infrared. Two different oscillators can be seen one at about 10 μm wavelength and the other about 22 μm wavelength.

3.5 Light interaction with surfaces

When light rays encounter material surfaces some of the light's energy is reflected and some is transmitted. The laws of reflection and refraction discussed in chapter 2 only provide the direction of the light rays. The wave picture of reflection and refraction allows us to obtain reflected and transmitted fields and therefore the amount of light flux reflected and transmitted at boundaries between two media as well as multilayer films and interference filters.

3.5.1 Single surface reflection, refraction, and transmission

Figure 3.16 shows an incident k-vector approaching a boundary between two different media. We take the plane of the boundary to be the x–y-plane. The actual orientation of the surface does not matter since we can always reorient our coordinate system so that all the k-vectors lie in a single plane of incidence, in this case the x–z-plane. For all negative z values the refractive index is N_0 and for positive z values the refractive index is N_1. The normal vector to the surface of the

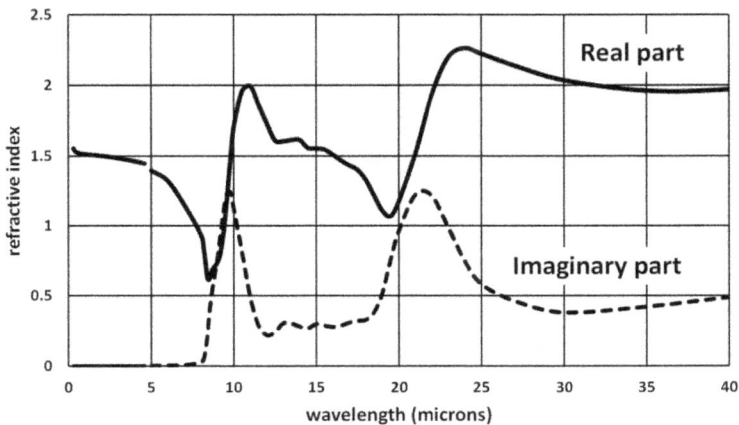

Figure 3.15. Refractive index data for clear soda lime glass showing the presence of two oscillators as predicted by the extended dispersion theory.

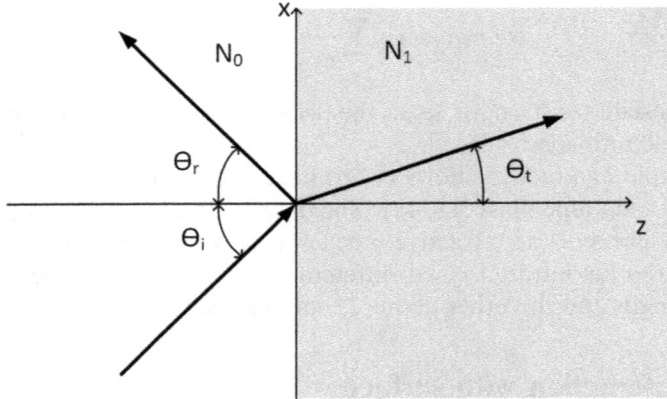

Figure 3.16. A wave with incident wave vector $\overline{k}_i$ propagating toward the boundary. A fraction of the wave reflects in the $\overline{k}_r$-direction and the remainder is transmitted in the $\overline{k}_t$-direction. Note that $\theta_i = \theta_r$ and $N_0 \sin \theta_i = N_1 \sin \theta_t$.

boundary is defined as $\hat{n} = \hat{z}$. The incident fields are defined as, $\overline{E}_i$ and $\overline{H}_i$, the reflected fields as, $\overline{E}_r$ and $\overline{H}_r$, and the transmitted fields as, $\overline{E}_t$ and $\overline{H}_t$.

Writing a general k-vector in terms of its magnitude and direction, with $\hat{k}$ the unit vector in the k-direction we have,

$$\vec{k} = \frac{2\pi N}{\lambda}\hat{k}$$

and,

$$\frac{2\pi N}{\lambda}(\hat{k} \times \vec{E}) = 2\pi\upsilon\mu\vec{H}$$

resulting in a general relation between the electric field, magnetic field, and propagation vector to be,

$$\frac{N}{c\mu}(\hat{k} \times \vec{E}) = \vec{H}$$

where again the fields and the k-vector form a mutually perpendicular set of vectors. To simplify notation, we define a value $\eta = N/c\mu$, so that,

$$\eta(\hat{k} \times \vec{E}) = \vec{H}$$

Now apply this general relation to each of the wave vectors in figure 3.16. We related the fields on each side of the boundary by applying the appropriate boundary conditions from electromagnetic theory [4, 5]. These conditions state that the tangential components of the electric fields and the tangential components of the magnetic field must be continuous across the boundary.

To obtain a relationship between the field components we will start by analyzing an incident wave vector at normal incidence to the boundary. This is shown in figure 3.17. Although this may seem only applicable to this special case, we will see

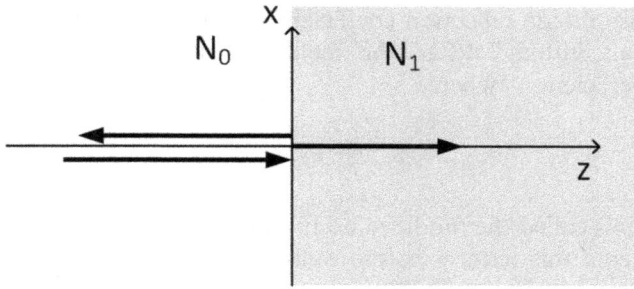

Figure 3.17. Wave vectors at normal incidence to a boundary. The incident and reflected wave vectors have been displaced slightly to avoid overlap in the drawing.

that it is relatively easy to extend the outcome of our calculation and show that it can apply to any oblique incidence angle, as in figure 3.16.

Since the electric field vectors and the magnetic field vectors for each wave are perpendicular to the k-vector, for this case, both fields are themselves tangential to the boundary. For example,

$$\overline{E_i} = \overline{E_i}_{\text{tan}} \text{ and } \overline{H_i} = \overline{H_i}_{\text{tan}}$$

So, in the N_0 medium, we have,

$$\eta_0(\hat{z} \times \overline{E_i}) = \overline{H_i}$$
$$\eta_0(-\hat{z} \times \overline{E_r}) = \overline{H_r}$$

and the N_1 medium,

$$\eta_1(\hat{z} \times \overline{E_t}) = \overline{H_t}$$

Now applying boundary conditions at $z = 0$, we obtain a relation between the magnitudes of the electric fields as,

$$E_i + E_r = E_t$$

and

$$\overline{H_i} + \overline{H_r} = \overline{H_t}$$

Substituting from above,

$$\eta_0(\hat{z} \times \overline{E_i}) - \eta_0(\hat{z} \times \overline{E_r}) = \eta_1(\hat{z} \times \overline{E_t})$$

Simplifying and eliminating the transmitted field using the electric field boundary condition yields,

$$\eta_0(E_i - E_r) = \eta_1(E_i + E_r)$$

Solving for the reflected field in terms of the incident field we obtain the ratio,

$$\frac{E_r}{E_i} = \frac{\eta_0 - \eta_1}{\eta_0 + \eta_1} \equiv r$$

where r is the amplitude reflection coefficient. A similar definition is made for the ratio of the transmitted field to the incident field to arrive at the amplitude transmission coefficient, t, where,

$$\frac{E_t}{E_i} \equiv t$$

The parameter η is called the modified admittance or the impedance [13]. Note that for dielectric media this reduces to the well-known equation for reflection from a boundary at normal incidence [2, 3],

$$r = \frac{n_0 - n_1}{n_0 + n_1}$$

Our general formulation also allows us to compute the reflectance from a metallic surface where the refractive index is a complex quantity. Again, for simplicity, we will assume normal incidence so that,

$$r = \frac{N_0 - N_1}{N_0 + N_1}$$

This relation also shows that for the situation where $n_0 < n_1$, r is negative. In fact, r can in general be a complex value since the impedance terms can be complex, as with a metal. Another way to write the refection coefficient using complex exponential notation as,

$$r = |r|e^{j\varphi}$$

where the value φ is a phase shift term. So, in our simple case for a reflection of a wave from the boundary between two dielectrics, r is negative when $\varphi = \pi$. Therefore, the physical interpretation of a negative reflection coefficient is that there is a 180° or π phase shift on reflection. These phase shifts can be a more complicated function of the refractive indices of the media and the incident angle and will be discussed later [2, 3].

Now suppose that the rays are reversed for propagation from the N_1 to the N_0 medium. There is again a reflection back into the N_1 medium and transmission to the N_0 medium. However, the reflection and transmission coefficients are different. We will designate these values as r' for the reflection coefficient and t' for the transmission coefficient. Because of conservation laws it is relatively easy to show that these coefficients are related to each other and are known as Stokes relations [3]. The Stokes relations are,

$$tt' + r^2 = 1$$
$$r = -r' = r'e^{j\pi}$$

where again we see that the difference between reflection coefficients is a π phase shift.

Since the light flux is proportional to the square of the electric field amplitude, and the incident light and reflected light are in the same medium, the ratio of the amount of reflected light flux to the incident light flux is,

$$\left| \frac{E_r}{E_i} \right|^2 = \left| \frac{\eta_0 - \eta_1}{\eta_0 + \eta_1} \right|^2 = |r|^2$$

Conservation of energy requires that the energy in the incident light be equal to the sum of the energy in the reflected light and the transmitted light. From the Poynting vector definition, the light flux is proportional to the square of the electric field amplitude. Using this definition, the ratio of the reflected light power to the incident light power is defined as the reflectance R and is,

$$R \equiv \frac{N_0 \cos(\theta_r)}{N_0 \cos(\theta_i)} \left| \frac{E_r}{E_i} \right|^2$$

Since $\theta_i = \theta_r$ with incident and reflected waves in the same medium this means that the reflectance is simply,

$$R = \left| \frac{E_r}{E_i} \right|^2 = |r|^2 = \left| \frac{\eta_0 - \eta_1}{\eta_0 + \eta_1} \right|^2$$

In a similar way, we define a value T called the transmittance, as the ratio of the transmitted power to the incident power,

$$T \equiv \frac{N_1 \cos(\theta_t)}{N_0 \cos(\theta_i)} \left| \frac{E_t}{E_i} \right|^2 = \frac{N_1 \cos(\theta_t)}{N_0 \cos(\theta_i)} |t|^2$$

It is straightforward to show that,

$$R + T = 1$$

which is just a statement of conservation of energy assuming no other absorption or scattering effects.

3.5.2 Fresnel equations

Since electromagnetic fields are vectors, it is useful to describe some common nomenclature used in describing the orientation of fields with respect to a boundary. Figure 3.18 shows vector orientations for two possible conditions that can occur when an electromagnetic wave approaches a boundary with wave vector $\vec{k}$ at an angle θ with respect to the surface normal. The x–z-plane shown in the figure is the plane of incidence. Figure 3.18(a) is the transverse magnetic, TM, orientation. That is, the magnetic field vector points in the $+y$-direction (out of the page), transverse or perpendicular to the plane of incidence and the electric field is in the plane of incidence and perpendicular to $\vec{H}$ and $\vec{k}$. Similarly, figure 3.18(b) illustrates the transverse electric, TE, orientation with the electric field vector pointing in the $+y$-direction, and the magnetic field in the plane of incidence and perpendicular to

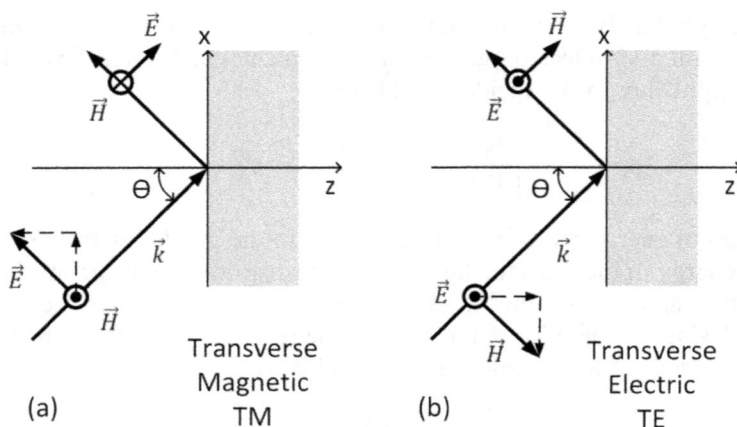

Figure 3.18. Definitions and vector diagrams of (a) transverse magnetic, TM waves and (b) transverse electric, TE waves incident on the boundary between two media. Tangential and normal components of the incident fields are shown as dashed vectors.

$\vec{E}$ and $\vec{k}$. The importance of these orientations is that any single incident wave can be written as a linear combination of TE and TM waves. Upon reflection at the boundary the field orientations are defined as shown in the figure.

Now, let us extend the results of the special case for near normal incidence to find the amplitude reflection coefficient for any oblique incidence angle. Most general texts treat each case separately to arrive at a set of relations which are the Fresnel equations [2, 3]. However, both field orientations can be obtained from one set of equations by making some simple definitions relating to the impedance terms. The key to this result is to notice that the tangential component of the electric field of a TM wave is $E_x = E \cos\theta$, and the tangential component of a TE wave is $H_x = -H \cos\theta$ as shown in figure 3.18. We must use these components when applying boundary conditions between fields on each side of the boundary. We know that in general,

$$\frac{N}{c\mu}(\hat{k} \times \vec{E}) = \vec{H}$$

For a TM wave,

$$\vec{H}_{\tan} = \vec{H} \text{ and } \vec{E}_{\tan} = \vec{E} \cos(\theta)$$

If we substitute these quantities into the equation above, we obtain a relation between only the tangential components of the fields.

$$\frac{N}{c\mu \cos(\theta)}(\hat{z} \times \vec{E}_{\tan}) = \vec{H}_{\tan}$$

As before we define the impedance term for TM waves to be, η_{TM},

$$\eta_{\text{TM}} \equiv \frac{N}{c\mu \cos(\theta)}$$

resulting in,

$$\eta_{\text{TM}}(\hat{z} \times \vec{E}_{\text{tan}}) = \vec{H}_{\text{tan}}$$

which is the same form of the equation used for the case of normal incidence and meets the requirements that the tangential components of the fields be continuous across the boundary. In fact, normal incidence is when $\theta = 0$. We can repeat this process for the TE wave and show that we arrive at the same relation but with, η_{TE} defined as,

$$\eta_{\text{TE}} \equiv \frac{N}{c\mu}\cos(\theta)$$

The specific angle θ to be used is different in each medium. So, in the N_0 medium we would use $\theta = \theta_i$ and in the N_1 medium we would use $\theta = \theta_t$. Of course, θ_t, can be found in terms of θ_i using Snell's law. Therefore, a general form relating the tangential components of the electric field and magnetic field across a boundary is,

$$\eta(\hat{z} \times \vec{E}_{\text{tan}}) = \vec{H}_{\text{tan}}$$

where η is the appropriate impedance for either TE or TM waves.

As an example, for a boundary between two dielectric media with refractive index n_0 and n_1 at incident angle θ_i and transmitted angle θ_t the amplitude reflection coefficient is [2, 3],

$$r_{\text{TE}} = \frac{\eta_0 - \eta_1}{\eta_0 + \eta_1} = \frac{n_0 \cos \theta_i - n_1 \cos \theta_t}{n_0 \cos \theta_i + n_1 \cos \theta_t}$$

This relation is the well-known TE reflection coefficient form of the Fresnel equations [2, 3]. We can also obtain the TM reflection coefficient just applying our formalism to this case,

$$r_{\text{TM}} = \frac{\dfrac{n_0}{\cos \theta_i} - \dfrac{n_1}{\cos \theta_t}}{\dfrac{n_0}{\cos \theta_i} + \dfrac{n_1}{\cos \theta_t}}$$

Figure 3.19 shows graphs of the transmission and reflection coefficients for both TE and TM cases. And figure 3.20 is a plot of the reflectance for the TE and TM cases.

The reflection coefficient for the TM wave can become zero at a specific angle as seen in figures 3.19 and 3.20. Setting the numerator to zero and using Snell's law leads to a condition that $\theta_i + \theta_t = 90°$ and defines the particular angle known as Brewster's angle or the polarization angle, where $\theta_i = \theta_B$. This angle is,

$$\theta_B = \tan^{-1}\left(\frac{n_1}{n_0}\right)$$

When natural light (randomly polarized light) is incident at this angle the reflected TM wave is zero and the only reflection is a TE polarization state, a linearly polarized reflection.

Phase shifts on reflection at a boundary between two media as a function of the incident angle may also be obtained from this simple relation for various cases of the

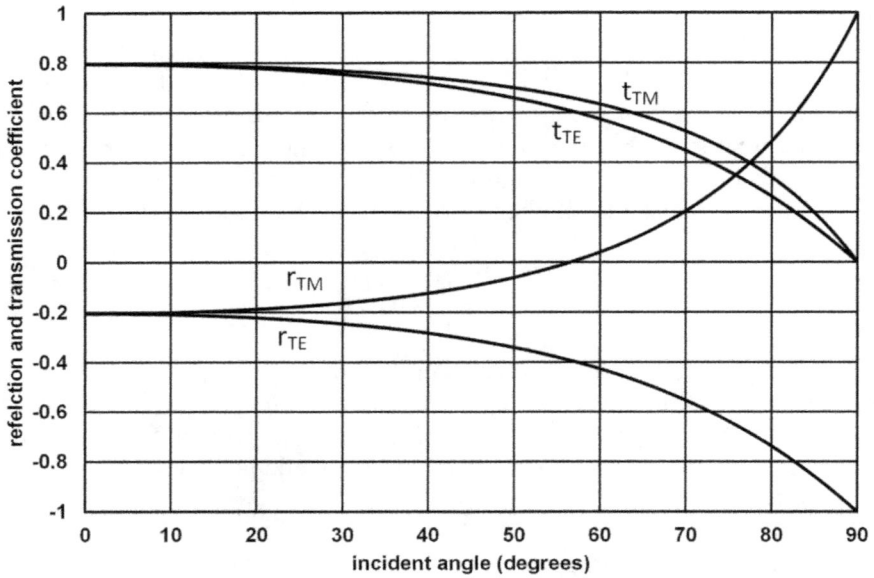

Figure 3.19. Reflection and transmission coefficients for TE and TM waves as a function of the incident angle for a HeNe source incident onto a NBK7 substrate (refractive index 1.5151).

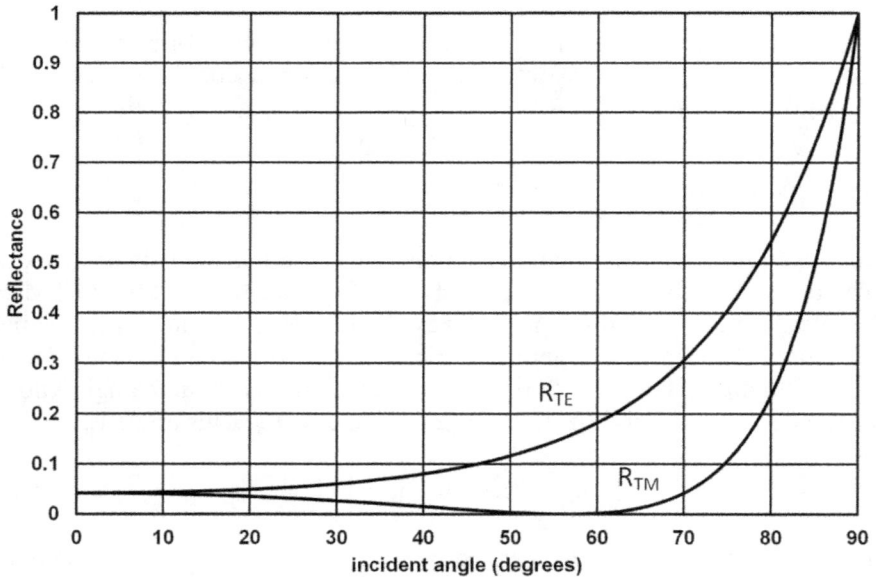

Figure 3.20. Reflectance for both the TE and TM waves as a function of the incident angle for a HeNe source incident onto a NBK7 substrate (refractive index 1.5151).

relative index of refraction between the two media and for either the TE or TM mode. To determine phase shifts, φ, we return to the definition of the reflection coefficient and its complex polar representation,

$$r = \frac{\eta_0 - \eta_1}{\eta_0 + \eta_1} = |r|e^{j\varphi}$$

To simplify notation, we define a relative admittance η as,

$$\eta = \frac{\eta_1}{\eta_0}$$

First, we examine the case of an incident TE wave on the boundary. From the relations above, and making use of the law of refraction, the relative admittance for the TE wave is,

$$\eta = \frac{n_1 \cos \theta_t}{n_0 \cos \theta_i} = \frac{\sqrt{n_1^2 - n_0^2 \sin^2 \theta_i}}{n_0 \cos \theta_i}$$

Now there are two possibilities, (1) $n_0 < n_1$ where the wave is propagating from a less dense to a more dense medium, or (2) $n_0 > n_1$, propagation from a more dense to a less dense medium. For the first possibility where $n_0 < n_1$, η is positive, real, and greater than one for all incident angles. Thus, the reflection coefficient,

$$r = \frac{1 - \eta}{1 + \eta} \leqslant 0$$

for all incident angles. Since r is negative, and $-1 = \exp(j\pi)$, the phase shift on reflection will be equal to π for all incident angles.

The second possibility $n_0 > n_1$, where there may be internal reflections, requires more investigation. Again examining the relative admittance for TE waves, we see that η is positive and real as long as the incident angle is less than the critical angle, $\theta_i \leqslant \sin^{-1} \frac{n_1}{n_0} \equiv \theta_c$. However, in this situation $0 < \eta < 1$ implying that r is positive and the phase shift on reflection, $\varphi = 0$. For incident angles greater than the critical angle we have total internal reflection so the amplitude of the reflected wave must be unity. But there is a phase shift, and it depends on the incident angle. From the definition of the relative admittance we see that the quantity under the radical in the numerator is always negative meaning that the relative admittance is purely imaginary. Rewriting the admittance in terms of a real positive value defined as 'a', we have,

$$\eta = -ja = -j\frac{\sqrt{n_0^2 \sin^2 \theta_i - n_1^2}}{n_0 \cos \theta_i}$$

for $\theta_i > \theta_c$. The choice of the negative sign is needed in order for the transmission coefficient to decrease with increasing distance into the transmission medium. The resulting reflection coefficient of,

$$r = \frac{1 + ja}{1 - ja} = \frac{\sqrt{1 + a^2}\, e^{+j\epsilon}}{\sqrt{1 + a^2}\, e^{-j\epsilon}} = e^{j2\epsilon}$$

where the angle $\epsilon = \tan^{-1}(a)$. As expected, the magnitude of the reflection coefficient is unity for total internal reflection,

$$|R| = 1$$

but the phase angle as a function of the angle of incidence is,

$$\varphi_{TE} = 2 \tan^{-1}\left[\frac{\sqrt{n_0^2 \sin^2 \theta_i - n_1^2}}{n_0 \cos \theta_i} \right]$$

for all $\theta_i > \theta_c$. Figure 3.21 shows sample plots of the phase shifts on reflection for TE waves incident on a boundary according to the relative media.

A similar procedure for the TM wave results in the following relations for the phase shift on reflection for, $n_0 < n_1$,

$$\varphi_{TM} = \begin{cases} \pi, & 0 \leqslant \theta_i < \theta_B \\ 0, & \theta_B < \theta_i < \dfrac{\pi}{2} \end{cases}$$

where $\theta_B \equiv \tan^{-1}\left(\frac{n_1}{n_0}\right)$ which is the polarization angle or Brewster's angle. When $n_0 > n_1$, then the phase shift on reflection for the TM wave is,

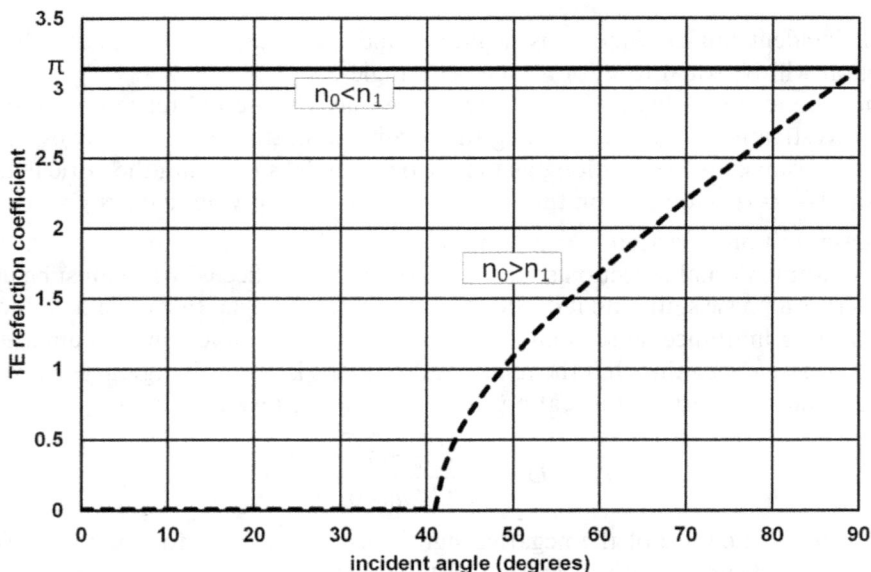

Figure 3.21. Phase shifts on reflection for TE waves as a function of the incident angle for a HeNe source incident onto a NBK7 substrate (refractive index 1.5151) and for waves internal to the NBK7 substrate into a vacuum.

$$\varphi_{\text{TM}} = \begin{cases} 0, & 0 \leqslant \theta_i < \theta'_B \\ \pi, & \theta'_B < \theta_i < \theta_c \\ \pi - 2\dfrac{\sqrt{n_0^2 \sin^2 \theta_i - n_1^2}}{\left(\dfrac{n_1}{n_0}\right)^2 \cos \theta_i}, & \theta_c < \theta_i < \dfrac{\pi}{2} \end{cases}$$

where $\theta'_B \equiv \tan^{-1}\left(\frac{n_0}{n_1}\right)$ is another polarization angle for the internally reflected waves. Note that θ_B and θ'_B are complementary angles. Plots of these functions are shown in figure 3.22.

In summary, the amplitude reflection coefficients and phase shifts on reflection can be obtained for a wave incident on a boundary between any two media (refractive index N_0 and N_1) using,

$$r = \frac{\eta_0 - \eta_1}{\eta_0 + \eta_1}$$

with appropriate selection of the refractive indices and incident vector state, using,

$$N_0 \sin(\theta_i) = N_1 \sin(\theta_t) \text{ and } \eta_{\text{TE}} \equiv \frac{N}{c\mu}\cos(\theta) \text{ or } \eta_{\text{TM}} \equiv \frac{N}{c\mu \cos(\theta)}$$

where θ is either θ_i or θ_t depending on the medium. Again, since the impedance factors can be complex r can also be complex. This is not surprising because of the known phase shifts for waves reflected from a boundary [1].

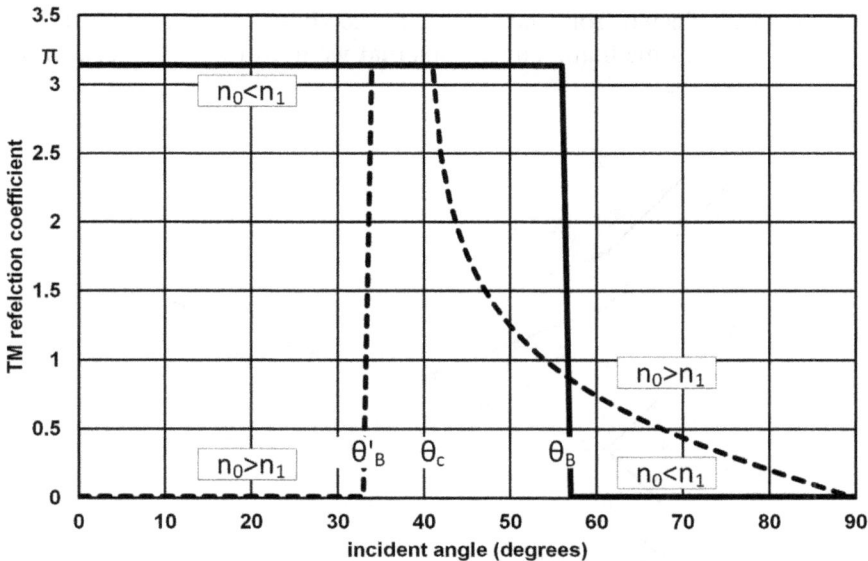

Figure 3.22. Phase shifts on reflection for TM waves as a function of the incident angle for a HeNe source incident onto a NBK7 substrate (refractive index 1.5151) and for waves internal to the NBK7 substrate into a vacuum.

3-35

3.5.3 Coatings on surfaces and thin film interference

Coatings placed on optical surfaces alter the reflectance from the surface and transmittance through the boundary. Single thick coatings act as if there are two different media with two boundaries that can reflect the incident radiation. Adding more boundary layers provides even more reflections back into the incident medium. Figure 3.23 shows several coating layers with a few rays traced into the medium. Whenever a ray intersects a boundary it splits into a reflected ray and a transmitted ray creating multiple reflections. All of the reflections at every boundary eventually contribute to the overall wave reflectance in the incident medium. A thick coating simply acts like a slab of material and displaces the incident and reflected beam.

If the coating is thin, then not only will a ray be displaced but the reflected waves associated with this ray will interfere with one another due to the superposition principle as long as the reflected waves have some degree of coherence with respect to each other. Suppose that there is a plane wave incident on the first surface boundary between the N_0 and N_1 medium at an angle θ_i. A ray diagram is shown in figure 3.24 representing this situation. At point A on the surface, the incident wave splits into two rays. Ray#1 is reflected back into the N_0 medium and Ray#2 is transmitted into the N_1 medium at angle θ_t. This wave (Ray#2) transmits to point B on the next boundary separating the N_1 and N_2 medium where it is reflected at angle θ_t toward point C on the surface between the N_1 and N_0 medium. At point C, Ray#2 is refracted at angle θ_i into the N_0 medium. The waves associated with Ray#1 and Ray#2 can interfere as long as they satisfy the interference condition.

To apply the interference condition we must compute the phase difference between the reflected wave from the first surface boundary and the wave transmitted into the coating, reflected from the second surface boundary and then transmitted back into the incident medium. This means that we need the OPD between Ray#1

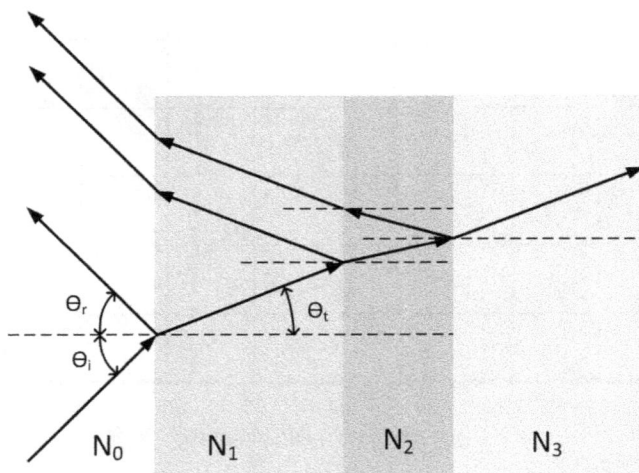

Figure 3.23. Several coating layers with multiple surfaces. A few selected rays are traced but they do show that all boundaries contribute to the reflectance in the N_0 medium.

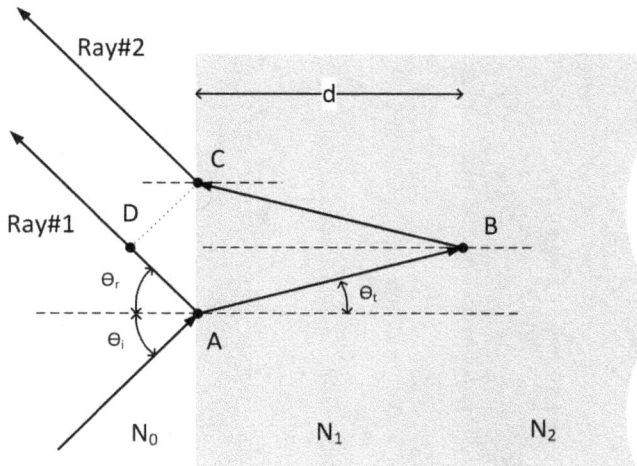

Figure 3.24. Ray diagram used to compute the OPD between a ray reflected at the top surface of a coating and the transmitted ray that is reflected from the bottom surface. The thickness of the coating is d.

and Ray#2 as they emerge together into the N_0 medium. The OPD is the difference in the optical path length taken by Ray#2 and the optical path length of Ray#1. Using the distances from figure 3.24 in terms of line segments we find,

$$OPD = N_1 \overline{ABC} - N_0 \overline{AD}$$

Point D in the figure is found by drawing a line segment perpendicular to Ray#1 and Ray#2 and intersecting with point C. Making use of Snell's law and geometry, this relation becomes,

$$OPD = 2 N_1 d \cos \theta_t$$

At normal incidence this reduces to, $OPD = 2 N_1 d$.

Continuing the discussion of a single layer structure, we will look at several special cases. The first is a further examination of multiple reflections. This occurs in both a thin slab of material and in two reflecting surfaces separated by a thin gap as illustrated in figure 3.25. These multiple reflections continue throughout no matter how large the plate as seen in figure 3.25(a).

To obtain the total reflected electric field we sum the scalar amplitudes from all the multiply reflected rays. Taking E_0 as the amplitude of the incident wave the field of the initial reflection is $E_1 = rE_0$ where r is the reflection coefficient for this surface. To obtain the next field we use the definitions of the reflection and transmission coefficients from the Stokes relations. The amplitude of the transmitted field is reduced by the transmission into the slab, t, reflection at the back surface, r', and transmission into the incident medium, t'. Remember from the Stokes relations that there is a phase shift of π for reflection from a less dense to a more dense medium and a zero phase shift on reflection for a wave traveling from a more dense to a less dense medium, $r' = -r$. There is also a phase delay with respect to the incident field that

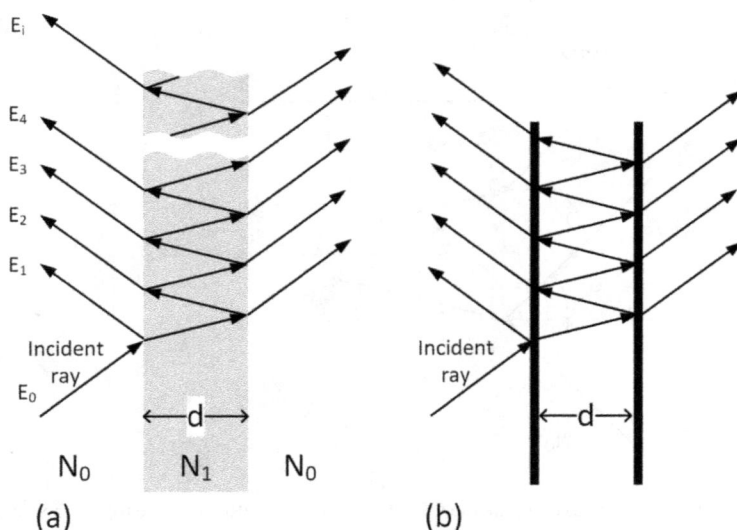

Figure 3.25. Multiple reflections occurring in (a) a thin parallel plate, and (b) between two mirror surfaces.

must be included because of the increased optical path of each successive reflected ray. Thus, the reflected field for the next reflected wave is,

$$E_2 = tt'r'E_0e^{-je}$$

where $\epsilon = \frac{4\pi}{\lambda}N_1\,d\,\cos\theta_t$. Successive fields can then be written using this process with all ith field components added to obtain the total reflected field, E_R as,

$$E_R = E_0\{r + tt'r'e^{-je} + tt'r'^3e^{-j2\epsilon} + tt'r'^5e^{-j3\epsilon} + \dots\}$$

Recognizing that a portion of the infinite series is a power series, this equation can be rewritten as,

$$E_R = E_0\left\{r + \frac{tt'r'e^{-je}}{1 - r'^2e^{-je}}\right\}$$

and using the Stokes relations,

$$E_R = E_0\left\{r + \frac{r(1 - e^{-je})}{1 - r^2e^{-je}}\right\}$$

The reflectance can then be obtained by taking the absolute square of the ratio of the reflected field to the incident field. This results in,

$$R = \frac{F\,\sin^2\left(\frac{\epsilon}{2}\right)}{1 + F\,\sin^2\left(\frac{\epsilon}{2}\right)}$$

3-38

where the parameter F is the coefficient of Finesse and is defined as,

$$F = \left(\frac{2r}{1 - r^2}\right)^2$$

From conservation of energy, assuming no absorption in the medium, the transmittance is,

$$T = \frac{1}{1 + F \sin^2\left(\frac{e}{2}\right)}$$

which is known as the Airy function [2, 3]. A graph of the transmittance as a function of the phase angle is shown in figure 3.26.

Transmittance is a maximum for phase angles that are an even multiple of π and a minimum for odd multiples of π. The larger the reflection coefficient (larger the coefficient of finesse) the lower the value of the minimum transmittance. A device formed in this manner is called an etalon when the distance d is fixed but is also referred to as a Fabry–Perot interferometer depending on the application. The Fabry–Perot interferometer can be employed as a high-resolution spectrometer since the phase term is a function of wavelength. It also forms the basic resonator used in a laser cavity [2, 3].

The second coating situation to examine is the reflectance from a thin single layer deposited on a substrate surface [14]. This is shown in figure 3.27.

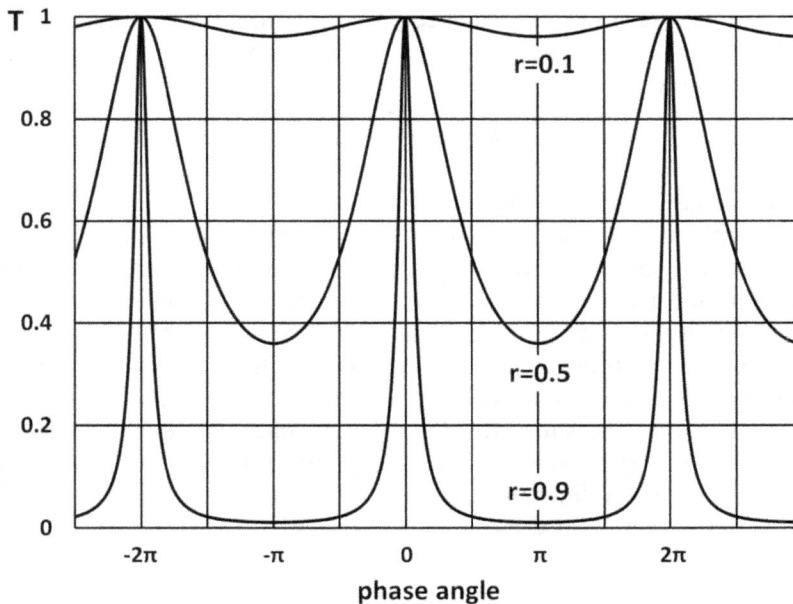

Figure 3.26. Transmittance through a thin separation between two reflecting surfaces as a function of the phase for different values of reflection.

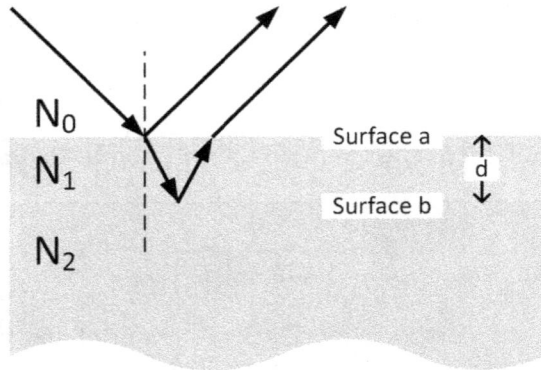

Figure 3.27. Wave incident on a film coating (N_1 medium) deposited on a substrate, N_2.

For simplicity, we will assume that the incident and reflected waves are at normal incidence. As we saw before, we can later extend the analysis to cover oblique incident angles. Under this assumption the OPD between the first and second reflected waves is just $2N_1d$. From our previous discussion, the condition for constructive interference is that the phase difference is and even multiple of π, and for destructive interference the phase difference is an odd multiple of π. These conditions are summarized as,

$$\Delta\phi_{\text{constructive}} = 2m\pi \ \text{ and } \ \Delta\phi_{\text{destructive}} = (2m+1)\pi$$

where m is an integer, $m = 0, 1, 2, 3, \ldots$ Computing the phase for any one wavefront, the phase shift with respect to a reference phase location will include both the phase associated with an OPD as well as any phase shifts on reflection. In general, any phase shift can be written,

$$\Delta\phi = \frac{2\pi}{\lambda}\text{OPD} + \Delta\epsilon$$

where $\Delta\epsilon$ is a phase shift on reflection.

As an example, take the situation of figure 3.27 and assume that the values of the refractive index are such that, $N_0 < N_1 < N_2$, which is a typical case. This could be a film deposited on a glass substrate. Phase shifts are determined with respect to the point at which the incident ray interacts with the surface. The first reflection at Surface a, has zero path difference but there is a π phase shift on reflection at this point since $N_0 < N_1$. The second wave travels through the film, reflects at Surface b, between the film and the substrate, propagating back through the film where it interferes with the first reflected wave. The phase difference for this wave is the optical path taken in the film and the phase shift on reflection. The overall phase difference between these two wavefronts is then,

$$\Delta\phi = \phi_2 - \phi_1 = \left[\frac{2\pi}{\lambda}(2N_1 d) + \pi\right] - [\pi] = \frac{4\pi}{\lambda}N_1 d$$

where λ is the wavelength in the N_0 medium.

Suppose we desire a coating that produces destructive interference or an anti-reflection (AR) coating [14]. Applying the destructive interference condition results in a requirement on the thickness of the film of,

$$d = \frac{(2m + 1)}{4} \frac{\lambda}{N_1} = (2m + 1)\frac{\lambda_1}{4}$$

where $\lambda_1 = \lambda/N_1$ which is the wavelength of the light in the film medium N_1. Thus, d must be an odd multiple of a quarter-wavelength in the film. Quarter-wave thin film layers become building block for many multilayer film structures including narrow-band interference filters [15].

For optimal performance of an anti-reflection coating, it is desirable for the reflectance at each surface be similar. Since we are assuming normal incidence, the reflectance at the top surface (surface a) is,

$$R_a = \left| \frac{N_0 - N_1}{N_0 + N_1} \right|^2$$

and the reflectance at surface b, is,

$$R_b = \left| \frac{N_1 - N_2}{N_1 + N_2} \right|^2$$

Equating these quantities results in a condition between refractive indices of,

$$N_1 = \sqrt{N_0 N_2}$$

3.5.4 Reflectance from multilayer films

While the observations in the preceding section provide physical insight into interference coatings and reflectance from layered structures a more formal treatment from electromagnetic theory provides a method to model the optical performance of multilayer thin films. This is accomplished by extending the single surface reflection formalism to include multiple surfaces and oblique incidence angles.

Figure 3.28 shows a wave impinging from external medium N_0 onto a film of index N_1 deposited onto a medium with refractive index N_2, here taken as a substrate or final medium. The thickness of the film is d_1. Two boundaries are identified, boundary a between the external medium and the film and boundary b between the film and the substrate. The z-axis is taking perpendicular to each boundary with origin at boundary a. The incident wave is also illustrated as a propagating wave at a general k-vector direction with wavelength λ, in the N_0 medium.

From before, we found a general relation for the tangential fields at a boundary. That is,

$$\eta(\hat{z} \times \vec{E}_{\text{tan}}) = \vec{H}_{\text{tan}}$$

To obtain the reflectance from this system we must relate the electromagnetic fields in medium N_2 about boundary b to the fields in medium N_0 about boundary a.

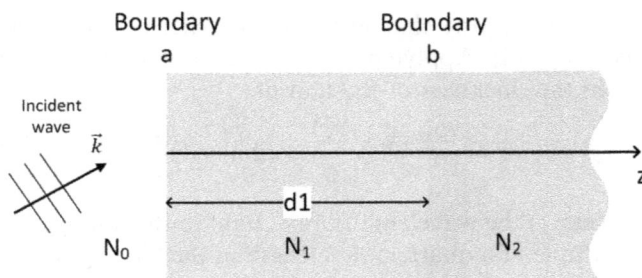

Figure 3.28. Film of index N_1 deposited onto another medium N_2 with boundaries between media specified as boundary a and boundary b.

To keep track of the direction of the reflected and transmitted fields in any region we will define $a +$ superscript to be a field propagating in the $+z$-direction and $a -$ superscript to be a field propagating in the $-z$-direction. We will also drop the tangential subscript since all field components are considered to be tangential. Thus, at boundary b, there will be a positive propagating wave and a negative propagating (reflected) wave in medium N_1 and we apply the boundary conditions to determine the field in N_2, that is, the tangential components of the electric and magnetic fields must be continuous across the boundary. This implies that,

$$\vec{H}_b = \vec{H}_{1b}^+ + \vec{H}_{1b}^- = \eta_1\left(\hat{z} \times \vec{E}_{1b}^+\right) - \eta_1(\hat{z} \times \vec{E}_{1b}^-)$$

$$\vec{E}_b = \vec{E}_{1b}^+ + \vec{E}_{1b}^-$$

These relations can be solved for the positive and negative going fields in medium N_1 in terms of the substrate fields at boundary b to be,

$$2\left(\hat{z} \times \vec{E}_{1b}^+\right) = \frac{\vec{H}_b}{\eta_1} + (\hat{z} \times \vec{E}_b)$$

$$2(\hat{z} \times \vec{E}_{1b}^-) = -\frac{\vec{H}_b}{\eta_1} + (\hat{z} \times \vec{E}_b)$$

The fields at boundary a are related to the fields at boundary b in the N_1 medium by a phase factor, δ, related to the optical path between the boundaries, or,

$$\delta = \frac{2\pi N}{\lambda}\cos\theta\, d$$

where N is the refractive of the medium, θ is the angle in the medium, and d is the thickness of the film for any medium. In other words, for our case of the N_1 medium,

$$\delta_1 = \frac{2\pi N_1}{\lambda}\cos\theta_1\, d_1$$

The angle θ is obtained from Snell's law for any ith medium since,

$$N_0 \sin\theta_0 = N_1 \sin\theta_1 = N_2 \sin\theta_2 = N_i \sin\theta_i$$

Writing the fields at boundary a, propagating in the N_1 medium we have,

$$\vec{E}_{1a}^{+} = \vec{E}_{1b}^{+} e^{j\delta_1} \text{ and } \vec{E}_{1a}^{-} = \vec{E}_{1b}^{-} e^{-j\delta_1}$$
$$\vec{H}_{1a}^{+} = \vec{H}_{1b}^{+} e^{j\delta_1} \text{ and } \vec{H}_{1a}^{-} = \vec{H}_{1b}^{-} e^{-j\delta_1}$$

Applying the boundary conditions at boundary a, we obtain a relation between these fields and the fields in the N_0 medium as,

$$\vec{H}_a = \vec{H}_{1a}^{+} + \vec{H}_{1a}^{-}$$
$$\vec{E}_a = \vec{E}_{1a}^{+} + \vec{E}_{1a}^{-}$$

Again, to obtain the reflectance we must relate the electromagnetic fields in medium N_2 about boundary b to the fields in medium N_0 about boundary a. We can now solve for this relationship. From the fields in N_1 at boundary a, we can take the cross product of the electric fields and substitute the cross-product relations for boundary b to obtain,

$$\left(\hat{z} \times \vec{E}_{1a}^{+} \right) = \frac{1}{2} \left[\frac{\vec{H}_b}{\eta_1} + (\hat{z} \times \vec{E}_b) \right] e^{j\delta_1}$$

$$(\hat{z} \times \vec{E}_{1a}^{-}) = \frac{1}{2} \left[-\frac{\vec{H}_b}{\eta_1} + (\hat{z} \times \vec{E}_b) \right] e^{-j\delta_1}$$

Similarly, for the magnetic field functions but using their relations to the electric fields, we find,

$$\vec{H}_{1a}^{+} = \eta_1 \left(\hat{z} \times \vec{E}_{1a}^{+} \right) = \frac{\eta_1}{2} \left[\frac{\vec{H}_b}{\eta_1} + (\hat{z} \times \vec{E}_b) \right] e^{j\delta_1}$$

$$\vec{H}_{1a}^{-} = -\eta_1 (\hat{z} \times \vec{E}_{1a}^{-}) = -\frac{\eta_1}{2} \left[-\frac{\vec{H}_b}{\eta_1} + (\hat{z} \times \vec{E}_b) \right] e^{-j\delta_1}$$

Taking the cross product of the electric filed boundary condition relation for the electric fields at boundary a, and substituting from above we find,

$$(\hat{z} \times \vec{E}_a) = \left[\frac{\vec{H}_b}{\eta_1} + (\hat{z} \times \vec{E}_b) \right] \frac{e^{j\delta_1}}{2} + \left[-\frac{\vec{H}_b}{\eta_1} + (\hat{z} \times \vec{E}_b) \right] \frac{e^{-j\delta_1}}{2}$$

which reduces to,

$$(\hat{z} \times \vec{E}_a) = (\hat{z} \times \vec{E}_b)[\cos \delta_1] + \vec{H}_b \left[\frac{j \sin \delta_1}{\eta_1} \right]$$

and repeating for the magnetic field boundary condition equation,

$$\vec{H}_a = (\hat{z} \times \vec{E}_b)[j\eta_1 \sin \delta_1] + \vec{H}_b[\cos \delta_1]$$

These two equations can be written in a matrix form as,

$$
\begin{bmatrix} (\hat{z} \times \vec{E}_a) \\ \vec{H}_a \end{bmatrix} = \begin{bmatrix} \cos \delta_1 & \frac{j \sin \delta_1}{\eta_1} \\ j\eta_1 \sin \delta_1 & \cos \delta_1 \end{bmatrix} \begin{bmatrix} (\hat{z} \times \vec{E}_b) \\ \vec{H}_b \end{bmatrix}
$$

to relate the fields at boundary a to boundary b. This matrix only contains information about the thin film layer and is called the characteristic matrix of the thin film [2, 3, 13]. Note that from our earlier definition,

$$
\eta_2(\hat{z} \times \vec{E}_b) = \vec{H}_b
$$

A similar relationship must also be valid for the fields in medium N_0 except the impedance value and therefore the reflectance are modified by the surfaces of the film as we saw in the discussion of multiple reflections from a thin film. We define a new parameter, Γ, which is the modified impedance or admittance of the assembly (assembly of film layers) so that,

$$
\Gamma(\hat{z} \times \vec{E}_a) = \vec{H}_a
$$

The admittance of the assembly acts essentially as the refractive index of a single effective medium and the reflection coefficient for the thin film system can be written as,

$$
r = \frac{\eta_0 - \Gamma}{\eta_0 + \Gamma}
$$

Using the characteristic matrix formulation for the film with only scalar components (albeit in general complex),

$$
\begin{bmatrix} B \\ C \end{bmatrix} = \begin{bmatrix} \cos \delta_1 & \frac{j \sin \delta_1}{\eta_1} \\ j\eta_1 \sin \delta_1 & \cos \delta_1 \end{bmatrix} \begin{bmatrix} 1 \\ \eta_2 \end{bmatrix}
$$

where, the scalar values B and C allow the value of Γ to be calculated as,

$$
\Gamma \equiv \frac{C}{B}
$$

From this value the reflection coefficient, and reflectance from the system may be found. Also, since the phase term δ_1 and the impedance term η_1 are function of the refractive index, wavelength, and angular variables this formalism allows reflectance to be determined as a function of any of these variables.

Let us now apply this formalism to determine the reflectance spectra for a single film coating onto a substrate. We will begin with a glass substrate of NBK7. From our earlier analysis we found that a design requirement for the refractive index of an anti-reflection film on a substrate ($N_2 = 1.5185$, NBK7 at 550 nm wavelength) with air as an external medium ($N_0 = 1$), is,

$$
N_1 = \sqrt{N_0 N_2} = \sqrt{(1)(1.5185)} = 1.233
$$

However, no suitable material having both physical and optical characteristics exists for a film coating with this refractive index. The best coating choice is magnesium fluoride (MgF_2) which has a refractive index of 1.3785 at 550 nm. Using these parameters, figure 3.29 shows the computed reflectance at normal incidence from a single film deposited onto a substrate of NBK7 glass ($N_2 = 1.5185$) with different thicknesses. Dispersion of the materials has not been considered in these plots. Also plotted on the graph is reflectance of the substrate with no film coatings.

Any coating thickness reduces the reflectance from the nominal reflectance of the substrate. The quarter-wave thickness indeed shows an absolute minimum at the center wavelength of 0.55 μm as predicted from destructive interference. Also, half-wave and full-wave thicknesses have a maximum at the center wavelength equal to the bare substrate reflectance, as if the coating were absent at this wavelength [14].

The formalism developed above allows us to extend the analysis of coating to multiple thin film layers. Since the application of one coating onto a substrate produces an equivalent single medium with a modified admittance, more layers can be considered by the successive application of the characteristic matrix for these layers. Therefore, in general, we simply multiply matrices to obtain a single characteristic matrix, M, for a set of multiple thin film layers. The necessary relations for calculating the reflection from a film stack of 'm' film layers deposited on a substrate of refractive index N_S from an incident N_0 medium are summarized as,

$$M = M_1 M_2 \cdots M_m = \prod_{i=1}^{m} \begin{bmatrix} \cos \delta_i & \dfrac{j \sin \delta_i}{\eta_i} \\ j\eta_i \sin \delta_i & \cos \delta_i \end{bmatrix}$$

$$\begin{bmatrix} B \\ C \end{bmatrix} = M \begin{bmatrix} 1 \\ \eta_S \end{bmatrix} \text{ and } \Gamma \equiv \frac{C}{B} \text{ so that } r = \frac{\eta_0 - \Gamma}{\eta_0 + \Gamma}$$

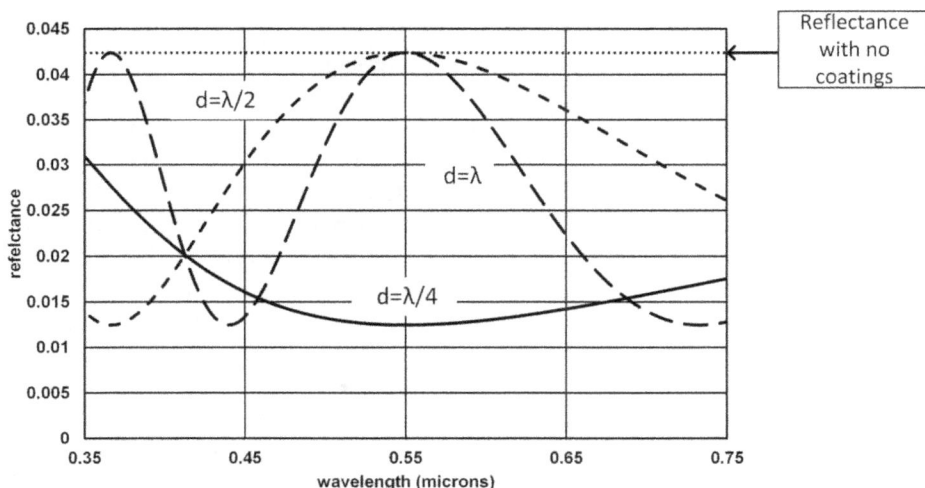

Figure 3.29. Single film anti-reflection coating of MgF2 on NBK7. The three plots correspond to the thickness of the film chosen to be quarter-wave, half-wave, and full-wave at 550 nm wavelength.

with

$$\delta_i = \frac{2\pi N_i}{\lambda}\cos\theta_i\, d_i$$

$$N_0\sin(\theta_0) = N_i\sin(\theta_i) = N_S\sin(\theta_S)$$

$$\eta_{TE} \equiv \frac{N_i}{c\mu_i}\cos(\theta_i) \ \text{ or } \ \eta_{TM} \equiv \frac{N_i}{c\mu_i\cos(\theta_i)}$$

As we have seen before the TE or TM waves must be considered separately using the above relations for incident oblique angles. In fact, some care must be taken because the values are in general complex, particularly for the modified admittance quantities. Alternate relations for the modified admittance values can be obtained by appropriately applying Snell's law. This results in,

$$\eta_{TE} = \frac{1}{c\mu_i}\sqrt{N_i^2 - N_0^2\sin^2\theta_0} \ \text{ and } \ \eta_{TM} = \frac{N_i^2}{\eta_{TE}}$$

While this analysis is an excellent tool for calculations and allows predictions to be made on the performance of a set of films and coatings applied to optical components, we need other insight to use in a design process. Quarter-wave layers provide a minimum in reflectance for a single layer. Suppose we have two layers of different materials each with a quarter wave thickness. We will find that these structures form the building blocks for multiple coatings and provide design insight [14]. Since the materials are different, we designate one a low index, N_L, layer as [L] and the other high index, N_H, layer as [H]. The quarter-wave thickness about a central wavelength, λ_0, in each film is then,

$$[L] \rightarrow d_L = \frac{\lambda_0}{4N_L} \ \text{ and } \ [H] \rightarrow d_H = \frac{\lambda_0}{4N_H}$$

As before, each layer will have its own characteristic matrix and to compute the characteristic matrix for a two-layer structure we would multiply these two matrices. For this case there are several possibilities. First, we could have two successive quarter wave layers of the same material resulting in a half-wave layer. In terms of the quarter-wave designation this would be written as either, [HH] or [LL] depending on the material. For the particular case where $\lambda = \lambda_0$, at the central wavelength, the phase term δ is a multiple of π. The characteristic matrix at this wavelength becomes the unity matrix. Thus, a structure of two similar quarter-wave layers deposited back-to-back is called an absentee layer or spacer layer [16]. Remember we saw this situation in figure 3.29 at the 0.550 µm wavelength where the half-wave layer was a maximum.

The second possibility for combining two quarter-wave layers is [HL]. Examining the phase term at the central wavelength we find that δ is an odd multiple of $\pi/2$. Substituting this value into the characteristic matrices for each single layer we find that they are,

$$[H] = \begin{bmatrix} 0 & \dfrac{j}{\eta_H} \\ j\eta_H & 0 \end{bmatrix} \text{ and } [L] = \begin{bmatrix} 0 & \dfrac{j}{\eta_L} \\ j\eta_L & 0 \end{bmatrix}$$

Multiplying these two matrices to obtain the characteristic matrix of the two-film structure we have,

$$[HL] = \begin{bmatrix} -\dfrac{\eta_L}{\eta_H} & 0 \\ 0 & -\dfrac{\eta_H}{\eta_L} \end{bmatrix}$$

and from this matrix the reflectance may be computed. An example graph for the reflectance at normal incidence from an $[HL]$ structure is shown in figure 3.30 for titanium dioxide and silicon oxide deposited onto a substrate of NBK7. In a similar manner, take the layer structure to be reversed as $[LH]$. The characteristic matrix for this structure is now,

$$[LH] = \begin{bmatrix} -\dfrac{\eta_H}{\eta_L} & 0 \\ 0 & -\dfrac{\eta_L}{\eta_H} \end{bmatrix}$$

The reflectance from the $[LH]$ structure is also shown in figure 3.30. The low-high structure provides a minimum reflectance over a wide portion of this wavelength, chosen especially to cover the visible portion of the spectrum. Film structures that are designed to minimize the reflectance over a wide spectral range are designated as BBAR for broad-band anti-reflection coatings [14].

Multiple quarter-wave layers form the building blocks for multilayer films. To designate a particular design the convention is to write a line associated with a given

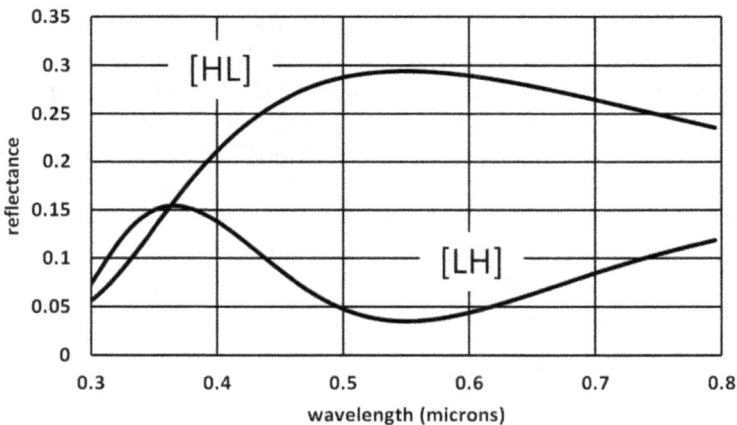

Figure 3.30. Reflectance as a function of wavelength for high and low quarter-wave layers $[HL]$ at 550 nm deposited onto a substrate of NBK7 and for the $[LH]$ structure. The high layer has a refractive index of 2.16 (TiO$_2$) and the low layer refractive index is 1.45 (SiO$_2$).

design out into a series of layers of the system. For example, six layers of alternating high and low index films in air covering a substrate would be designated as,

$$\text{air}|HLHLHL|\text{substrate} \quad \text{or} \quad \text{air}|[HL]^3|\text{substrate}$$

From the example above, we saw that a maximum reflectance occurred about the central wavelength with an $[HL]$ film layer deposited onto the substrate. Increasing the number of layers will continue to increase the reflectance at this wavelength. We can analyze the concept from the characteristic matrix formulation. From above, we found the $[HL]$ matrix, so multiplying two of the matrices together will form a structure of four layers or,

$$[HLHL] = [HL]^2 = \begin{bmatrix} \left(-\frac{\eta_L}{\eta_H}\right)^2 & 0 \\ 0 & \left(-\frac{\eta_H}{\eta_L}\right)^2 \end{bmatrix}$$

which can easily be extended to a total of m-layers. Computing the modified admittance of the entire assembly on a substrate of refractive index N_S results in,

$$\Gamma = \eta_S \left(-\frac{\eta_H}{\eta_L}\right)^{2m}$$

and since $N_H > N_L$, $\Gamma > 1$, and increases the larger the value of m. Assuming normal incidence from air the reflectance from an m-layer structure becomes,

$$R = \left| \frac{1-\Gamma}{1+\Gamma} \right|^2 = \left| \frac{\frac{1}{\Gamma} - 1}{\frac{1}{\Gamma} + 1} \right|^2$$

This result implies that R approaches unity as the modified admittance becomes large. In practice, this condition can be reached with only a few layers. Figure 3.31 shows the spectral reflectance from different $[HL]$ stacks using the same values as in figure 3.30. Even more layers will increase the reflectance and over a fairly broad spectral range about the central design wavelength. Thus, a component made from a set of high-low index layers deposited on a substrate acts as a reflector over a wavelength band becoming a dielectric mirror.

In the discussion of multiple reflections from surfaces separated by a thickness we examined the Fabry–Perot structure comprised of two mirrors. Using dielectric mirrors, we can develop a similar concept. Take as an example the structure,

$$\text{air}|[HL]^m HH[LH]^m|\text{substrate}$$

This is equivalent to two mirrors separated by a gap where the gap is the $[HH]$ half-wave separation or spacer layer. Analysis of the complete characteristic matrix for this type of structure at the central wavelength λ_0 shows again that we arrive at a unity matrix indicating a minimum reflectance or maximum transmittance but only

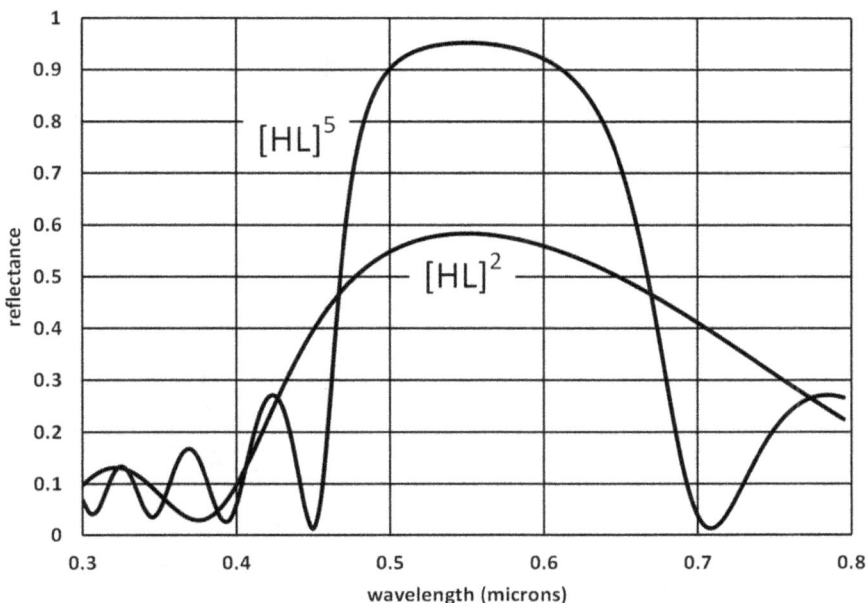

Figure 3.31. Reflectance as a function of wavelength for multiple high-low index film structures with increasing number of layers centered about 550 nm and deposited onto a substrate of NBK7. The same film materials were used as in figure 3.30.

for this specific wavelength. This is illustrated in figure 3.32 where a plot of the reflectance as a function of wavelength is shown for

$$\text{air}|[HL]^4HH[LH]^4|\text{substrate}$$

The more layers on each side of the spacer layer that are added the narrower the transmittance peak about the central wavelength. These types of components are narrowband spectral filters. They are designed about a specific wavelength and can be made narrower by increasing layers or made broader by altering the thicknesses of surrounding layers [14, 17]. Designs for edge filters that allow low wavelength to pass or high wavelengths to pass can also be made using similar multilayer structures [16, 17].

In specifying filters for a system design some general terminology is often used. Many spectral filters provide an average transmittance at the peak along with some measure of spectral width. The region of the spectrum with low transmittance (or high reflectance) is called the stop band. Likewise, high transmittance spectral regions are referred to as the pass band. Figure 3.33 is an illustration of the transmittance of a filter and some typical values that might be given as part of the specification. These include the center wavelength, average transmittance within the pass band, the full width at half-maximum (FWHM), and the wavelength span at 10% of the average transmittance. The width of the stop band and the blocking level may also be specified. Blocking level is the value of the transmittance (also given in terms of optical density, OD) within the stop band.

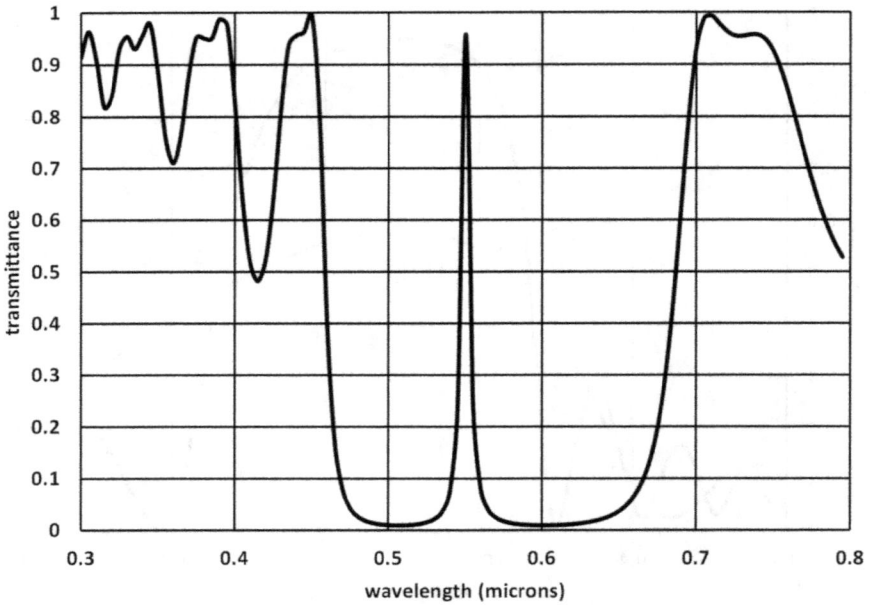

Figure 3.32. Narrowband spectral filter example using a stack of four multiple high-low layers separated by a high-index spacer layer. The substrate used for this calculation is NBK7. The high layer has a refractive index of 2.16 (TiO_2) and the low layer refractive index is 1.45 (SiO_2).

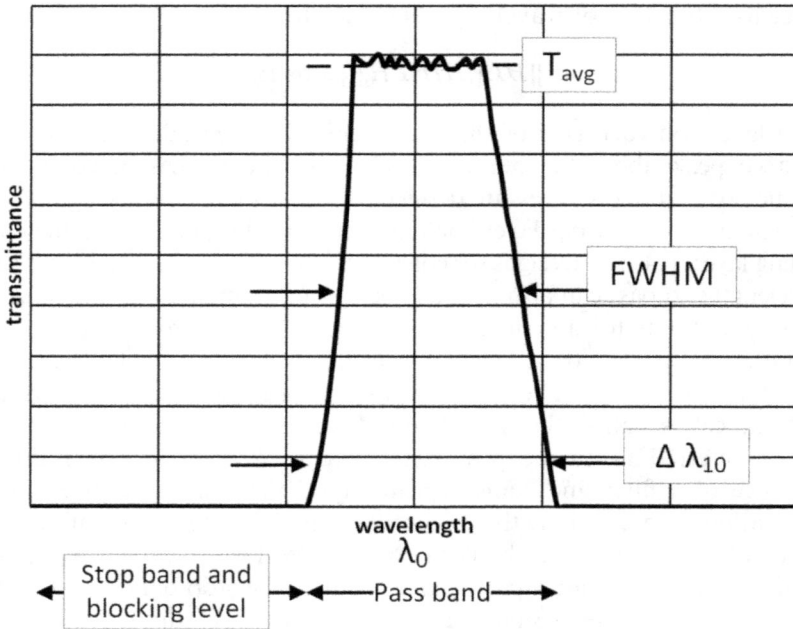

Figure 3.33. Selected parameters used in specifying a spectral filter.

Although we have used dielectric materials so far in our discussion and calculations, the formalism allows for metallic materials which have complex refractive indices. As an example, suppose that a single pair of high-low dielectric layers is deposited on a metallic substrate with complex refractive index $N_S = n_R + jn_I$. If we examine the reflectance into an $N_0 = 1$ external medium at normal incidence for just the central wavelength, we find that the modified admittance is,

$$\Gamma = (n_R + jn_I)\left(-\frac{n_H}{n_L}\right)^2$$

Resulting in a reflectance value of,

$$R = \left|\frac{1-\Gamma}{1+\Gamma}\right|^2 = \frac{1 - \left[\dfrac{2\left(\frac{n_H}{n_L}\right)^2 n_R}{1 + \left(\frac{n_H}{n_L}\right)^4\left(n_R^2 + n_I^2\right)}\right]}{1 + \left[\dfrac{2\left(\frac{n_H}{n_L}\right)^2 n_R}{1 + \left(\frac{n_H}{n_L}\right)^4\left(n_R^2 + n_I^2\right)}\right]}$$

One condition for this value to be larger than the reflectance from bare metal arises when,

$$\left(\frac{n_H}{n_L}\right)^2 > 1$$

which is the case for the high-low coating pairs. Figure 3.34 shows the reflectance spectra for aluminum and for an aluminum surface coated as air $|HLHL|$ Aluminum. The reflectance of bare metal has been computed from refractive index data [12, 18]. The high layer has a refractive index of 2.16 (TiO_2) and the low layer refractive index is 1.45 (SiO_2) with layer thicknesses centered at 550 nm. As predicted, the reflectance is higher than that of bare metal in the vicinity of the central wavelength.

As the figure illustrates, coatings can boost the reflectance of metallic mirrors but only within a particular spectral range. Metallic mirrors are not only coated for boosted reflectance but to protect the surfaces from physical damage and degradation.

3.6 Polarization

The propagating electromagnetic wave illustrated in figure 3.3 assumes that the electric field vector oscillates but remains in the x–z-plane. This also means that the magnetic field vector must remain in the y–z-plane. A wave that meets this condition is said to be linearly polarized or plane polarized. The vector nature of the propagating waves is called polarization.

Maxwell's equations require that the electric field and magnetic field be perpendicular to one another as an electromagnetic wave propagates in space. So, in general the electric field $\vec{E_0}$ for the wave propagating in the z-direction could be written as,

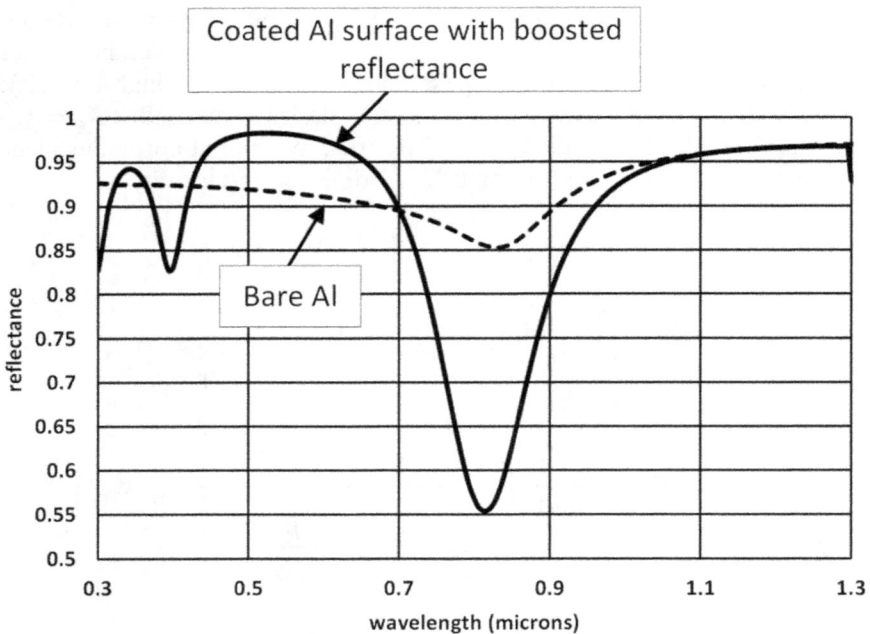

Figure 3.34. Reflectance spectra from an aluminum surface and an aluminum surface coated with quarter wave layers illustrating boosted reflectance.

$$\vec{E_0} = E_{x0}e^{j\varphi_x}\hat{x} + E_{y0}e^{j\varphi_y}\hat{y}$$

where E_{x0} and E_{y0} are real amplitudes and φ_x and φ_y are the corresponding phase constants of each vector component. Because the electric field and the magnetic field are related through Maxwell's equations with vector directions perpendicular to one another we only need to consider one field in analyzing polarized waves [3, 4].

3.6.1 Jones vectors

A convenient way to analyze the state of polarized fields as a wave propagates through an optical system is to use Jones vectors [2, 3, 5]. A Jones vector is formed as an x-axis and y-axis column vector for a wave propagating in the z-axis direction. The vector from above in algebraic notation can then be easily represented as a column vector,

$$\vec{E_0} = \begin{bmatrix} E_{x0}e^{j\varphi_x} \\ E_{y0}e^{j\varphi_y} \end{bmatrix}$$

As an example, return to figure 3.3 for a propagating electromagnetic wave where the electric field vector is oriented along the x-axis direction. This polarization state for the case where, $E_{x0} = E_0$, $E_{y0} = 0$, and $\varphi_x = \varphi_y = 0$, can be described by a Jones vector as,

$$\overrightarrow{E_0} = E_0 \begin{bmatrix} 1 \\ 0 \end{bmatrix}$$

which represents a linearly polarized state oriented along the x-direction.

All field vectors will have some amplitude that is constant for both components, like the E_0 value above. In using this notation, it is convenient to divide out this value since we are just determining the state of polarization and not computing field magnitudes. However, the complete field can always be found. With this in mind, we define a complex quantity,

$$\tilde{E} = \frac{\overrightarrow{E_0}}{E_0}$$

Then any polarized vector field state can be rewritten in a general form, as,

$$\tilde{E} = A \begin{bmatrix} a_x \\ a_y e^{j\varepsilon} \end{bmatrix}$$

where A is a normalization constant, a_x and a_y are constant amplitudes and ε is the phase difference between the y-component and the x-component of the fields [3, 19].

For a wave where there is a phase shift between components, assume that $E_{x0} = E_{y0} = E_0$, $\varphi_x = 0$, and $\varphi_y = \pi/2$. For this case, the Jones vector is then,

$$\tilde{E} = A \begin{bmatrix} 1 \\ e^{j\frac{\pi}{2}} \end{bmatrix}$$

where A is a normalization constant to be determined because the magnitude of the vector must be maintained as unity. From Euler's identity, $e^{j\frac{\pi}{2}} = j$. The magnitude of the vector, the vector multiplied by its complex conjugate is, $A^2 + A^2 j(-j) = 1$, resulting in $A = \frac{1}{\sqrt{2}}$. With the values the vector that describes this polarization state is,

$$\tilde{E} = \frac{1}{\sqrt{2}} \begin{bmatrix} 1 \\ j \end{bmatrix}$$

This type of state is known as left circular polarization. Analysis of this propagating wave shows that the vector direction of the electric field rotates about the z-axis as time progresses. Viewed along the propagation direction, the tip of the vector will process in time through a circle since the length of the vector remains constant.

Another convenient form of the Jones vector useful for computations and identification of polarization states, is to use complex notation. It is easy to show that,

$$\tilde{E} = \frac{1}{\sqrt{a^2 + b^2 + c^2}} \begin{bmatrix} a \\ b + jc \end{bmatrix}$$

where $\varepsilon = \varphi_y - \varphi_x$, a, b, and c are real constants. In arriving at this result, we also used $\tan \varepsilon = c/b$. The general form for a polarization state is elliptically polarized light [2, 3, 5]. Table 3.1 lists Jones vectors for several different polarization states.

Table 3.1. List of selected Jones vectors. The parameters: a, b, and c used in the table are positive, real constants.

Property	Equation	x–y projection in time
Linear—horizontal	$\begin{bmatrix} 1 \\ 0 \end{bmatrix}$	
Linear—vertical	$\begin{bmatrix} 0 \\ 1 \end{bmatrix}$	
Linear—angle θ from x-axis	$\begin{bmatrix} \cos\theta \\ \sin\theta \end{bmatrix}$	
Circular—left	$\frac{1}{\sqrt{2}}\begin{bmatrix} 1 \\ j \end{bmatrix}$	
Circular—right	$\frac{1}{\sqrt{2}}\begin{bmatrix} 1 \\ -j \end{bmatrix}$	
Elliptical—$a < c$ Left + Right −	$\frac{1}{\sqrt{a^2+c^2}}\begin{bmatrix} a \\ \pm jc \end{bmatrix}$	

Elliptical—$a > c$
Left $+$

Right $-$

$$\frac{1}{\sqrt{a^2 + c^2}}\begin{bmatrix} a \\ \pm jc \end{bmatrix}$$

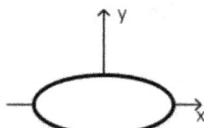

Elliptical
Left $+$

Right $-$

$$\frac{1}{\sqrt{a^2 + b^2 + c^2}}\begin{bmatrix} a \\ b \pm jc \end{bmatrix}$$

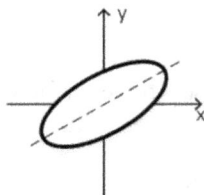

3.6.2 Natural light and polarizing mechanisms

Jones vectors only describe polarized light; however, they can be used when natural light is incident on the system. This special condition requires that only polarization states will be analyzed as the wave propagates through the system. We used this earlier in arriving at the Fresnel equations. In that case the incident radiation was represented as two orthogonal, linearly polarized waves, the TE and TM modes. If Jones vectors were used in this situation, we would just describe the vector as equal oscillations in both the x- and y-directions. So, natural light is just written as,

$$\tilde{E} = \frac{1}{\sqrt{2}}\begin{bmatrix} 1 \\ 1 \end{bmatrix}$$

Note that this is the same polarization state as a linear polarization oriented at 45°. Again, Jones vectors are only for polarized light so if this vector is a resultant state, the light is plane polarized at 45°.

Suppose that natural light is incident on a surface as described above. Then the Jones vector is a linear combination of TE and TM waves. Using our results for the reflection coefficient, we found earlier that there is an incident angle at which the TM wave reflection is zero. (This can be observed in the plots of figures 3.19 and 3.20.) This occurs when,

$$r_{\text{TM}} = \frac{\dfrac{n_0}{\cos \theta_i} - \dfrac{n_1}{\cos \theta_t}}{\dfrac{n_0}{\cos \theta_i} + \dfrac{n_1}{\cos \theta_t}} = 0$$

Leading to the definition of the polarizing angle or Brewster's angle,

$$\theta_B = \tan^{-1}\left(\frac{n_1}{n_0}\right)$$

When natural light is incident at the polarization (Brewster's angle) the amplitude of the reflected TM wave is zero and the only reflection is a TE polarization state, producing a linearly polarized reflection. Thus, the observed phenomena of polarization by reflection is one of the physical mechanisms that produces polarized light. In addition to polarization by reflection, polarized light is observed in a number of situations where light waves interact with materials [20]. Natural light scattered from small particles in suspension in a medium polarizes the scattered light in preferential directions. The light from the Sun is polarized by scattering from atmospheric molecules to generate polarized skylight.

Certain crystalline materials exhibit birefringence (refractive index is different along different crystalline axes) creating double refraction. There are three classifications used to describe crystalline materials in terms of their refractive index in three dimensions. An isotropic material has the same refractive index in all directions within the crystal. When one of the refractive index values is different than the other two, we call this a uniaxial crystal. The direction of the unique index (n_e, the extraordinary index) is the optic axis and is an axis of symmetry. The other two indices are called the ordinary refractive index, n_o. When all the refractive indices are different in three different directions the material is a biaxial crystal.

An incident beam of light will refract differently depending on the crystal orientation and these refractions are polarized [2, 3, 19]. When an incident beam propagates along the crystalline axis within the birefringent material there is a phase delay between the polarization states because there is a different refractive index for each orthogonal state. The direction with the largest refractive index is referred to as the slow axis, the smallest refractive index is the fast axis. A uniaxial crystal cut into a slab of thickness L so that its optic axis lies in the plane of incidence is a phase retarder or waveplate. This can be easily seen by computing the optical path taken by two different polarization states, one parallel to the optic axis. The OPD of the wave is,

$$\text{OPD} = (n_e - n_o)L \equiv \Delta\varphi\frac{\lambda}{2\pi}$$

When the beam emerges, there will be a phase shift between the two polarization directions altering the type of polarization. When $\Delta\varphi = \pi/2$ it is a quarter waveplate (QWP). When $\Delta\varphi = \pi$ the component is a half waveplate (HWP). Of course, this condition is for one particular wavelength and does depend on the two refractive indices at that wavelength.

Polarization by absorption is observed in some materials when one polarized component is allowed to transmit, and all other components are absorbed. We can use Jones vectors to analyze this situation. Take a linearly polarized beam with its electric field vector oriented at a general angle θ to the x-axis. The vector for this state of incident light is,

$$\vec{E_i} = E_0\begin{bmatrix} \cos\theta \\ \sin\theta \end{bmatrix}$$

Suppose this beam is sent into an absorbing material that absorbs light in all directions except along the vertical (x-axis). The concept is illustrated in figure 3.35.

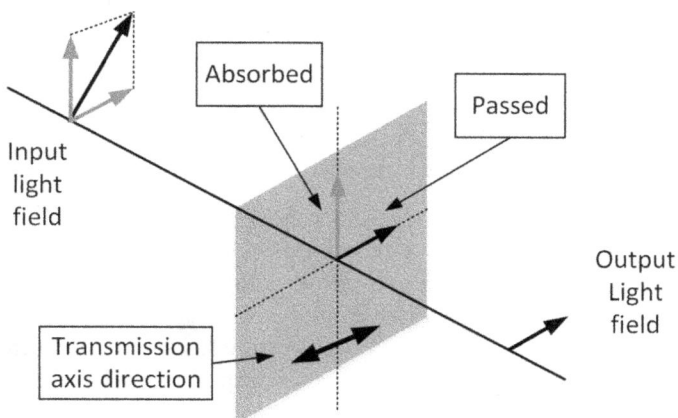

Figure 3.35. Absorbing polarizer. Only the component of the incident light along the direction of the polarizer's transmission axis is passed.

The resultant transmitted light field will only be the x-component of the incident light, or,

$$\vec{E_t} = E_0 \begin{bmatrix} \cos\theta \\ 0 \end{bmatrix}$$

Since the irradiance of a beam is proportional to the square of the amplitude of the field, using the Poynting vector average flux, the transmitted irradiance is,

$$S_{\text{ave}} = \frac{1}{2}\varepsilon_0 c E_t^2 = S_{0\,\text{avg}}\cos^2\theta$$

this relationship is known as Malus's law. The angle θ is not a general angle but the angle between the incident polarized vector and the axis of orientation of the absorber. Polarization by absorption is seen in ordinary polarized sunglasses and polarizing sheets and precision polarizer optical components [20].

3.6.3 Jones matrices

Jones vectors specify the state of polarization at some point along a propagating wave. This can be applied to optical systems using a general systems approach. As with any systems approach, the output can be related to the input of the system through a transformation. In a similar manner used to follow ray vectors through a system using system matrices, we can follow the polarization vectors through a system using Jones matrices. This system concept is shown in figure 3.36. Since the incident and transmitted fields are vectors, they are related through a 2×2 system matrix. The matrix equation relating an incident column vector and the resultant transmitted field vector is,

$$\vec{E_t} = M_{\text{sys}}\vec{E_i}$$

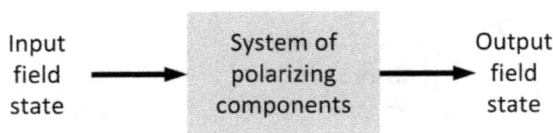

Figure 3.36. Systems approach to using Jones matrices.

where M_{sys} is the final system matrix. To use this in a systems model we need to know various system matrices for polarizing components comprising the system.

To determine the form of the polarization component matrices we use cases of incident input to a general matrix knowing an expected output. To begin, suppose we want the matrix describing a linear polarizer with a vertical transmission axis (TA). Whatever the input to the system the output must have a linear vertical polarization state. So, for an input natural light beam, the matrix equation is,

$$\begin{bmatrix} 0 \\ 1 \end{bmatrix} = \begin{bmatrix} a & b \\ c & d \end{bmatrix}\begin{bmatrix} 1 \\ 1 \end{bmatrix}$$

which gives a set of equations in terms of the general a, b, c, and d values in the matrix. For self-consistency, another input filed should result in the same output. So,

$$\begin{bmatrix} 0 \\ 1 \end{bmatrix} = \begin{bmatrix} a & b \\ c & d \end{bmatrix}\begin{bmatrix} 0 \\ 1 \end{bmatrix}$$

From these conditions, we have four equations and four unknowns so that all of the values may be determined resulting in a matrix for a vertical polarizer component,

$$M_V = \begin{bmatrix} 0 & 0 \\ 0 & 1 \end{bmatrix}$$

This process can be followed for all types of components, polarizers, rotators, phase retarders, and waveplates [3, 19]. Table 3.2 lists the Jones matrices for a set of selected polarizing components.

Determining the polarization state of a beam after propagating several polarizing elements is as simple as multiplying matrices for each component. The final matrix for the polarizing system of n components is,

$$M_{sys} = M_n M_{n-1} \dots M_4 M_3 M_2 M_1$$

3.7 Applications: refractive index measurements using wave optics

3.7.1 Refractive index of air using interferometry

An assumption often made in analyzing optical systems is that the index of refraction of air is the same as that of a vacuum and is taken as unity. This assumption is valid for most systems but of course depends on the level of accuracy and precision needed. When the effects of temperature are an important consideration then the temperature coefficient dn/dT, for the materials is required. However, absolute measurements of the temperature coefficients must also include the temperature coefficient air since measurements are most often made in air [21].

Table 3.2. List of selected Jones matrices for several selected polarizing components.

Component type	Matrix
Linear—TA horizontal	$\begin{bmatrix} 1 & 0 \\ 0 & 0 \end{bmatrix}$
Linear—TA vertical	$\begin{bmatrix} 0 & 0 \\ 0 & 1 \end{bmatrix}$
Linear—TA at angle θ from x-axis	$\begin{bmatrix} \cos^2\theta & \sin\theta\cos\theta \\ \sin\theta\cos\theta & \sin^2\theta \end{bmatrix}$
Rotator θ to $\theta+\beta$	$\begin{bmatrix} \cos\beta & -\sin\beta \\ \sin\beta & \cos\beta \end{bmatrix}$
Phase retarder—general	$\begin{bmatrix} e^{j\varepsilon_x} & 0 \\ 0 & e^{j\varepsilon_y} \end{bmatrix}$
Phase retarder—quarter waveplate $\Delta\varepsilon = \pi/2$ Slow axis—horizontal	$e^{j\pi/4}\begin{bmatrix} 1 & 0 \\ 0 & -j \end{bmatrix}$
Phase retarder—quarter waveplate $\Delta\varepsilon = \pi/2$ Slow axis—vertical	$e^{-j\pi/4}\begin{bmatrix} 1 & 0 \\ 0 & j \end{bmatrix}$
Phase retarder—half waveplate $\Delta\varepsilon = \pi$ Slow axis—horizontal	$e^{j\pi/2}\begin{bmatrix} 1 & 0 \\ 0 & -1 \end{bmatrix}$
Phase retarder—half waveplate $\Delta\varepsilon = \pi$ Slow axis—vertical	$e^{-j\pi/2}\begin{bmatrix} 1 & 0 \\ 0 & -1 \end{bmatrix}$

Interferometry is an accurate and straightforward method to employ for a direct measure the refractive index, especially for gasses and thin transparent plates [2, 22].

A Michelson interferometer was used to measure the refractive index of air at a room temperature of 21.7 °C. Figure 3.37 shows the arrangement where a vacuum cell of length $d = 7.97$ cm with transparent windows was placed in one of the arms of the interferometer. A HeNe laser ($\lambda = 632.8$ nm) was used as the source. The interference condition arises when the OPD traveled by the beam changes. In this case, the refractive index of the gas within the vacuum cell is changing. Thus, m fringes will be observed to pass by the center of the interference pattern for every wavelength change in OPD.

$$m\lambda = \text{OPD} = 2d(n - n_0)$$

where n is the refractive index at pressure P and n_0 is the refractive at pressure P_0. The factor of two is included since the beam makes two passes through the cell.

To compute the OPD we assume that the refractive index is proportional to the gas pressure for a range near atmospheric pressure and at room temperature [23]. With this assumption, the refractive index, n, can be determined from,

$$n = 1 + \frac{m\lambda P}{2d\Delta P}$$

where $\Delta P = P - P_0$, the change in nominal pressure applied to the cell.

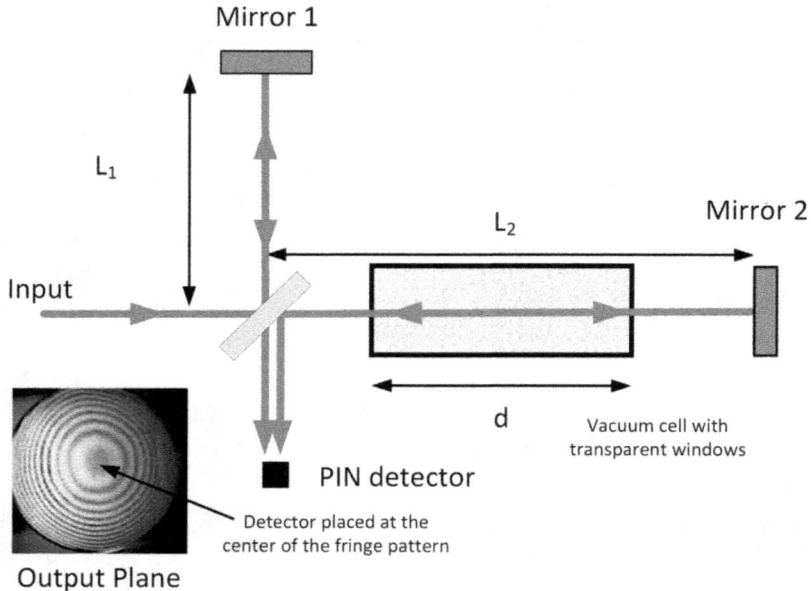

Figure 3.37. Michelson interferometer arrangement used for measuring the refractive index of a gas in the vacuum cell of length d. The photo inset shows a typical fringe pattern observed in the plane in which a small detector is placed.

To measure the refractive index of air, the pressure in the cell was lowered from nominal atmospheric pressure using a vacuum pump. When the cell is opened and the pressure within the cell returns to atmospheric pressure the fringe pattern shifts due to the changing refractive index of air in the cell. The central fringe of the pattern alternated from bright to dark as the OPD changed. Placing a small detector at the center of the fringe pattern allows a signal to be collected for the number of fringes as a function of time. A typical signal is shown in figure 3.38 recorded by a digital storage oscilloscope.

Simply counting fringes from the oscilloscope trance allows the refractive index to be computed. Atmospheric pressure was measured to be 99.7 kPa ± 0.5 kPa and a range of gauge pressures were set for ΔP from 50 to 80 kPa. Finally, the refractive index of air was measured to be 1.000 27 ± 0.000 03 based on the change in pressure within a vacuum cell.

3.7.2 Refractive index of glass from polarized reflectance measurements

Another technique for determining the refractive index of solids is to use the Fresnel reflectance of polarized light. Figure 3.39 shows the layout of the test setup. The sample under test is required to have a flat polished surface so that incident light from a polarized laser beam will be reflected. The test sample is placed on a precision rotation stage with its surface centered on the axis of rotation of the stage and the incident beam. The incident beam on the sample is aligned so that it crosses the rotation axis with a state of polarization in either the TE or TM mode with respect

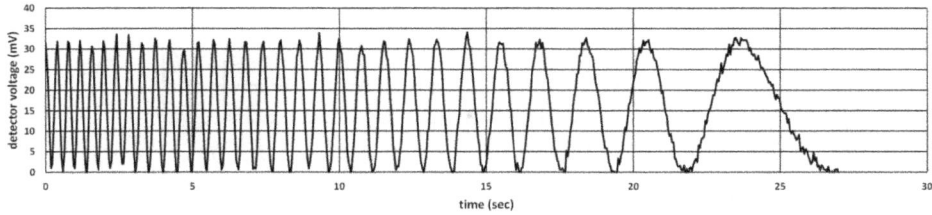

Figure 3.38. Signal recorded on a storage oscilloscope from a small detector placed at the position of the central fringe of the interference pattern as the OPD changed in time.

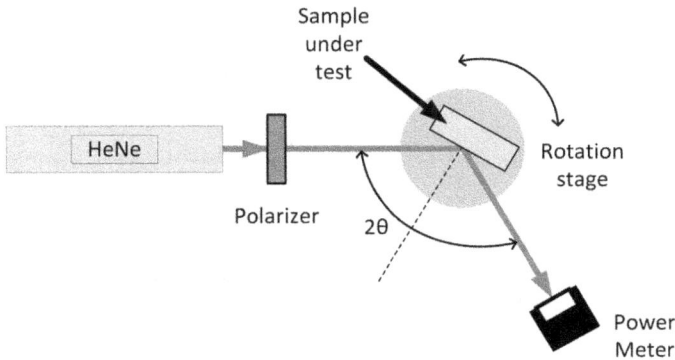

Figure 3.39. Test layout for the experimental measurement of reflectance.

to the sample surface. Rotating the stage by the incident angle θ produces a reflected beam at angle 2θ. The power of the reflected light can then be measured as a function of the angle of incidence.

In a particular test, an unpolarized HeNe laser was used as the source. Power of the reflected light from the sample was measured as a function of the angle of incidence for both the TE and TM mode set by the polarizer. Figure 3.40 shows both sets of data along with a best fit curve for each. The average value of the refractive index obtained from this fit was 1.53 ± 0.02. This value is consistent with the sample of NBK7. A separate measure of Brewster's angle was also made by measuring the angle at which the TM reflected beam was minimized and found to be $56.7° \pm 0.1°$ resulting in a refractive index of 1.52 ± 0.01. Although this method is not as accurate as others, it is a simple test and can provide useful information about unknown materials and can be used as a quality control test for identifying materials.

Exercises and problems

1. The wavefunction describes a traveling wave in a region of space, where x is in units of meters and t is in seconds.

$$\psi(x, t) = 10 \cos\left(4x - 2t + \frac{\pi}{4}\right)$$

Compute the phase velocity of the wave and the initial phase angle?

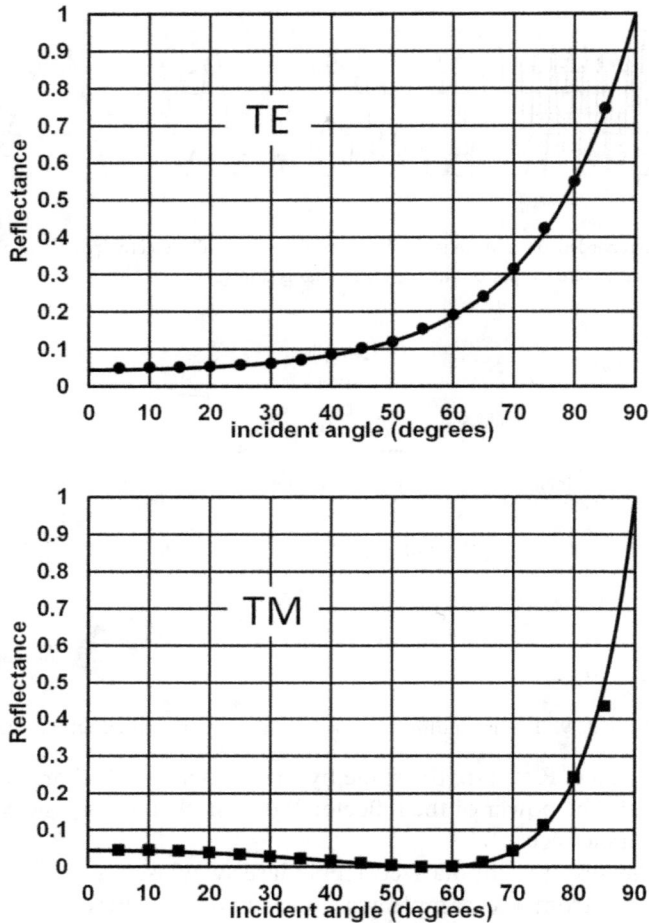

Figure 3.40. TE and TM mode reflectance from a test sample and fit to experimental data to determine the refractive index of the sample under test.

2. Determine the direction and magnitude of the k-vector for the plane electromagnetic wave with wavelength of 600 nm traveling in vacuum given by the wavefunction below. In this case x, y, z are in units of meters and t is in seconds.

$$\psi(x, y, z, t) = 10 \text{ V m}^{-1} \sin\left(\frac{k}{\sqrt{14}}x + \frac{2k}{\sqrt{14}}y + \frac{3k}{\sqrt{14}}z - (3.1415 \times 10^{15} \text{ Hz})t + \frac{\pi}{4}\right)$$

3. A plane wave is incident on a planar interface between two dielectric media (incident medium, n_0 and transmitted medium, n_1) at angle θ_i. The surface lies in the x–y-plane and the normal to the boundary between the two media is in the z-direction. The propagation vector of the incident wave is:

$$\vec{k_i} = \frac{2\pi}{\lambda_0}(1.08\hat{x} + 1.44\hat{z})$$

where λ_0 is the wavelength in vacuum. Determine the incident angle with respect to the surface normal and the refractive index of the incident medium.

4. The magnetic field component of a plane sinusoidal electromagnetic wave traveling in vacuum has a maximum amplitude value of 10^{-8} Tesla whose vector direction is along the $+x$-axis. (a) Find the maximum amplitude of the electric field. (b) What is the direction of the energy flow if the electric field component is in the $+z$-direction? (c) Find the average power per unit area carried by the electromagnetic wave.

5. A plane polarized electromagnetic wave is traveling through vacuum with its electric field described by:

$$\vec{E}(y, t) = (10 \text{ V m}^{-1})\hat{x} \sin[(1.00 \times 10^7 \text{ rad m}^{-1})y + (3.00 \times 10^{15} \text{ rad s}^{-1})t]$$

(a) What is the direction of propagation of the wave? (b) What is the speed of this wave? (c) What is the wavelength of this wave? (d) What is the maximum amplitude of the magnetic field? (e) What is the direction of the magnetic field vector? (f) Compute the magnitude of the average Poynting vector associated with this wave?

6. Compute the complex magnetic field $\vec{H}(\vec{r})$ for plane wave propagating in free space and having electric fields:
   (a) $\vec{E}(\vec{r}) = \hat{y}e^{-jkz}$
   (b) $\vec{E}(\vec{r}) = \hat{y}e^{-j(k_x x + k_z z)}$
   (c) $\vec{E}(\vec{r}) = (\hat{x}k_z - \hat{z}k_x)e^{-j(k_x x + k_z z)}$

7. (a) What is the coherence length of a source of wavelength 500 nm with a 1 nm linewidth? (b) What source linewidth would be required to produce a source with a 1 m coherence length?

8. A narrow spectral band of light centered around a mean wavelength of 520 nm is chopped by a shutter at a frequency of 40 MHz. Determine the linewidth (in nm) of the resulting light.

9. Determine the linewidth for a HeNe laser (632.8 nm wavelength) with a 10.0 m coherence length.

10. Compute the coherence time for a visible white light source with wavelengths equally distributed between 390 nm and 780 nm.

11. A liquid dye is dissolved in water. Light from a green HeNe laser (wavelength 543 nm) is sent through a 1.0 cm cuvette containing only water and detected by a large area photodiode that collects all of the beam. The detector voltage circuit gives a signal of 68.8 mV. Now the water in the cell is replaced with the dye and the detector signal is now found to be 55.1 mV. What is the absorption coefficient of the dye?

12. A commercial narrowband-pass filter specification sheet says that the filter transmits wavelengths in a 4.0 nm wide band centered about 600.0 nm.

When this filter is placed behind a white light lamp, what is the coherence time of this source?

13. Plane plates of glass (20 cm square) are in contact along one side and held apart by a wire 0.05 mm in diameter, parallel to the edge in contact. Interference fringes are observed when filtered green mercury light (546 nm wavelength) is directed normally on the plates from above. How many dark fringes appear between the edge and the wire?

14. A typical Young's double slit arrangement is shown in figure P3.14.

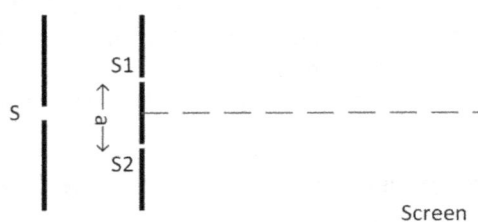

Figure P3.14. Young's double slit arrangement.

(a) In this case, source S is a small pinhole illuminated by light from a ruby laser (694.3 nm wavelength). A fringe pattern is observed on the distant screen with a visibility of 1. The fourth bright fringe is found to be 1.0° above the central axis. What is the slit separation, a?

(b) A thin absorber was placed in front of slit S1 so that the light transmitted by S1 is 50% of the amount of light transmitted by slit S2. What is the fringe visibility?

(c) Would fringes be observed if the source S was replaced by a light bulb with a long thin filament? Why or why not?

(d) Would fringes be observed if the two slit sources (S1 and S2) were each replaced by a long filament light bulb? Why or why not?

15. An unknown monochromatic source was tested using a standard Young's double slit arrangement (figure P3.14). The slit width $a = 0.10$ mm. The fringes produced were captured with a camera (0–250 gray-levels) at the distant screen location. The resulting irradiance pattern from the camera is shown as gray-level versus pixel number in figure P3.15.

(a) What is the magnitude of the complex degree of coherence for this source?

(b) A thin attenuator filter with transmittance of 0.29 was placed over slit S1 producing a new irradiance pattern. What is the visibility of this new pattern?

16. A perfectly flat piece of glass ($n = 1.5$) is placed over another perfectly flat piece of plastic ($n = 1.2$) as shown in figure P3.16. They touch on the left side but are separated on the right side. Light of wavelength 600 nm is incident normally from above (incident angle 0°).

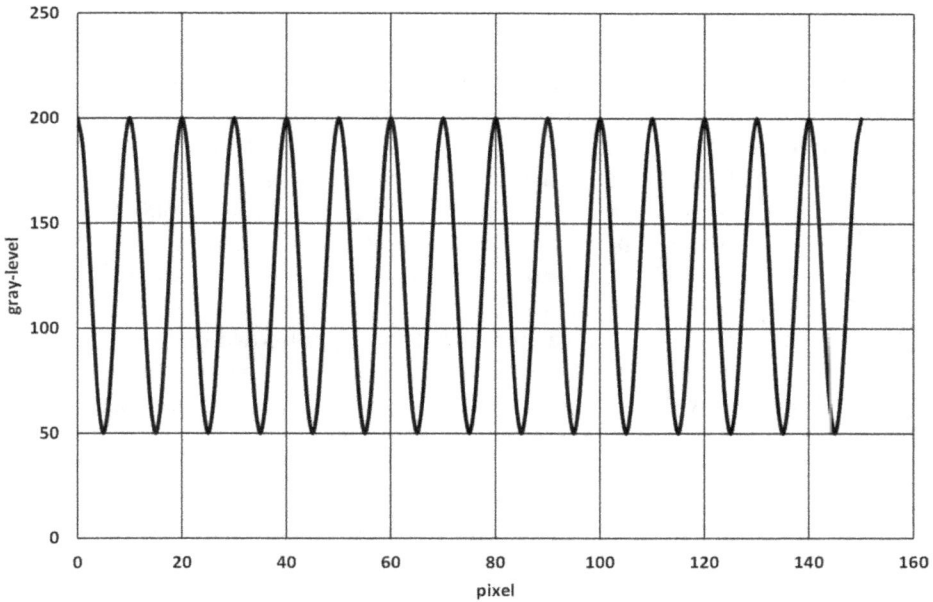

Figure P3.15. Irradiance pattern of fringes observed in a Young's double slit experiment and captured by a camera.

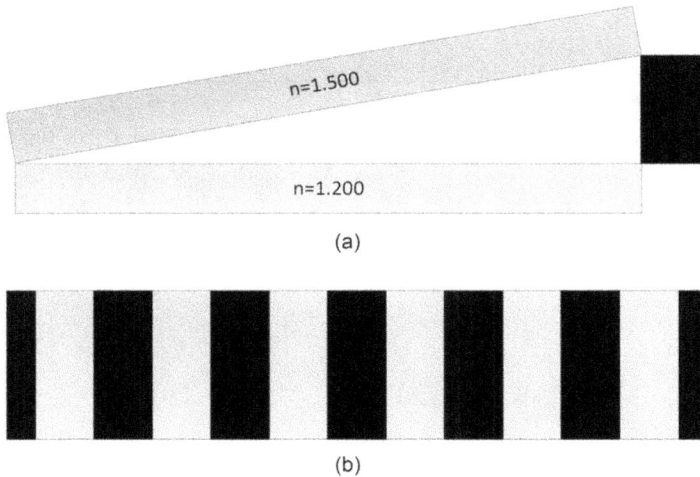

(a)

(b)

Figure P3.16. (a) Side view of two plates separated by a thin strip at one edge. (b) Top view showing fringes observed from above.

(a) A thin metal strip is placed between the glass and plastic forming an interference pattern as observed from above. A photo of the pattern is shown below. The black bands are the dark fringes. How thick is the metal strip?

(b) The space between the glass and the plastic is now filled with water ($n = 1.333$). Will the fringe observed on the left side where the two pieces touch be bright or dark? Explain.

17. Fringes are observed with monochromatic light incident on a Michelson interferometer. When the movable mirror is translated 0.073 mm, a shift of 300 fringes is observed. (a) What is the wavelength of the light from this monochromatic source? (b) When a thin glass slab of thickness 0.005 mm is placed into one of the arms of the interferometer, a shift of 10.5 fringes was observed. Find the refractive index of the glass slab.

18. A single thin film ($n_f = 1.38$) with physical thickness of 600 nm covers a polymer optical window of refractive index 1.47. Determine the vacuum wavelengths of light that are not reflected when the window is illuminated from above with sunlight (500 nm wavelength).

19. The reflectance spectra was measured for an unknown glass material in air as a function of wavelength as shown in the table below.
    (a) What will be the phase shift on reflection for incident light reflecting from the glass surface?
    (b) Complete the table and find the refractive index of the material as a function of wavelength.
    (c) Sketch a graph of refractive index versus wavelength and state whether or not this material has normal or anomalous dispersion.

Wavelength (nm)	Reflectance	Refractive index
400	0.0497	
500	0.0452	
600	0.0428	
700	0.0414	

20. A laser beam from a laser pointer is directed toward the flat surface of a D-shaped glass element of refractive index 1.50 from below through the circular section. See figure P3.20. Compute the TE and TM amplitude reflection coefficients, r, and reflectance R at the glass-air surface for incident angles of $0°$, $30°$, $60°$.

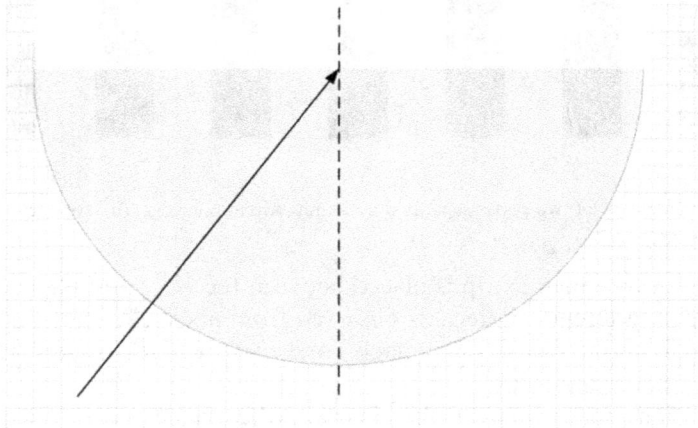

Figure P3.20. Laser beam incident onto the flat surface of a D-shaped glass element.

21. Compute the real and imaginary parts of the refractive index of a conductor in terms of the permittivity ε, conductivity σ, permeability μ, and angular frequency of light ω.

22. Compute the real and imaginary parts of the refractive index of a metal assuming that $\omega \gg \gamma$ (this is known as the high frequency limit).

23. Light is normally incident from a medium of refractive index N_0 onto the surface of a medium with a complex refractive index specified as $N_R + jN_I$. Find an equation for the reflectance from this surface.

24. The website http://refractiveindex.info/ gives information about the refractive index of many materials. Go to the website and find the refractive index of aluminum as a function of wavelength. From the available data plot the reflectance at normal incidence as a function of wavelength for a bare aluminum surface in the region 200–1000 nm. Comment on the shape of the plot.

25. Given that H and L represent the characteristic matrix of high and low quarter wave layers, respectively, is the reflectance from a [air $|HL|$ glass] system the same as an [air $|LH|$ glass] system? Prove your answer by computing the transfer matrix for each case.

26. An unpolarized helium–neon laser (wavelength of 633 nm) was used to measure various effects of polarization. The output power of the beam was measured to be 5.0 mW using a laser power meter.

    (a) The critical angle for total internal reflection within a particular liquid was determined to be 45°. What is the refractive index of the liquid?

    (b) If the unpolarized laser was directed toward the surface of the liquid from above in air, at what incident angle would the reflected light be completely polarized?

    (c) In another experiment, the light from the laser was incident normal to a sheet of polarizing film. What would be the measured power of the beam after transmission through the Polaroid sheet?

    (d) A second polaroid was placed behind the first with its transmission axis at 30° with respect to the first sheet. Now, what would be the measured power of the beam after transmission through the second sheet?

27. Light linearly polarized with a horizontal transmission axis is sent through another linear polarizer with TA at 45° and then through a QWP with SA horizontal. Use the Jones matrix technique to determine and describe the polarization state of the transmitted light.

28. Calculate the reflection and transmission coefficients for the TE mode of light incident from air onto a glass plate at 30° to the surface normal. The refractive index of the glass is 1.60.

29. At a wavelength of 1.00 μm the refractive index of gold is known to be 0.270 $+ j7.07$. (a) Determine the absorption coefficient at 1.0 μm. (b) Compute the real and imaginary parts of the susceptibility at 1.0 μm. (c) What is the reflectance at normal incidence at 1.0 μm?

30. At a wavelength of $\lambda = 1.0$ μm, amorphous carbon has a complex refractive index of: $N = 1.4 + j0.6$. (a) Calculate the absorption coefficient. (b) Find the value for the dielectric constant. (c) What is the susceptibility?

31. Indium Tin Oxide (ITO) may be considered as a Drude metal with low conduction electron Density of $N = 3 \times 10^{26}$ m^{-3}. Compute the plasma frequency for ITO.

32. Semiconductor lasers can be made using the two cleaved surfaces along crystal planes to act as mirrored reflectors forming a plane parallel cavity. Assume normal incidence for a crystal in air with a refractive index of 3.5 with a 1.50 mm length. Take the losses in the cavity to be associated with the surface reflections, a loss due to diffraction of 0.40 cm^{-1}, and losses due to absorption of 2.0 cm^{-1}. (a) Find the total distributed loss coefficient for this laser. (b) Compute the coefficient of finesse.

33. Design a single AR coating for a germanium lens for use in the infrared at a wavelength of 5.00 μm. The refractive index of germanium at this wavelength is 4.016 19.

   (a) What is the refractive index of the ideal material that you would need for the thin film coating?
   (b) The coating material of choice for this application is diamond (refractive index of 2.380 67) due to its physical properties. What thickness would you pick for this film?
   (c) Compute the reflectance from the coated surface at normal incidence at the 5.00 μm wavelength.

34. Find the refractive index of the substrate required for the reflectance from a structure: air $[LH]\, n_S$ to be zero. Assume that the low refractive index film is cerium trifluoride ($n_L = 1.650$) and the high refractive index film is zirconium dioxide ($n_H = 2.100$).

35. Evaporated aluminum on a front surface mirror has a complex refractive index of $1.32 + j7.61$ at 633 nm wavelength. (a) What is the absorption coefficient for aluminum at this wavelength? (b) What is the reflectance from the mirror at this wavelength for a normal incidence reflection? (c) Is there a way to increase this reflectance? How?

36. Plot both TE and TM reflectance versus wavelength over the visible spectrum (350–700 nm) for light at (a) normal incidence, and (b) incident at 20° to the normal for the following system

$$\text{Air } [HLHL]\text{glass}$$

where $N_L = 1.38$, $N_H = 1.70$ and the index of the glass substrate is 1.51. Take the thickness of the H layers to be 81 nm and the L layers to be 100 nm.

References

[1] Born M and Wolf E 1975 *Principles of Optics* (New York: Pergamon)
[2] Hecht E 2002 *Optics* 4th edn (Reading, MA: Addison-Wesley)

[3] Pedrotti F L, Pedrotti L S and Pedrotti L M 2007 *Introduction to Optics* 3rd edn (Englewood Cliffs, NJ: Prentice Hall)

[4] Wangsness R K 1986 *Electromagnetic Fields* 2nd edn (New York: Wiley)

[5] Klein M V and Furtak T E 1986 *Optics* 2nd edn (New York: Wiley)

[6] Young H D, Freedman R A, Ford A L and Sears F W 2004 *Sears and Zemansky's University Physics: With Modern Physics* (San Francisco, CA: Pearson Addison-Wesley)

[7] Carter W H 1995 Coherence theory *Handbook of Optics, Volume I: Fundamentals, Techniques, and Design* 2nd edn, ed M Bass, E W Van Stryland, D R Williams and W L Wolfe (New York: McGraw-Hill) ch 4

[8] Goodman J W 2017 *Introduction to Fourier Optics* 4th edn (New York: W H Freeman)

[9] Fox M 2010 *Optical Properties of Solids* 2nd edn (Oxford: Oxford University Press)

[10] Johnson P B and Christy R W 1972 Optical constants of the noble metals *Phys. Rev.* B **6** 4370–79

[11] Rubin M 1985 Optical properties of soda lime silica glasses *Sol. Energy Mater.* **12** 275–88

[12] Polyanskiy M N *Refractive Index Database* https://refractiveindex.info (Accessed 13 November 2020)

[13] Saleh B E A and Teich M C 2007 *Fundamentals of Photonics* 2nd edn (New York: Wiley)

[14] Willey R R 2006 *Field Guide to Optical Thin Films* (Bellingham, WA: SPIE)

[15] Heavens O S 1991 *Optical Properties of Thin Solid Films* 2nd edn (New York: Dover Publications)

[16] Macleod H A 2010 *Thin-film Optical Filters* (Boca Raton, FL: CRC Press/Taylor & Francis)

[17] Thelen A 1989 *Design of Optical Interference Coatings* (New York: McGraw-Hill)

[18] McPeak K M, Jayanti S V, Kress S J P, Meyer S, Iotti S, Rossinelli A and Norris D J 2015 Plasmonic films can easily be better: rules and recipes *ACS Photon.* **2** 326–33

[19] Smith D G 2012 *Field Guide to Physical Optics* (Bellingham, WA: SPIE)

[20] Bennett J M 1995 Polarization *Handbook of Optics, Volume I: Fundamentals, Techniques, and Design* 2nd edn, ed M Bass, E W Van Stryland, D R Williams and W L Wolfe (New York: McGraw-Hill) ch 5

[21] SCHOTT Technical Information TIE-29 *Refractive Index and Dispersion* (https://us.schott.com/)

[22] Jenkins F A and White H E 1976 *Fundamentals of Optics* (New York: McGraw-Hill)

[23] Owens J C 1967 Optical refractive index of air: dependence on pressure, temperature and composition *Appl. Opt.* **6** 51–9

IOP Publishing

Optical Systems Design Detection Essentials

Radiometry, photometry, colorimetry, noise, and measurements

Robert M Bunch

Chapter 4

Photon optics and sources of photons

4.1 Introduction

Ray optics is used to describe light propagation through relatively large macroscopic systems of lenses, mirrors, and other components. As the interaction of light with objects becomes on the order of the wavelength, wave optics adequately predicts effects such as interference, diffraction, and quality of the images formed in coherent and incoherent illumination systems. Wave optics also describes the relation to the wave amplitude and the flow of energy in a system.

As light interacts with matter on an even smaller microscopic scale, the size of atoms and molecules, there are phenomena that wave optics alone cannot predict. Instead of a continuous spectrum, spectroscopic measurements of the emission from gas discharge tubes showed a pattern of discrete wavelengths or spectral lines with different patterns for different elements [1, 2]. Blackbody or thermal radiation spectra deviated from the classical theory of cavity radiation as the frequency of the light increased, often referred to as the ultraviolet catastrophe [1]. Experiments of the photoelectric effect, emission of electrons when a metallic surface is illuminated with a beam of light, indicated that the kinetic energy of the emitted electrons did not depend on the amount of incident light but on the wavelength of the incident light [1]. What does depend on the amount of light is the number of electrons at a specific energy. These experiments along with other evidence eventually led to our understanding of the particle-like nature of light or the photon model, quantized energy states, and the birth of quantum mechanics.

So, is light a ray, wave, or photon? The answer is yes. Light can be considered any of these models depending on the size and character of the interaction, often called the principle of complementarity. Attempts to understand the nature of light leads to many philosophical discussions about the photon which is beyond the scope of this text. We can detect individual photons by measuring their energy transfer and measure the properties of photons of different energies or frequencies. From a systems perspective when we detect the energy of light and its interaction with materials, such as detectors, it is convenient to use the photon concept. This will be the approach used here.

4.2 Basic properties of photons

Spectral lines are observed due to the discrete or quantized energy levels in atoms and molecules. When electrons are in two different levels in an atom, we say that the atom has two states. In moving from one state to the other the atom undergoes a transition and the energy associated with this transition is through the photon that is either emitted or absorbed. This concept is illustrated in figure 4.1 for the simple case of two energy states. There are actually three different possibilities for transitions to occur, spontaneous absorption, spontaneous emission, and stimulated emission [1, 3]. The difference in energy between the energy levels of the two states is equal to the energy of a photon.

As we shall see later each photon associated with the light has an energy given by,

$$\mathcal{E} = h\nu = \frac{hc}{\lambda}$$

where λ is the wavelength, ν is the frequency, and h is a constant called Planck's constant with a value of 6.6261×10^{-34} J s. Photons have no mass as other material particles and travel at the speed of light, c. As the stimulated emission process also indicates, photons are indistinguishable from each other and are not observed directly but only by being either created or annihilated through an energy transition process. Also, photons in different states would be represented by two different monochromatic waves. So, a single monochromatic plane electromagnetic wave could just be considered as a large number of propagating photons occupying the

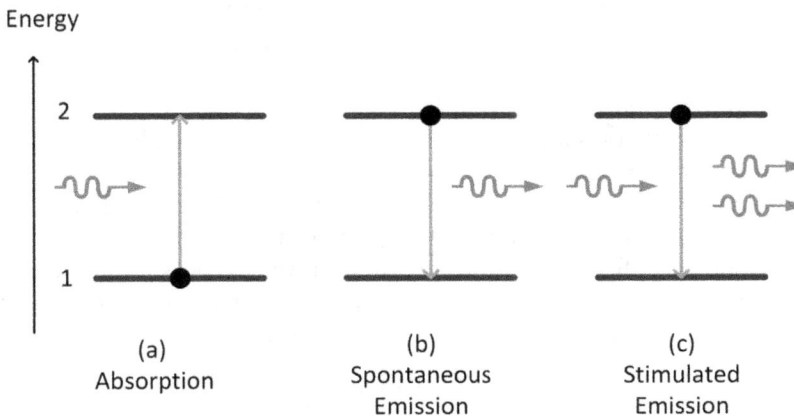

Figure 4.1. (a) Absorption of an incident photon resulting in the electron occupying state 1 to make a transition to state 2. (b) Emission of a photon when an electron already in state 2 makes a transition to state 1. (c) Stimulated emission when an incident photon of energy equal to the energy difference between state 1 and state 2 causes a transition to occur releasing another photon of equal energy.

same state where the discrete nature of an individual photon can no longer be observed.

4.3 Photon statistics

To better understand the interaction of photons with matter we must investigate the discrete nature of the photon as well as the type of quantum system in which the photon interacts. Because of the large number of particles as well as the different characteristics of the materials involved in this analysis, we must resort to statistical methods in order to understand the behavior of various systems. We need to determine the average behavior of a system of particles and how these particles are distributed in energy within the various available states.

4.3.1 Photon flux and photon numbers

From the previous chapter we found that the irradiance, E_e, on a surface from an incident wavefront was given by the time average Poynting vector and is proportional to the square of the amplitude of the electric field of the electromagnetic wave. To connect this concept with the photon picture we take the ratio of the irradiance to the energy of a single photon to obtain,

$$\frac{E_e}{h\nu} = \frac{E_e}{hc}\lambda$$

which is a value specifying the number of photons per unit surface area per unit time. So, to measure the optical power within a photon stream; for example, falling on an optical detector, the detector by its nature performs a time average of the large number of photons imparting energy and arriving within some time interval.

The question remains whether or not we can detect individual photons and count the number of photons in a photon stream. As one might expect, the distribution of photons depends on the type of source, coherent or incoherent [3–5]. Straightforward experiments have been developed to illustrate the properties of photon statistics by counting photons [6].

One important class of photon statistics is when the incident photons form a set of independent random events with a photon flux proportional to the optical power. We also assume that the arrival of a photon is independent of a previous photon arrival. To obtain the distribution function we consider a small interval of time Δt at some time t, shown in figure 4.2. In this case we have two possibilities, (1) a photon count in time t and none in Δt, or (2) zero counts in time t and one count in Δt.

We define the probability of zero counts in t and one in Δt as,

$$P(0, t)(\alpha \, \Delta t)$$

where the quantity α is a constant of proportionality. Likewise, we define the probability of one count in t and zero counts in Δt as,

$$P(1, t)(1 - \alpha \, \Delta t)$$

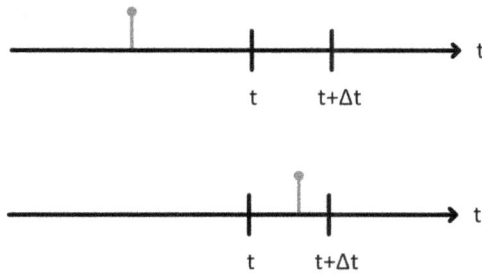

Figure 4.2. Schematic illustration of possibilities of photon arrival within a time t and Δt.

Now, the probability of detecting a one count in t and Δt is,

$$P(1, t + \Delta t) = P(1, t)(1 - \alpha \, \Delta t) + P(0, t)(\alpha \, \Delta t)$$

In the limit at Δt approaches zero, we obtain a differential equation relating the probabilities.

$$\frac{dP(1, t)}{dt} = -\alpha P(1, t) + \alpha P(0, t)$$

Using the same argument, the probability of counting zero photons in time t leads to the differential equation,

$$\frac{dP(0, t)}{dt} = -\alpha P(0, t)$$

whose solution is,

$$P(0, t) = \exp(-\alpha t)$$

Substituting this result into the relation above and solving the differential equation results in,

$$P(1, t) = \alpha t \exp(-\alpha t)$$

where $P(1, 0) = 0$.

Extending to n photons results in a differential equation for the probability,

$$\frac{dP(n, t)}{dt} = -\alpha P(n, t) + \alpha P(n - 1, t)$$

whose solution is the well-known Poisson distribution function [3, 4],

$$P(n, t) = \frac{(\alpha t)^2 e^{-at}}{n!}$$

To test this distribution function, we compute the sum of all probabilities which should be unity. By substitution of variables, let $u = \alpha t$. Then the summation is,

$$\sum_{n=0}^{\infty} P(n, t) = \sum_{n=0}^{\infty} \frac{u^2 e^{-u}}{n!} = e^{-u} \sum_{n=0}^{\infty} \frac{u^2}{n!} = e^{-u} e^{+u} = 1$$

Next, compute the mean value of n photon arrivals. We use the notation of an average quantity as the quantity or expression is a bar over that quantity. The definition of the mean or expectation value of n ($\bar{n}$) is,

$$\bar{n} \equiv \sum_{n=0}^{\infty} nP(n) = \sum_{n=0}^{\infty} n \frac{u^2 e^{-u}}{n!} = u$$

This result provides an alternative relationship for the Poisson distribution function of,

$$P(n) = \frac{(\bar{n})^2 e^{-\bar{n}}}{n!}$$

Another statistical parameter to compute for a probability distribution is the variance. The variance, σ^2 is defined as,

$$\sigma^2 \equiv \overline{(n - \bar{n})^2} = \overline{n^2} - \bar{n}^2$$

Using the Poisson distribution, it can be shown that the variance is $\sigma^2 = \bar{n}$. This statistical distribution and its properties will be used later in discussing detection of light and noise issues associated with detection [7, 8].

4.3.2 Signal-to-noise ratio (SNR)

In transmitting a signal by detecting photon arrivals the randomness of the events introduce noise. A general optical signal as a function of time incident on a detector will cause measured electrical fluctuations as shown in figure 4.3. The electrical measurement may be a voltage or current. Characteristic measures of this signal are the mean value and the variance associated with the statistics of the fluctuations.

The signal-to-noise ratio for such an electrical signal is defined as,

$$\text{SNR}_{\text{electrical}} \equiv \frac{(\text{mean value})^2}{\text{variance}}$$

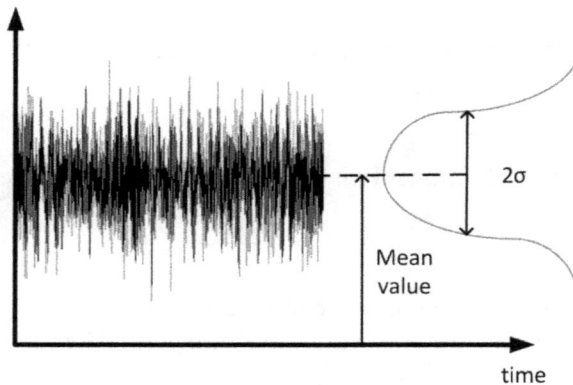

Figure 4.3. Noisy electrical signal from a detector as a function of time. The signal is characterized by an average value and a variance.

So, if the signal follows Poisson statistics this value reduces to,

$$\mathrm{SNR_{electrical}} = \bar{n}$$

For a detection system conforming to Poisson statistics, the electrical SNR linearly increases with the average value of the mean number of photons detected. We shall see this appear later in discussion of specific detection systems.

In general, the SNR is defined as the ratio of the signal power to the noise power. The emphasis here on the signal being an electrical signal is to make a specific distinction between an electrical signal and an optical signal. In an electronic system, the electrical power is proportional to the square of the current or voltage. However, in optical engineering the optical signal power is directly proportional to the detected current or voltage and not the square. Thus,

$$\mathrm{SNR_{electrical}} = \mathrm{SNR_{optical}^2}$$

Care must be taken when specifying an SNR to be clear whether or not it is the optical or electrical value [9]. This will be discussed later as a metric of detection systems.

4.3.3 Statistical distribution functions

There are three specific statistical distribution functions that are of importance for photon interactions with matter at the microscopic level. All these distribution functions and their properties will be discussed in turn; however, we will not derive them here. When the particles are distinguishable then the classical case of Maxwell–Boltzmann statistics can be applied. Such a case is that of a gas at some temperature. Photons and electrons are both indistinguishable particles, but they obey different statistical distributions. The distinction between the statistics for the population of states between them depends on another principle, the Pauli exclusion principle [1, 2]. This principle tells us that there cannot be more than one electron in the same quantum state in a multi-electron atom. Thus, electrons are distributed within states according to Fermi–Dirac statistics, so they are referred to as fermions. On the other hand, many photons can be in the same state as we have already discussed and are an example of particles called bosons which follow Bose–Einstein statistics. The common characteristic between all of these statistical distributions is that the probable number of particles in a state at energy $\mathcal{E}$ depends directly on the temperature of the system.

The Maxwell–Boltzmann distribution arises from an analysis of the probability of finding a particle in given state between energy $\mathcal{E}$ and $\mathcal{E} + \Delta\mathcal{E}$. The number of energy states is also assumed to be independent of $\mathcal{E}$ so the system contains uniformly distributed states. This leads to the result that the probable number of distinguishable particles as a function of the energy is,

$$n_{\mathrm{MB}}(\mathcal{E}) = Ae^{-\mathcal{E}/kT}$$

where T is the temperature, k is Boltzmann's constant 1.381×10^{-23} J K^{-1}, and, A, is a normalization constant dependent on the total number of particles in the system. It is easy to show that the average energy of the system is,

$$\overline{\mathcal{E}} \equiv \frac{\int_0^\infty \mathcal{E} n_{\mathrm{MB}}(\mathcal{E}) d\mathcal{E}}{\int_0^\infty n_{\mathrm{MB}}(\mathcal{E}) d\mathcal{E}} = kT$$

which is the law of equipartition of energy [1]. Note that for some systems the number of energy states is not independent of $\mathcal{E}$. For these situations, the number of particles must be modified by a density of states factor related to the degeneracy of states at a particular energy. Degeneracy just means that a system may have different ways to have the same energy [2].

The probable number of indistinguishable particles in an energy state $\mathcal{E}$ in thermal equilibrium at temperature T and for particles that do not obey the Pauli exclusion principle is given by the Bose–Einstein distribution.

$$n_{\mathrm{BE}}(\mathcal{E}) = \frac{1}{A e^{\mathcal{E}/kT} - 1}$$

where again, A, is a normalization factor that is not dependent (or at best weakly dependent) on temperature. Note that for large A and for $\mathcal{E} > kT$ the first term in the denominator dominates resulting in an approximation of the Maxwell–Boltzmann distribution.

The Fermi–Dirac distribution function is,

$$n_{\mathrm{FD}}(\mathcal{E}) = \frac{1}{A e^{\mathcal{E}/kT} + 1}$$

For this case, because of the exclusion principle, the constant A does depend on temperature. The Fermi–Dirac distribution function is often written as,

$$n_{\mathrm{FD}}(\mathcal{E}) = \frac{1}{e^{(\mathcal{E} - \mathcal{E}_{\mathrm{F}})/kT} + 1}$$

where $\mathcal{E}_{\mathrm{F}}$ is called the Fermi energy. For energy states less than the Fermi energy, the exponential factor is negative and the value of $n_{\mathrm{FD}} \approx 1$. Thus, all states below the Fermi energy can be occupied by only one particle.

Each of the three distribution functions are plotted as a function energy in figure 4.4. The energy axis is relative scale in comparison to kT but it is the same on all plots.

As one can see, all the distribution functions approach zero for large values of energy or when $\mathcal{E} \gg kT$. This means as energy increases the probability of a state being occupied is small. For small energy, the Maxwell–Boltzmann distribution has a finite value (normalized to one), however, the Bose–Einstein distribution can become rather large indicating that many particles can be in the same state at lower energy. On the other hand, the Fermi–Dirac distribution is always one or less for lower energies less than the Fermi energy. In fact, for low temperature, the value is

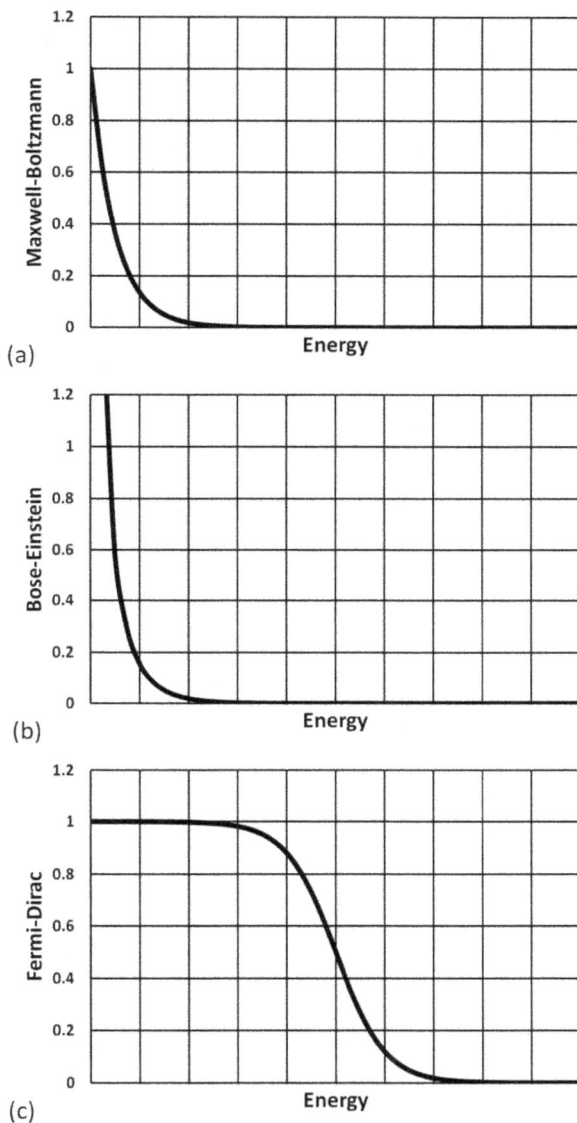

Figure 4.4. (a) Maxwell–Boltzmann, (b) Bose–Einstein, and (c) Fermi–Dirac distribution functions plotted as a function energy.

equal to one for all energies less than the Fermi energy. We will see the implications of the Fermi energy later in relation to detection of light by semiconductor detectors.

4.4 Thermal sources and blackbody radiation

An object hotter than its environment will emit radiation and cool. If the object is cooler than its environment it will absorb energy and heat up. Thermal equilibrium occurs when the rate of emission and the rate of absorption are equal. Thermal sources are objects that emit according to their temperature.

The experimental observations of thermal sources and our understanding of the theoretical foundation of these sources has a long history [1, 2, 4]. From the human observer perspective, hot objects are called self-luminous since they emit radiation in the visible portion of the spectrum. In particular, the spectral radiation of an object as a function of temperature was found to follow some predictable trends. For consistency, a hollow object was heated, and a small opening allowed radiation to escape forming what was called cavity radiation. Another term used is blackbody radiation since a perfect emitter of radiation would also be a perfect absorber.

One observation is called Wein's displacement law, that is, the product of the peak emission wavelength and the temperature of a cavity radiator is a constant.

$$\lambda_{\text{peak}} T = \text{constant} = 2.898 \times 10^{-3} \text{ m K}$$

When the complete spectra were collected for various temperatures and the total spectral radiant flux per unit area of the opening was measured, the result showed that this value was proportional to the fourth power of the temperature. This relation is summarized as Stefan's law,

$$\frac{\Phi_e}{A} = \sigma T^4$$

where the constant $\sigma = 5.67 \times 10^{-8}$ W m^{-2} K^{-4}. Examples of cavity radiation spectra are shown in figure 4.5.

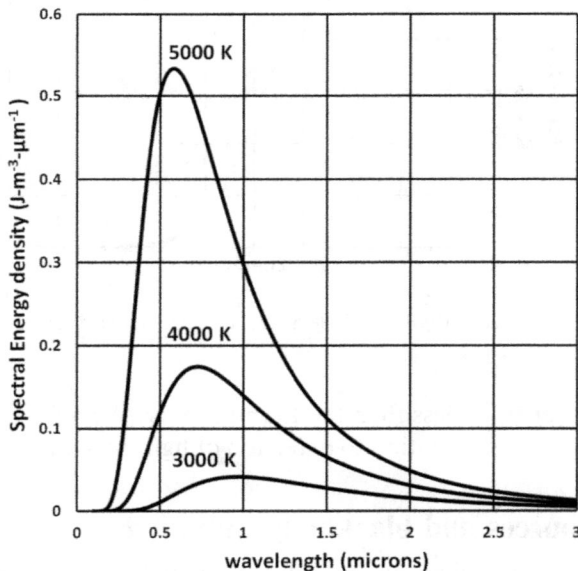

Figure 4.5. Blackbody spectral curves for three different temperatures.

The classical theory of cavity radiation was one of the first attempts to understand the observed experimental evidence of blackbody radiation. This is done by examining electromagnetic waves confined within a cavity and counting the number of modes of radiation within a given frequency interval [1, 2]. In addition, using the law of the equipartition of energy, where the average energy of each mode would be kT, the energy density within a frequency interval is,

$$w_e(\nu)d\nu = \frac{8\pi\nu^2 kT}{c^3}d\nu$$

This relation is known as the Rayleigh–Jeans formula for blackbody radiation [1].

Figure 4.6 shows a plot of the Rayleigh–Jeans formula along with that of an experimental blackbody spectrum for a fixed temperature. It is obvious that these two curves do not match, especially in the ultraviolet region of the spectra (high frequency range). There must be something incorrect about the theoretical assumptions necessary to describe the experiments.

The resolution to the ultraviolet catastrophe was provided by Max Planck. Planck hypothesized that energy in an oscillating atom of the cavity was absorbed and emitted as discrete amounts of energy, quanta [1, 2]. He also assumed that this energy was proportional to the frequency of oscillation and that no individual wave could contain more energy than kT. With these assumptions, the average energy of the system could be computed when the energy was,

$$E = nh\nu$$

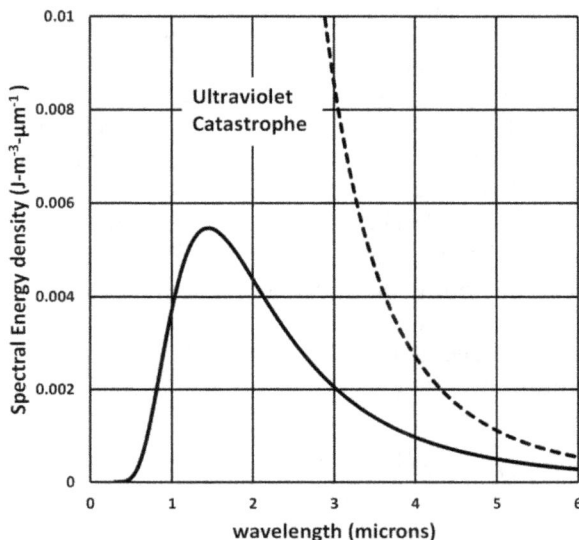

Figure 4.6. Classical theory of cavity radiation (Rayleigh–Jeans formula) for a temperature $T = 2000$ K (dashed line) compared to an experimental blackbody spectrum (solid line) of the same temperature as a function of wavelength illustrating the ultraviolet catastrophe.

where n is an integer and h is Planck's constant 6.6261×10^{-34} J s. Calculating the average energy using a discrete summation is accomplished with,

$$\overline{\mathcal{E}} \equiv \frac{\sum\limits_{n=0}^{\infty} \mathcal{E} n_{MB}(\mathcal{E})d\mathcal{E}}{\sum\limits_{n=0}^{\infty} n_{MB}(\mathcal{E})d\mathcal{E}} = \frac{\sum\limits_{n=0}^{\infty}\left(\frac{nh\nu}{kT}\right)e^{-\frac{nh\nu}{kT}}}{\sum\limits_{n=0}^{\infty}\left(\frac{1}{kT}\right)e^{-\frac{nh\nu}{kT}}}$$

Resulting in,

$$\overline{\mathcal{E}} = \frac{h\nu}{e^{+\frac{h\nu}{kT}} - 1}$$

Using this average energy instead of the classical value of kT leads to a relation for the energy density of cavity radiation,

$$w_e(\nu)d\nu = \frac{8\pi\nu^2}{c^3}\frac{h\nu}{e^{+\frac{h\nu}{kT}} - 1}d\nu$$

which is the frequency form of Planck's blackbody distribution function. Converting this to a function of wavelength results in an alternate form,

$$w_e(\lambda)d\lambda = \frac{8\pi hc}{\lambda^5}\frac{1}{e^{+\frac{hc}{\lambda kT}} - 1}d\lambda$$

Planck's prediction of the blackbody radiation spectra and the resolution of the ultraviolet catastrophe provided further evidence that photons were discrete amounts of energy. From a practical perspective, blackbody radiation provides a known spectrum that only depends on the measurement of one parameter, the temperature. Thus, this provides a source that can be used to calibrate spectral instruments or use spectra to determine temperature. Planck's quantum hypothesis was eventually carried further to show that this concept was fundamental in understanding the microscopic properties of atoms leading to quantum mechanics.

The temperature defining the blackbody radiation spectral distribution is called the radiation temperature [10]. The radiation temperature should not be confused with the value know as color temperature. The color temperature is determined from the temperature of a blackbody that most closely matches the color coordinate of a source. We will discuss this again in another chapter.

In calculating the total energy density of a blackbody spectrum at all wavelengths for a given temperature, we find,

$$w_e = \int\limits_0^{\infty} w_e(\lambda)d\lambda = \int\limits_0^{\infty} \frac{8\pi hc}{\lambda^5}\frac{1}{e^{+\frac{hc}{\lambda kT}} - 1}d\lambda = \frac{4}{c}\sigma T^4$$

which predicts the same temperature to the fourth power dependence as found in Stefan's law.

Surfaces that are not perfect blackbodies are called graybodies if a constant scaling factor can be used to match the spectrum of the radiant flux per unit wavelength of a blackbody. White-light sources such as traditional incandescent bulbs may in some cases be considered graybodies. Reflected light from surfaces of materials can also exhibit a graybody effect.

Another way to describe the energy transmitted is in terms of photon flux. This type of description is called actinometry [11]. As described earlier, we recognize that the radiant energy in a photon stream is,

$$Q_e = N_\lambda \frac{hc}{\lambda}$$

where N_λ is the total number of photons at wavelength λ. Taking the time derivative of this relation gives us a flux, or,

$$\frac{dQ_e}{dt} = \frac{dN_\lambda}{dt} \frac{hc}{\lambda} \equiv \Phi_e$$

We define the photon flux or quantum flux with a symbol Φ_q, then,

$$\Phi_q = \frac{dN_\lambda}{dt} = \frac{\lambda}{hc} \Phi_e$$

For a monochromatic radiation there is a simple conversion between radiometric and actinometric values. If the radiant flux is a function of wavelength then as above, the total photon flux is,

$$\Phi_{q,\,tot} = \frac{1}{hc} \int_0^\infty \lambda \Phi_\lambda d\lambda$$

Photon quantities are commonly used in photobiology and biomedical optics disciplines.

The mean value of the energy calculated above for a thermal source shows that they are described statistically by the Bose–Einstein probability distribution function whose mean value is,

$$\bar{n} = \frac{1}{e^{+\frac{h\nu}{kT}} - 1}$$

As we did earlier for the Poisson distribution, let us calculate the variance according to this distribution. The result is,

$$\sigma^2 = \bar{n} + \bar{n}^2$$

Suppose we want thermal light to carry a signal. The signal-to-noise ratio for this source would be,

$$\text{SNR}_{electrical} = \frac{\bar{n}}{n + 1}$$

which is always less than unity indicating this type of source is not a candidate for a communication system. However, thermal sources are often a part of a system as background and can contribute to the noise of the system.

4.5 Photon interactions with atoms, molecules, and solids

Light as an electromagnetic wave interacts with materials because the electric field oscillates in the light wave interact with electrons. Microscopically, photon interactions with materials are observed due to an energy transfer mechanism as discussed previously. Detection and emission measurements of photons require an understanding of photon interactions with materials.

Besides thermal excitation, photon emission can arise when atoms are excited by electronic collisions or other incident photons. There are numerous physical processes that result in the emission of photons generally referred to as luminescence [3]. Examples of types of luminescence are when photons of one wavelength can be generated by the absorption of higher energy photons into a different energy level (photoluminescence), excitation from collisions with an electron beam (cathode luminescence), and excitation from an externally applied electric field (electro-luminescence) [12]. Typical fluorescent lamps have a gaseous discharge within a low-pressure mercury gas producing an ultraviolet wavelength spectral emission that excites a phosphor coating that results in a near continuum of visible radiation. White-light LEDs (light emitting diodes) also use a phosphor material excited by a blue LED emitter to produce the white-light continuum. All these processes are used in the design of a variety of display devices [12].

Line spectra provided one of the first indications that atoms had quantized energy states. These spectra were first obtained by examining the emitted light from arcs in a gas tube or flames vaporizing a solid material [2, 13]. The applicability of emission spectra is that it provides a unique determination of the chemical elements comprising the material. The wavelength associated with each line is a value related to an electronic transition from one state to another for an atom releasing a photon whose energy is the difference between the energy of the two states. This is shown in the spontaneous emission property of figure 4.1(b).

4.5.1 Photoelectric effect

The photoelectric effect occurs when light is incident on a metallic surface (photo-cathode) releasing electrons. Figure 4.7 shows a schematic illustration of a system for observing the photoelectric effect. Incident light falls on a photocathode within an evacuated tube, releasing electrons that are collected by the anode with an output current measured by an ammeter the in the external circuit. If the anode is at a higher potential than the photocathode the negatively charged electrons will be attracted by the positive potential of the anode providing a current flow. The rate of emitted electrons depends directly on the amount of incident light. This effect is actually one of the first systems for the electronic detection of light.

As the applied potential of the anode becomes more negative with respect to the photocathode the detected current will decrease since the energy of the electrons will not be sufficient to make it to the anode. At some negative potential, V_{stop}, the current goes to zero. This is called the stopping potential since all electron current has stopped. Thus, the highest kinetic energy electrons from the photoelectric effect interaction have stopped so that the value of the kinetic energy is $K_{\text{max}} = eV_{\text{stop}}$.

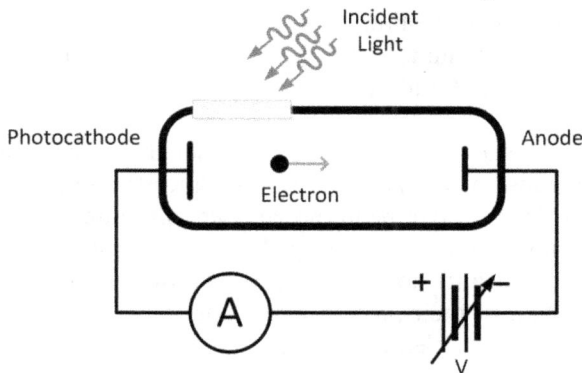

Figure 4.7. Schematic of a photoelectric effect experimental apparatus.

One of the curious results of these experiments is that the stopping potential does not depend on the amount of incident light but only on the wavelength of the incident light and the material of the photocathode.

Einstein explained the results of the photoelectric effect by assuming that the incident light was in the form of photons, as postulated by Planck in analyzing blackbody radiation. He assumed that an incident photon with energy quantum $h\nu$ is absorbed by an electron in the photocathode and ejected from the material surface if it had sufficient energy. This was a simple conservation of energy relationship,

$$K_{\max} = h\nu - W_0$$

where $K_{\max}$ is the kinetic energy of the electron released from the surface and W_0 is the work done by the electron to release it from the metallic surface, a material dependent value known as the work function. Any photon with energy greater than the work function will release electrons and provide them with kinetic energy to be detected. Any photon with energy less than the work function does not have sufficient energy for an electron to be released and detected. There is a critical frequency (or wavelength) called cutoff where $K_{\max} = 0$. Every photocathode material is characterized by this cutoff value. The cutoff wavelength for a material with work function W_0 is,

$$\lambda_c = \frac{hc}{W_0}$$

4.5.2 Band structure of solids—review

Photon interactions within an isolated atom occur when the energy of the photon is absorbed exciting an electron to an upper energy level or emitted as an excited electron makes a transition from an upper energy level to a lower energy level. A similar process occurs for molecules, but the energy levels arise from potentials related to interatomic binding forces. These levels can be classified as rotational and vibrational modes of the atoms making up the molecule [1, 3].

The photoelectric effect is observed from photon interactions with a metallic surface. In order to further understand this interaction, we must review the energy level structure of solid materials and how we use this to classify their physical properties.

If two isolated atoms are brought closer together the energy levels of the system must deviate from those of the isolated atoms. Since the Pauli exclusion principle tells us that no two electrons can be in the same energy state, the energy of this state is said to split creating two different energy levels. Figure 4.8 illustrates the concept of a single energy level and how it splits when four atoms are placed in close proximity as a function of their atomic separation.

As more atoms are assembled to form a solid material each of the energy levels will split and the atoms separation will decrease to the equilibrium lattice spacing of the material. The more atoms the more states are split creating almost a continuum of levels at a particular atomic separation. The near continuum is a band. Figure 4.9 shows two levels and creation of bands associated with those levels.

As one can see, when the bands of allowed states are formed there is a region of energy where there are no allowed states which is the energy band gap. In general, the lower energy bands are filled first with the number of electrons. The highest energy band in a solid is the conduction band and energy band just below is called the valence band. The valence band contains states populated by electrons and is the highest energy band that is completely filled. For the case of metals, these materials have a partially filled conduction band which is responsible for their relatively high electrical conductivity.

In an insulator material, the conduction band contains no filled electron states and is separated from the valence band by a large energy gap. Even though there are

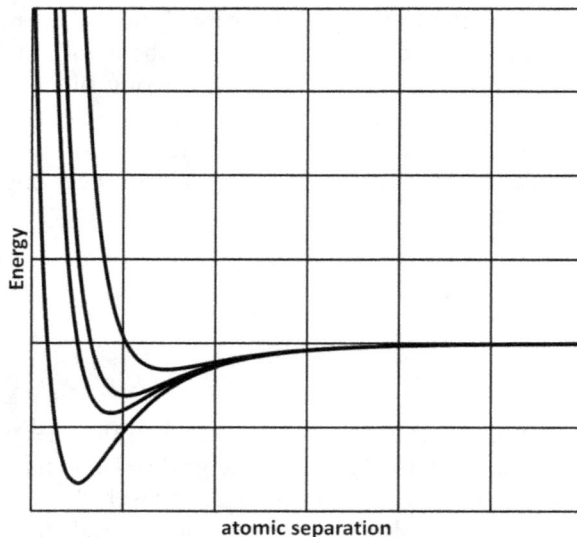

Figure 4.8. Splitting of a single atomic energy level with atomic separation between four atoms (all scales relative).

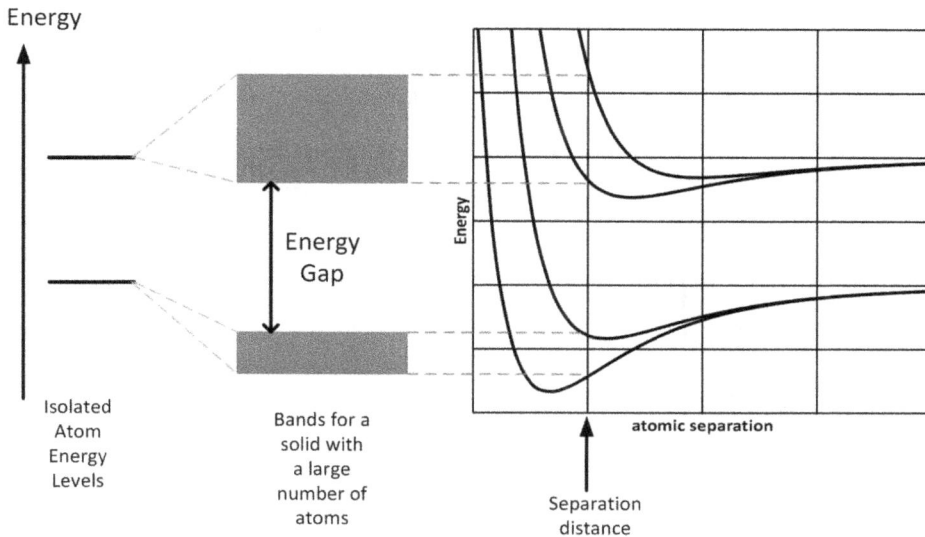

Figure 4.9. Illustration of two energy levels splitting with decreasing atomic separation and the formation of two bands of allowed states (all scales relative).

many available states in the conduction band, the energy required to promote an electron into one of these states is very high, so such materials do not conduct electricity. This concept is illustrated in figure 4.10. Also shown on this diagram is a plot of the Fermi–Dirac distribution indicating probable population of electrons. The Fermi energy lies somewhere within the energy gap.

Semiconductors have a similar but subtle difference from insulators in terms of their band structure. This is shown in figure 4.11. like an insulator there is an energy gap between the conduction band and the valence band, but the size of the energy gap is quite small. This means that there are some empty but more probable states that are available in the conduction band, as illustrated by the Fermi–Dirac distribution of figure 4.11 (shaded area of the conduction band). Given sufficient energy, greater than the gap energy, an electron can be promoted into a state in the conduction band.

Understanding the basic optical and electrical properties of materials is just a starting point. To engineer components and devices based on the materials usually requires that the material properties (optical, thermal, electronic, and mechanical) be tailored to provide a specific characteristic. Purposely introducing impurity atoms into the base material, or doping, is one of the primary ways that engineers can alter material properties [14].

Doped insulators are employed as the host for many laser materials. The dopant becomes the atoms that provide the desired optical property while the insulator material is chosen for its other physical properties. All solid-state lasers are designed along these criteria. Examples of these type of lasers include Nd:YAG (neodymium ions doped in the host material yttrium aluminum garnet) ruby (chromium ions

Energy

Conduction Band

$\mathscr{E}_F$

Energy Gap

Valence Band

$n_{FD}(\mathcal{E})$

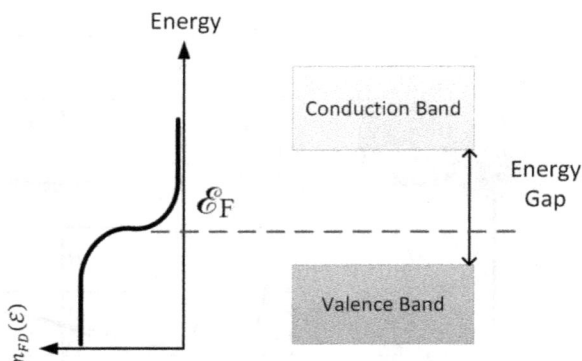

Figure 4.10. Band structure characteristics of an insulating material represented by a large band gap and filled valence band.

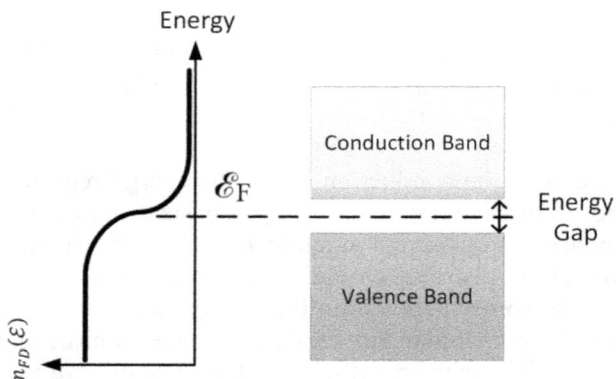

Energy

Conduction Band

$\mathscr{E}_F$

Energy Gap

Valence Band

$n_{FD}(\mathcal{E})$

Figure 4.11. Band structure characteristics of a semiconductor material represented by a relatively small energy gap. Shaded area in the conduction band show the possibility of some available states.

doped into aluminum oxide), fiber lasers (rare earth ions doped into the glass core of a fiber) [15].

Semiconductors are also intentionally doped to alter the number of available charge carriers. These are classified into two basic types of doped materials called n-type or p-type. The n-type semiconductor is doped with an impurity that provides an excess electron to the overall crystal lattice. This makes electrons the majority carrier and is called a donor impurity since it donates an extra charge carrier [16, 17]. The p-type materials are doped with impurity atoms having one less electron than the host semiconductor. These dopants are called acceptors since they need an electron to fill a vacant chemical bond. The missing electron site is termed a 'hole'. Electrically, holes (positive charges) can move through the crystal as an effective majority carrier. Note that doped materials, as a whole, are still completely electrically neutral because the donor and acceptor atoms are replacing an atom of the base semiconductor within the lattice. An example is germanium, atomic number $Z = 32$. The next element in the periodic table, arsenic $Z = 33$, is an appropriate donor impurity.

4.5.3 Introduction to the p–n junction

The purpose of this section is to provide a phenomenological description of a homojunction semiconductor device, the p–n junction. The importance of this device to optics and photonics applications such as detectors, LEDs, laser diodes, solar cells merits a rudimentary understanding of the physical characteristics as well as the electrical properties. The detailed physics of this device will not be covered here. Applications using the concepts from this section will be used later.

As discussed in the previous section, doped semiconductors form p-type or n-type materials depending on whether the material is doped with acceptor or donor atoms. From a band theory perspective, the doped atoms have energy states that are at energies within the bandgap since they are either below the bottom of the conduction band for donors or higher than the top of the valance band for acceptors. Doped semiconductor materials are called extrinsic semiconductors as opposed to the undoped intrinsic semiconductor. For purposes of discussion, it is useful to make some definitions for the various energy levels. We define, $\mathcal{E}_g$ as the bandgap energy, $\mathcal{E}_C$ as the energy level of the bottom of the conduction band, $\mathcal{E}_V$ as the energy level at the top of the valance band. Thus, the value of the energy gap is,

$$\mathcal{E}_g = \mathcal{E}_C - \mathcal{E}_V$$

Figure 4.12 shows the concept of donor states for n-type materials and acceptor states for p-type materials. Take the energy difference between $\mathcal{E}_C$ and the energy of the donor level to be $\mathcal{E}_D$ and the energy difference between $\mathcal{E}_V$ and the energy of the acceptor level as $\mathcal{E}_A$. In most cases at room temperature $\mathcal{E}_D$ is small enough that there is sufficient thermal energy to promote many electrons into the conduction band substantially increasing the electrical conductivity of n-type materials. A similar effect occurs in p-type semiconductors.

Another energy level that can be identified is the Fermi energy level. Since the Fermi energy is found from the statistics of the population of states, the Fermi energy is different for n-type and p-type materials. The specific value of the Fermi energy depends on impurity concentrations and temperature and is shown in figure 4.12 only to represent its relation between material types.

Numerous techniques exist for fabricating p–n junction devices [16, 17]. For simplicity, we are just going to assume that two materials are joined with an abrupt junction. This is shown schematically in figure 4.13. When this occurs diffusion of both holes and electrons create regions near the junction where ions of positively charged donor atoms are left in the n-type material and negatively charged acceptor ions are left in the p-type material. When an equilibrium charge distribution has been established diffusion stops. This region is called the depletion region or depletion layer since it has been depleted of majority carriers in both slabs of the materials.

Energy

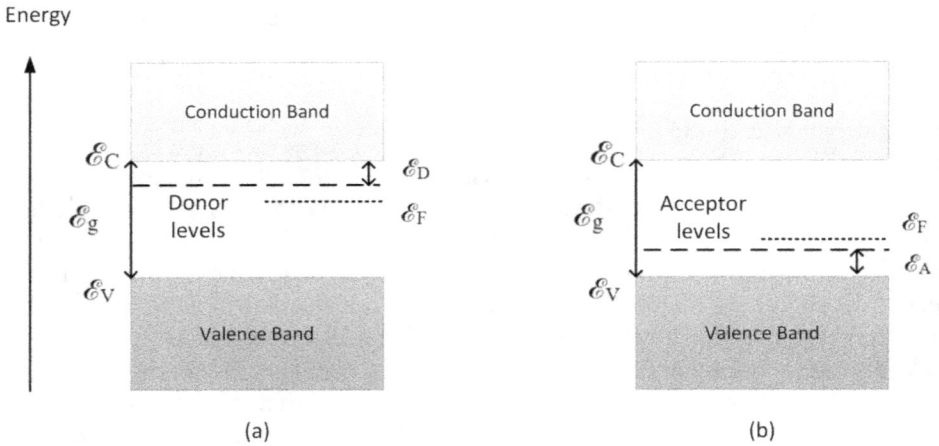

Figure 4.12. Band structure of a semiconductor with (a) donor levels in an n-type material and (b) acceptor levels in a p-type material.

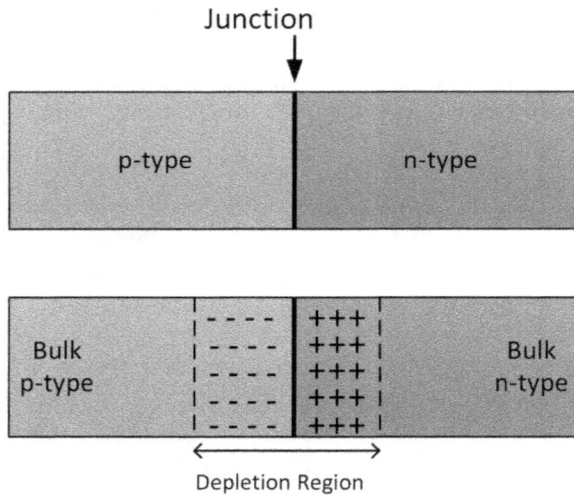

Figure 4.13. Schematic of the abrupt joining of n-type and p-type materials forming the depletion layer around the junction.

Another consequence of the diffusion process means that there is a separation of charge across the junction (space charge layer) setting up an electric field across the junction throughout the depletion region. This also means that there is an electric potential gradient created across the depletion region. Returning to the band structure energy level approach, figure 4.14, we can easily see this effect by mapping energy at various positions across the entire p–n structure. The common level between the two materials as they are joined is the Fermi energy.

The total potential across the junction is the equilibrium contact potential, identified as V_0. Note that the potential difference is directly related to the energy

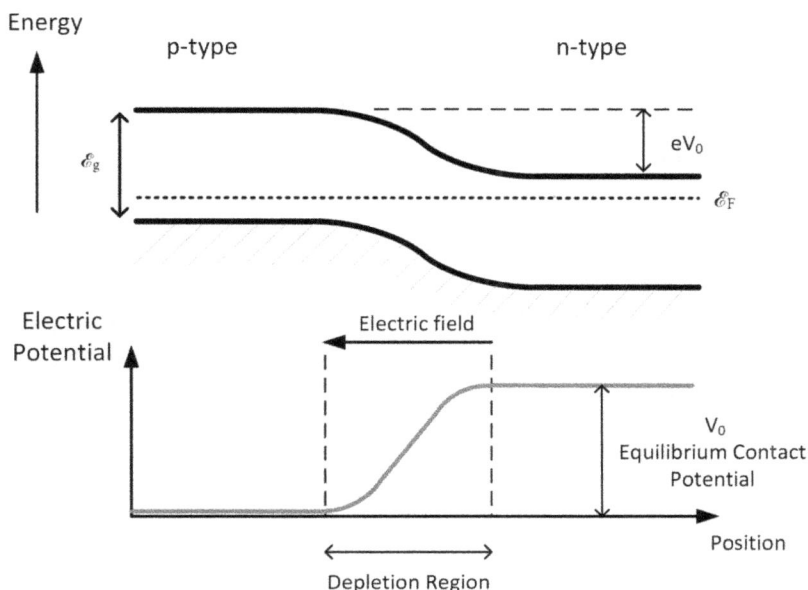

Figure 4.14. Energy band structure of a p–n junction in equilibrium after the junction has been formed and the depletion region has been established. An electric field is created across the junction and a potential gradient across the depletion region as a function of the position through the device.

difference eV_0 or the energy level difference between the conduction band energies of the p-type and n-type material referenced to the Fermi energy.

4.5.4 p–n junction in reverse and forward bias

In equilibrium the total current through the junction must be zero. (Actually, current density but we will only use current for simplicity assuming a constant cross-sectional area.) We can calculate the current by adding all contributions of the carrier flow due to both electrons and holes. There are two contributions for each carrier. One contribution is the diffusion current already discussed. Since the electric field setup across the junction will sweep minority carriers across the junction a drift or generation current arises in the opposite direction to the diffusion currents. This can be written as,

$$i = i_{pd} - i_{pg} + i_{nd} - i_{ng}$$

where i_{pd} and i_{pg} are the diffusion and generation currents for holes and i_{nd} and i_{ng} are the diffusion and generation currents for electrons. Which is again zero for the equilibrium situation since both diffusion currents and generation currents for each carrier must be equal and opposite.

Now let us apply an external potential across the junction in the forward bias configuration. This circuit is shown in figure 4.15 along with the resulting band structure schematic. The Fermi energy must shift by the energy associated with the applied potential. Since the height of the potential barrier across the junction has

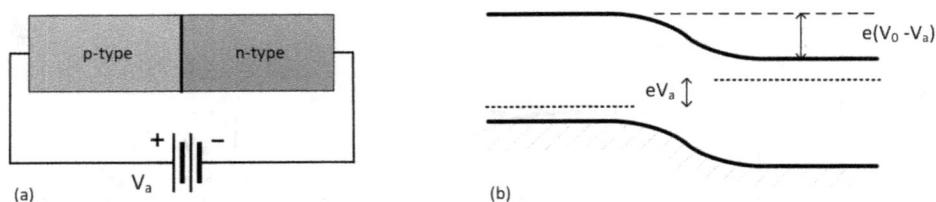

Figure 4.15. (a) p–n junction in forward bias from a simple circuit with externally applied potential and (b) band structure.

been reduced, majority carriers are said to be injected across the junction [16, 17]. The resulting diffusion currents become substantially larger than the generation currents. Detailed analysis of the carrier concentrations results in relationships for the forward bias drift currents which are simply [3, 17],

$$i_{pd} = i_{pd,\,0}\exp\left(\frac{eV_a}{kT}\right) \text{ and } i_{nd} = i_{nd,\,0}\exp\left(\frac{eV_a}{kT}\right)$$

where $i_{pd,0}$ and $i_{nd,0}$ are the drift currents in equilibrium when $V_a = 0$.

Using these two relations the total current as a function of applied voltage becomes,

$$i = (i_{pd,\,0} + i_{nd,\,0})\exp\left(\frac{eV_a}{kT}\right) - (i_{pg} + i_{ng})$$

When $V_a = 0$ $i = 0$ so $\left(i_{pd,\,0} + i_{nd,\,0}\right) = (i_{pg} + i_{ng}) \equiv i_{sat}$ the reverse saturation current. For negative values of V_a, the reverse bias configuration, results in a small negative total current. As V_a becomes more negative the total current asymptotically approaches i_{sat}.

The general current–voltage characteristic equation for a p–n junction with $V = V_a$ is then,

$$i = i_{sat}\left[\exp\left(\frac{eV}{kT}\right) - 1\right]$$

plotted in figure 4.16(a) and is a reasonably good approximation to experimental measurements for most bias voltages. Electrically, the p–n junction acts as a diode, allowing current to flow in the forward direction and not in the reverse direction depending on the bias current. The insert in figure 4.16(a) shows the symbol used in electronic circuit diagrams to represent a diode. The arrow points in the direction of forward current flow from the anode to the cathode.

While figure 4.16(a) is an ideal i–V curve for a diode, in practice there are physical limits on the applied potential that a single device can withstand. Small changes in the forward voltages generate large changes in current. Large currents provide the need for more power dissipation causing heat generation which will damage the device. Reverse bias voltages are limited by a device parameter called the reverse breakdown voltage; a value provided on any specification sheet. Any potential

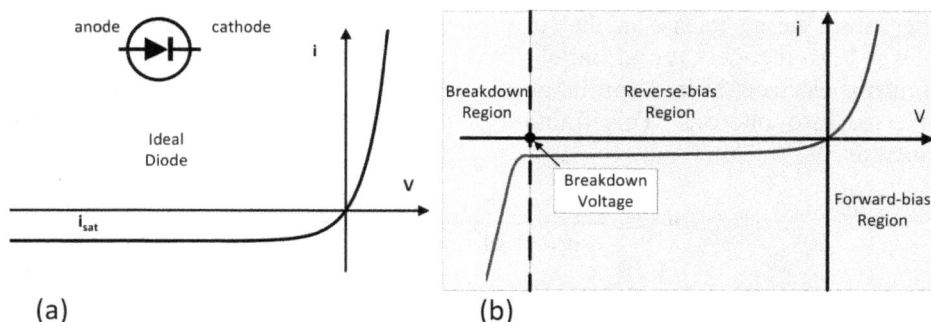

Figure 4.16. (a) Current–voltage characteristic curve for an ideal p–n junction. The inset to this figure also shows the electronic circuit symbol used to represent a diode. (b) Typical characteristic curve for a diode with three bias regions highlighted.

applied with a magnitude greater than this voltage begins to alter the crystal structure of the device material allowing a large negative current to flow. As with a large forward current, the diode can be permanently damaged due to excessive heat. The three regions of operation for a typical diode are highlighted in figure 4.16(b) with the reverse breakdown shown by the dashed line.

4.6 Photonic devices: sources

All optical systems require a source of photons. The system may be using natural light from the Sun or reflected from objects illuminated by a natural source. The target may be the thermal radiation inherently produced by a target object in the infrared portion of the spectrum. Specifying the source is one of the first issues a designer must face. Issues such as wavelength/wavelength band, minimum power, and continuous wave or pulsed/modulated must be determined. Traditional white-light sources such as tungsten-halogen sources may be used however, newer photonic devices are becoming more common and these will be discussed in this section.

Light emitting diodes (LEDs), lasers, and laser diodes are devices used in a wide variety of enabling applications [18]. In an LED, the electrical current from a forward biased p–n junction injects an electron which recombines with a hole releasing a photon. Lasers operate because of the stimulated emission process in an active material media. While we will not delve into a rigorous treatment of these devices here, because of their importance, we will briefly discuss some of their basic principles and distinguishing characteristics so that we may use them in later discussions of applications.

4.6.1 Light emitting diode (LED)

LEDs are semiconductor devices that are designed to emit photons when current is injected causing electrons to recombine with holes releasing energy as a photon. The difference between LEDs and an ordinary diode is optimization for light output rather than energy release in the form of heat dissipated. The light emitted from LEDs is both incoherent and unpolarized.

During electron–hole recombination only a fraction of the injected charge is converted into photons. This fraction is the internal quantum efficiency and is defined as,

$$\eta_{\text{int}} \equiv \frac{\text{number of photons}}{\text{number of injected electrons}}$$

This internal quantum efficiency depends on the device structure and active materials which are governed by both the radiative (r_{rad}) and nonradiative (r_{nr}) rates of recombination. The radiative rate is just the number of photons generated per unit time. So, another way to write the quantum efficiency is,

$$\eta_{\text{int}} = \frac{r_{\text{rad}}}{r_{\text{rad}} + r_{\text{nr}}}$$

Since all injected electrons will recombine through either radiative or nonradiative processes. In addition, the total rate of recombination must be the rate of flow of charge carriers, or $r_{\text{rad}} + r_{\text{nr}} = \frac{i}{e}$, so that,

$$r_{\text{rad}} = \eta_{\text{int}} \frac{i}{e}$$

And thus, the internally generated flux of an LED source of wavelength λ is,

$$\Phi_{e\,\text{int}} = \eta_{\text{int}} \frac{hc}{\lambda e} i$$

Which tells us that the optical power generated internally is directly proportional to the injected current into the device. However, this is the internal efficiency and not the total efficiency since the usable light flux must still be extracted from the LED. Also, not included here is a limitation that the current cannot be increased past a certain maximum value due to saturation effects and device breakdown.

Spectral characteristics of LEDs can be obtained using what we have learned from band structure definitions and quantum statistics. Since the photon energy arises from electron–hole recombination, we will use a simple model for their energy. Take $\mathcal{E}_e$ as the energy of an electron (effective mass m_e) and $\mathcal{E}_h$ as the energy of a hole (effective mass m_h). These energies can be written as,

$$\mathcal{E}_e = \mathcal{E}_C + \frac{p_e^2}{2m_e} \text{ and } \mathcal{E}_h = \mathcal{E}_V - \frac{p_h^2}{2m_h}$$

where $\mathcal{E}_C$ is the potential energy of the conduction band, $\mathcal{E}_V$ the potential energy of the valance band, and p_e and p_h are the electron and hole momenta. On recombination, applying conservation of momentum in the effective collision results an equal momentum for each particle or $p_e = p_h \approx p$. resulting in a photon energy of,

$$\mathcal{E} = \mathcal{E}_e - \mathcal{E}_h = \mathcal{E}_g + \frac{p^2}{2m} = h\nu$$

where a new effective mass has been defined as $\frac{1}{m} = \frac{1}{m_e} + \frac{1}{m_h}$.

This result shows us that photon energy depends both on the value of the band gap energy and an effective kinetic energy in the form of available momentum states near the band edges. It can be shown that an approximate relation for the spontaneous emission rate of a transition from these states is proportional to [3],

$$\sqrt{\mathcal{E} - \mathcal{E}_g} \exp\left(-\frac{\mathcal{E}}{kT}\right)$$

To find the peak photon energy for LED output we find the value of $\mathcal{E} = h\nu$ which maximizes this emission rate. This results in,

$$\mathcal{E}_{max} = \mathcal{E}_g + \frac{kT}{2} = h\nu_{max}$$

And since most thermal energies are smaller than the band gap energy an approximate peak wavelength value is,

$$\lambda_{max} \approx \frac{hc}{\mathcal{E}_g}$$

Figure 4.17 shows a plot of the measured spectra from a 640 nm peak LED along with the simple theoretical model described here. As one can see, there is reasonable agreement for the higher energy portion of the spectra up to the peak. In this model there is a sharp cutoff at the band edge because details of states residing within the bandgap at lower energies are not included. In addition, some further analysis of this simple model shows that the spectral width (full-width half-maximum, FWHM) can be predicted to be,

$$\Delta\lambda \approx \frac{1.795 \, kT\lambda^2}{hc}$$

For the data of figure 4.17 the measured FWHM is 18.8 ± 1.5 nm and the predicted value from this model is 15.5 nm at $T = 300$ K.

There are a wide variety of LED types that depend on the material properties of the semiconductors, the doping levels within the materials, growth methods used to form the junctions, and device packaging for light collection. Two of the most common structures used in fabricating the LED die are classified as (1) surface emitting and (2) edge emitting. In a surface emitting structure, the light output is perpendicular to the plane of the junction region and is often a spatially broad output pattern. An edge emitting LED provides light output along a line parallel with the junction. Some are also in the form of waveguides to increase output which is the special case known as a superluminescent diode (SLD). Edge emitters generally have a narrow spatial output, at least along one direction depending on the structure of the device. A schematic cross-section of a surface and edge emitter LED structure is provided in figure 4.18.

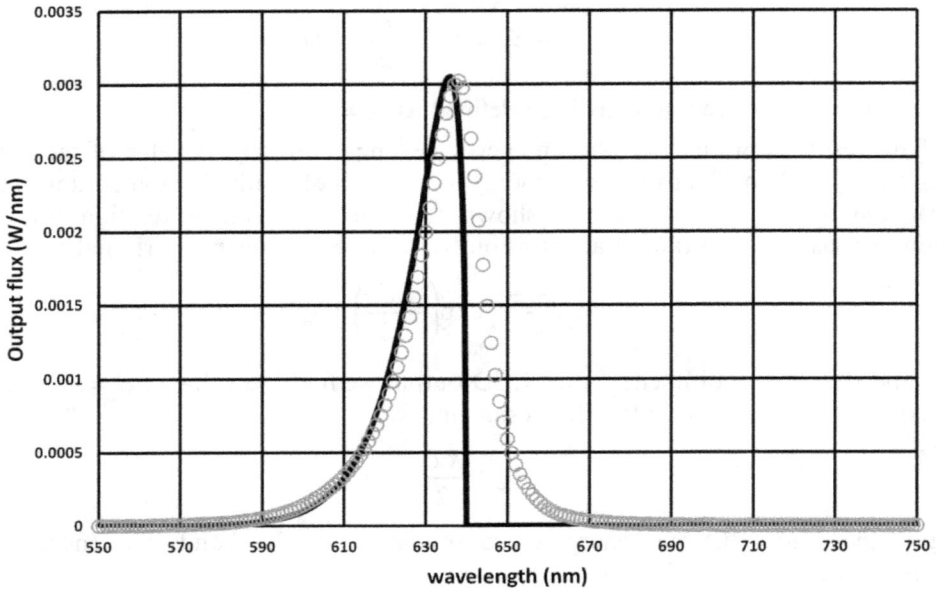

Figure 4.17. Measured spectra from a 640 nm peak LED (circled data points) along with the simple model for LED output (solid line).

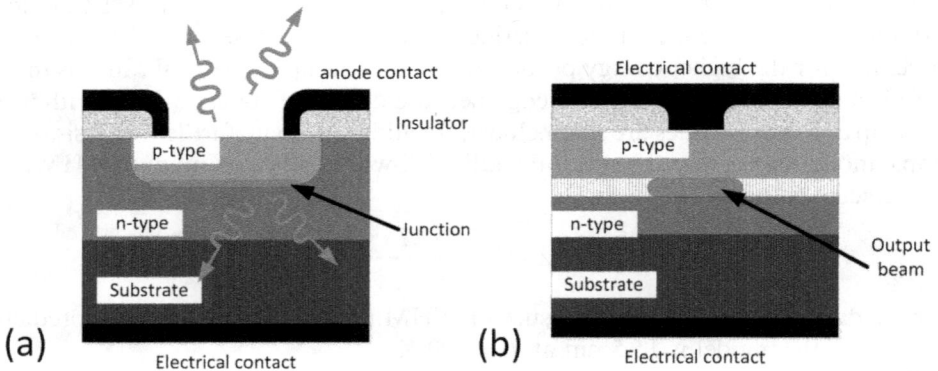

Figure 4.18. LED basic structures for a (a) surface emitter, and (b) edge emitter.

Examples of substrate and active materials include Gallium Arsenide (GaAs), Gallium Phosphide (GaP), Gallium Arsenide Phosphide combinations, Aluminum Gallium Phosphide (AlGaP), and Silicon Carbide (SiC). These materials are chosen for their band gap energy and thus the dominant wavelength. Fabrication of layers and structures are done using processes such as impurity diffusion.

LEDs have been used as indicator lights and alphanumeric displays for many years [18, 19]. Development of new materials for LEDs and detectors has been driven in part by the needs of fiber optic communications [20]. The discovery of blue and ultraviolet higher energy LEDs is driving the lighting industry. All these applications have led manufacturers to improve the light output and overall efficiency of these sources.

The internal quantum efficiency was computed earlier and depends on device structure and materials. The total efficiency into the external environment is also controlled by the efficiency of extraction. We can model extraction from LEDs by simply taking a source of flux embedded at the junction and determine how much of the internally generated flux will escape.

For example, using a surface emitting LED, we place a point source at the junction as shown in figure 4.19.

Light from the point source emits in all directions but only the fraction of rays with angles less than the critical angle leave the surface. The fractional trans-mittance, T_f, is the ratio of the solid angle cone in medium n_1 to the solid angle of the hemispherical radiation into region n_2. This value is,

$$T_f = \frac{\pi \sin^2 \theta_C}{2\pi} = \frac{1}{2}\left(\frac{n_2}{n_1}\right)^2$$

Another contribution to the extraction efficiency is Fresnel losses at the boundary between the two media. These values can be computed for all incident angles, but a good approximation is to use near normal incidence, or

$$T_{\text{Fresnel}} = \frac{4n_1 n_2}{(n_1 + n_2)^2}$$

In addition, there will be attenuation of each ray as it propagates through medium 1 which can be approximated as,

$$T_{\text{loss}} = \exp(-\alpha l_1)$$

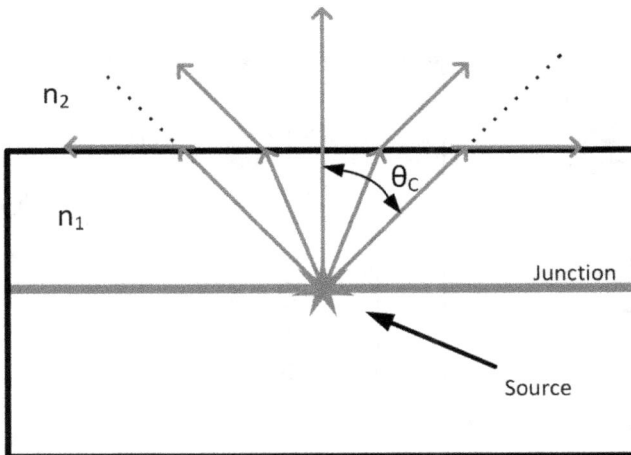

Figure 4.19. Model for light extraction efficiency of a surface emitting LED structure.

where α is the absorption coefficient at wavelength λ in medium 1 and l_1 is the thickness of medium 1 above the junction. When all of these transmittance values are included then a model for the surface emitting LED output flux is,

$$\Phi_{eout} = \eta_{\text{int}} \frac{2hc}{\lambda e} \left(\frac{n_2}{n_1} \right)^2 \frac{n_1 n_2}{(n_1 + n_2)^2} \exp(-\alpha l_1) i$$

This model indicates several parameters that can be used to maximize the output flux from LED devices. One is to reduce the thickness of the region of medium n_1 by etching away some of the material [20]. Other techniques include encapsulating the device in a dome shaped medium to increase the value of the ray angles that escape to an external air medium. Many indicator LEDs have this type of encapsulation.

Common current driver circuits for DC operation are shown in figure 4.20. LEDs are current driven devices, so controlling the current flow is necessary. For DC operation a simple voltage divider circuit containing a series resistor is used to limit the current, figure 4.20(a). Figure 4.20(b) shows a transistor controlling current flow through the application of a bias voltage applied to the base. An AC signal applied to the base can be used to modulate the current flow and LED output power, figure 4.20(c).

For the voltage divider of figure 4.20(a) the required series resistance is,

$$R = \frac{V - V_d}{i}$$

In designing the circuit, the desired current is usually provided based on the output power required. The battery voltage is chosen not to exceed a maximum applied voltage given on the LED specification sheet. In AC operation the bias voltage is chosen so that the amplitude of the input voltage oscillation is about one-half of the maximum voltage value. This allows the light output to follow the input oscillation. Of course, this modulation output is only proportional to the input as long and the LED and transistor are operating in their linear regions.

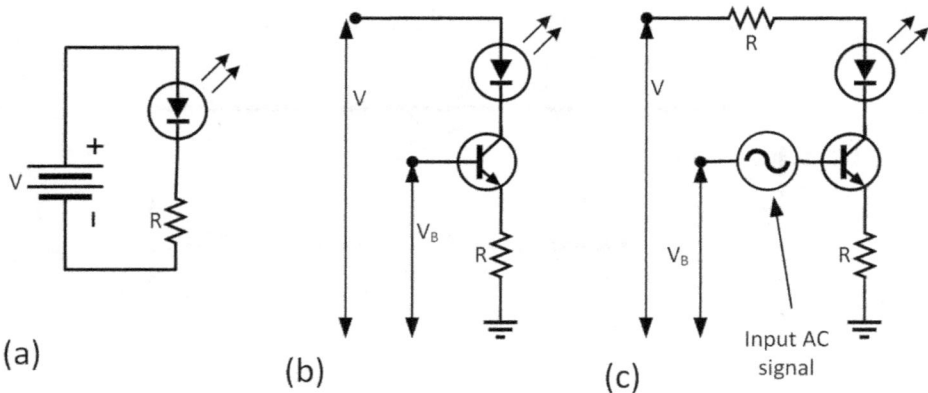

Figure 4.20. Some simple LED drive circuits (a) simple voltage divider, (b) biased transistor for current control, and (c) alternating current used to modulate the LED output.

4.6.2 Basics of laser operation

Laser sources have several unique properties such as being nearly monochromatic, coherent, and highly directional light sources. The processes of emission and absorption of photons between two energy levels as shown in figure 4.1 form the basis for our discussion of laser operation.

For a laser to operate there are three basic ingredients required. The first is an active medium, also called the gain medium, which is a source of atoms or molecules whose energy levels can be prepared to allow for a higher probability of stimulated emission to occur. Second, an input energy source often called a pump source is required to provide energy for the system to achieve population inversion, more upper states filled than lower energy states. Third, an optical feedback mechanism which most often is in the form of a cavity resonator to allow for controlled amplification at a particular frequency/wavelength.

Take $\mathcal{E}_1$ to be the energy of level 1 with population N_1 atoms per unit volume and $\mathcal{E}_2$ to be the energy of level 2 with population N_2 atoms per unit volume. The difference between these two states is then the energy of the photon that is emitted or absorbed when transitions occur. The ratio of the populations is given by the Boltzmann distribution,

$$\frac{N_2}{N_1} = e^{-(\mathcal{E}_2 - \mathcal{E}_1)/kT} = e^{-h\nu/kT}$$

We define a rate of transition as the number of atoms in one state that make a transition to the other state per second. Another term used to describe these transitions is in terms of the lifetime which is just the reciprocal of the rate. The rate of spontaneous emission from state 2 to state 1 is just, $R_{\text{spon}} = N_2 A_{21}$ where A_{21} is a constant. The rates of spontaneous absorption and stimulated emission require a radiation field with energy density w_e. Those rates are defined as, $R_{\text{abs}} = N_1 B_{12} w_e(\nu)$ and $R_{\text{stim}} = N_2 B_{21} w_e(\nu)$, respectively, where B_{12} and B_{21} are constants. The constants of proportionality are the Einstein A and B coefficients.

If the system is in thermal equilibrium, then the rate of emission must equal the rate of absorption [1–3, 21]. This detailed balancing of rates is,

$$N_1 B_{12} w_e(\nu) = N_2 A_{21} + N_2 B_{21} w_e(\nu)$$

Using the ratio of N_2/N_1 from above and solving for w_e, we obtain,

$$w_e(\nu) = \frac{A_{21}}{B_{12}} \frac{1}{e^{+\frac{h\nu}{kT}} - \frac{B_{21}}{B_{12}}}$$

Since this must be equal to the same energy density from Planck radiation then we find,

$$\frac{B_{21}}{B_{12}} = 1 \text{ and } \frac{A_{21}}{B_{12}} = \frac{8\pi h \nu^3}{c^3}$$

Another way to think about this energy density is that it represents the number of photons per unit volume, N, with energy $h\nu$, or, $w_e = Nh\nu$. We saw earlier that the irradiance of a wavefront propagating through a material along a distance z was,

$$E_e(z) = E_e(0)e^{-\alpha z}$$

where the coefficient α is the macroscopic absorption coefficient and $E_e(0)$ is the incident irradiance at $z = 0$. Because of this absorption process the change in irradiance within a small element dz is,

$$dE_e(z) = -\alpha E_e(z)dz$$

Since the irradiance is the energy per unit area per unit time, this quantity can also be written in terms of numbers of photons as, $E_e = Nh\nu\frac{c}{n}$ where the quantity c/n is the speed of light in the medium of refractive index n. This process is illustrated in figure 4.21 for a set of incident photons.

Our detailed balancing argument tells us that if we want to increase the rate of stimulated emission then we must increase the radiation density and the population N_2. From a photon picture, we model the net rate of loss of photons per unit volume in this beam as,

$$-\frac{dN}{dt} = N_1 B_{12} w_e(\nu) - N_2 B_{21} w_e(\nu)$$

In this simplified situation we have not included spontaneous emission since it does not contribute to the beam. Writing this equation in terms of irradiance along the z-direction we obtain,

$$\frac{dE_e}{dz} = \left[(N_1 - N_2)\frac{B_{21}h\nu}{(c/n)} \right]E_e$$

Or the absorption coefficient is,

$$\alpha = (N_1 - N_2)\frac{B_{21}h\nu}{(c/n)}$$

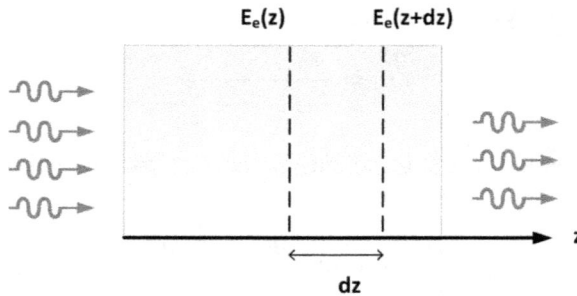

Figure 4.21. Irradiance as a function of position for a wavefront (photon stream) passing through a distance dz of a material medium.

Therefore, an atomic system with two levels in thermal equilibrium has a high probability of absorbing photons rather than emitting photons. For any gain or emission to occur then the coefficient α must become negative. This requires that the population difference between the two levels be negative or $N_2 > N_1$. This requirement runs counter to the population of energy levels in the normal situation of thermal equilibrium. However, population inversion can take place by including one or more energy levels in the system operation.

Figure 4.22(a) shows a bar graph of the energy of three separate energy states and the population of those states for a system in thermal equilibrium. To induce population inversion, suppose that the dynamics of the system are such that the third energy level, energy $\mathcal{E}_3 > \mathcal{E}_2$ is employed so that a large energy input or pump will place electrons in this state. If the third state allows a fast nonradiative transition to level $\mathcal{E}_2$, and if the lifetime of this state is fairly long (a metastable state), then population inversion can be achieved, figure 4.22(b) energy is input to the system the population.

Three-level and four-level schemes are typical models used in understanding laser energy levels. Illustration of the energy level diagrams for each is provided in figure 4.23. For each system, the laser emission is occurring between energy level 2 and energy level 1. The ground state (lowest energy level) for the four-level scheme is N_0. The excitation pump promotes electrons to an upper energy state in the pump band where there is a fast decay to the state with energy $\mathcal{E}_2$. An incident photon with energy, $\mathcal{E}_2 - \mathcal{E}_1 = h\upsilon$ passing through the active medium stimulates the system to produce another photon to achieve amplification.

When population inversion occurs, the small-signal gain coefficient $\beta = -\alpha$, becomes,

$$\beta = -\alpha = (N_2 - N_1)\frac{B_{21}h\upsilon}{(c/n)}$$

And the irradiance of the propagating wavefront after traveling through a distance z is,

$$E_e(z) = E_e(0)e^{+\beta z}$$

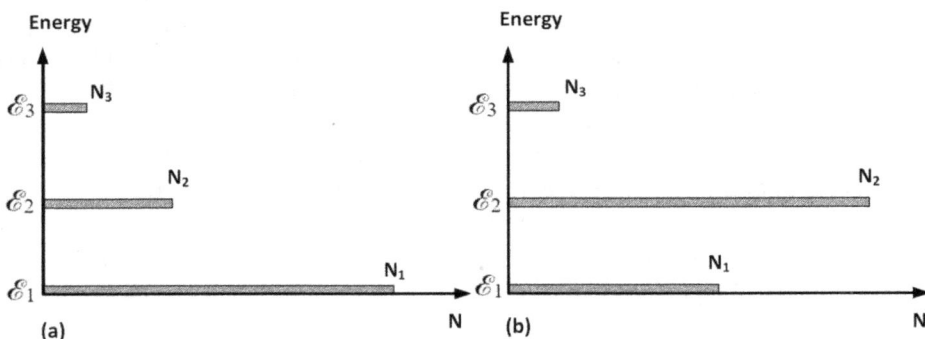

Figure 4.22. Energy level diagrams for a system (a) in thermal equilibrium, and (b) a system that has been pumped to induce a population inversion.

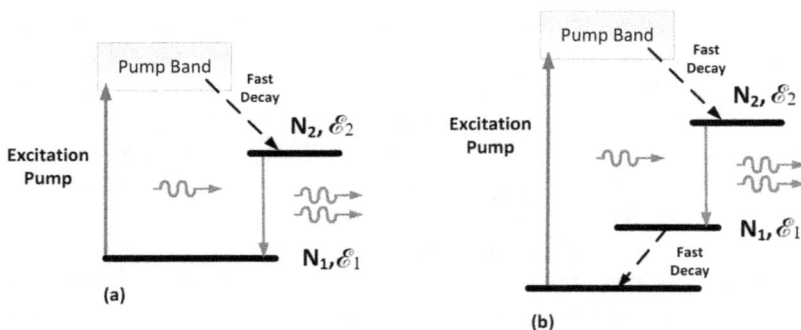

Figure 4.23. (a) Three-level energy diagram and (b) four-level energy diagram showing a pump source.

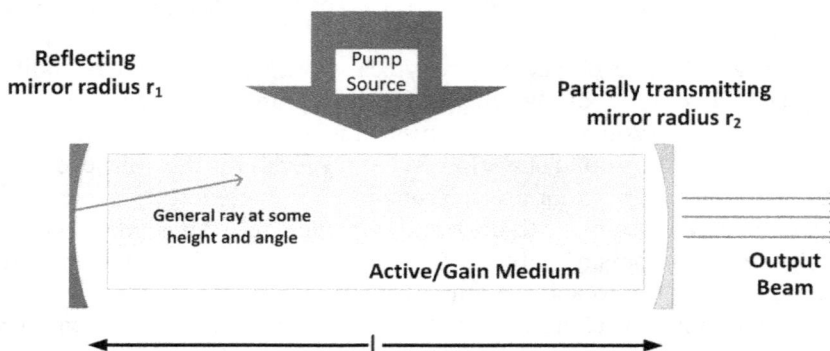

Figure 4.24. Laser configuration. One mirror has only a partial transmittance so that the beam can emerge.

In practice, amplification occurs only when the overall gain of a system becomes greater than all other absorption losses in the system. However, this light is still not a laser but only an amplified photon stream.

The final component for laser operation is a cavity resonator which provides controlled feedback to the gain medium. Figure 4.24 shows the basic configuration of a laser comprising two mirrors forming the cavity, the gain medium, and some type of pump source. In the design of the laser this resonator must be stable, and the round-trip gain must be greater than the losses incurred on the beam. We can use some simple ray tracing to obtain criteria for cavity stability. Start with a general ray propagating at some arbitrary ray height and angle, between two mirrors of radius of curvature, r_1 and r_2 separated by a length L as shown in figure 4.24.

Using ray transfer matrices, the system matrix for one round trip pass for the ray beginning at the surface of mirror 1 at height y_0 and angle u_0 and ending reflecting from mirror 1 with height y_1 and angle u_1 is,

$$\begin{pmatrix} y_1 \\ u_1 \end{pmatrix} = \mathcal{R}_1 T_L \mathcal{R}_2 T_L \begin{pmatrix} y_0 \\ u_0 \end{pmatrix} = \begin{bmatrix} A & B \\ C & D \end{bmatrix} \begin{pmatrix} y_0 \\ u_0 \end{pmatrix}$$

where the coefficients A, B, C, and D can be easily determined through matrix multiplication. But we need this ray transfer for a large number of passes. An identify from linear algebra says, that a 2×2 matrix to the Mth power is,

$$\begin{bmatrix} A & B \\ C & D \end{bmatrix}^M = \frac{1}{\sin\theta} \begin{bmatrix} A\sin M\theta - \sin(M-1)\theta & B\sin M\theta \\ C\sin M\theta & D\sin M\theta - \sin(M-1)\theta \end{bmatrix}$$

where

$$\cos\theta = \frac{1}{2}(A + D)$$

The values of A and D from above can be substituted into the relation for the cosine function and since the cosine must be real, this places a condition on the parameters so that,

$$0 \leqslant \left(1 - \frac{L}{r_1}\right)\left(1 - \frac{L}{r_2}\right) \leqslant 1$$

which is the stability criterion for laser resonators. As one can see, if both mirrors are planar the stability criterion is still possible but at an extremum.

As mentioned before, another practical criterion for laser operation is that the overall gain must exceed all other losses in the laser system. Issues such as absorption and scattering at the mirror surfaces, scattering in the laser active media, and diffraction losses all contribute to system loss.

If we include a single loss parameter α to account for all losses except those due to reflectance at the mirrors then we modify the irradiance of the beam within the cavity as,

$$E_e(z) = E_e(0)e^{+(\beta-\alpha)z}$$

After a complete round trip through the cavity of length L where the beam reflects from mirror 2 and then mirror 1, the ratio of the final irradiance to the initial irradiance is,

$$R_1 R_2 \exp[2(\beta - \alpha)L]$$

where R_1 and R_2 are the reflectance values for mirrors 1 and 2, respectively. As long as this value is greater than unity then the small system gain will overcome other losses producing laser output. When this value is unity, we define a gain threshold, β_{th}, the value of the small-signal gain necessary to overcome losses. For this situation, we have,

$$\beta_{th} = \alpha + \frac{1}{2L}\ln\left(\frac{1}{R_1 R_2}\right)$$

Note that this gain coefficient has units of inverse length. Increasing the length will increase the overall gain of the laser.

Even though our discussion of laser operation is not completely rigorous it does explain the important characteristics of laser systems. Because the photons generated in a laser active medium are due to transitions between energy levels the laser is nearly monochromatic. Amplification due to the stimulated emission process results in the high degree of coherence of the laser light since the stimulated emission photon is in phase with the incident photon. Finally, the cavity resonator that provides controlled feedback forms a system leading to the directionality of the output laser beam.

4.6.3 Laser diode

The active medium of a semiconductor laser is the p–n junction, similar to an LED. For lasing action there must also be population inversion and optical feedback. The structure of most semiconductor diode lasers is a series of layers forming what is called a double heterostructure [22]. This is similar to an edge emitter where an active medium layer lies between two outer layers (cladding layers). The cladding layers are the doped p and n materials forming the junction in the active layer. Figure 4.25 shows a schematic of the basic concept for the double heterostructure system. When the refractive index of the cladding layers is lower than the active layer, a waveguide is formed confining the photons generated to be within the active gain medium. The laser cavity is typically created by cleaving the end of the structure along a facet of the material.

The refractive index of most semiconductor materials is relatively high compared to glass. A material such as AlGaAs has an index of refraction of about 3.7 at 780 nm wavelength. At normal incidence into air the reflectance at the output face of a laser made from this material is 0.33. Some lasers are coated to alter their reflectance to optimize output at a given wavelength.

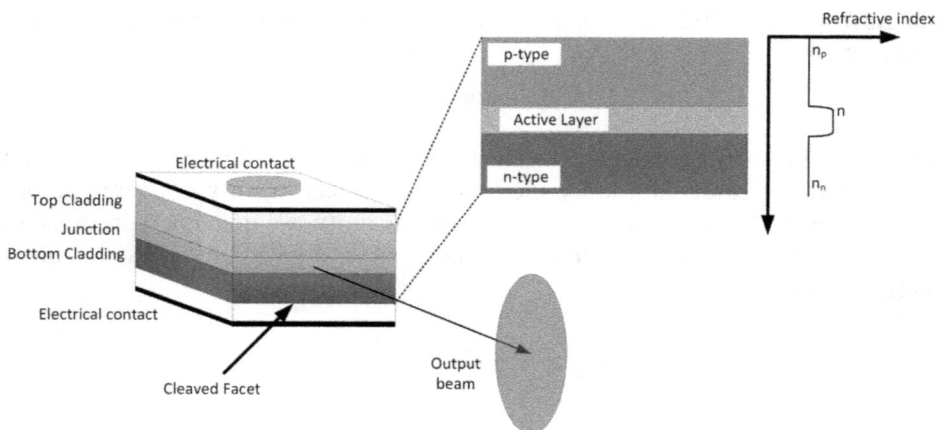

Figure 4.25. Laser diode schematic double heterostructure.

Along with a high refractive index comes absorption which limits gain. We found earlier that the internal photon flux generated through radiative recombination of electrons and holes was proportional to the injection current. This implies that there at some value of current, the overall gain will exceed the threshold gain. This value of current is called the threshold current. At low values of injection current the system will have radiative output as with an LED dominated by spontaneous emission of photons. But above the threshold current, stimulated emission of photons occur providing laser output. The characteristic light output for a diode laser as a function of input current is shown in figure 4.26. The value of threshold current is identified by extrapolating the linear region of the laser output back to the current axis.

4.7 Application: determining laser diode threshold current

In some design situations it may be important to characterize a laser device to determine its operation specifications. In particular, for a laser diode, the threshold current value from a specification sheet must be verified. The threshold current is highly dependent on the environmental temperature and device temperature so a test under a specific set of operating conditions may be necessary. Figure 4.26 shows a graph of a typical light output versus injection current for a laser diode and indication of the threshold current. This value was obtained by fitting the linear region of the laser output to a line and extrapolating the line to find the current axis

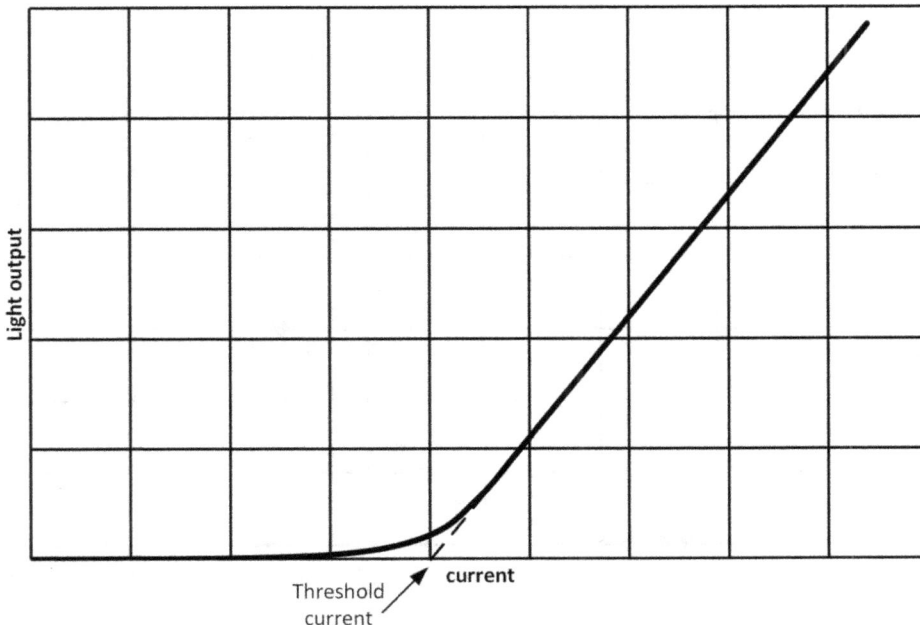

Figure 4.26. Characteristic light output curve for a diode laser. The threshold current value is indicated.

intercept. This is only one of several methods used to determine the threshold current value from experimental measurements [23, 24].

Laboratory measurements for the light output as a function of drive current are shown in figure 4.27. The data follows the general characteristic curve as expected. The light output was measured with a detector whose output voltage was proportional to the incident light flux. Errors in measurement were less than the size of the data points. Using a linear line fit to the data from 24 mA up resulted in a line whose slope was 82.3 ± 0.22 mV mA^{-1} with a y-intercept of -1960.0 ± 5.90 mV. Computing the threshold current is performed by setting the voltage equal to zero and solving for that current value to obtain ith $= 23.8 \pm 0.095$ mA. While this calculation method is often used, it may not provide a good comparison to a threshold current value obtained from fundamental laser physics theory and has some drawbacks in that it depends on the specific data chosen by the user for the fit [24].

Two other methods of data analysis may offer a better representation for determining the threshold current. These two methods rely on first-order and second-order differences/derivatives. The same data of figure 4.27 is replotted by taking the numerical derivative at each of the equally spaced consecutive data points for each Δi interval resulting in figure 4.28(a). The computation used for consistent current values is a centering difference derivative a standard technique used in image processing [25]

$$\frac{dV}{di} \approx \frac{V(i + \Delta i) - V(i - \Delta i)}{2\Delta i}$$

A similar calculation can also be done to obtain the numerical second derivative. This relation is given by,

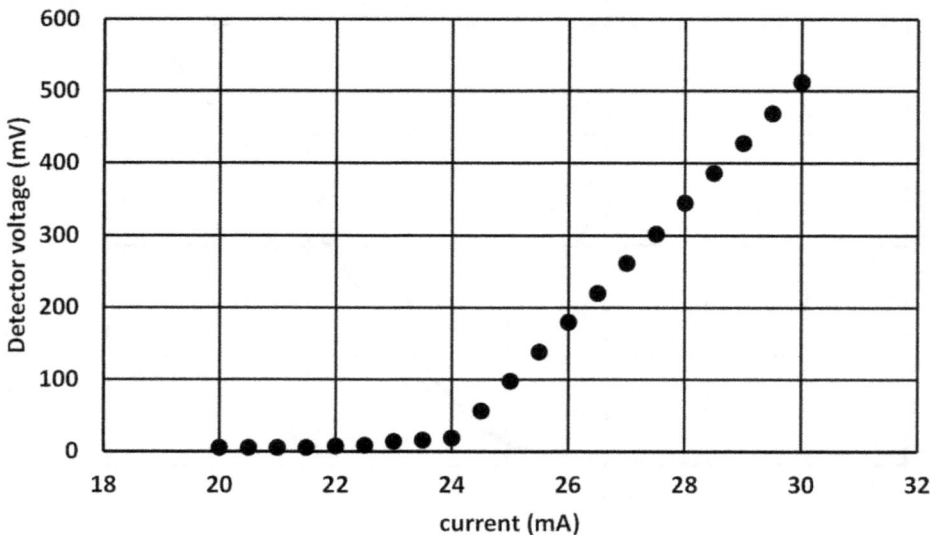

Figure 4.27. Experimental measurement of the light output versus drive current for a laser diode.

(a)

(b)

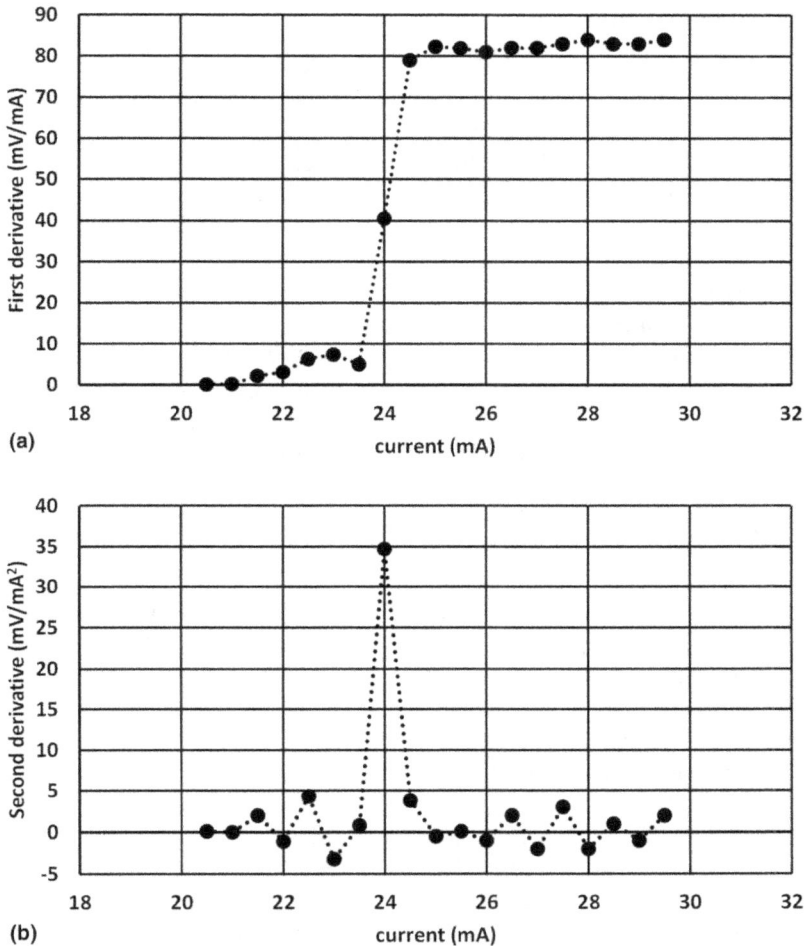

Figure 4.28. First and second numerical derivatives of the threshold current data from figure 4.27.

$$\frac{d^2V}{di^2} \approx \frac{V(i + \Delta i) + V(i - \Delta i) - 2V(i)}{\Delta i^2}$$

Figure 4.28(b) shows a plot of the numerical second derivative.

A general definition for obtaining the threshold current from first-derivative data is the current value when for the light output reaches half its maximum value. The criteria used in the second derivative analysis are the current related to the maximum value. In both cases, a value for the threshold current is identified as 24 mA. Unfortunately, this data set does not contain enough data points near the region of the threshold current to allow a proper error analysis. In summary, several methods for determining the threshold current have been examined. One must take care obtaining a sufficient number of data points in the region of the threshold current if this quantity is needed to a given accuracy.

Exercises and problems

1. A photon of energy 2.81 eV enters the depletion region of a p–n junction with a bandgap energy of 1.1 eV and is totally absorbed. On average, how many electron hole pairs are created?

2. A crystalline material becomes transparent for wavelengths above 300 nm and is opaque for all lower wavelengths. Find the energy of the energy gap for this material.

3. A small sodium vapor lamp (589 nm wavelength) emits 10 W of optical power into a spherical volume. If you stand 10 m away from this source, how many photons enter your eye (4.0 mm diameter pupil) each second?

4. Solar radiation is incident on the Earth's surface at an average of 1000 W m^{-2} on a surface normal to the rays. For a mean wavelength of 550 nm, calculate the number of photons falling on a 1 cm^2 of the surface each second.

5. The average output power of a Nd:YAG pulsed laser is 33 W. What is the peak output power for this laser running with a duty cycle of 5.0×10^{-8}?

6. A pulsed laser with 6.0 mJ output energy has a peak power of 3000 W. What is the pulse width of the laser?

7. We observe optical spectra emitted by atoms when an electron in an atom makes a transition from one energy level to another. Find the wavelength of the photon emitted when an electron makes a transition from an upper state to a lower state with energy difference of 3.0 eV.

8. An atom has two energy levels with a transition wavelength of 694.3 nm. What percentage of the atoms are in the upper state at room temperature ($T = 300$ K)? Assume that all of the atoms are in either of these two energy states.

9. An atom has only two populated energy levels with a transition wavelength of 580 nm between them. At room temperature ($T = 300$ K) there are $6.0 \times 10^{+20}$ atoms in the lower state. (a) Compute the energy of a photon released when one of these atoms makes a transition from the upper state to the lower state. (b) How many atoms occupy the upper state at room temperature under conditions of thermal equilibrium? (c) Now suppose that $4.0 \times 10^{+20}$ atoms are pumped into the upper state with $2.0 \times 10^{+20}$ atoms remaining in the lower state. How much total energy would be released until the states reach equilibrium?

10. A total of 10^{16} noninteracting distinguishable atoms are distributed between two energy states. The excited state is 0.22 eV above the ground state energy. At a temperature of 400 K, how many atoms are in the excited state and how many are in the ground state?

11. The threshold wavelength for the photoemission of electrons from a gold surface is 257.3 nm. (a) Does the threshold wavelength depend on the intensity of the light? (b) Does the rate at which electrons are emitted depend on the intensity of the light? (c) What is the work function for gold? (d) What is the stopping voltage when ultraviolet light of 100.0 nm is incident on gold?

12. The following set of data was recorded in a photoelectric effect experiment.

Stopping potential (V)	Incident light frequency (10^{14} Hz)
-1	2.42
0	4.84
$+1$	7.25
$+2$	9.67
$+3$	12.09

   (a) What is the threshold wavelength (also called cutoff wavelength)?

   (b) What is the work function of the metal used as the target for this experiment? (Express your answer in eV).

   (c) What is the maximum kinetic energy of photoelectrons when the stopping potential is 3 Volts?

   (d) If you were to plot the data above as stopping potential versus frequency, what would the slope of the curve represent? Give a numerical value for this result.

   (e) Does the cutoff wavelength depend on the intensity of the light?

   (f) Does the rate at which electrons are emitted depend on the intensity of the light?

13. A block of steel is initially at room temperature (20 °C). Assume that the steel block acts as an ideal blackbody radiator. (a) At room temperature, what wavelength does this block emit the maximum electromagnetic radiation? (b) If we heat the block to 1000 K, how many times more total thermal radiation does the block now emit? (c) What temperature do we need to heat up the block so that it glows red with a peak wavelength of 650 nm?

14. Compute the total radiant exitance and peak wavelength of a blackbody source whose temperature is (a) 1000 K, (b) 2000 K, and (c) 6000 K.

15. At what temperature is the total radiant exitance of a graybody with emissivity of 0.12 equal to the total radiant exitance of a blackbody of temperature 1000 K?

16. Why are at least three energy levels needed in an active medium for lasing action to occur?

17. Compute the amount of population inversion required per unit volume per unit frequency interval for a CO_2 laser (10.6 μm wavelength) to be 0.5 m^{-1}. Take the Einstein A coefficient for this laser to be 200 s^{-1}.

18. Given a simple two-level atomic system with ground state energy level identified as Level #0 and an upper energy level identified as level #1.

   (a) If $B_{10} = 10^{19}$ m^3 W^{-1} s^{-3}, what is A_{10} and the corresponding spontaneous lifetime τ_{10} for a transition wavelength of 500 nm? (b) What temperature would be necessary for a 10% population inversion of a simple two-level system at 500 nm? Assume that all atoms are in either of the two states?

References

[1] Eisberg R M and Resnick R 1985 *Quantum Physics of Atoms, Molecules, Solids, Nuclei, and Particles* 2nd edn (New York: Wiley)

[2] Krane K S 2019 *Modern Physics* 4th edn (New York: Wiley)

[3] Saleh B E A and Teich M C 2007 *Fundamentals of Photonics* 2nd edn (New York: Wiley)

[4] Hecht E 2002 *Optics* 4th edn (Reading, MA: Addison-Wesley)

[5] Arecchi F T 1965 Measurement of the statistical distribution of Gaussian and laser sources *Phys. Rev. Lett.* **15** 912–16

[6] Koczyk P, Wiewior P and Radzewicz C 1996 Photon counting statistics—undergraduate experiment *Am. J. Phys.* **64** 240–45

[7] Davenport W B Jr and Root W L 1987 *An Introduction to the Theory of Random Signals and Noise* (New York: Wiley-IEEE Press)

[8] van Etten W 2006 *Introduction to Random Signals and Noise* (Chichester: Wiley)

[9] Hobbs P C D 2000 *Building Electro-optical Systems: Making It all Work* (New York: Wiley)

[10] Boyd R W 1983 *Radiometry and the Detection of Optical Radiation* (New York: Wiley)

[11] Zalewski E F 1995 Radiometry and photometry *Handbook of Optics, Second Edition Volume II: Devices, Measurements, and Properties* ed M Bass, E W Van Stryland, D R Williams and W L Wolfe (New York: McGraw-Hill) ch 24

[12] Wilson J and Hawkes J F B 1998 *Optoelectronics: An introduction* (London: Prentice Hall Europe)

[13] Jenkins F A and White H E 1976 *Fundamentals of Optics* (New York: McGraw-Hill)

[14] Kittel C 2004 *Introduction to Solid State Physics* 8th edn (New York: Wiley)

[15] Sennaroglu A 2010 *Photonics and Laser Engineering: Principles, Devices, and Applications* (New York: McGraw-Hill)

[16] Sze S M and Ng K K 2006 *Physics of Semiconductor Devices* 3rd edn (New York: Wiley)

[17] Streetman B G and Banerjee S 2016 *Solid State Electronic Devices* 7th edn (Harlow: Pearson)

[18] National Research Council 1998 *Harnessing Light: Optical Science and Engineering for the 21st Century* (Washington, DC: The National Academies Press)

[19] Haitz R H, Crawford M G and Weissman R H 1995 Light emitting diodes *Handbook of Optics, Second Edition Volume II: Devices, Measurements, and Properties* ed M Bass *et al* (New York: McGraw-Hill) ch 12

[20] Keiser G 2010 *Optical Fiber Communications* 4th edn (New York: McGraw-Hill)

[21] Einstein A 1917 Zur Quantentheorie de Strahlung *Phys. Z* **18** 121–8

[22] Derry P L, Figueroa L and Hong C 1995 Semiconductor lasers *Handbook of Optics, Second Edition Volume II: Devices, Measurements, and Properties* ed M Bass, E W Van Stryland, D R Williams and W L Wolfe (New York: McGraw-Hill) ch 13

[23] Bellcore standard *Introduction to Reliability of Laser Diodes and Modules Special Report SR-TSY-001369* http://telecom-info.telcordia.com/ido/AUX/SR_TSY_001369_TOC.i01.pdf

[24] *The Differences Between Threshold Current Calculation Methods* https://newport.com/medias/sys_master/images/images/hc9/hd1/9680121757726/AN-12-REV03-Differences-Between-Threshold-Current-Calculations.pdf

[25] Gonzalez R C and Woods R E 2018 *Digital Image Processing* 4th edn (Upper Saddle River, NJ: Pearson/Prentice Hall)

Optical Systems Design Detection Essentials
Radiometry, photometry, colorimetry, noise, and measurements
Robert M Bunch

Chapter 5

Linear optics and optical system functions

5.1 Introduction

In this chapter the linear systems approach to wave propagation in optical systems is discussed. We consider the spatial distribution of a source field as a functional input to a general system and use linear systems theory to follow this field through a system to characterize optical system performance limitations [1, 2]. Applying a linear systems model for wave propagation in systems, results in the same predictions for output light fields as obtained from classical interference and diffraction. These phenomena will be identified in turn as we encounter various examples of particular input field functions. We will continue to use the geometry and notation introduced in chapter 1 and shown in figure 1.5 to obtain information about the electromagnetic fields in each plane of a system. Only one-dimensional functions are used in some cases for simplicity in understanding basic properties, however, all results can be easily extended to two-dimensional functions and system operators. Optical system limitations and imaging performance issues are also covered including diffraction limits, resolution, image quality, and the optical transfer function (OTF).

5.2 Linear optical systems basics

5.2.1 One-dimensional systems properties

In chapter 1 we introduced the operator formalism for obtaining the output response of a system with output function $g(x)$ in terms of the input function represented by $f(x)$. The system operator $\mathcal{H}$ maps or transforms the output from the input as,

$$\mathcal{H}\{f(x)\} = g(x)$$

Again, $f(x)$ and $g(x)$ are model mathematical functions representing some physical phenomena. The models must then have characteristics representative of a physical

doi:10.1088/978-0-7503-2252-2ch5

mechanism. So, the functions and the system operator must have properties that correspond to those observed in nature.

Three specific system properties that are commonly needed for physical systems are causality, shift-invariance, and linearity. The system property of causality means that no output occurs before there is an input. This can be stated mathematically as,

$$\text{If } f(x) = 0 \text{ for all } x < x_0 \text{ then } g(x) = \mathcal{H}\{f(x)\} = 0 \text{ for all } x < x_0$$

Shift-invariance of a system operator states that for a fixed value x_0,

$$\text{If } g(x) = \mathcal{H}\{f(x)\} \text{ then } g(x+x_0) = \mathcal{H}\{f(x+x_0)\}$$

(This is also called space invariance if x represents a spatial coordinate or time invariance if the function is in the temporal domain). Finally, linearity is the property such that,

$$\text{If } g_i(x) = \mathcal{H}\{f_i(x)\} \text{ for all } f_i(x) \text{ elements of } f(x), \text{ then}$$
$$\mathcal{H}\{af_1(x) + bf_2(x))\} = a\mathcal{H}\{f_1(x)\} + b\mathcal{H}\{f_2(x)\} = ag_1(x) + bg_2(x)$$

where a and b are arbitrary constants. As we will see, these systems properties relate to most optical systems that we are interested in modeling.

5.2.2 Impulse response function and point source model

The simplest system mapping that can be done is to take only a single value input and determine the resulting output. The mathematical model used to symbolically describe the quantity of a single point input is the Dirac delta function [3]. The definitions and mathematical theory of the delta function is described by Lighthill [4] and summarized in *Principles of Optics*, appendix IV by Born and Wolf [5].

The delta function is defined through its properties as,

$$\delta(x) = 0 \text{ for all } x \neq 0$$
$$\int_{-\infty}^{+\infty} \delta(x)dx = 1$$

This definition also leads to the following property,

$$\int_{-\infty}^{+\infty} f(x)\delta(x - x_0)dx = f(x_0)$$

where $f(x)$ is a continuous function of x and x_0 is a value within the x domain. (This property is often called the sifting-property of the delta function.) Again, the use of the delta function is to symbolically describe the quantity of a function at a single point. Figure 5.1 shows plots of a one-dimensional delta function and a two-dimensional function. It is conventional to illustrate the result of the sifting-property by a line with height proportional to the value of the function at the coordinate, as in these cases, $f(x_0)$ and $f(x_0, y_0)$.

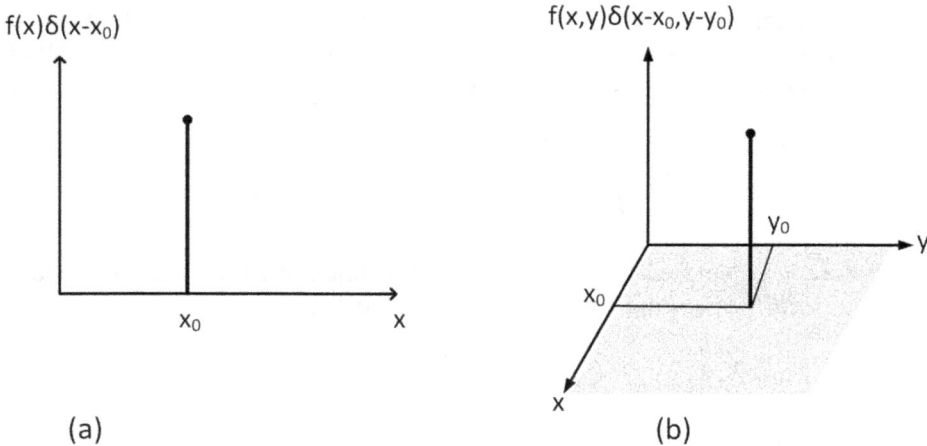

Figure 5.1. Plots of delta function representations for (a) one-dimensional function centered at a point x_0, and (b) two-dimensional function centered at the coordinate (x_0, y_0).

This representation of a delta function is also called an impulse function. The concept of impulse arises when describing system inputs in the time domain [6]. In the spatial domain the impulse function models our definition of an optical point source.

As a quick review of our earlier discussion in the introduction to systems, take a single point source as the input to a general linear system. (For simplicity, we are assuming only one-dimensional functions.) The point source input function is then defined at a general point w so that,

$$f(x) = \delta(x - w)$$

and the magnitude of the function is one. The output of a system with this input impulse is then,

$$g(x) = \mathcal{H}\{f(x)\} = \mathcal{H}\{\delta(x - w)\} \equiv h(x, w)$$

Thus, the new function $h(x, w)$ is defined as the impulse response function or system function and is the response of the linear system to a unit impulse input at any general point where $x = w$.

5.2.3 Convolution and the convolution theorem

In the linear systems formalism, we start with an input spatial function. Applying this approach to optical systems, this function is the mathematical model for a general object optical field or intensity function. Using the properties of the delta function, a general function $f(x, y)$ can be written as,

$$f(x, y) = \iint\limits_{-\infty}^{+\infty} f(x_0, y_0)\delta(x - x_0, y - y_0)dx_0 dy_0$$

5-3

where the integration is over the (x_0, y_0) input plane. The resulting output function $g(x, y)$ is obtained by operating on $f(x, y)$ with the system operator, $\mathcal{H}$. Note that $\mathcal{H}$ only affects functions of x and y and treats the x_0 and y_0 variables as constants. Thus,

$$g(x, y) = \int\!\!\!\int_{-\infty}^{+\infty} f(x_0, y_0)\mathcal{H}\{\delta(x - x_0, y - y_0)\}dx_0dy_0$$

From before we identified the impulse response function as the function resulting from the operator operating on a delta function so that,

$$g(x, y) = \int\!\!\!\int_{-\infty}^{+\infty} f(x_0, y_0)h(x - x_0, y - y_0)dx_0dy_0$$

This integral function is a form called the convolution. It states that the output function of a linear system is completely characterized by the convolution of the input with the system impulse response.

For simplicity, in one dimension we define the convolution integral as,

$$g(x) = \int_{-\infty}^{+\infty} f(w)h(x - w)dw$$

where $h(x - w) \equiv \mathcal{H}\{\delta(x - w)\}$. The parameter w used in the above integral is a dummy variable of integration. The convolution operation is also mathematically written in a shorthand operator notation as,

$$g(x) = f(x) \circledast h(x)$$

So far all of the analysis we have done has been in the spatial domain. From chapter 1 we know that we can also express the field functions in the spatial frequency domain. This was done using the Fourier transform defined by the integral relation,

$$F(f_x) = \int_{-\infty}^{+\infty} f(x)e^{-j2\pi f_x x}dx \equiv \mathcal{F}\{f(x)\}$$

where $F(f_x)$ is the Fourier transform of the function $f(x)$, and the variable f_x is the spatial frequency. Since the transformation function is obtained through an integral operation, we have also used operator notion defining the Fourier transform operator as $\mathcal{F}$.

Using this definition, we can also obtain the spatial frequency function of the output, $g(x)$ by taking the Fourier transform, so that,

$$G(f_x) = \int_{-\infty}^{+\infty} g(x)e^{-j2\pi f_x x}dx \equiv \mathcal{F}\{g(x)\}$$

But this is the Fourier transform of the convolution of the input function with the system function, or,

$$G(f_x) = \mathcal{F}\{f(x) \circledast h(x)\}$$

If we define $H(f_x)$ as the Fourier transform of the system function $h(x)$, $H(f_x) \equiv \mathcal{F}\{h(x)\}$, then it can be shown that,

$$G(f_x) = F(f_x)H(f_x)$$

The proof of this result is known as the convolution theorem. This is often stated as, the Fourier transform of a convolution between two functions is the product of the Fourier transforms of each individual function. In optical systems this can be useful when representing a function in the spatial frequency domain.

5.2.4 Coherent and incoherent imaging system functions

In chapter 2 we modeled an object as a combination of many point sources each emitting rays of light. Using geometrical optics principles, we traced rays of light through various optical systems to determine image locations. This can also be done using a wave optics approach. In addition, we will find that the results of wave optics provide us with information about the quality of images formed as well as locations.

An optical imaging system is an example of the mapping of one single point in the object plane (input plane) to points in the image plane (output plane) through the system impulse response function. As discussed in the section on coherence, assumptions must always be made related to the sources used and their degree of coherence. If the amplitude and phase of the point objects must be considered, then the input and output functions are linear in amplitude and represent field functions. The impulse response function for this type of system is a coherent system function and describes what is called coherent imaging. Incoherent imaging arises when spatially adjacent points in the object and image have no phase relationship. In this situation, a linear systems approach can also be used except that the input and output functions now represent a light flux proportional to the absolute square of the field amplitude. The impulse response function for an incoherent imaging system describes the incoherent mapping between all object points to points in the image plane [2].

Figure 5.2 illustrates the imaging concept for both coherent and incoherent systems where a single point source at a point (x_{obj}, y_{obj}) in the (x_0, y_0) plane is imaged in the (x_1, y_1) plane about point (x_{img}, y_{img}). The physical optical system is comprised of both a combination of optical elements and the spaces between elements where waves propagate. The character of the function in the image plane is the result of the system function operating on the point source input, or the impulse response function $g(x_1, y_1) = h(x_1 - x_{img}, y_1 - y_{img})$.

Of course, ideally the image would also be a point source. Since the function $g(x_1, y_1)$ is not a delta function, the effect of the optical system is to spread out the light from the point source image location by the impulse response function.

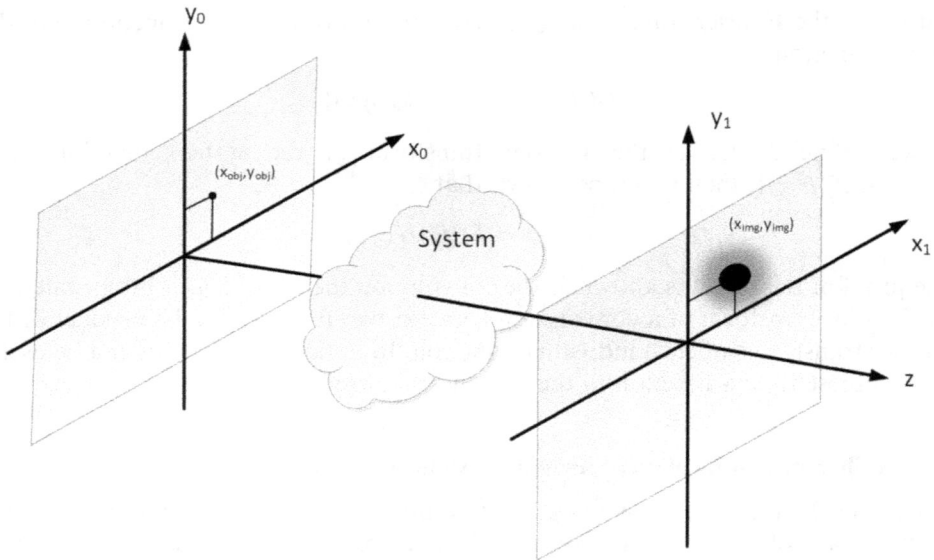

Figure 5.2. Point source imaging by a general optical system. The input function is, $\delta(x_0 - x_{obj}, y_0 - y_{obj})$ resulting in an output function that is the system impulse response function $g(x_1, y_1) = h(x_1 - x_{img}, y_1 - y_{img})$.

For coherent systems this function is called the amplitude point spread function (PSF) and for incoherent systems the function is the intensity PSF [1, 2].

Examples and calculations of the impulse response functions for both coherent and incoherent imaging systems will be given in a later section.

5.3 Diffraction

As light propagates through optical systems it diffracts. Optical diffraction is any deviation from geometrical optics predictions caused by the wave nature of light. The historical foundation of the theory of diffraction has been well documented by many authors [5–8]. However, we are more interested here in the implications of diffraction on optical systems design. The major consequence of diffraction phenomena is that it imposes limits on the overall performance of optical systems [2]. Two such effects caused by diffraction are limited resolution and degradation of image quality [1].

5.3.1 Fresnel diffraction integral and system function of space

We know from Maxwell's equations that the wave field arriving at a plane a distance z from a general point source is a spherical wave. This geometry is shown in figure 5.3. We want to compute the value of the field from contributions at all points in the input plane. The magnitude of the field u_1, at a general point P in plane 1 is of the form,

$$du_1 = \frac{u_0}{j\lambda\rho}e^{j(\vec{k}\cdot\vec{p})}dx_0 dy_0$$

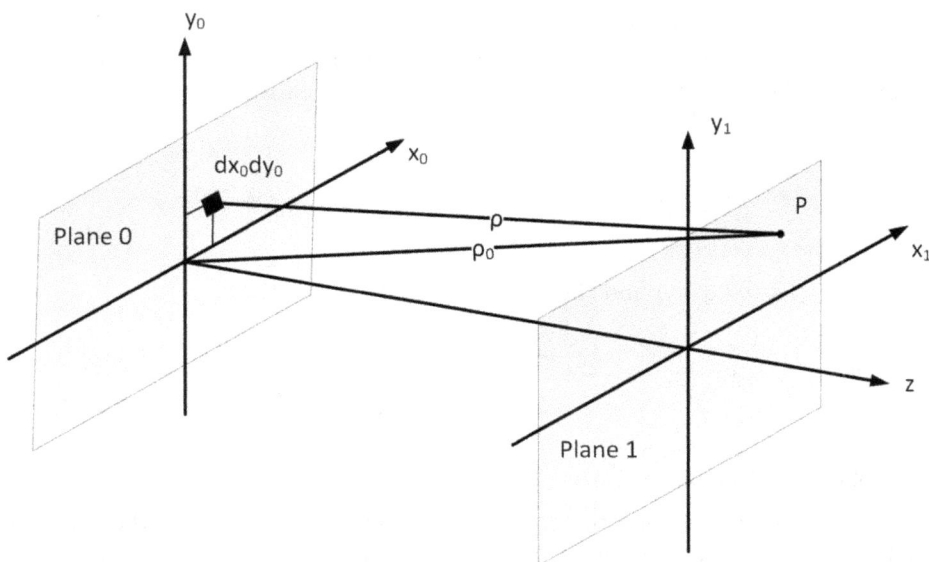

Figure 5.3. Point object in plane 0 with a wave propagating to point P in plane 1. The distance from the origin of plane 0 to point P is ρ_0.

where u_0 is the value of the field at a point in plane 0, $\vec{k}$ is the wave propagation vector, and $\vec{\rho}$ is the vector from the source point to the point in the output plane. Remember that, $|k| = \frac{2\pi}{\lambda}$. Also shown in figure 5.3 is a value ρ_0 which is the distance from the origin of the input plane to point P in the output plane.

From this geometry we can write,

$$\rho^2 = z^2 + (x_1 - x_0)^2 + (y_1 - y_0)^2 \text{ and } \rho_0^2 = z^2 + x_1^2 + y_1^2$$

We also make the approximations that we restrict the observation point P to be relatively near the z-axis, $z \gg \sqrt{x_1^2 + y_1^2}$, and $\rho_0 \approx z$. Expanding ρ in a Taylor series and keeping the first two terms leads to,

$$\rho \cong z\left\{1 + \frac{1}{2}\left[\left(\frac{x_1 - x_0}{z}\right)^2 + \left(\frac{y_1 - y_0}{z}\right)^2\right]\right\}$$

The distance ρ is a quadratic form in x and y. Now substituting this into the equation for the field at point P results in,

$$du_1(x_1, y_1) = \frac{u_0(x_0, y_0)}{j\lambda z}e^{jkz}\exp\left\{+\frac{jkz}{2}\left[\left(\frac{x_1 - x_0}{z}\right)^2 + \left(\frac{y_1 - y_0}{z}\right)^2\right]\right\}dx_0 dy_0$$

And the value of the field at a general point P in plane 1 is the Fresnel diffraction integral,

$$u_1(x_1, y_1) = \frac{1}{j\lambda z}e^{jkz}\iint\limits_{-\infty}^{+\infty} u_0(x_0, y_0)\exp\left\{+\frac{jkz}{2}\left[\left(\frac{x_1 - x_0}{z}\right)^2 + \left(\frac{y_1 - y_0}{z}\right)^2\right]\right\}dx_0 dy_0$$

If we assign a general function $f(x_0, y_0)$ as the light field leaving plane 0 due to any incident light impinging on plane 0 and the amplitude transmittance function of any aperture in plane 0 then the form of the Fresnel diffraction integral can be recognized as a convolution operation [2, 9].

$$g(x_1, y_1) = \int\!\!\!\int_{-\infty}^{+\infty} f(x_0, y_0)h(x_1 - x_0, y_1 - y_0)dx_0dy_0$$

where $g(x_1, y_1) \equiv u_1(x_1, y_1)$ and the system function is,

$$h(x_1 - x_0, y_1 - y_0) = \frac{1}{j\lambda z}e^{jkz} \exp\left\{ +\frac{jkz}{2}\left[\left(\frac{x_1 - x_0}{z}\right)^2 + \left(\frac{y_1 - y_0}{z}\right)^2\right]\right\}$$

5.3.2 Spherical phase wavefronts

Suppose that we want to find the output field from a point source at the origin. We model the point source function in plane 0 as an impulse function, or delta function. Using the properties of a delta function, the output field in plane 1 is,

$$u_1(x_1, y_1) = \frac{A}{j\lambda z}e^{j\frac{2\pi}{\lambda}z} \exp\left\{ +\frac{j\pi}{\lambda z}(x_1^2 + y_1^2)\right\}$$

where A is a constant proportional to the light field. We interpret the form of this field as a wave of wavelength λ propagating in the $+z$-direction with a wavefront that has a quadratic phase in any x_1 and y_1 plane. A quadratic wavefront is a spherical wave. Since the power per unit area (irradiance) in the output plane is proportional to the absolute square of the field through the Poynting vector we see that,

$$|u_1|^2 \sim \frac{1}{z^2}$$

This is the commonly known inverse square law.

5.3.3 Quadratic phase function

Returning to the quadratic phase term in the Fresnel diffraction integral, we find,

$$\exp\left\{ +\frac{jkz}{2}\left[\left(\frac{x_1 - x_0}{z}\right)^2 + \left(\frac{y_1 - y_0}{z}\right)^2\right]\right\} = \exp\left\{ +\frac{j\pi}{\lambda x}[(x_1 - x_0)^2 + (y_1 - y_0)^2]\right\}$$

Expanding this equation and grouping like terms results in a product of three exponential functions,

$$\exp\left[+\frac{j\pi}{\lambda z}(x_1^2 + y_1^2)\right]\exp\left[+\frac{j\pi}{\lambda z}(x_0^2 + y_0^2)\right]\exp\left[-\frac{j2\pi}{\lambda z}(x_0x_1 + y_0y_1)\right]$$

If we define two new variables, $f_x \equiv \frac{x_1}{\lambda z}$ and $f_y \equiv \frac{y_1}{\lambda z}$ as the spatial frequency variables for the output plane coordinates (plane 1), then this equation becomes,

$$\exp\left[+\frac{j\pi}{\lambda z}(x_1^2 + y_1^2)\right]\exp\left[+\frac{j\pi}{\lambda z}(x_0^2 + y_0^2)\right]\exp[-j2\pi(x_0 f_x + y_0 f_y)]$$

The third exponential term is identical to the Fourier transform kernel function in the definition of the two-dimensional Fourier transform,

$$F(f_x, f_y) = \int\!\!\!\int\limits_{-\infty}^{+\infty} f(x, y)e^{-j2\pi(f_x x + f_y y)}dxdy$$

Also, defining a new function, the quadratic phase factor (or signal) as [6, 10],

$$Q(x, y; z) \equiv \exp\left[+\frac{j\pi}{\lambda z}(x^2 + y^2)\right]$$

The Fresnel diffraction integral can be simplified [2, 9, 10],

$$u_1(x_1, y_1) = \frac{1}{j\lambda z}e^{j\frac{2\pi}{\lambda}z}Q(x_1, y_1; z)\int\!\!\!\int\limits_{-\infty}^{+\infty} f(x_0, y_0)Q(x_0, y_0; z)$$
$$\times \exp[-j2\pi(x_0 f_x + y_0 f_y)]dx_0 dy_0$$

And using operator notation, the Fresnel diffraction integral can also be written as,

$$u_1(x_1, y_1) = \frac{1}{j\lambda z}e^{j\frac{2\pi}{\lambda}z}Q(x_1, y_1; z)\mathcal{F}\{f(x_0, y_0)Q(x_0, y_0; z)\}$$

The use of quadratic phase functions can assist in performing diffraction calculations and interpreting their results. This is due to the particular mathematical properties of these functions. A table of these functions and their properties are provided in the appendix and will be used throughout.

5.3.4 Fraunhofer diffraction

If z is large enough in comparison to the wavelength of light and the region of interest (ROI) in the input plane, a further approximation can be made for the diffraction field in the output plane (plane 1). This approximation criterion is,

$$\frac{(x_0^2 + y_0^2)}{\lambda z} \ll 1$$

In other words, the ratio of the area of interest in plane 0 to the λz product must be small. This is often referred to as the far-field approximation. Applying this approximation means that the quadratic phase function $Q(x_0, y_0; z)$ is approximately one. Thus, the resulting diffraction field in plane 1 is,

$$u_1(x_1, y_1) = \frac{1}{j\lambda z}e^{j\frac{2\pi}{\lambda}z}Q(x_1, y_1; z)\mathcal{F}\{f(x_0, y_0)\}$$

Or, in the Fraunhofer diffraction limit, the electric field of the diffracted wave is proportional to the Fourier transform of the field function in the input plane. Again, in applying the Fourier transform operator the spatial frequency variables are, $f_x \equiv \frac{x_1}{\lambda z}$ and $f_y \equiv \frac{y_1}{\lambda z}$.

As a simple test of this result, suppose that the input field function is a point source so that

$$f(x_0, y_0) = A\delta(x_0, y_0)$$

where A is the amplitude of the point source field. The Fourier transform of the function delta function is unity so that

$$\mathcal{F}\{f(x_0, y_0)\} = A$$

Resulting in the far-field diffraction function,

$$u_1(x_1, y_1) = \frac{A}{j\lambda z}e^{j\frac{2\pi}{\lambda}z}Q(x_1, y_1; z)$$

Thus, the Fraunhofer diffraction field at distance z from a point source is a spherical wave. But as z approaches infinity the quadratic phase factor approaches a value of unity. This means that the wave is essentially a plane wave. However, as seen before the irradiance distribution of the wavefront still follows the inverse square law.

Another example of the application of the linear systems approach to diffraction that of a single slit. In this case, the input field function for a slit of width, a, along the x-axis, height b along the y-axis, and wave amplitude, A, can be written as [2, 11]

$$f(x_0, y_0) = A \operatorname{rect}\left(\frac{x_0}{a}, \frac{y_0}{b}\right)$$

Following the same process used above we calculate the Fourier transform of $f(x_0, y_0)$ and obtain,

$$\mathcal{F}\{f(x_0, y_0)\} = A|a||b|\operatorname{sinc}\left(\frac{ax_1}{\lambda z}\right)\operatorname{sinc}\left(\frac{by_1}{\lambda z}\right)$$

This is the familiar function describing the Fraunhofer diffraction field of a rectangular slit [2, 8, 11].

5.3.5 Interference-system view

Interference phenomena results as a consequence of the wave nature of light and the superposition of coherent waves. We saw previously that when spatial phase is important then we will use spatial functions of the complex electromagnetic field amplitude and assume complete spatial coherence. Using the linear systems approach to diffraction we can analyze the interaction between two electromagnetic wave fields resulting in classical interference.

We found that we could model a single point source as a delta function and that the wavefront propagating from this point source results in a spherical wave.

Suppose that the input function at an input plane is composed of two identical point sources separated by a general distance $2a$, each with field amplitude A. Again, for simplicity, only a one-dimensional function is used. The input plane function for the electric fields of each point source is,

$$f(x_0) = A\delta(x_0 - a) + A\delta(x_0 + a)$$

This function is shown in figure 5.4. So, the input plane contains two displaced point sources. In classical optics this is the functional representation for a double slit with a very narrow slit width.

Now, using the linear system approach under the far-field approximation, the resulting field in the output plane (x_1 coordinate) a distance z away from the input is,

$$u_1(x_1) = \frac{A}{j\lambda z}e^{j\frac{2\pi}{\lambda}z}e^{j\frac{\pi}{\lambda z}x_1^2}\mathcal{F}\{\delta(x_0 - a) + \delta(x_0 + a)\}$$

Resulting in,

$$u_1(x_1) = \frac{A}{j\lambda z}e^{j\frac{2\pi}{\lambda}z}e^{j\frac{\pi}{\lambda z}x_1^2}\left\{\exp\left[+\frac{j2\pi}{\lambda z}ax_1\right] + \exp\left[-\frac{j2\pi}{\lambda z}ax_1\right]\right\}$$

Remembering that, $f_x \equiv \frac{x_1}{\lambda z}$ in the Fourier transform integral definition, this equation reduces to a single cosine function,

$$u_1(x_1) = 2\frac{A}{j\lambda z}e^{j\frac{2\pi}{\lambda}z}e^{j\frac{\pi}{\lambda z}x_1^2}\cos\left[+2\pi\frac{a}{\lambda z}x_1\right]$$

In interference we are interested in the irradiance pattern across the output plane, $E_e(x_1)$. We can obtain this function by taking the absolute square of the field function, or,

$$E_e(x_1) = 4E_e(0)\cos^2\left[+2\pi\frac{a}{\lambda z}x_1\right]$$

where $E_e(0)$ is the irradiance that would be measured from a single source. This function is the same as we obtained in classical optics from superposition of two, point source waves for the fringe pattern observed in a Young's double slit experiment [7, 8]. Note that our result also predicts the factor of four at the

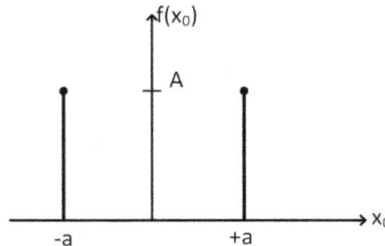

Figure 5.4. Function representing two identical point sources in the input plane of an optical system (two displaced delta functions).

$x_1 = 0$ position in the output plane between the two, point sources arising from constructive interference. This relationship is identical to the form found earlier using the addition of plane waves.

5.4 Fourier optics

In chapter 1 we saw that input functions to optical systems can be described in both the time/space domain as well as the temporal/spatial frequency domain. The domains are related through the Fourier transform operation. Our systems approach to the propagation of light waves also resulted in applying a Fourier transform of the input field to obtain the output field. In fact, it has long been recognized that many phenomena in wave optics can be described using some type of Fourier analysis [2, 6, 10, 12]. This recognition is the origin of the term Fourier optics. In addition to simply describing physical optics phenomena, Fourier optics analysis can predict image quality through calculation and measurement of the OTF of a system [1].

5.4.1 Phase transformation function of a lens

The field propagating from a plane containing an optical point source was computed in the previous section using Fresnel diffraction. The resulting wavefront was shown to be a spherical wave centered at the point source position. Suppose the spherical wave from a point source is collected by a thin lens. From ray optics and the laws of geometrical optics we know that if the point source is located at the front focal point of the lens then the rays from the point source will be refracted and travel parallel to the optical axis. In wave optics this means that the spherical wavefront emitted from the point source will be converted to a plane wave on leaving the plane containing the lens. This concept is shown in figure 5.5.

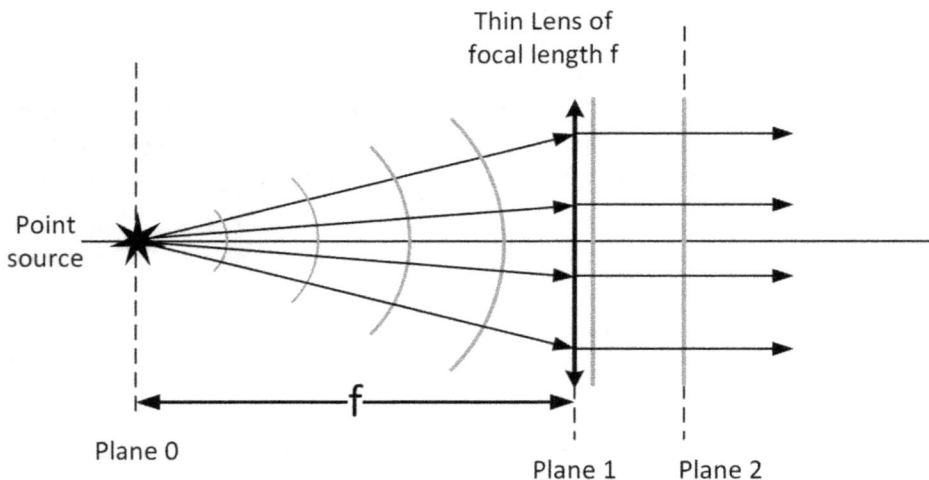

Figure 5.5. Light from a point source collected by a thin lens showing both rays and wavefronts.

Thus, the field leaving plane 1, u_1' located just after the thin lens (also in plane 1) is the product,

$$u_1'(x_1, y_1) = u_1(x_1, y_1)u_{\text{lens}}(x_1, y_1)$$

where u_{lens} is the phase transformation introduced by the thin lens. Using results from above for the field incident on plane 1 after propagating over distance $z = +f$ becomes,

$$\frac{A}{j\lambda f}e^{j\frac{2\pi}{\lambda}f} = \frac{A}{j\lambda f}e^{j\frac{2\pi}{\lambda}f}\exp\left\{+\frac{j\pi}{\lambda f}(x_1^2 + y_1^2)\right\}u_{\text{lens}}(x_1, y_1)$$

Again, A is the field amplitude of the point source in plane 0.

Solving for the lens phase transformation function we find,

$$u_{\text{lens}}(x_1, y_1) = \exp\left\{-\frac{j\pi}{\lambda f}(x_1^2 + y_1^2)\right\} \equiv Q(x_1, y_1; -f)$$

Therefore, the phase transformation of an ideal thin lens of focal length $+f$ is simply a quadratic phase factor function.

5.4.2 Coherent imaging

Applying a wave optics approach to an optical imaging system is an example of the application of the linear systems formalism. In this situation the input and output functions represent the object and image field spatial distributions. As before, we take $u_0(x_0, y_0)$ as the input or object field in plane 0, and $u_1(x_1, y_1)$ to be the field in the output or image plane. We relate the image field and the object field using the convolution with the system impulse response function h. This is called coherent imaging since the input and output functions are proportional to the complex light field where each point in the object has a relative phase relationship.

$$u_1(x_1, y_1) = \int\!\!\!\int_{-\infty}^{+\infty} u_0(x_0, y_0)h(x_1 - x_0, y_1 - y_0)dx_0dy_0 = u_0 \circledast h$$

As shown in figure 5.2, the effect of the system on a point source object is to spread out the image through a PSF. In order to use this relation to compute the output field both the input function and the system impulse response function must be known completely, which is not always the case. Thus, the convolution approach is important from a theoretical understanding of various optical effects within a system but not always for computations.

Another approach is to directly compute the field at planes successively through the system. We are able to do this because we know the system function for free space as obtained from Fresnel diffraction. Since a lens introduces a phase transformation on the propagating wave, waves from a point leaving the object plane (input plane) must be transformed by the entire system to result in the corresponding image point in the image plane (output plane). To do this we combine the Fresnel diffraction propagation with the phase transformation introduced by the lens successively from plane-to-plane through the system to arrive at the condition for imaging.

As a simple example, take a point source object at the origin of the input plane of a lens system as shown in figure 5.6. We will then follow the wave propagation through the system (comprised of a large size single thin lens) to the output/image plane and analyze the resulting field.

First, we write down the field function at the thin lens plane (x, y) using Fresnel diffraction over the distance z_0 for a general input function $f(x_0, y_0)$, to obtain,

$$u(x, y) = \frac{1}{j\lambda z_0} e^{j\frac{2\pi}{\lambda} z_0} Q(x, y; z_0) \mathcal{F}\{f(x_0, y_0) Q(x_0, y_0; z_0)\}$$

The field function for a point source at the origin is,

$$f(x_0, y_0) = A\delta(x_0, y_0)$$

resulting in the field function just encountering the lens in the (x, y) plane.

$$u(x, y) = \frac{A}{j\lambda z_0} e^{j\frac{2\pi}{\lambda} z_0} Q(x, y; z_0)$$

The field $u'(x, y)$ leaving the (x, y) plan is the product of the incoming field $u(x, y)$ and the phase transformation introduced by the thin lens of focal length $+f$, or,

$$u'(x, y) = \frac{A}{j\lambda z_0} e^{j\frac{2\pi}{\lambda} z_0} Q(x, y; z_0) Q(x, y; -f)$$

Now, to obtain the field in the image plane we must propagate the field $u'(x, y)$ again using Fresnel diffraction, or,

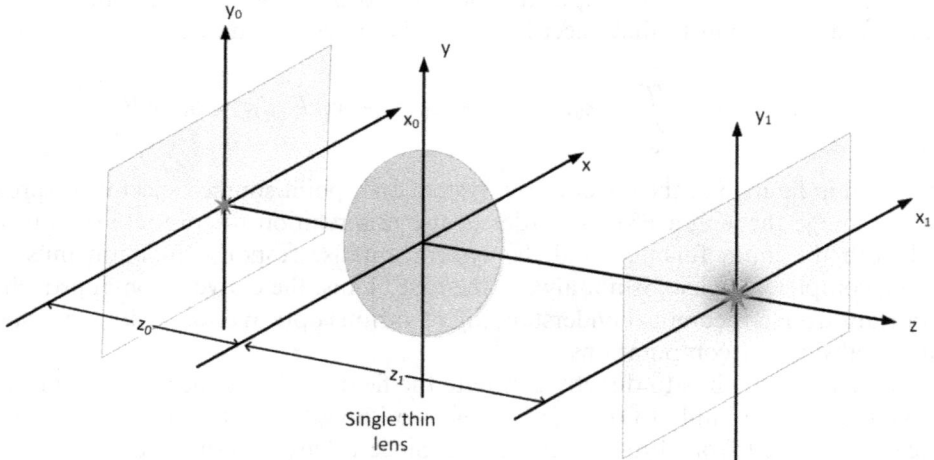

Figure 5.6. Imaging a point source by a thin lens using wave propagation properties. The input function is a point source object in plane 0, coordinates (x_0, y_0), the lens is in a general intermediate plane specified by coordinates (x, y), and the image plane (or output plane) is specified by coordinates (x_1, y_1).

$$u_1(x_1, y_1) = \frac{1}{j\lambda z_1} e^{j\frac{2\pi}{\lambda} z_1} Q(x_1, y_1; z_1) \mathcal{F}\{u'(x, y)q(x, y; z_1)\}$$

Substituting the value of u' into this equation gives,

$$u_1(x_1, y_1) = \frac{-A}{\lambda^2 z_0 z_1} e^{j\frac{2\pi(z_0+z_1)}{\lambda}} Q(x_1, y_1; z_1) \mathcal{F}\{Q(x, y; z_0)Q(x, y; -f)Q(x, y; z_1)\}$$

Now examine the product of the three quadratic phase functions under the Fourier transform operator. Using properties of the quadratic phase factors the Fourier transform portion of the function reduces to,

$$\mathcal{F}\{Q(x, y; z_0)Q(x, y; -f)Q(x, y; z_1)\} = \mathcal{F}\left\{\exp\left[\frac{j\pi}{\lambda}(x^2 + y^2)\right]\left[\frac{1}{z_0} + \frac{1}{z_1} - \frac{1}{f}\right]\right\}$$

Notice that when $z_0 = -t$, the object distance and when $z_1 = t'$ the image distance that,

$$\frac{1}{t'} - \frac{1}{t} - \frac{1}{f} = 0$$

which is just the statement of the imaging condition we obtained from ray optics with thin lenses. When we apply the imaging condition, the Fourier transform function becomes a delta function since the argument of the Fourier transform is unity. But the Fourier transform operation results in a delta function with variables of the spatial frequency $f_x \equiv \frac{x_1}{\lambda z_1}$ and $f_y \equiv \frac{y_1}{\lambda z_1}$. Therefore, the field in the output plane under the imaging condition is,

$$u_1(x_1, y_1) = \frac{-A}{\lambda^2 z_0 z_1} e^{j\frac{2\pi(z_0+z_1)}{\lambda}} Q(x_1, y_1; z_1)\delta\left(\frac{x_1}{\lambda z_1}, \frac{y_1}{\lambda z_1}\right)$$

This field is a point source at the origin of the image plane but in a scaled coordinate. Using properties of the delta function this term can be rewritten as,

$$\delta\left(\frac{x_1}{\lambda z_1}, \frac{y_1}{\lambda z_1}\right) = \lambda^2 t^2 \delta\left(-\frac{t}{t'}x_1, -\frac{t}{t'}y_1\right) = \lambda^2 t^2 \delta\left(-\frac{x_1}{M}, -\frac{y_1}{M}\right)$$

where we recognize that the magnification of the system is $M = t'/t$. Thus, the field function in the image plane is finally,

$$u_1(x_1, y_1) = A e^{j\frac{2\pi(z_0+z_1)}{\lambda}} Q(x_1, y_1; z_1)\frac{1}{M}\delta\left(-\frac{x_1}{M}, -\frac{y_1}{M}\right)$$

which is the field in the image plane as predicted by geometrical optics with proper scaling by the system magnification. Note the negative sign in the coordinate also predicts that the image will be inverted for a positive focal length lens that was assumed in this case. This same result can also be obtained using a linear systems/convolution approach instead of the Fresnel diffraction method used here using the coherent impulse response function [2, 10].

5.4.3 Incoherent imaging

From electromagnetic theory the flow of energy is described by the complex Poynting vector. In the image plane (or when the light falls on a detector) the irradiance at each point is a time average of the incident electromagnetic fields. When we only consider sources where the spatial phases are uncorrelated this is called incoherent imaging.

Following a similar linear systems procedure used for coherent imaging we take $u_0(x_0, y_0, t)$ as the input or object field in plane 0, and $u_1(x_1, y_1, t)$ to be the field in the output or image plane. In this case the time dependence of the field functions are shown explicitly since we must perform a time average to obtain the light flux. Thus, irradiance in the image plane (x_1, y_1) is,

$$E_{e1}(x_1, y_1) = \langle u_1(x_1, y_1, t)u_1^*(x_1, y_1, t) \rangle$$

where the brackets represent a time average operation. Likewise, the irradiance in the object plane is,

$$E_{e0}(x_0, y_0) = \langle u_0(x_0, y_0, t)u_0^*(x_0, y_0, t) \rangle$$

Now substituting the integral function from above for u_1 but using dummy variables of integration within each integral in order to keep track of the functional variables we find,

$$E_{e1}(x_1, y_1) = \left\langle \iint\limits_{-\infty}^{+\infty} u_0(w, s, t)h(x_1 - w, y_1 - s)dwds \right.$$
$$\left. \times \iint\limits_{-\infty}^{+\infty} u_0^*(w', s', t)h^*(x_1 - w', y_1 - s')dw'ds' \right\rangle$$

Note that the only time dependences are associated with the object field functions. So, rewriting this equation results in,

$$E_{e1}(x_1, y_1) = \int\limits_{-\infty}^{+\infty} \int\limits_{-\infty}^{+\infty} \int\limits_{-\infty}^{+\infty} \int\limits_{-\infty}^{+\infty} h(x_1 - w, y_1 - s)h^*(x_1 - w', y_1 - s')$$
$$\times \langle u_0(w, s, t)u_0^*(w', s', t) \rangle dwdsdw'ds'$$

But this simply reduces to,

$$E_{e1}(x_1, y_1) = \iint\limits_{-\infty}^{+\infty} |h(x_1 - x_0, y_1 - y_0)|^2 E_{e0}(x_0, y_0)dx_0dy_0$$

As seen, this equation is again in the form of a convolution integral with the input function as the irradiance distribution of the object in plane 0 and the output function as the irradiance distribution in the image plane. This also shows that for incoherent imaging, the incoherent impulse response function is just the absolute square of the coherent impulse response function. Therefore, once the complex coherent system function is determined, the incoherent system function can be computed.

For the simple example of point source imaging above, the image irradiance distribution is just

$$E_{el}(x_1, y_1) = \frac{A^2}{M^2} \delta\left(-\frac{x_1}{M}, -\frac{y_1}{M}\right)$$

Or the image irradiance predicted by geometrical optics for imaging a point source by a large thin lens.

5.4.4 Pupil function and resolution limits

Earlier we assumed that the lens used in forming the image of the object was a large diameter thin lens similar to the assumptions made when using ray optics. Using wave optics, we have the opportunity to analyze systems in which the lens has a finite extent as well as phase relations that deviate from the ideal lens phase trans-formation. This is accomplished by introducing a new function called the pupil function that describes these effects. However, we will still assume that the lens is thin and that the pupil function is described in the thin lens plane. Thus, a more general lens phase transformation function in an (x, y) lens plane can be written as,

$$u_{lens}(x, y) = P(x, y)Q(x, y; -f)$$

where $P(x, y)$ is the pupil function. Since this is a field function, the pupil function can in fact be complex and describe not only amplitude effects, lens sizes and apodization, but additional phase effects or aberrations [2]. A general form for the pupil function would be,

$$P(x, y) = A(x, y) \exp\left[j\frac{2\pi}{\lambda} W(x, y)\right]$$

where $A(x, y)$ is the amplitude function and $W(x, y)$ is a function representing the wave aberrations that may exist in a particular system [2]. While this is important to realize, a detailed analysis of systems with apodized pupils or specific aberrations is beyond the scope of our discussion.

A typical singlet lens with diameter d and focal length $+f$ would have a pupil function specified by the circ(r) function with $r^2 = x^2 + y^2$, or

$$P(x, y) = \text{circ}\left(\frac{r}{d/2}\right)$$

To examine the effects of this simple pupil function, we consider a plane wave incident on a single lens and examine the field in the back focal plane. This system is illustrated in figure 5.7. Take the incident field on the lens to be a plane wave of amplitude A. The field $u'(x, y)$ just behind the lens is then simply,

$$u'(x, y) = A \, \text{circ}\left(\frac{r}{d/2}\right) Q(x, y; -f)$$

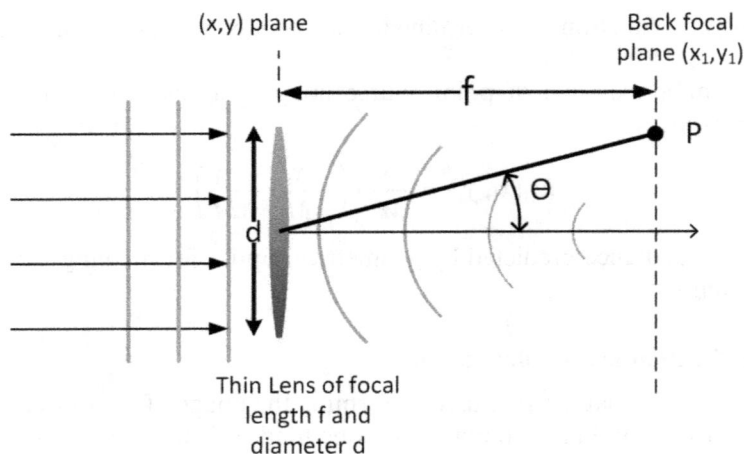

(x,y) plane

Back focal
plane (x_1, y_1)

f

P

θ

d

Thin Lens of focal
length f and
diameter d

Figure 5.7. Field in the back focal length of a lens with diameter d, focal length $+f$ with an incident plane wave. A general point in the back focal plane at point P with coordinate (x_1, y_1) can also be identified by an angle θ to that point because of circular symmetry. Because of the pupil, the wavefronts behind the lens are not exactly spherical but do converge toward the back focal plane.

To determine the field in the back focal plane we apply the Fresnel transformation of the $u'(x, y)$ function over the focal length distance f. This results in a field function in the (x_1, y_1) plane of,

$$u_1(x_1, y_1) = \frac{A}{j\lambda f}e^{+j\frac{2\pi}{\lambda}f}Q(x_1, y_1; +f)\mathcal{F}\left\{\text{circ}\left(\frac{r}{d/2}\right)Q(x, y; -f)Q(x, y; +f)\right\}$$

Note that the product of the two quadratic phase factors within the Fourier transform operator are unity. Using the Fourier transform of the $\text{circ}(r)$ function gives,

$$u_1(x_1, y_1) = \frac{A}{j\lambda f}e^{+j\frac{2\pi}{\lambda}f}Q(x_1, y_1; +f)\left[\frac{d}{2}\frac{J_1(\pi\alpha d)}{\alpha}\right]$$

where the variable $\alpha^2 = f_x^2 + f_y^2$, and as usual $f_x \equiv \frac{x_1}{\lambda f}$ and $f_y \equiv \frac{y_1}{\lambda f}$ resulting in the final field of,

$$u_1(x_1, y_1) = \frac{Ad}{j2}e^{+j\frac{2\pi}{\lambda}f}Q(x_1, y_1; +f)\left[\frac{J_1\left(\frac{\pi d}{\lambda f}\sqrt{x_1^2 + y_1^2}\right)}{\sqrt{x_1^2 + y_1^2}}\right]$$

Using the angle θ shown in figure 5.7, and knowing that this system is circularly symmetric, we can define another variable $\gamma \equiv \frac{\pi d}{\lambda}\sin\theta = \frac{\pi d}{\lambda f}\sqrt{x_1^2 + y_1^2}$

Then the irradiance distribution in the back focal plane can be found to be,

$$E_e(x_1, y_1) = E_e(0)\left[\frac{2J_1(\gamma)}{\gamma}\right]^2$$

where $E_e(0,0)$ is the irradiance for angle $\theta = 0$ or on the axis at coordinates $(0,0)$. This irradiance distribution function is the same Airy disk function found in diffraction from circular apertures [7, 8]. Essentially, the distribution in the back focal plane of a lens of diameter d is the Fraunhofer diffraction pattern of the circular aperture of the lens and is symmetrical about the optical axis.

Figure 5.8 is a plot of the normalized Airy disk irradiance distribution function. This function is characterized by a central maximum surrounded by circular rings of minima and maxima. The first zero of the normalized function occurs at a value of $\gamma = 3.832$. From the above definition of γ we compute the approximate angle out to the first minima as,

$$\theta_1 = \frac{1.22 \lambda}{d}$$

often referred to as the angular half width of the radius of the central spot.

Basic optics shows us that this diffraction limits the ability of an optical system to resolve the images of two closely separated points [7, 8]. The angular separation is a statement of Rayleigh's criterion that describes a condition when two adjacent point sources are said to be just resolved. In other words, the maxima of the image of one point source overlaps the minima of the image of the other point source. Therefore, the pupil function is the critical physical element that limits the ultimate resolution of any optical system.

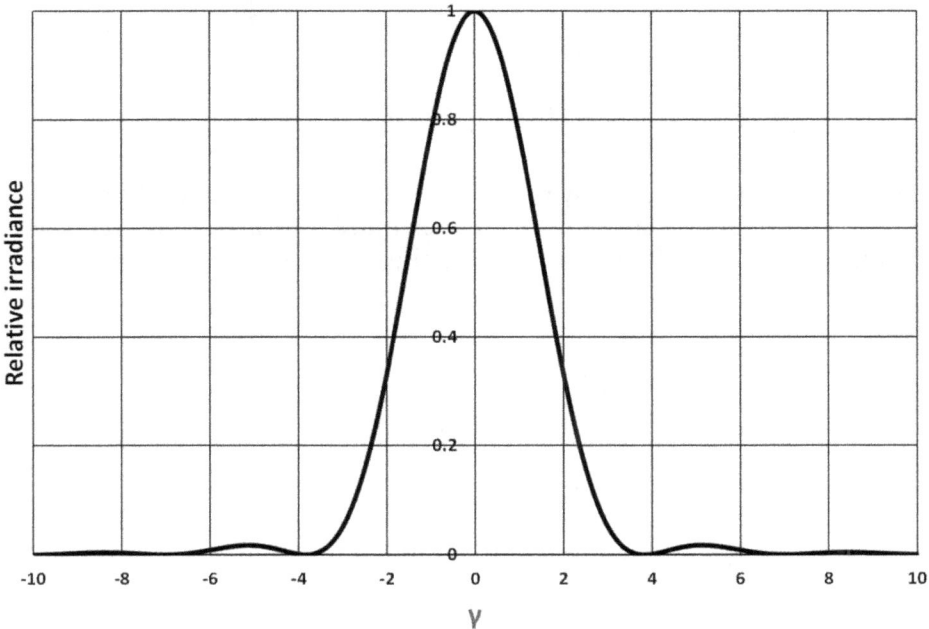

Figure 5.8. Plot of the normalized Airy disk irradiance distribution as a function of the general parameter, $\gamma \equiv \frac{\pi d}{\lambda}\sin\theta$.

5.4.5 Pupil function and the imaging system response function

Since the pupil function dictates resolution limits of an optical system, we will now return to the linear systems approach to imaging to obtain a more general relation for the response function of an imaging system. As before, we take $u_0(x_0, y_0)$ as the field in the object plane and $u_1(x_1, y_1)$ to be the field in the image plane. The notation used is shown in figure 5.9 with distance from the object plane to the lens plane is z_0 and the distance from the lens plane to the image plane is z_1. The lens of focal length $+f$ has a pupil function specified by $P(x, y)$.

Linear systems analysis allows us to compute the image field and the object field using a convolution with the system impulse response function. Now we will rewrite the field function in a way that allows us to extract the response function for this imaging system and relate it to the pupil function. In general,

$$u_1(x_1, y_1) = \int\!\!\int_{-\infty}^{+\infty} u_0(x_0, y_0)h(x_1 - x_0, y_1 - y_0)dx_0dy_0$$

From Fresnel diffraction over the distance z_0, the field just entering the lens plane is,

$$u(x, y) = \frac{1}{j\lambda z_0}e^{j\frac{2\pi}{\lambda}z_0}A(x, y; z_0)\mathcal{F}\{u_0(x_0, y_0)Q(x_0, y_0; z_0)\}$$

The field just leaving the lens plane is then simply,

$$u'(x, y) = u(x, y)P(x, y)Q(x, y; -f)$$

And the field on the image plane is just,

$$u_1(x_1, y_1) = \frac{1}{j\lambda z_1}e^{+j\frac{2\pi}{\lambda}z_1}Q(x_1, y_1; z_1)\mathcal{F}\{u'(x, y)Q(x, y; z_1))\}$$

Although these functions are easy to write down, they are algebraically complicated since we have Fourier transforms of Fourier transforms and must keep track of appropriate variables of integration. Substituting these relations into the last

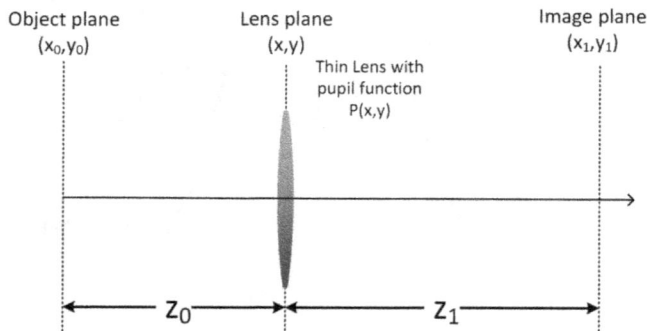

Figure 5.9. Single lens imaging system. The lens has pupil function $P(x, y)$.

equation for u_1, writing out the Fourier transforms in their integral representations, and applying the imaging condition results in,

$$u_1(x_1, y_1) = \frac{e^{+j\frac{2\pi}{\lambda}(z_1+z_1)}}{-\lambda^2 z_0 z_1} Q(x_1, y_1; z_1) \iint\limits_{-\infty}^{+\infty} u_0(x_0, y_0) Q(x_0, y_0; z_0)$$

$$\times \left[\iint\limits_{-\infty}^{+\infty} P(x, y) \exp\left\{ -j2\pi \left[x\left(\frac{x_0}{\lambda z_0} + \frac{x_1}{\lambda z_1}\right) + y\left(\frac{y_0}{\lambda z_0} + \frac{y_1}{\lambda z_1}\right) \right] \right\} dxdy \right] dx_0 dy_0$$

Note that the integral within the bracket is of the form of a Fourier transform of the pupil function but in a set of coordinates where $f_x = (\frac{x_0}{\lambda z_0} + \frac{x_1}{\lambda z_1})$ and $f_y = (\frac{y_0}{\lambda z_0} + \frac{y_1}{\lambda z_1})$. If this integral were evaluated, then the remaining integral is in the form of a convolution of the field u_0 and the remaining functions. To simplify this function, we define a set of reduced coordinates [2],

$$x_0' = Mx_0 \quad \text{and} \quad y_0' = My_0$$

where M is the magnification of the system as discussed in the coherent imaging section. With this addition, the field in the image plane can be written as a single convolution,

$$u_1(x_1, y_1) = \iint\limits_{-\infty}^{+\infty} h(x_1 - x_0', y_1 - y_0') \, u_g(x_0', y_0') dx_0' dy_0'$$

where u_g is the image predicted by geometrical optics. This is the more general statement of the earlier example of a point source. Therefore, the imaging system amplitude response function is,

$$h(x_1, y_1) = \frac{1}{\lambda^2 z_1^2} \iint\limits_{-\infty}^{+\infty} P(x, y) \exp\left\{ -j\frac{2\pi}{\lambda z_1}[xx_1 + yy_1] \right\} dxdy$$

As seen before, the amplitude response function is related to the Fourier transform of the pupil function [2].

5.4.6 Amplitude transfer function or coherent transfer function

From the analysis in the previous sections we saw that the convolution of the input function/object function with the appropriate impulse response function allows us to compute the output field for coherent imaging or the output irradiance for incoherent imaging systems. The convolution theorem tells us that the Fourier transform of the convolution of two functions is just the product of the Fourier transform of each function. For example, in coherent imaging, the Fourier transform of an object field yields a function that physically represents the spatial frequency characteristics of the object. So, let us now examine the various imaging system functions and their relationships in the spatial frequency domain.

First, we will start with coherent imaging. In this case we define the function, U_g, U_1, and H, which are the Fourier transforms of u_g, u_1, and h, respectively.

$$U_g(f_x, f_y) = \int\!\!\int_{-\infty}^{+\infty} u_g(x, y)\, \exp[-j2\pi(f_x x + f_y y)]dxdy \equiv \mathcal{F}\{u_g(x, y)\}$$

$$U_1(f_x, f_y) = \int\!\!\int_{-\infty}^{+\infty} u_1(x, y)\, \exp[-j2\pi(f_x x + f_y y)]dxdy \equiv \mathcal{F}\{u_1(x, y)\}$$

$$H(f_x, f_y) = \int\!\!\int_{-\infty}^{+\infty} h(x, y)\, \exp[-j2\pi(f_x x + f_y y)]dxdy \equiv \mathcal{F}\{h(x, y)\}$$

Thus, from the convolution theorem,

$$U_1(f_x, f_y) = H(f_x, f_y) U_g(f_x, f_y)$$

However, since we know the coherent imaging impulse response function $h(x_1, y_1)$, then

$$H(f_x, f_y) = \int\!\!\int_{-\infty}^{+\infty} \left[\frac{1}{\lambda^2 z_1^2} \int\!\!\int_{-\infty}^{+\infty} P(x, y)\, \exp\left\{ -j\frac{2\pi}{\lambda z_1}[xx_1 + yy_1] \right\} dxdy \right]$$
$$\times \exp[-j2\pi(f_x x_1 + f_y y_1)]dx_1 dy_1 \equiv \mathcal{F}\{h(x_1, y_1)\}$$

which simply reduces to,

$$H(f_x, f_y) = P(-\lambda z_1 f_x, -\lambda z_1 f_y)$$

Thus, the coherent transfer function or amplitude transfer function (ATF) $H(f_x, f_y)$ is just the pupil function expressed in a set of scaled spatial frequency coordinates. Therefore, if we know the pupil function then we can simply write down the transfer function for coherent imaging [2, 6, 10].

To better understand the physical meaning of the ATF we will return to an ideal system and then review an example of a particular system with a known input function. An ideal imaging system would have a pupil function $P(x, y) = 1$. This is the assumption made in most simple ray optics approaches to imaging. So, if the pupil function is unity for all values of the coordinates then, the ATF would also be 1 for all spatial frequencies, or, $H(f_x, f_y) = 1$. The physical meaning of this assumption is that spatial frequencies present in the object will be passed to the image.

5.4.7 Example: ATF of a square pupil

Now, let us examine a particular example system. In this case we will assume a square pupil function. This function is of the form,

$$P(x, y) = \text{rect}\left(\frac{x}{l}\right)\text{rect}\left(\frac{y}{l}\right)$$

A plot of the pupil function in the (x, y) coordinate plane is shown in figure 5.10 where the value is unity everywhere with the bounds of the square and zero outside.

From the analysis above we can easily obtain the ATF as,

$$H(f_x, f_y) = \text{rect}\left(\frac{\lambda z_1 f_x}{l}\right)\text{rect}\left(\frac{\lambda z_1 f_y}{l}\right)$$

In this equation the negative signs on the coordinates have been dropped since the function is symmetric about the coordinates. A cross-section of the (f_x, f_y) coordinate plane for this function is shown in figure 5.11(a) and a one-dimensional cut of the function along the f_x-axis is shown in figure 5.11(b).

For a numerical example, let us look at a one-dimensional case for simplicity. Suppose this lens is used to image an input field function of a cosine grating with a pitch of 50 lines/mm. The amplitude function of this grating is then,

$$f(x) = \cos\left[2\pi\left(50\frac{\text{lines}}{\text{mm}}\right)x\right] \quad \text{where } x \text{ in mm units}$$

The imaging system is a $2f$ unit magnification system with lens of focal length of 100 mm, square pupil with sides of length 25 mm at 0.500 μm wavelength. Because this is a $2f$ system, then the object distance and image distance are 200 mm. With these values we can compute the ATF and the Fourier transform of the image field, $F(f_x)$, to obtain,

$$F(f_x) = \frac{1}{2}\left[\delta\left(f_x - 50\frac{\text{lines}}{\text{mm}}\right) + \delta\left(f_x + 50\frac{\text{lines}}{\text{mm}}\right)\right]$$

$$H(f_x) = \text{rect}\left(\frac{f_x}{250 \text{ mm}^{-1}}\right)$$

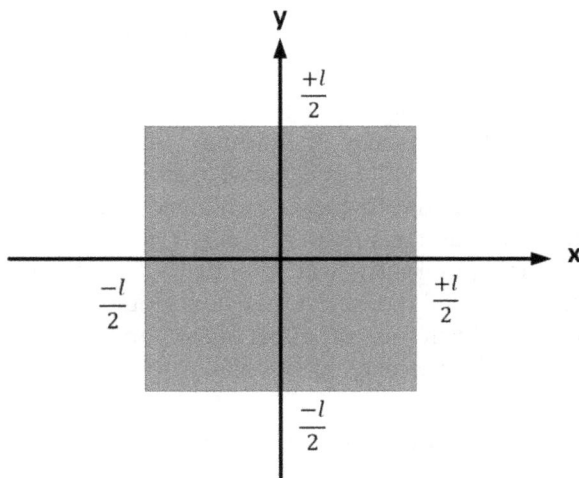

Figure 5.10. Pupil function of a square lens of side l.

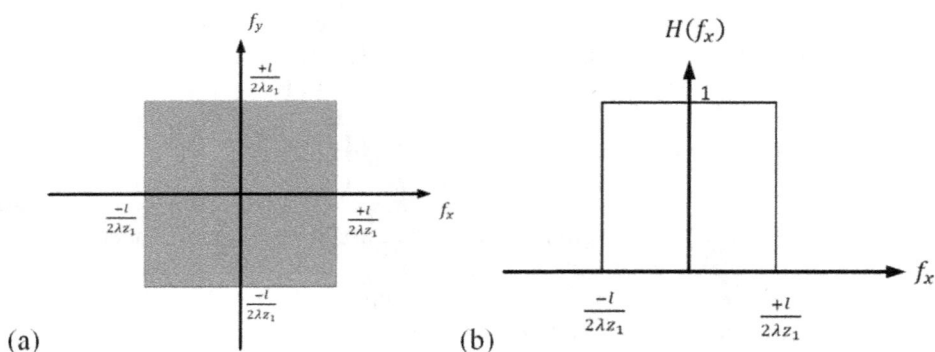

Figure 5.11. (a) Amplitude transfer function of a lens with a square pupil function of side l at an image distance z_1 with light of wavelength λ. (b) $H(f_x)$, a one-dimensional cut of the function.

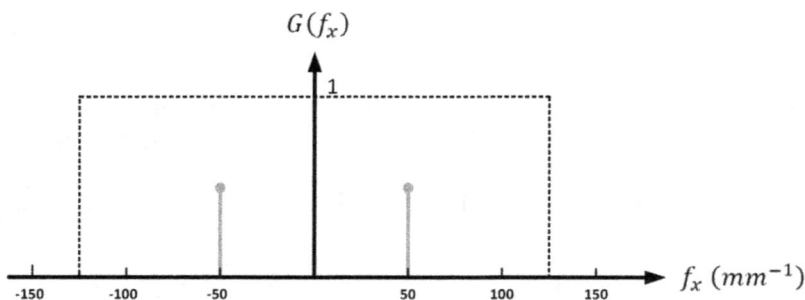

Figure 5.12. Plot of the output field function of a cosine grating as the object in a $2f$ imaging system. The 100 mm focal length lens has a square pupil function with sides of length 25 mm. The wavelength of light is 0.500 μm.

From the convolution theorem the Fourier transform of the output field is the product of these two functions, or,

$$G(f_x) = F(f_x)H(f_x) = \frac{1}{2}\left[\delta\left(f_x - 50\frac{\text{lines}}{\text{mm}}\right) + \delta\left(f_x + 50\frac{\text{lines}}{\text{mm}}\right)\right]\text{rect}\left(\frac{f_x}{250 \text{ mm}^{-1}}\right)$$

A plot of the function $G(f_x)$ is shown in figure 5.12. An outline of the ATF is shown indicating that at a spatial frequency of 125 mm^{-1} there is a sharp cutoff beyond which all values are zero. This value is called the cutoff frequency.

For this case, since the spatial frequency contained in the object is 50 mm^{-1} the output image is the same as the input with unit transfer. However, suppose the spatial frequency of the input grating was 150 mm^{-1}. Then the product of the ATF and the Fourier transform of the object function would be zero, resulting in an image with zero contrast. Therefore, the effect of the pupil function is to limit spatial frequency transfer through any imaging system.

5.4.8 Optical transfer function or incoherent transfer function

In this section we will continue our analysis of the spatial frequency characteristics of imaging of images formed using incoherent light. For incoherent imaging systems the system impulse response function is the absolute square of the coherent system impulse response function. This function is often referred to as the PSF since it would be the observed output irradiance distribution for an input point source object [2, 6, 12]. For this case, from before, the irradiance distribution in the image plane is,

$$E_{e1}(x_1, y_1) = \int\!\!\!\int_{-\infty}^{+\infty} |h(x_1 - x_0, y_1 - y_0)|^2 E_{e0}(x_0, y_0) dx_0 dy_0$$

Again, we must take the Fourier transforms of these functions to examine their spatial frequency characteristics. Since this equation is a form of the convolution, we know that,

$$\mathcal{F}\{E_{e1}\} = \mathcal{F}\{|h|^2\}\mathcal{F}\{E_{e0}\}$$

Since the irradiance distributions must be real and non-negative their Fourier transforms, or spatial spectra will have a maximum value when $f_x = 0$ and $f_y = 0$. For example, if the object had a uniform illumination then the Fourier transform of the constant illumination would be a delta function at the origin multiplied by the value of the constant, uniform illumination. So, the value of the spectrum at the origin in the spatial frequency domain indicates the relative amount of the total illumination of the object or image. As we will see, this quantity relates to the overall contrast of the image or object [2].

With this in mind we define a set of normalized functions to use in analyzing incoherent imaging systems. These definitions are [2],

$$\mathcal{G}_g(f_x, f_y) = \frac{\int\!\!\!\int_{-\infty}^{+\infty} E_{e0}(x, y) \exp[-j2\pi(f_x x + f_y y)] dx dy}{\int\!\!\!\int_{-\infty}^{+\infty} E_{e0}(x, y) dx dy}$$

$$\mathcal{G}_1(f_x, f_y) = \frac{\int\!\!\!\int_{-\infty}^{+\infty} E_{e1}(x, y) \exp[-j2\pi(f_x x + f_y y)] dx dy}{\int\!\!\!\int_{-\infty}^{+\infty} E_{e1}(x, y) dx dy}$$

$$\mathcal{H}(f_x, f_y) = \frac{\int\!\!\!\int_{-\infty}^{+\infty} |h(x, y)|^2 \exp[-j2\pi(f_x x + f_y y)] dx dy}{\int\!\!\!\int_{-\infty}^{+\infty} |h(x, y)|^2 dx dy}$$

Resulting in a normalized form for frequency domain relation for incoherent imaging of,

$$\mathcal{G}_1(f_x, f_y) = \mathcal{G}_g(f_x, f_y)\mathcal{H}(f_x, f_y)$$

The function $\mathcal{H}(f_x, f_y)$ is called the OTF. Since the OTF can in general be a complex function it can be expressed in terms of an amplitude and phase, as,

$$\text{OTF}(f_x, f_y) = \text{MTF}(f_x, f_y) \exp[-j\text{PTF}(f_x, f_y)]$$

where MTF is the modulation transfer function and PTF is the phase transfer function [1]. To illustrate the utility of this definition, suppose that we have a real object with a particular spatial frequency modulation. The image of this object will have a modulation at this spatial frequency modified by the MTF of the imaging system. If we could measure the input and output modulation or image contrast at this spatial frequency then the MTF would be,

$$\text{MTF} = \frac{M_{\text{image}}}{M_{\text{object}}}$$

where the modulations or contrast are defined as [1, 8],

$$M_{\text{image}} = \frac{E_{e\,1\,\text{max}} - E_{e\,1\,\text{min}}}{E_{e\,1\,\text{max}} + E_{e\,1\,\text{min}}} \quad \text{and} \quad M_{\text{image}} = \frac{E_{e\,0\,\text{max}} - E_{e\,0\,\text{min}}}{E_{e\,0\,\text{max}} + E_{e\,0\,\text{min}}}$$

Based on this discussion, it is important to note that the MTF is a system MTF. In other words, the system may be composed of many elements, components, or operations. Each part of the system alters the transfer of spatial frequency information in turn. Therefore, the overall MTF of a system is the product of all the MTF functions of each individual part of the system. For a system with n contributions for each ith MTF function, the total MTF is,

$$\text{MTF} = \prod_{i=1}^{n} \text{MTF}_i$$

An alternative relationship for computing the OTF from the ATF is useful since the ATF is related to the pupil function. From the definition of the OTF above, we can rewrite the function using operator notation, and making use of the convolution theorem,

$$\mathcal{H}(f_x, f_y) = \frac{\mathcal{F}\{|h(x, y)|^2\}}{\int\!\!\int_{-\infty}^{+\infty} |h(x, y)|^2 \, dxdy} = \frac{\mathcal{F}\{h(x, y)\} \circledast \mathcal{F}\{h^*(x, y)\}}{\int\!\!\int_{-\infty}^{+\infty} |h(x, y)|^2 \, dxdy}$$

But since, $\mathcal{F}\{h(x, y)\} = H(f_x, f_y) = P(-\lambda z_1 f_x, -\lambda z_1 f_y)$

$$\mathcal{H}(f_x, f_y) = \frac{H \circledast H^*}{\int\!\!\int_{-\infty}^{+\infty} |h(x, y)|^2 \, dxdy}$$

Taking this to one dimension for simplicity (it can be easily extended to two dimensions) writing out the convolution integral, and making an appropriate change of variables of integration yields,

$$\mathcal{H}(f_x) = \frac{\int_{-\infty}^{+\infty} H\left(w + \frac{f_x}{2}\right)H^*\left(w - \frac{f_x}{2}\right)dw}{\int_{-\infty}^{+\infty} |H(w)|^2\, dw}$$

Therefore, we see that the OTF as a function of spatial frequency can be computed from the overlap integral between two displaced pupil functions [2].

5.4.9 Example: OTF of a square pupil

As with the case of coherent imaging systems, we will compute the OTF for an imaging system with a square pupil function. In this example we will only use one dimension and then extend the result to two dimensions since the functions are separable. This one-dimensional square pupil function of side l is,

$$P(x, y) = \text{rect}\left(\frac{x}{l}\right)$$

So, the ATF is then,

$$H(f_x) = \text{rect}\left(\frac{\lambda z_1}{l}f_x\right)$$

Using the overlap integral from above with this function results in,

$$\mathcal{H}(f_x) = \frac{\int_{-\infty}^{+\infty} \text{rect}\left[\frac{\lambda z_1}{l}\left(w + \frac{f_x}{2}\right)\right]\text{rect}\left[\frac{\lambda z_1}{l}\left(w - \frac{f_x}{2}\right)\right]dw}{\int_{-\infty}^{+\infty} \left| \text{rect}\left[\frac{\lambda z_1}{l}w\right]\right|^2 dw}$$

In order to perform the necessary integrals, it is useful to examine a plot of the overlapping functions as shown in figure 5.13. On the plots the bounds of each overlap are shown to obtain the necessary limits of integration.

From figure 5.13 the integrations in the denominator for the OTF can be evaluated as,

$$\int_{-\infty}^{+\infty} \left| \text{rect}\left[\frac{\lambda z_1}{l}w\right]\right|^2 dw = \int_{\frac{-l}{2\lambda z_1}}^{\frac{+l}{2\lambda z_1}} dw = \left(\frac{l}{\lambda z_1}\right)$$

And the integral in the numerator

$$\int_{-\infty}^{+\infty} \text{rect}\left[\frac{\lambda z_1}{l}\left(w + \frac{f_x}{2}\right)\right]\text{rect}\left[\frac{\lambda z_1}{l}\left(w - \frac{f_x}{2}\right)\right]dw = \int_{\frac{-l}{2\lambda z_1}+\frac{f_x}{2}}^{\frac{+l}{2\lambda z_1}-\frac{f_x}{2}} dw = \left(\frac{l}{\lambda z_1} - f_x\right)$$

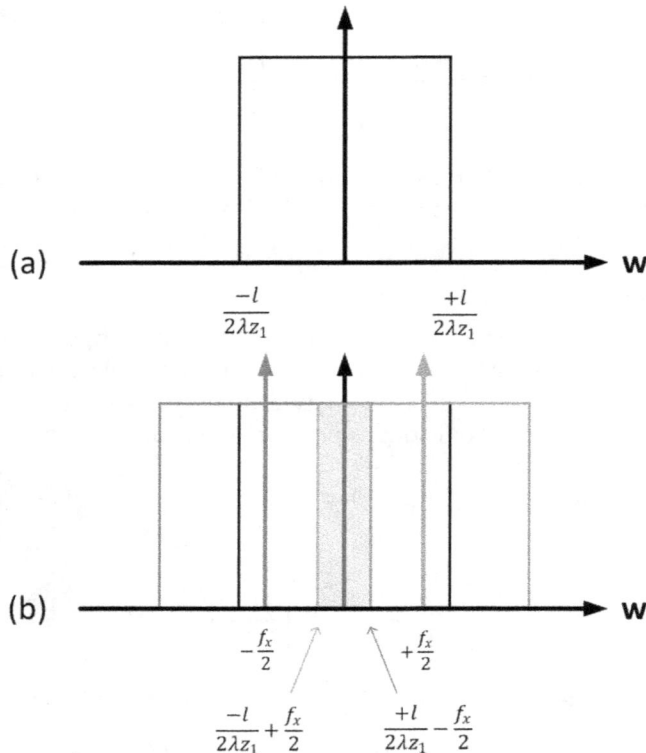

Figure 5.13. Plots of (a) the rectangle function in the denominator as a function of the dummy integration variable w, and (b) the two displaced overlapping rectangle functions for the integral in the numerator.

Resulting in the one-dimensional OTF to be,

$$\mathcal{H}(f_x) = \left(1 - \frac{f_x}{\left(\frac{l}{\lambda z_1}\right)}\right) = \text{tri}\left(\frac{f_x}{\frac{l}{\lambda z_1}}\right)$$

Since this function is real the OTF is equal to the MTF.

For a numerical example we will use the same system example discussed for coherent imaging. As before, the imaging system is a $2f$ unit magnification system with lens of focal length of 100 mm, square pupil with sides of length 25 mm at 0.500 μm wavelength. Because this is a $2f$ system, then the object distance and image distance are 200 mm. The MTF plot for this system is shown in figure 5.14.

Note that the value of the quantity $(l/\lambda z_1) = 250$ mm^{-1}. This spatial frequency value is the cutoff frequency for incoherent illumination for this system since all spatial frequencies greater than this value will have a modulation of zero and will have no contrast. Note that this value is twice the value of the cutoff frequency for coherent illumination. Even though the cutoff is larger for incoherent illumination all spatial frequencies less than the cutoff for coherent illumination are transferred equally. That is not the case for incoherent illumination imaging systems.

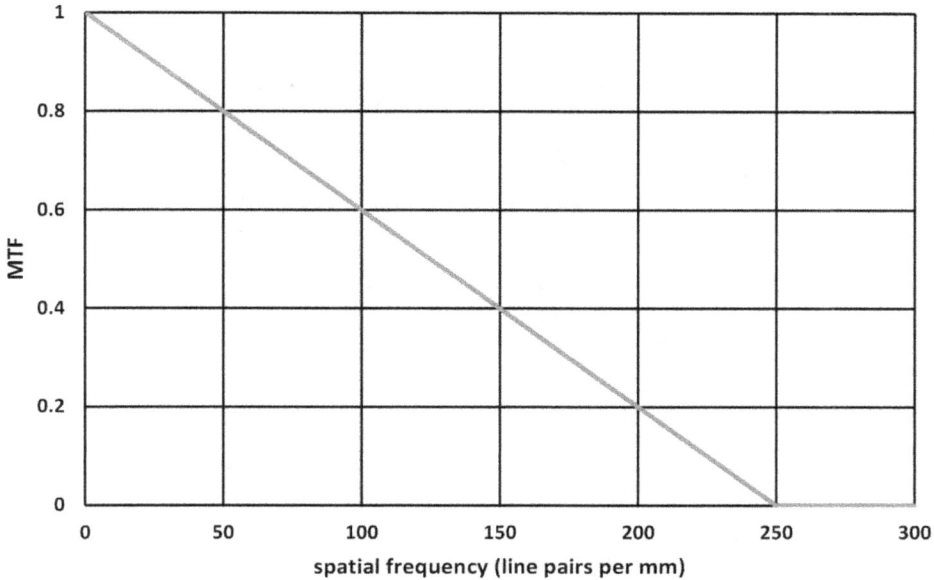

Figure 5.14. MTF plot of a lens with a square pupil function and cutoff frequency of 250 mm^{-1}.

As before, if the input object is a cosine modulation then the output function will also be a cosine modulation. For the 50 lines/mm grating of the earlier example of a coherent system, we found that the image contrast was the same as the object. However, with an incoherent system, the MTF at 50 lines/mm is 0.80 so that the contrast is reduced in the image plane. Figure 5.15 shows the input and resulting output functions for different spatial frequencies illustrating the effect of the MTF on a particular modulation.

5.4.10 Circular pupil function

Earlier we examined the circular pupil function and its effect of limiting resolution in a system. Since most typical off-the-shelf lenses are circular because of their physical properties, knowing the OTF for a lens with circular aperture is important to understanding resulting imaging quality. The OTF of a circular pupil function of diameter d with no aberrations is,

$$\mathcal{H}(\alpha) = \frac{2}{\pi}\left[\cos^{-1}\left(\frac{\alpha}{2\alpha_0}\right) - \frac{\alpha}{2\alpha_0}\sqrt{1 - \left(\frac{\alpha}{2\alpha_0}\right)^2}\right]$$

where $\alpha^2 = f_x^2 + f_y^2$ and $2\alpha_0 = \frac{d}{\lambda z_1}$. The value of $2\alpha_0$ is the cutoff spatial frequency for this system [2].

As a comparison, we will again examine a $2f$ unit magnification system with lens of focal length of 100 mm at 0.500 μm wavelength, but now evaluate the MTF for a circular pupil of diameter $d = 25$ mm. The MTF plot for this system, along one

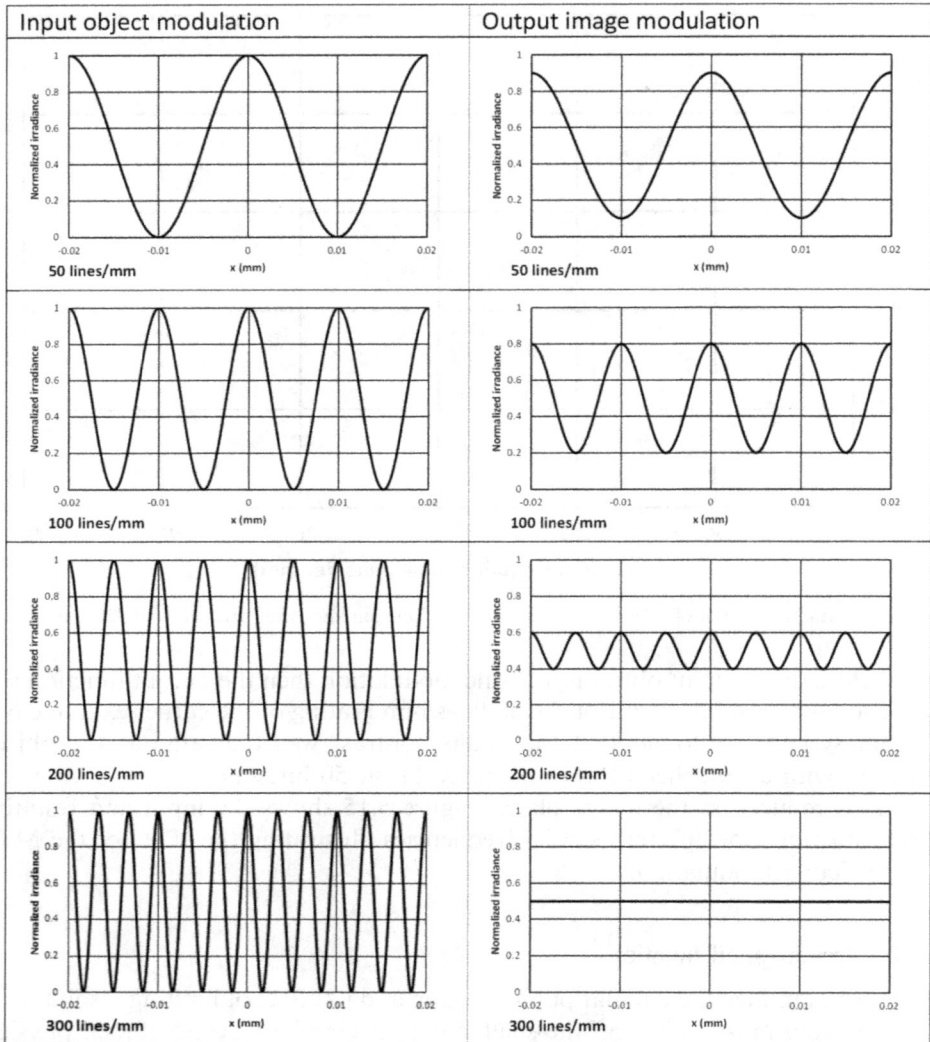

Figure 5.15. Object and image contrasts for four different input functions at 50, 100, 200, 300 lines/mm modulation modified by the square pupil function MTF of figure 5.14.

spatial frequency axis is shown in figure 5.16 along with the graph of a square pupil for comparison.

5.4.11 Relationships between response functions and transfer functions

In the previous sections we have defined response functions and transfer functions for both coherent and incoherent imaging systems. These functions describe the characteristics of images produced by systems with a known pupil function. Thus, if we know the pupil function then all other system functions may be obtained.

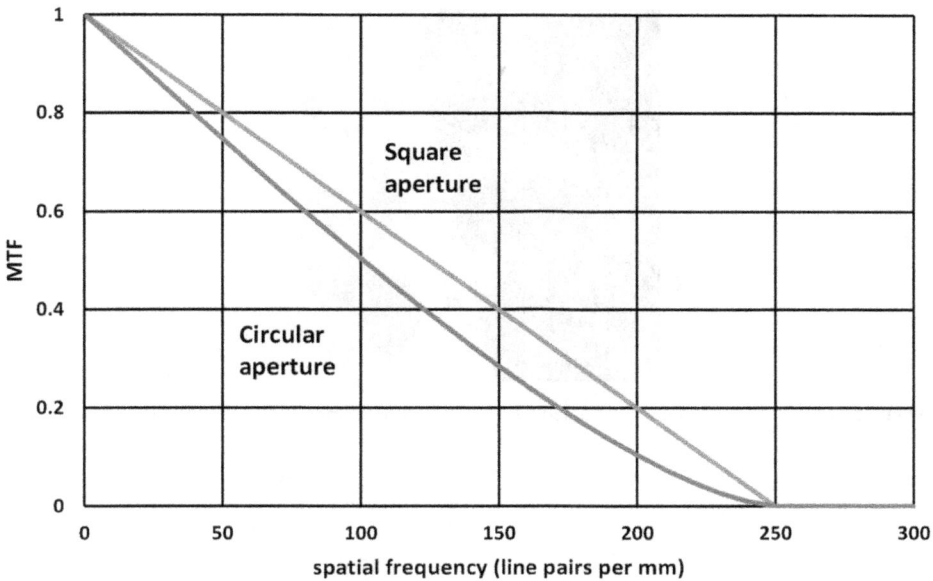

Figure 5.16. MTF plots along the f_x spatial frequency axis for a square pupil function of side 25 mm and a circular aperture of diameter 25 mm.

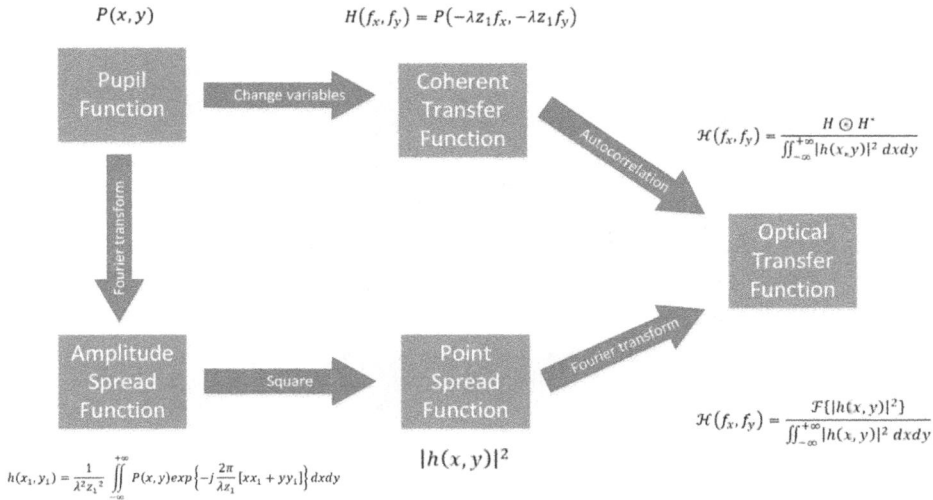

Figure 5.17. Flowchart for computing various system and transfer function knowing the pupil function.

Figure 5.17 provides a flowchart mapping of the interrelations between all of these various system functions. This flowchart also shows that if one measures the PSF then the OTF/MTF may be computed. In fact, this is one of the methods used to experimentally determine the MTF for a system [1].

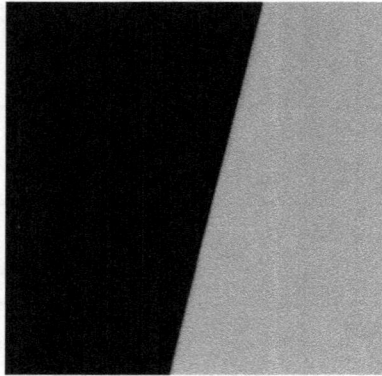

Figure 5.18. A region of interest extracted from a slanted edge image collected by an optical system.

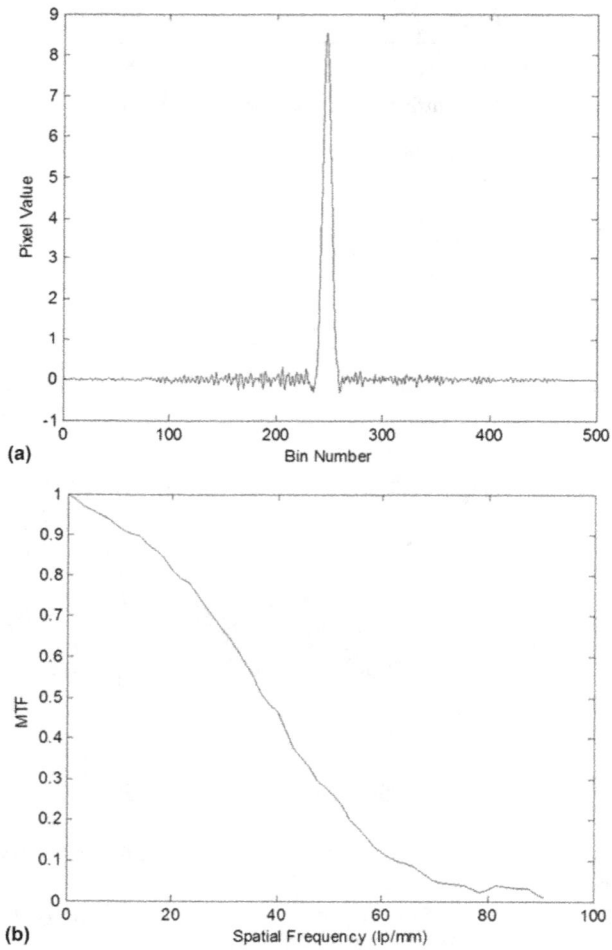

(a)

(b)

Figure 5.19. (a) The LSF and (b) resulting MTF calculated from the ESF.

5.5 Application: experimental determination of MTF

There are numerous methods and standards-based techniques to obtain the OTF experimentally from an image collected by an imaging system [13–16]. From the theoretical discussion above, knowing the PSF provides a method for obtaining the OTF. However, an accurate, low noise PSF is difficult to measure. Capturing the image of a slanted edge and using image processing techniques to obtain a low noise, high resolution line spread function (LSF) has proven to be an excellent means of determining a system MTF [14, 17–19]. The development of this slanted edge method has evolved into a standard for still images as ISO standard 12233 where the outcome results in what is known as the edge-based spatial frequency response (SFR) [18–20].

To summarize the SFR method, the image of a slanted edge object is captured by the system and a specific ROI is chosen. The detailed procedure is described fully in ISO standard 12233 [21]. Figure 5.18 shows an example of a captured slanted edge ROI. Using image processing methods, the centroid of each row of image data is obtained and a best-fit line through each of these centroids is calculated and used to obtain the slope of the line [22].

A super-sample binning process is applied to each row of image data and combined to obtain an average ESF [21]. At various points in the process Hamming windows can be applied to reduce noise [22]. The ESF can then be discretely differentiated to form the discrete LSF. The discrete Fourier transform of the ESF then results in the MTF plot. An example of the LSF and resulting MTF computed for a system is shown in figure 5.19.

Exercises and problems

1. A 50 slit aperture/transmission grating has a slit spacing of 50 μm and a slit width of 10 μm. Diffracted light from a laser normally incident on the grating is observed on a screen. What is the ratio of the irradiance of the primary maximum to that of the first secondary maximum?

2. A narrow single slit in an opaque screen is illuminated by a beam from a HeNe laser (632.8 nm), and it is found that the 10th dark band in the Fraunhofer pattern lies at an angle of 6.20° off the central axis. What is the slit width?

3. Plane waves with wavelength 550 nm are incident normally on a narrow slit in the (x_0, y_0) input plane having a width of 22.0 μm. (a) The Fraunhofer diffraction pattern is observed on a screen $z = 4.0$ m away. What is the width of the central maximum? (b) Now a lens of focal length 200.0 mm is placed just behind the slit. What distance z away from the lens is the Fraunhofer pattern observed?

4. Coherent light from a laser with a 442 nm wavelength is incident on a single slit of width 2.0×10^{-6} m. At what angle from the central maximum in the diffraction pattern is the first diffraction minimum observed?

5. Evaluate the convolution by <u>direct integration</u> and sketch a graph of the result.

$$f(x) = \text{rect}\left(\frac{x}{2}\right) \circledast \text{rect}\left(\frac{x}{2}\right)$$

6. Write down the Fourier transform of the function $g(x) = \text{sinc}^2(2x)$ and sketch the result in spatial frequency coordinates f_x. Is the function $g(x)$ a band-limited function? Is the Fourier transform of $g(x)$ a band-limited function?

7. Evaluate the Fourier transforms of the following functions.
   (a) $g_1(x) = \text{sinc}(x)\text{sinc}(\frac{x}{2})$
   (b) $g_2(x) = \cos(4\pi x)\text{rect}(\frac{x}{2})$

8. Evaluate the convolution of $f(x) = \cos(\pi x + \frac{\pi}{4})$ with the following functions and plot the results:
   (a) $h_1(x) = \text{rect}(x)$
   (b) $h_2(x) = \frac{1}{2}\text{rect}(\frac{x}{2})$

9. Evaluate the convolution of the rectangle functions.

$$g(x) = \text{rect}\left(\frac{x}{2}\right) \circledast \text{rect}\left(\frac{x}{2}\right)$$

10. Given the two functions $f_1(x)$ and $f_2(x)$, find the Fourier transform of their product,

    $G(f_x) = \mathcal{F}\{f_1(x)f_2(x)\}$. Sketch a graph of G, to scale, in spatial frequency coordinates f_x.
    (a) Take $f_1(x) = 2\cos(6\pi x)$ and $f_2(x) = \text{sinc}(x)$. The units of x are in mm.
    (b) Take $f_1(x) = \cos(10\pi x)$ and $f_2(x) = \text{rect}(x)$. The units of x are in mm.

11. An aperture is placed in the (x_0, y_0) plane specified by the function $\text{rect}(10x_0)\text{rect}(20y_0)$ where x_0 and y_0 are in millimeter units. The aperture is illuminated (in air) by a plane waves from a krypton ion laser (461.9 nm wavelength). If the observing screen is 1.0 m away, determine whether or not the resulting diffraction pattern is in the far-field or not. Prove your answer with a calculation.

12. A delta function input to an optical system produced the following impulse response function.

$$r(x) = \begin{cases} 0 & -\infty < x < +1 \\ 2 & +1 \leqslant x \leqslant +2 \\ 0 & +2 < x < +\infty \end{cases}$$

where x is in units of mm.

A signal $s(x)$ was then input to the system with the following function.

$$s(x) = \begin{cases} 0 & -\infty < x < +3 \\ 1 & +3 \leqslant x \leqslant +5 \\ 0 & +5 < x < +\infty \end{cases}$$

Compute and sketch the system output from this signal.

13. Sketch the 2-D function:

$$f(x, y) = \text{rect}\left(\frac{x}{2}, \frac{y}{2}\right) \circledast [\delta(x - 1, y - 1) + \delta(x + 1, y + 1)]$$

14. Sketch an (x, y) plane view of the amplitude transmission object given by the functions specified below. The units of the x and y coordinates are in millimeters. What specific optical component do each of these functions represent?

$$t_A(x, y) = \frac{1}{2}\left[\text{rect}(x) \circledast \text{comb}\left(\frac{x}{2}\right)\right]$$

$$t_A(x, y) = \frac{1}{2}[1 + \cos(ar^2)] \text{ where } r = \sqrt{x^2 + y^2}$$

15. Given the apertures shown below in figures P5.15 (a)–(d), write down the light field transmission function in (x, y) coordinates for each aperture. Grid spacings in figure P5.15 are in millimeters.

16. The spatial function represented by a particular Ronchi ruling is shown below in figure P5.16. The ruling object is illuminated by a laser pointer of wavelength 630 nm. (a) Find the dominant spatial frequency. (b) Sketch the pattern you would observe on a screen (i) placed just behind the ruling, and (ii) placed in a plane 5.0 m away along the axis of the ruling.

17. A sinusoidal transmission grating has an amplitude transmission function given by:

$$t(x, y) = \frac{1}{2}\{1 + \sin(8\pi x)\}$$

where the units of the x position are in microns. When illuminated with a plane parallel beam of light at 450 nm determine the irradiance profile that appears in the back focal plane of a lens with focal length 100 mm placed just behind the grating.

18. Three point sources are placed in the input plane of an optical system. One source is placed at the origin and the other two sources are placed on the x-axis a distance of 0.5 mm on each side of the origin. The source at the origin has an electric field amplitude of 2.0 V m^{-1} and the other sources have equal amplitudes of 1.0 V m^{-1}. (a) Provide an equation for the spatial

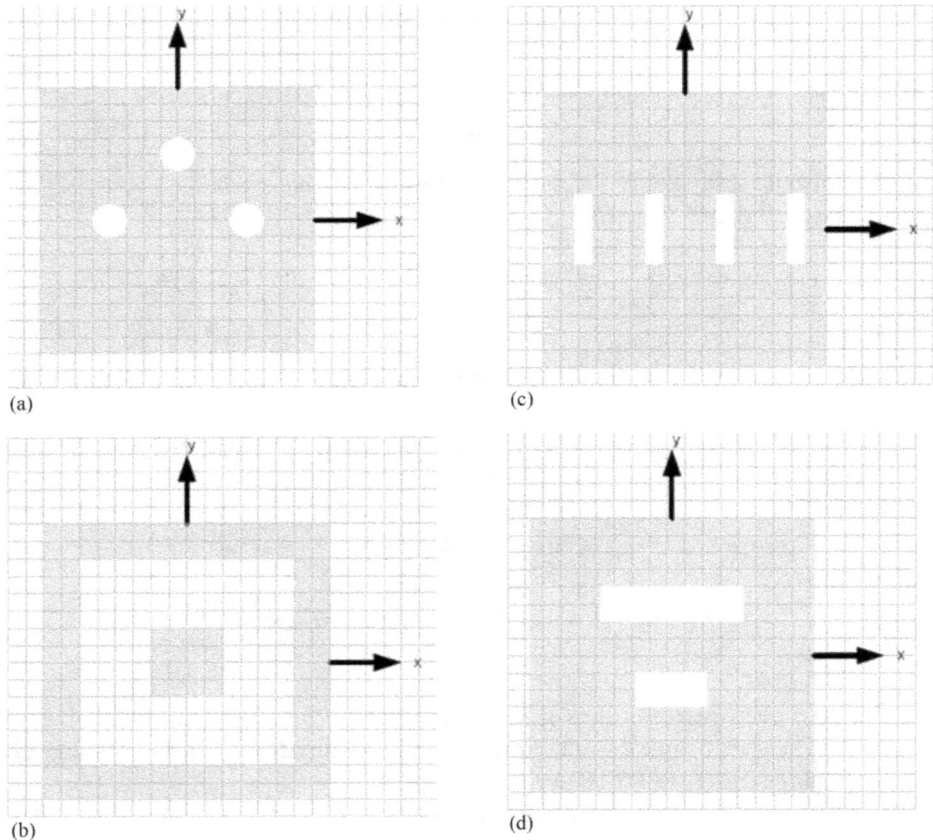

Figure P5.15. (a)–(d) Various two-dimensional apertures.

function of the light field, just following the plane containing the sources. (b) Compute the Fourier transform of this function to obtain the diffraction field function.

19. Two thin object transparencies are placed together in front of a lens of focal length 158 mm. Light from an expanded and collimated HeNe laser (633 nm wavelength) with unit amplitude illuminates the objects. The transmission (field) functions along the x coordinate axis for each object are: $t_1(x) = 2\cos(2\pi \cdot 5 \text{ mm}^{-1}x)$ and $t_2(x) = 2\text{sinc}^2(2 \text{ mm}^{-1}x)$ where the units of x are in mm and the function constants are in units of mm^{-1} as shown.

    (a) Compute the function for the field that would be in the back focal plane of the lens whose coordinate axis is x_f.

    (b) Sketch a profile of the <u>irradiance</u> pattern that you would see on a card placed in the back focal plane of this lens. The sketch should include appropriate dimensions of the x_f coordinate and the irradiance.

Figure P5.16. Overhead view of a Ronchi ruling.

20. A plane wave is incident from the left onto two thin lenses placed in contact. One lens has focal length $f_1 = 100$ mm the other has a focal length $f_2 = -50$ mm. Take the coordinate axis of the lenses plane to be (x, y).
    (a) What is the light field predicted just after passing through both lenses?
    (b) Compute the effective focal length of the two-lens combination.

21. A five-slit aperture function is shown in an (x_1, y_1) coordinate system with each slit a width of length a, height of length b and separated by length c (figure P5.21). (a) Write down the transmission function for this aperture. (b) A unit amplitude plane wave of wavelength λ is incident on the input plane. Determine the complex light field in the Fraunhofer diffraction plane a distance z from the (x_1, y_1) plane. Take the coordinates of the diffraction plane to be (x_0, y_0). (c) Plot a graph of the irradiance along the central axis in the x_0-direction for $a = 0.5$ microns, $b = 6.0$ microns and $c = 1.0$ microns with $\lambda = 400$ nm.

22. The aperture function for an object transparency placed in the front focal plane (P1) of a lens of focal length 100 mm and specified with coordinates (x_1, y_1) is given by the function,

$$t(x_1, y_1) = \sin(0.20\pi x_1)\cos(0.30\pi y_1)$$

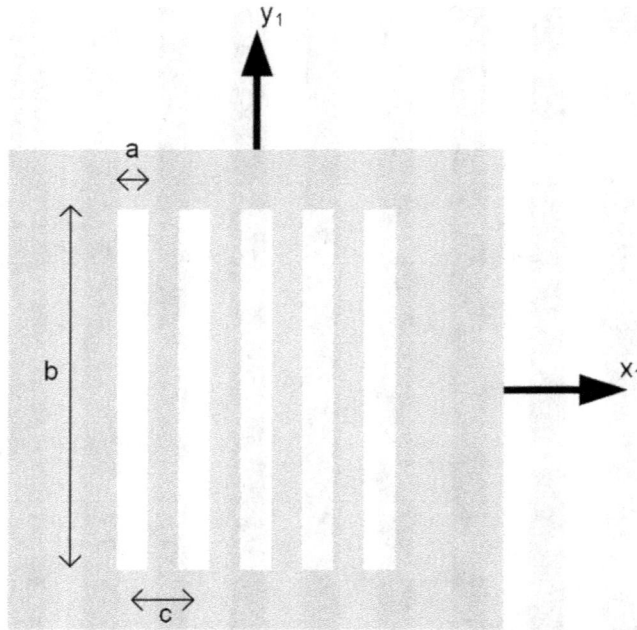

Figure P5.21. Five slits of width a, height b, and separated by a distance c.

where x_1 and y_1 are in units of microns. An incident plane wave of unit amplitude and wavelength 0.500 μm is incident on the transparency. Determine the complex light field observed in the back focal plane (P2) of this lens taking the coordinates to be (x_2, y_2). Make a sketch of the irradiance that you would observe in the P2 plane.

23. A square aperture of side, a, is illuminated by an <u>incident spherical wave</u> (not a plane wave) converging toward a point P located on the axis normal to the center of the aperture in a parallel plane a distance z behind the aperture. Find an analytical function for the <u>Fresnel</u> diffraction field in the plane (x_0, y_0) containing P.

24. Two coherent point sources (wavelength $\lambda = 500$ nm) lie in the (x, y) plane of a Cartesian coordinate system whose origin is at the midpoint between the sources. The two sources are separated by a distance "$2a = 10$ microns" and each has the same field amplitude, you can assume to be unity. The diffraction pattern from these sources is observed on a screen a distance $z = 10$ m away. The light emitted by one of the sources is shifted in phase from the other by a constant amount φ. (a) Sketch a diagram of the situation. (b) Write down a transmission function for the light field distribution in the (x, y) plane. (c) Take the screen coordinates to be a plane specified by (x_1, y_1) and determine the Fraunhofer diffraction pattern observed on the screen. What is the effect of this phase factor φ on the diffraction pattern when $\varphi = \pi/2$ compared to the pattern when the phase factor is zero?

25. A plane wave (coherent) of unit amplitude is incident on a long slit of width 2.0 mm placed in an (x, y) coordinate input plane. Instead of examining the diffraction pattern, a thin lens of focal length 10.0 cm images the slit in a plane 20 cm from the lens. The impulse response function of this lens along the 1D x-axis direction is given by the function,

$$h(x) = \begin{cases} 0 & x < 1.0 \text{ mm} \\ 2 & 1.0 \text{ mm} \leqslant x \leqslant 2.0 \text{ mm} \\ 0 & x > 2.0 \text{ mm} \end{cases}$$

(a) Write down the function $f(x)$ that describes the light field along the x-axis direction just past the (x, y) plane. (b) Determine the output function observed in the image plane and sketch the resulting field function. (Hint: because $h(x)$ is 1D you only need to consider a 1D solution.)

26. A long slit object of width $b = 2.0$ mm is the input to an optical system and centered on the optical axis of the system. See figure P5.26. The slit is illuminated by a plane wave from a HeNe laser (633 nm wavelength). The system impulse response function, $h(x)$ was measured to be:

$$h(x) = \text{rect}\left(\frac{x}{2}\right)$$

where x is in units of mm.

(a) Write down the function $f(x)$ that describes the light field distribution exiting the slit in the slit plane along the x-axis. Take the amplitude of the light to be normalized to 1.0 and the units of x to be in mm.

(b) Compute the output light field distribution that would be observed from this system in the output/image plane.

27. An object screen is placed in the input of a $4f$ optical processing system. The focal length of the first lens is 30 cm and the focal length of the second lens

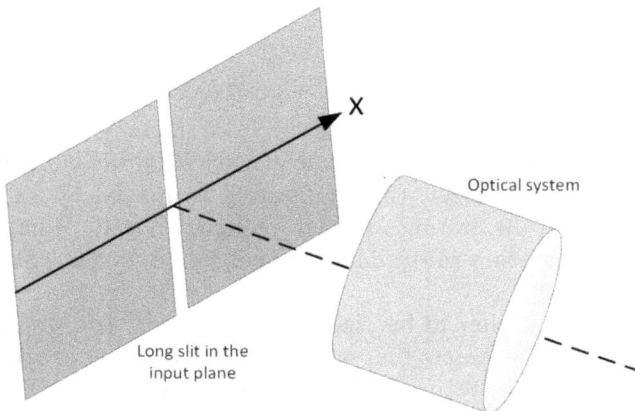

Figure P5.26. Light from a single long slit collected by an optical system.

is 15 cm. (a) What is the magnification of the image formed in the output plane? (b) The input object is then moved behind the first lens, between lens#1 and the transform plane. Plot a graph of the magnification of the image produced in the output plane as a function of the distance from the object screen to the transform plane.

28. A thin positive focal length lens has an apodized pupil function given by,

$$P(x, y) = \exp\left[\frac{-(x^2 + y^2)}{a^2}\right]$$

Compute the MTF of this lens when placed in an incoherent imaging system with wavelength λ and image distance d. Plot the MTF as a function of the spatial frequency f_x (lines/mm) for $\lambda = 600$ nm, $d = 300$ mm and $a = 1.8$ mm.

29. An incoherent imaging system is used to image a target using light from a $\lambda = 0.534$ μm wavelength source. The image distance is $d_i = 187$ mm. The 1D pupil function for the x-cross-section of the lens is:

$$P(x) = \text{rect}\left(\frac{x}{4}\right)$$

where the image plane coordinate is x_i. The units of x and x_i are in mm. Compute the 1D optical transfer function for this lens and determine the incoherent cutoff spatial frequency?

30. Compute the spatial impulse response, $h(x_i)$, the coherent system transfer function, $H(f_x)$, and the MTF for the following exit pupils in terms of the wavelength $\lambda = 0.500$ μm and image distance, $d_i = 100$ mm. The input coordinate x is in units of mm. In addition, plot the MTF and indicate the cutoff frequency value on your graph.
    (a) $P(x) = \text{rect}(2x)$
    (b) $P(x) = \text{rect}(\frac{x}{2}) \cos(\frac{\pi x}{2})$ (this is an apodized pupil).

31. A sinusoidal input object (an amplitude grating) has a form specified by the function:

$$f(x, y) = \frac{1}{2}[1 + \cos(60\pi x)]$$

where x and y are in units of millimeters. (a) What is the contrast and spatial frequency of this object? (b) The grating is imaged by an optical system whose MTF at the grating's spatial frequency is 0.45. What is the contrast and spatial frequency of the image formed by this system? Sketch the output image function.

32. Design a spatial filter to be placed in the Fourier transform plane of a $4f$ optical processing system that will highlight only the horizontal edges of an object that is in the form of a letter E as shown in figure P5.32. Sketch the filter.

Figure P5.32. Letter E object and the desired filtered image.

33. You are given the input object shown in figure P5.33 as a transparency in a 4*f* processing system. The focal lengths of the lenses are the same.
    (a) Sketch the two-dimensional optical transform of this object.
    (b) Design and sketch the filter that will produce only the letter **R** in the output plane.
    (c) Design and sketch the filter that will produce only the letter **H** in the output plane.

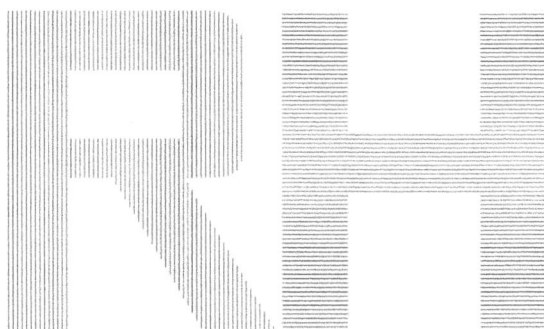

Figure P5.33. Object transparency for input to a 4*f* optical filtering system.

34. A laser beam has a Gaussian field distribution of amplitude A and beam waist w_0. At the output of the laser with coordinates (x_1, y_1) the laser light field is specified as,

$$u(x_1, y_1) = A \exp\left[\frac{-(x_1^2 + y_1^2)}{w_0^2}\right]$$

Using <u>Fresnel</u> diffraction principles, determine the field of this laser beam as it propagates over a distance z from the (x_1, y_1) plane to a new set of coordinates (x_0, y_0). Identify the new beam waist $w(z)$ in the (x_0, y_0) plane.

35. Four sinusoidal amplitude targets (A, B, C, D) shown in figures P5.35 (a)–(d) were used as objects and imaged by an unknown optical system setup in a 2*f* imaging configuration. Line profile data scans of each object and resulting image are shown below for a 0.10 mm section of the object and image. Using this information, fill in the table and sketch the MTF of the system labeling both coordinate axes including appropriate units using the grid shown in figure P5.35(e).

**Image
Target A**

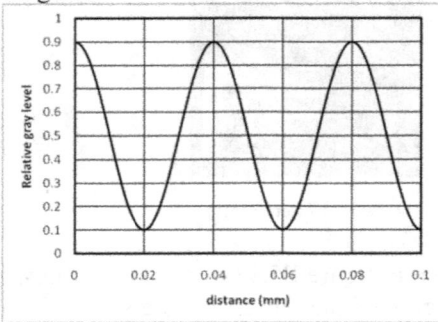

Object

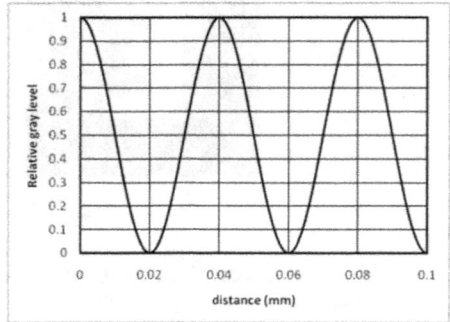

**Image
Target B**

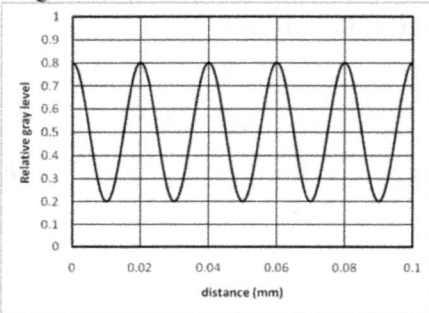

Object

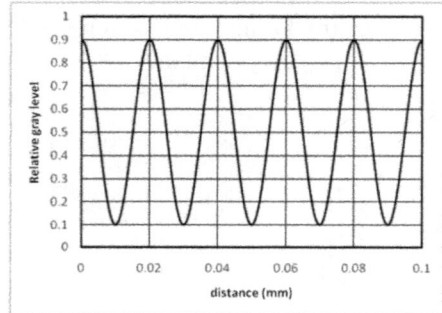

**Image
Target C**

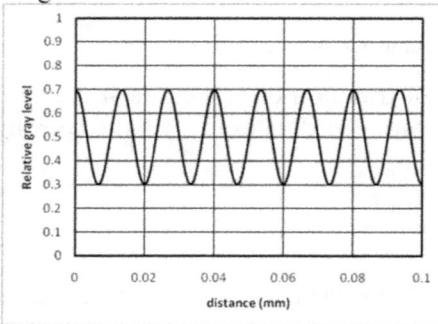

Object

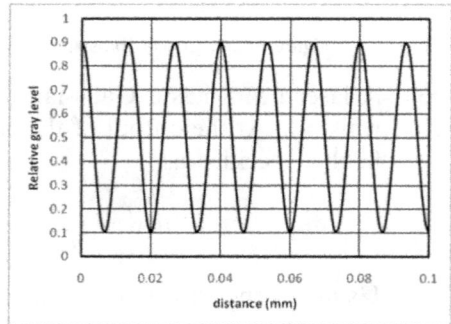

Figure P5.35. (a)–(d) Line profile data scans of the images formed by various sinusoidal object gratings.

Image
Target D

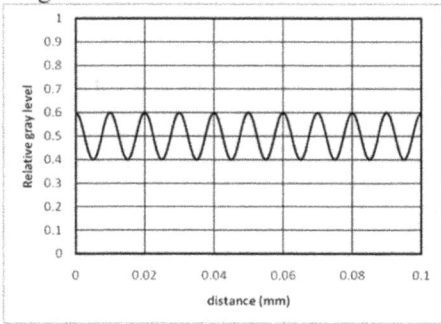

Object

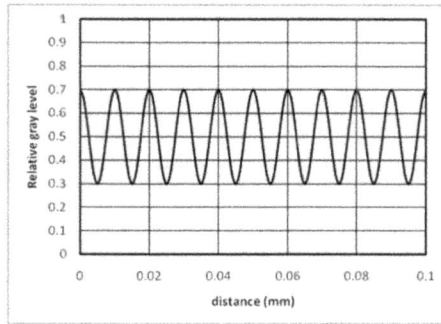

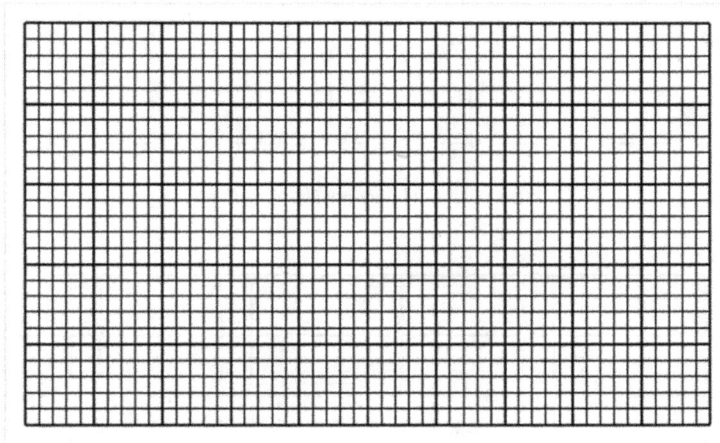

(e)

Figure P5.35. Grid used to plot the MTF of the imaging system.

Target	Spatial frequency	Image contrast	Object contrast	MTF
A				
B				
C				
D				

36. The graph below shows MTF curves for three lenses, A, B, and C. A sinusoidal transmission grating with a spatial frequency of 5.0 line pairs per millimeter was imaged by each of the three lenses using a $2f$ imaging system. The grating has an object contrast of 0.9 with an average gray level value of 100. Sketch three separate graphs showing a line-scan cross-section of the images produced by each lens on the axes below (figure P5.36).

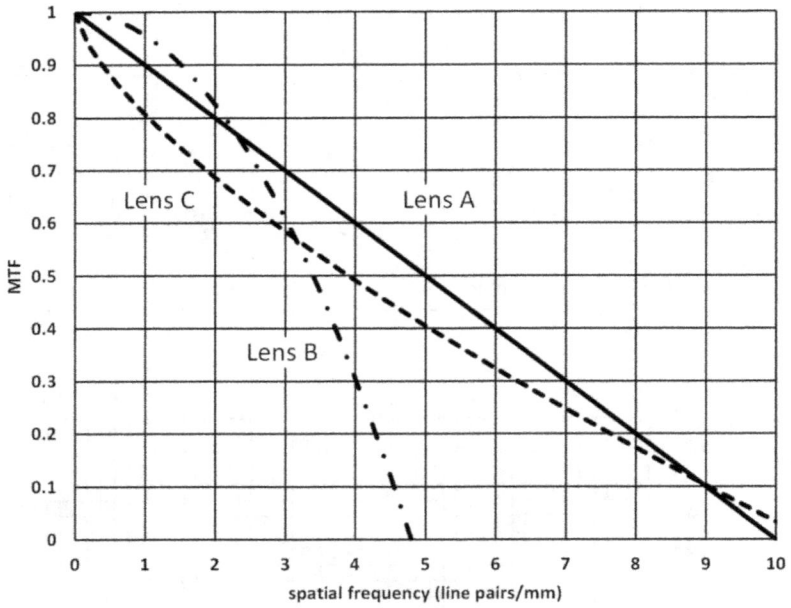

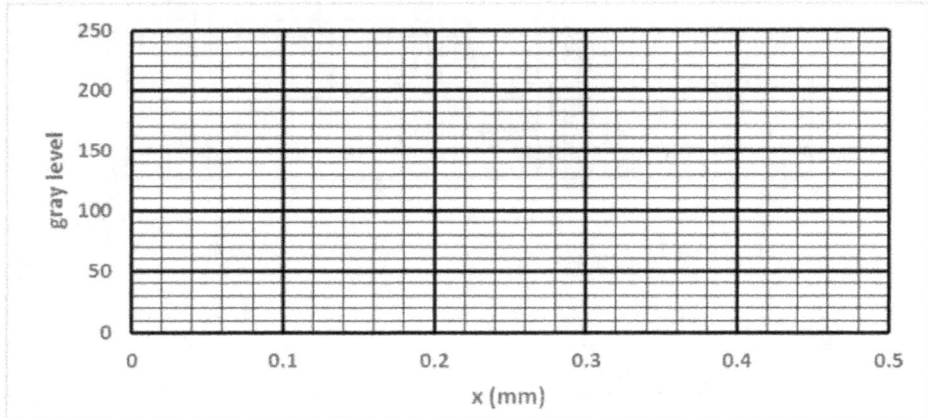

Figure P5.36. (a) MTF curves for three different lenses, and (b)–(d) axis grids to display the line scan images observed Lens A, Lens, B, and Lens C.

Lens B

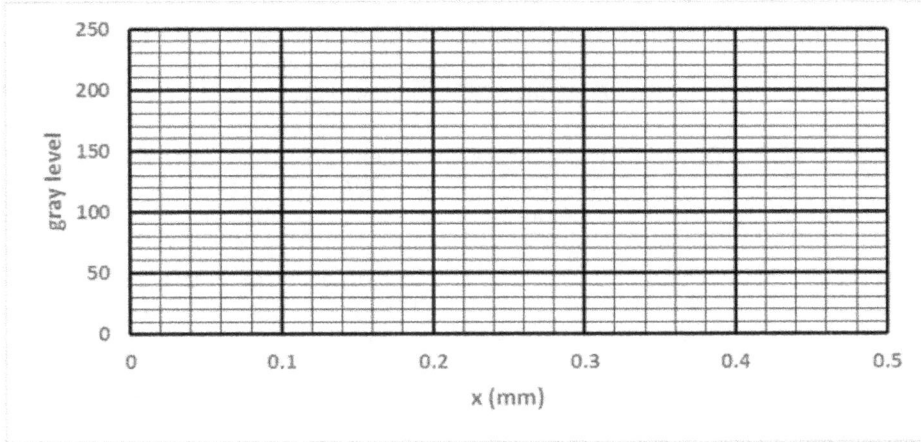

Lens C

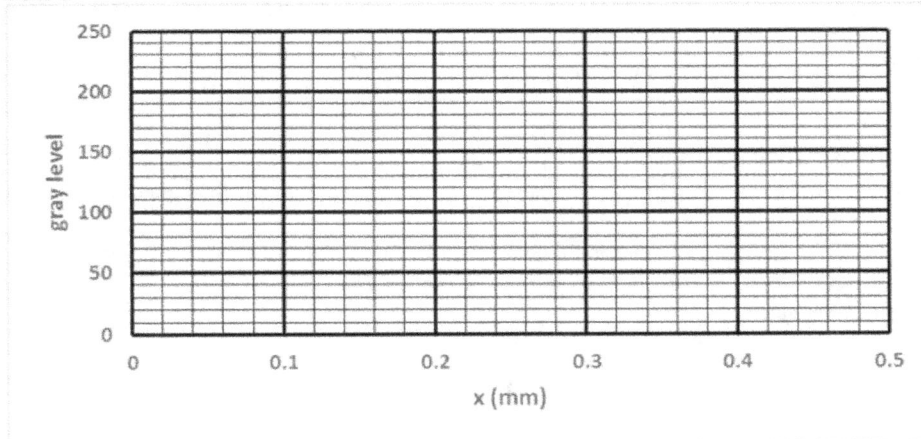

Figure P5.36. (Continued.)

References

[1] Boreman G D 2001 *Modulation Transfer Function in Optical and Electro-Optical Systems* (Bellingham, WA: SPIE Press)

[2] Goodman J W 2017 *Introduction to Fourier Optics* 4th edn (New York: WH Freeman)

[3] Dirac P 1930 *The Principles of Quantum Mechanics* (Oxford: Clarendon)

[4] Lighthill M J 1958 *An Introduction to Fourier Analysis and Generalized Functions* (Cambridge: Cambridge University Press)

[5] Born M and Wolf E 1975 *Principles of Optics* (New York: Pergamon)

[6] Gaskill J D 1978 *Linear Systems, Fourier Transforms and Optics* (New York: Wiley)

[7] Hecht E 2002 *Optics* 4th edn (Reading, MA: Addison-Wesley)

[8] Pedrotti F L, Pedrotti L S and Pedrotti L M 2007 *Introduction to Optics* 3rd edn (Englewood Cliffs, NJ: Prentice Hall)

[9] Smith D G 2012 *Field Guide to Physical Optics* (Bellingham, WA: SPIE Press)

[10] Cathey W T 1974 *Optical Information Processing and Holography* (New York: Wiley)

[11] Tyo J S and Alenin A S 2015 *Field Guide to Linear Systems in Optics* (Bellingham, WA: SPIE Press)

[12] Reynolds G O, DeVelis J B, Parrent G B and Thompson B J 1989 *The New Physical Optics Notebook: Tutorials in Fourier Optics* (Bellingham, WA: SPIE Press)

[13] International Organization for Standardization ISO 15529:2010, Optics and photonics— Optical transfer function—Principles of measurement of modulation transfer function (MTF) of sampled imaging systems https://www.iso.org/standard/56069.html

[14] Scott F, Scott R M and Shack R V 1963 The use of edge gradients in determining modulation-transfer function *Photogr. Sci. Eng.* **7** 64–8

[15] Williams T L 1998 *The Optical Transfer Function of Imaging Systems* (Bristol: Institute of Physics Publishing)

[16] Williams C S and Becklund O A 2002 *Introduction to the Optical Transfer Function* (Bellingham, WA: SPIE Press)

[17] Jones R A 1967 An automated technique for deriving MTFs from edge traces *Photogr. Sci. Eng.* **11** 102–6

[18] Williams D *et al* 2008 A pilot study of digital camera resolution metrology protocols proposed under ISO 12233 edition 2 *Proc. SPIE* **6808** 680804

[19] Burns P and Williams D 2008 Sampling efficiency in digital camera performance standards *Proc. SPIE* **6808** 680805

[20] Williams D and Burns P 2014 Evolution of slanted edge gradient SFR measurement *Proc. SPIE* **9016** 901605

[21] International Organization for Standardization ISO 12233:2017 Photography—Electronic still picture imaging—Resolution and spatial frequency responses https://www.iso.org/standard/71696.html

[22] Gonzalez R C and Woods R E 2018 *Digital Image Processing* 4th edn (New York: Pearson/ Prentice Hall)

IOP Publishing

Optical Systems Design Detection Essentials
Radiometry, photometry, colorimetry, noise, and measurements
Robert M Bunch

Chapter 6

Radiometry, photometry, and color

6.1 Introduction

Understanding the design and operation of optical systems requires a knowledge of the basic principles, terms, and definitions used to measure electromagnetic radiation. We have already used a few radiometric terms in previous chapters to describe propagating light and electromagnetic radiation. In this chapter we will concentrate on radiometry and use radiometric principles to predict the amount of radiation that propagates in optical systems.

Before discussing radiometry, we must differentiate between the subjects of radiometry and photometry, which both deal with measuring an amount of light. The particular measure of electromagnetic radiation depends on how the optical system will be used. Confusion in terminology has developed between the terms of radiometry and photometry over the years because of the historical development of light measurement coming from various scientific fields. For example, astronomers refer to light measurements as photometry because early observation of star light was in terms of magnitude units where the human eye was the detector [1]. Another example is the area of spectrophotometry which grew because selective spectral measurements of chemical compounds, usually in the visible part of the spectrum, provided scientists with information about chemical compounds [2]. As technology developed, optical detectors opened up the possibility of detecting radiation in regions outside the visible and these were exploited to extend classical photometric measurements. The current definition of photometry relates to light measurements restricted to the visible portion of the spectrum and attempts to emulate the response of a human observer [3–8]. Nevertheless, some fields continue to use the term photometry for measurements of radiation outside the visible.

Radiometry is the study of measurements based on electromagnetic energy content and includes radiation in all portions of the electromagnetic spectrum. Quantities involved in radiometry include source measurements as well as detection measurements. We will make some basic assumptions in our discussion of

doi:10.1088/978-0-7503-2252-2ch6   6-1

radiometry: (1) sources of radiation will be essentially incoherent sources so we will not deal with phase or interference phenomena, (2) propagation can be treated with ray optics, and (3) radiant energy is conserved in transparent non-absorbing media.

One of the difficulties in studying radiometry and photometry has been an inconsistent use of units of measurement and symbolic notation by various authors and publications. This has been recognized and addressed in numerous articles and is now part of a standard developed by the International Organization for Standardization (ISO) [9]. ISO 80 000–7:2019 standardizes names, symbols, definitions, and units for radiometric quantities in the wide wavelength range of approximately 1 nm to 1 mm. Some of the more typical units and symbols used here have been adopted by the CIE, International Commission on Illumination and American National Standards Institute (ANSI/IES RP-16–1980 updated from ANSI Z7.1–1967) [10]. Note that in defining radiometric terms and symbols a subscript 'e' is used with each quantity to emphasize that this value is an energy unit. In photometry a subscript 'v' is used to represent a visual quantity.

The phenomena we call color arises from combinations of spectra observed directly from sources, transmitted through filters, or reflected from objects and then detected by the eye. When we view sources of illumination and the reflected light from an object one of the characteristics the object is its color. Color has seen an increasing importance in the industry as part of comprehensive quality control measures. Just consider products such as paint, textiles, traffic lights, automotive signal lamps, color printers, and displays. Thus, several standards have been developed to measure colors and compare colors to one another. We will examine some color standards as well as topics such as color vision, chromaticity coordinates, radiation temperature, correlated color temperature (CCT), color rendering index, and color mixing.

6.2 Solid angle and geometrical concepts

In defining radiometric quantities, it is useful to describe some of the geometrical concepts used in radiometric definitions. From basic geometry a cone is defined as the volume swept through from a straight line at a vertex intersecting a closed curve. Figure 6.1(a) shows a closed spherical surface of radius R and an area on the sphere identified as A. The boundary lines from the center of the sphere around the area A define a volumetric cone. The solid angle, ω, is defined in this case as,

$$\omega = \frac{A}{R^2}$$

Note that the solid angle has no dimensions. In order to keep track of the quantity in calculations, we assign the unit of a solid angle as steradians (sr or str) in a similar manner to the radian unit used in defining the angle subtended by an arc length section of a circle. The solid angle subtended by an entire sphere of radius R would be the surface area of the sphere $4\pi R^2$ divided by R^2 or $\omega_{\text{sphere}} = 4\pi$ (sr).

Similarly, figure 6.1(b) shows a three-dimensional right-handed coordinate system with a differential surface element dA at a general radius r from the origin.

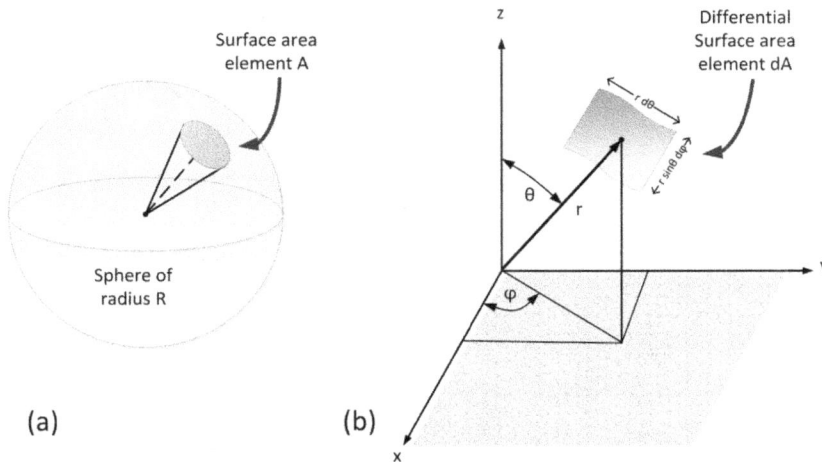

Figure 6.1. (a) Solid angle subtending a surface area element A on the surface of a sphere of radius R. (b) Three-dimensional coordinate geometry and solid angle.

The center of the element dA is at spherical coordinates (r, θ, φ). The length of each side of the area element is also shown on the figure. Like before, calculating the differential solid angle results in,

$$d\omega = \frac{dA}{r^2} = \sin \theta \; d\theta \; d\varphi$$

Now let us examine geometric possibilities for rays emanating from source areas and collected by a detector area. We have already used the concept of a point source and flux collection by a general detector area in previous chapters. With radiometry we can now formally make these definitions. The geometric concept of irradiance, as used before and shown in figure 6.2(a), defines the radiant flux acceptance by a surface area. All flux falling on this area independent of angle of incidence contributes to the irradiance. There are three geometries used to define various concepts used for describing optical sources, exitance, intensity, and radiance. These are illustrated in figure 6.2(b)–(d).

While figure 6.2 gives us a starting point for defining the geometry of source points and areas along with detection areas, in laying out an optical system the geometric relationship between the source and detector is also important. Take for example a point source illuminating a surface area. The orientation of the detecting surface area to the source must be included when determining the flux collected by this area. Figure 6.3 shows three different orientations of a detector area in relation to a point source with the detection area normal vector in the plane of the figure. When the surface area points toward the source then a maximum flux is received but when the surface normal vector is at right angles to the source then no flux is received. At angles between these extremes there is a received flux, but its value is reduced by the cosine of the angle between the surface normal and the line to the

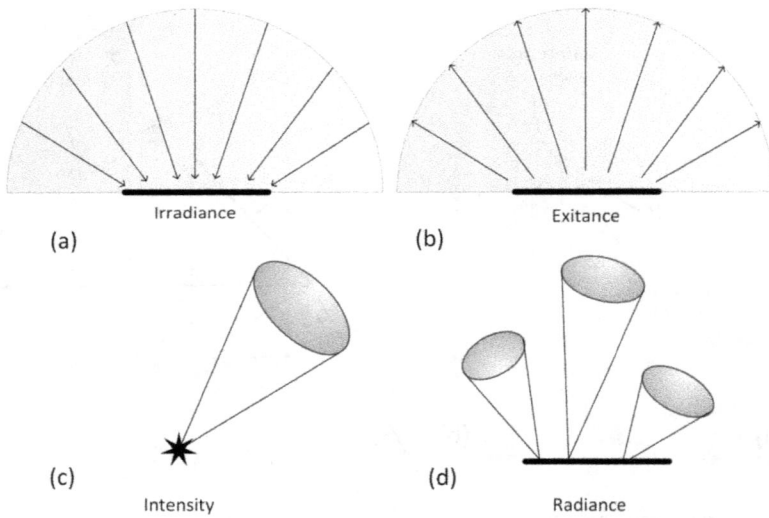

Figure 6.2. Geometrical concepts used in radiometry. (a) Irradiance of a detection area, (b) exitance from a source area, (c) intensity emanating from a point source into a solid angle, and (d) radiance from an extended source area.

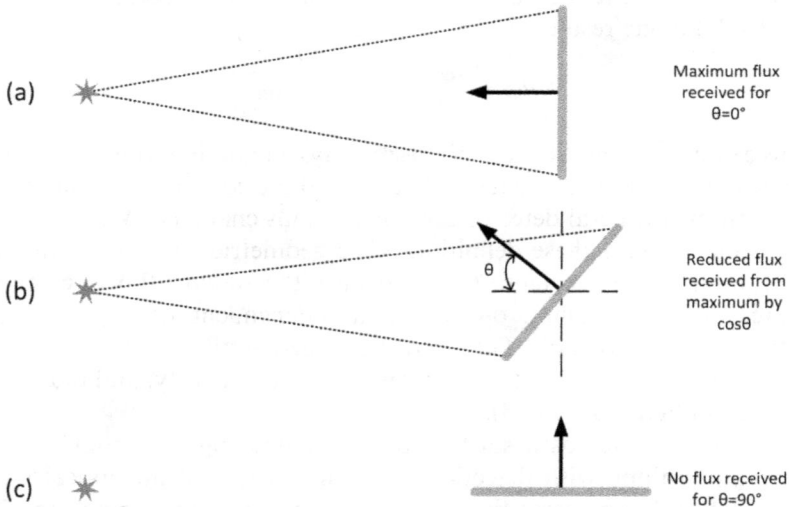

Figure 6.3. Projected area of a detector area element in relation to a point source. (a) Maximum flux received, (b) reduced flux received by the cosine of the angle of the surface normal with respect to the source, and (c) no flux received.

source. This is referred to as a projected area with respect to a source, $dA_{\text{proj}} = dA \cos \theta$. Although the area element is the same size in all cases its orientation reduces the received flux. Note also that the solid angle at the source and subtended by the rotated area element in figure 6.3(b) is smaller than the solid angle of figure 6.3(a) where a maximum flux is received.

Table 6.1. Radiometric quantities, symbols, and typical units of measure.

Quantity	Symbol	Unit (MKS)
Radiant energy	Q_e	J
Radiant energy density	w_e	J m^{-3}
Radiant flux (radiant power)	$\Phi_e = \frac{dQ_e}{dt}$	W
Irradiance (radiant flux areal density)	$E_e = \frac{d\Phi_e}{dA}$	W m^{-2}
Radiant exitance	$M_e = \frac{d\Phi_e}{dA}$	W m^{-2}
Radiant intensity	$I_e = \frac{d\Phi_e}{d\omega}$	W sr^{-1}
Radiance	$L_e = \frac{d^2\Phi_e}{dA\,d\omega\,\cos\theta}$	W m^{-2} sr^{-1}

Where t is time, A is the surface area receiving or emitting radiant flux, ω is the solid angle in the given direction at a point on the surface of a source, and θ is the angle between the surface normal and the direction of the solid angle. The subscripts e explicitly show that the quantities are related to energy units.

Another term important to the geometrical relationships found in radiometry is that of projected solid angle [8, 11]. This projection arises when an extended source surface normal is tilted with respect to the line of sight to the detector element. Like the projected area, the projected solid angle $d\Omega$ is defined as,

$$d\Omega \equiv d\omega \cos \varphi$$

where the angle φ is the angle between the surface normal of the source and the line of sight to the detection element. The quantity is not easy to immediately visualize except in a differential form for small detection area elements [11]. To determine a complete projected solid angle there is an integration over angles that must be included. When both source and detection area become large relative to each other a form factor or configuration factor can be calculated and used for certain geometries [12].

6.3 Radiometric quantities and definitions

Radiometry is based on the energy content of the electromagnetic radiation. All radiometric quantities relate back to radiant energy. Table 6.1 lists the various radiometric quantities along with symbols and units associated with each quantity [9]. We have already defined and used a subset of these physical quantities in discussions of electromagnetic radiation and basics of flux propagation.

It is useful to look at a few examples of the interrelationships between some of these quantities. Take a 1.0 mW continuous wave (constant radiant flux with time) laser pointer with a 1.0 mm^2 beam area. Using the definition of radiant flux, we know that the amount of energy delivered can be written as,

$$dQ_e = \Phi_e(t)\,dt$$

Since the flux with time is a constant the energy delivered by this beam in a 1.0 s time interval is $Q_e = (1.0\text{ mW})(1.0\text{ s}) = 1.0$ mJ. Now take a 1000 W pulsed laser source with a pulse width of 1.0 μs and pulse rate of one pulse every 10 s and with 1.0 mm^2 beam area. For this case, the energy deposited is $Q_e = (1000\text{ W})(1.0\text{ μs}) = 1.0$ mJ.

The same energy as the laser pointer and the same beam area. The critical difference is not just energy deposited but power density, the power delivered per unit area or irradiance.

As another example: What would be the difference in measuring the radiation from a 5.0 mW HeNe laser source and an ordinary incandescent light bulb using an optical power meter? As shown in figure 6.4(a), the detector collects all of the light from the small beam within the detector area element so the power reading on the instrument will provide the total power of the laser. However, for the light bulb of figure 6.4(b) only a fraction of the total light flux output from the bulb will be recorded by the meter. The ratio of the power read by the meter to the area of the detector is the irradiance.

These examples provide an initial illustration about the interrelationship between radiometric quantities and the importance of recognizing what values are being computed and/or measured in particular situations for different types of sources and detectors. In the sections that follow we will use the radiometric definitions to obtain some useful relationships for various source and detection models.

6.3.1 Point source radiometry

The definition of radiant intensity provides a source quantity that is the amount of radiant flux emitted by a source per unit solid angle. Since a point source has no extent it is only a model used for a small source radiating uniformly in all directions. The intensity of a point source is a constant because the amount of flux per unit solid angle from the source is the same in all directions. In general, the intensity of a source is a function of angles with respect to the direction of the line of sight to a detection surface and the source area.

The term intensity as used in most introductory physics textbooks is not the same as used here and in the standard definitions. Historically, intensity has been used to describe the time average of the Poynting vector, or power per unit area [13]. This is in fact the definition of irradiance.

Since the intensity of a point source is a constant it becomes a useful source model to use. In fact, we used this extensively in discussing rays and geometrical optics. Figure 6.5 shows a point source with intensity I_e and a detection are element dA at distance z from the source whose surface normal vector is along the line-of-sight

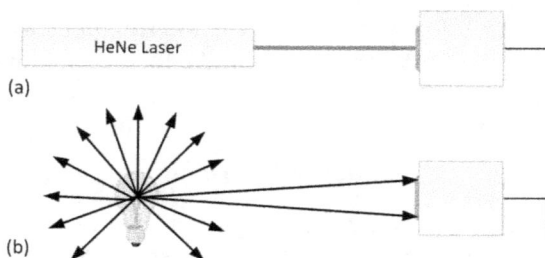

Figure 6.4. Comparison of optical power meter for two different sources. (a) Helium–neon laser source. (b) Ordinary incandescent light bulb.

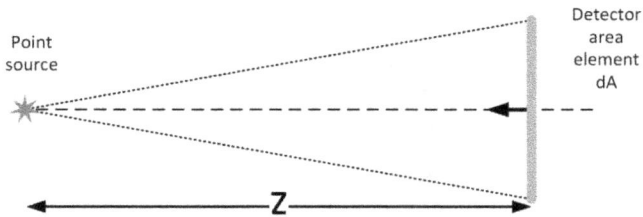

Figure 6.5. Irradiance from a point source.

direction. Using the definitions of intensity of the source and irradiance at the detection surface element dA, we equate the differential flux collected by the detector area subtended within in the solid angle $d\omega$ to obtain,

$$E_e dA = I_e d\omega$$

Since $d\omega = \frac{dA}{z^2}$, we find,

$$E_e = \frac{I_e}{z^2}$$

which is the familiar inverse square law for irradiance from point sources [3–6].

If the normal vector to the detector area element is not along the line of sight, then we must take into account the projected area in the irradiance calculation. Again, we will start with a point source with intensity I_e a perpendicular distance z from a detection surface area element. Now we want to calculate the irradiance falling on another surface area element in the same plane but at a distance d from the first element. Figure 6.6 shows both a cross-section and three-dimensional representation of this geometry.

As before, take the point source intensity as a constant I_e, then the irradiance a distance z from the source is, $E_e = \frac{I_e}{z^2}$. Define the irradiance at distance z' to be E_e'. From geometry, taking the angle $\theta = \tan^{-1}\left(\frac{d}{z}\right)$, then $z' = \frac{z}{\cos\theta}$. In addition, the detector area projection a distance z' is $A' = \frac{A}{\cos\theta}$. Using these values, we can now compute the irradiance E_e',

$$E_e' = \frac{I_e}{z'^2}\cos\theta = \frac{I_e}{z^2}\cos^3\theta = E_e\cos^3\theta$$

which is referred to as the cosine-cubed law for point sources. This provides a model for the reduction in irradiance observed from a source when a detector is translated in a perpendicular plane from a source. When there are multiple sources illuminating a detector area then the irradiance falling on the area is a simple sum of the irradiance from each source. As in this case the appropriate projected areas and distance from source to detector must be used.

Earlier we used point sources in imaging using both ray optics and wave optics. We will now examine point source imaging from the point of view of radiometry.

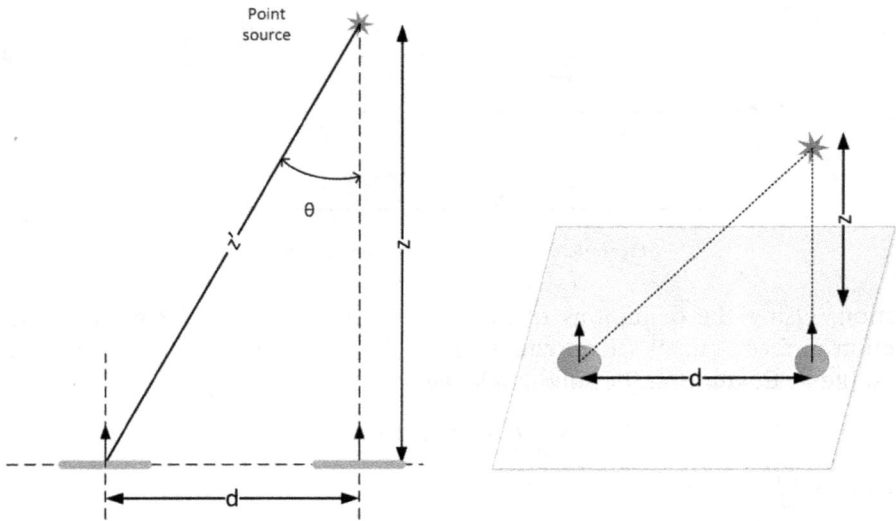

Figure 6.6. Irradiance falling on a detector area from a point source with normal vector along the line of sight and another area in the same plane but translated a distance d.

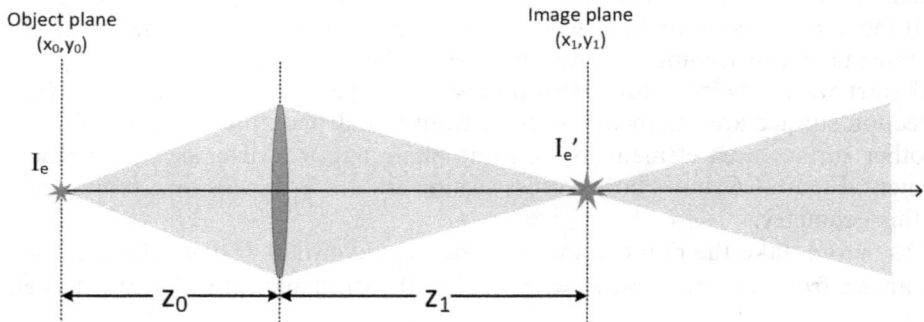

Figure 6.7. Imaging a point source with a thin lens of focal length f and diameter D. Solid angle cones in object space and image space are highlighted illustrating the flux collection of the lens.

The layout of a point source imaging system using a lens of focal length f and diameter D is shown in figure 6.7. A point source object with intensity I_e is imaged to form a point image.

The function of the lens is to collect the emitted radiation from the source, as we have seen earlier. From a radiometry perspective this means that the flux within a solid angle cone ω_0 in object space is directed to the image of the point source in image space with a new intensity I_e' within a solid angle cone ω_1. For a lossless system, the flux to the left of the lens must equal the flux to the right of the lens or,

$$I_e\omega_0 = I_e'\omega_1$$

But, $\omega_0 = \frac{A_{lens}}{z_0^2}$ and $\omega_1 = \frac{A_{lens}}{z_1^2}$ resulting in an image intensity of,

$$I_e' = \left(\frac{z_1}{z_0}\right)^2 I_e = m^2 I_e$$

where m is the magnification of the system. If the magnification is greater than one, then the image intensity is greater than the object intensity. However, energy conservation is not violated because the image intensity only delivers flux within the solid angle cones in image space whereas the object intensity projects radiation uniformly in all directions in object space.

6.3.2 Extended sources

When we discussed ray optics and the image formed by large objects, we used the concept that the object was comprised of a large number of point sources. We follow the same reasoning in radiometry for extended source objects. An extended source object with a given surface area is then built up from a large number of point sources each radiating from the surface with a flux per unit projected solid angle per unit area. The radiometric term that describes this process is radiance, L_e.

The final term from table 6.1 to be discussed is the extended source quantity radiant exitance. Radiant exitance is the power or flux per unit area of the source. The similarity of the definition to irradiance sometimes leads to confusion. A good way to think of radiant exitance is to take a large radiating surface area and have a small detector located a large distance away move closer to the surface. The irradiance falling on the small detector (area dA) will be from contributions at all points on the source. When the detector is placed directly on the source surface the irradiance will be the radiant exitance at that point on the surface within source area dA. Radiant exitance and radiance are directional quantities and depend on source characteristics and geometry.

Figure 6.8 shows an arbitrary shaped source of radiance L_e with a highlighted surface element dA. This element is radiating into solid angle $d\omega$. From the definition of radiance, the amount of flux from this small surface area is then,

$$d^2\Phi_e = L_e\, dA\, d\omega\, \cos\theta$$

where θ is the angle between the surface normal vector and the direction of observation.

Again, L_e is a directional quantity over the surface area so it can be a function of angular and surface area variables θ and φ, as shown in figure 6.8(b) where the normal vector of surface element, dA, is oriented along the z-axis direction. Using this definition if we integrate over the area of the source only then we obtain,

$$d\Phi_e = d\omega \int L_e \cos\theta\, dA$$

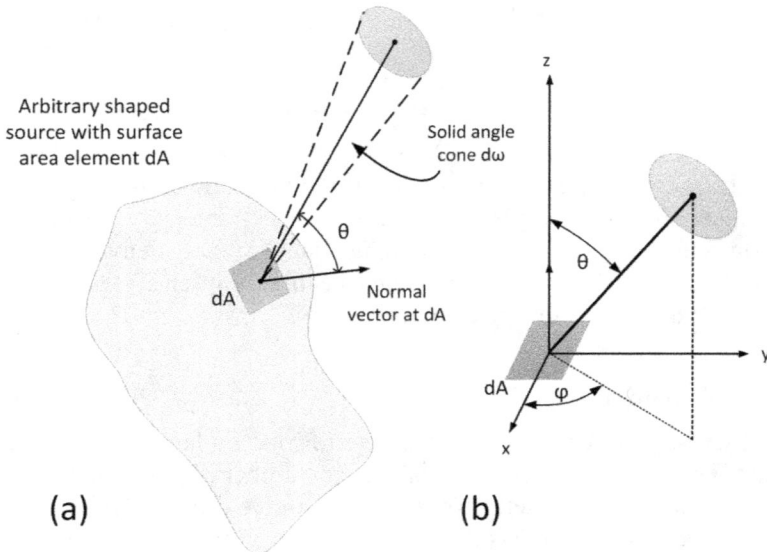

Figure 6.8. A source of radiance L_e radiating from a surface element into a solid angle. (a) General three-dimensional view, and (b) coordinate representation with surface normal vector pointing in a z-axis direction.

But from the definition of intensity as the flux per unit solid angle, we find,

$$I_e = \int L_e \cos \theta \, dA$$

So, the intensity of a source can be calculated knowing the radiance function. Similarly, if we integrate over the solid angles,

$$d\Phi_e = dA \int L_e \cos \theta \, d\omega$$

But the radiant flux per unit surface area element is the radiant exitance, so that,

$$M_e = \int L_e \cos \theta \, d\omega$$

In both relations the cosine factor has been retained within the integrand to emphasize that the solid angle integration is over projected solid angle, and the area integration is over the projected area.

6.3.3 Conservation of radiance and radiative transfer

Radiance quantities are not just related to sources but can also be used for reflective surfaces. In fact, radiance is even more fundamental and as a field quantity can be calculated at any point along some direction of propagation [14].

Suppose that we have a surface area dA_1 radiating into a second surface area dA_2 separated by a line-of-sight distance ρ all in the same lossless medium. Surface element 1 has a normal vector with respect to the line-of-sight θ_1 and surface 2 has its

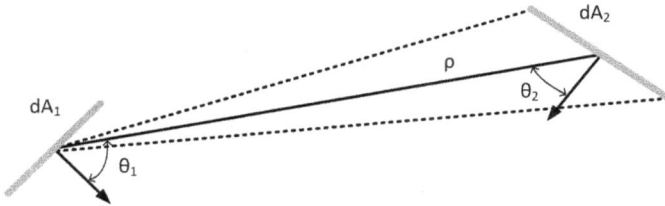

Figure 6.9. Radiance relationship between two surface areas in a uniform medium. The dotted lines from surface 1 to surface 2 indicate the solid angle subtended by dA_2 from dA_1.

surface normal vector at angle θ_2. This geometry is shown in a single plane in figure 6.9.

From our definitions of solid angles, we define the solid angle subtended by surface dA_2 at dA_1 to be,

$$d\omega_1 = \frac{dA_2 \cos \theta_2}{\rho^2}$$

And likewise, the solid angle subtended by surface dA_1 at dA_2 is,

$$d\omega_2 = \frac{dA_1 \cos \theta_1}{\rho^2}$$

Taking the radiance at surface 1 to be L_{e1} and using the definition of radiance, the differential flux from dA_1 is,

$$d^2\Phi_e = L_{e1} \, dA_1 \, d\omega_1 \cos \theta_1$$

Since radiance can also be associated with reflections from objects, we take the radiance from surface as L_{e2} and the flux received back to surface 1 is,

$$d^2\Phi_e = L_{e2} \, dA_2 \, d\omega_2 \cos \theta_2$$

Equating flux and substituting the definitions of each solid angle into the relation gives,

$$L_{e1} \, dA_1 \left(\frac{dA_2 \cos \theta_2}{\rho^2} \right) \cos \theta_1 = L_{e2} \, dA_2 \left(\frac{dA_1 \cos \theta_1}{\rho^2} \right) \cos \theta_2$$

Resulting in,

$$L_{e1} = L_{e2}$$

The radiance at each surface area is the same value, showing that radiance is conserved in a lossless optical system. Again, this is a consequence of the field concept of radiance [14]. In this calculation we have assumed that the intervening medium is the same or has the same refractive index which is the case in many situations. A more general conservation law, the radiance theorem shows that actually the radiance quantity L_e/N^2 is conserved where N is the refractive index of the medium. This conservation law must be applied when a beam propagates across a boundary between two different media [6].

Conservation of radiance is useful in many calculations since the radiance at points within an optical system remain the same value. It also shows that the flux received by a detection surface from an object with known radiance is,

$$d^2\Phi_e = L_e \frac{dA_1\, dA_2 \cos\theta_1 \cos\theta_2}{\rho^2}$$

Another consequence of conservation of radiance can be seen when we equate flux or,

$$d^2\Phi_e = L_{e1}\, dA_1\, d\omega_1 \cos\theta_1 = L_{e2}\, dA_2\, d\omega_2 \cos\theta_2$$

But now since radiance is conserved, we find that,

$$dA_1\, d\omega_1 \cos\theta_1 = dA_2\, d\omega_2 \cos\theta_2$$

or in terms of the projected solid angle,

$$dA_1\, d\Omega_1 = dA_2\, d\Omega_2$$

which is a quantity defined as the throughput, geometric extent, or étendue [3, 8, 12]. We will see this quantity again, particularly in discussions of illumination.

6.4 Lambertian sources and surfaces model

A particular model used in many radiometric analyses of optical systems is an assumption that the radiance is a constant value with angle and area. The term used to describe such objects is Lambertian [3–6]. When radiation is incident on a surface the reflected distribution is a constant then the surface is said to be Lambertian.

6.4.1 Intensity of a Lambertian source and Lambert's law

From the definition of radiance, we found that the intensity of a source could be computed if we know the radiance. This relation is,

$$I_e = \int L_e \cos\theta\, dA$$

If the radiance is a constant with area of the source then the radiance function may be factored out of the integral to obtain,

$$I_e(\theta) = L_e \cos\theta \int dA = L_e A_s \cos\theta$$

where A_s is the entire source area. This shows us that the intensity for such a source is angular dependent and is directly related to the cosine of the angle from the source normal. When the angle is $\theta = 0°$ then the intensity is simply, $I_e(0) = L_e A_s$, and,

$$I_e(\theta) = I_e(0) \cos\theta$$

which is known as Lambert's Law. Returning to the definition of radiance,

$$L_e(\theta) = \frac{d^2\Phi_e}{dA\,d\omega\,\cos\theta} = \frac{dI_e(\theta)}{dA\,\cos\theta} = L_e(0)$$

We see that the radiance at all angles is a constant value $L_e(0)$ or the radiance of this particular type of source is independent of viewing angle.

As we did with point sources, we can compute the irradiance falling on a surface area element a perpendicular distance z from a Lambertian extended source. We can also compute the irradiance falling on another surface area element in the same plane but at a distance d from the first element. Figure 6.10 shows both a cross-section and three-dimensional representation of this geometry.

For this case we wish to again find the irradiance as a function of the angle θ. Our previous result for point sources gives us the relation between irradiance and intensity where in this case the angular functions are shown explicitly for $\theta = 0°$.

$$E_e(0) = \frac{I_e(0)}{z^2}$$

Now the irradiance E_e' at a general angle θ becomes,

$$E_e'(\theta) = \frac{I_e(\theta)}{z'^2}\cos\theta = \frac{I_e(0)}{z^2}\cos^4\theta = E_e(0)\cos^4\theta$$

called the cosine-fourth law for Lambertian sources. The additional cosine factor from the cosine-cubed law arises from the projection of the extended source.

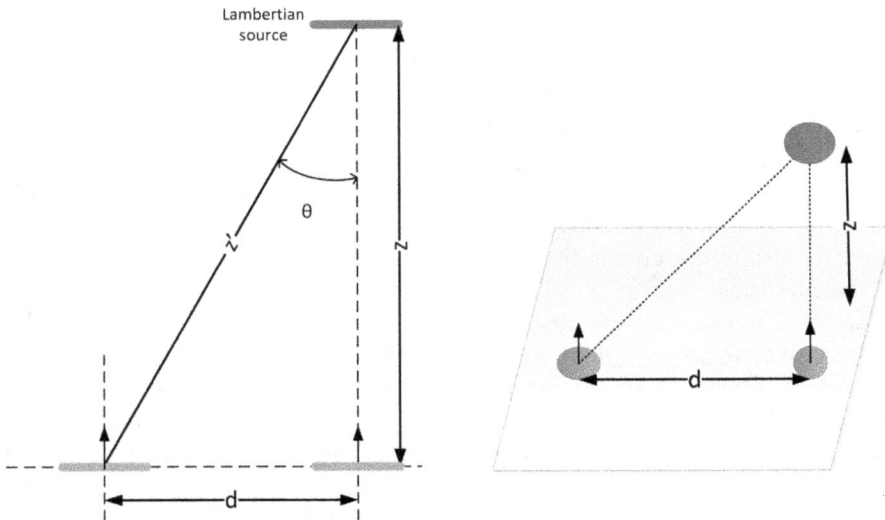

Figure 6.10. Irradiance falling on a detector area from a Lambertian extended source with normal vector along the line of sight and another area in the same plane but translated a distance d.

6.4.2 Radiant exitance of a Lambertian source

We can also compute the radiant exitance of a Lambertian source. Using the relationship from above,

$$M_e = \int L_e \cos \theta \, d\omega$$

From our definition of the differential solid angle, this relation becomes,

$$M_e = \iint L_e \cos \theta \sin \theta \, d\theta \, d\varphi$$

Since we desire to find the total radiant exitance from this source, the limits of integration must cover an entire hemisphere because the source radiates in all angles from the extended surface area. If we take the values of φ to be between 0 and 2π, then for a complete hemisphere the limits of the θ integration must be from 0 to $\pi/2$. Thus, for a Lambertian source with constant radiance, the radiant exitance reduces to,

$$M_e = 2\pi L_e \int_0^{\frac{\pi}{2}} \cos \theta \sin \theta \, d\theta = \pi L_e$$

One might think, from a units perspective, that the radiant exitance would just be the product of the radiance multiplied by the solid angle of a hemisphere, 2π. That would be true for an isotropic source, a source with some extent radiating like a point source. The cosine dependence of the projected solid angle again plays a role in this calculation.

6.4.3 Irradiance from a disk Lambertian source

The specific source geometry of a disk is one of the most common optical configurations as seen in lens cross-sections and apertures. Because of conservation of radiance, if a lens collects flux from an extended source then the radiance in the lens aperture will be the same radiance as the source. We will compute the irradiance falling on a small detection surface on-axis of a disk Lambertian source and then examine some limiting effects that are useful approximations in performing radiometric calculations.

Figure 6.11 shows the geometry of the disk source or radius, a, with respect to a small detector element (dA_{det}) on the disk axis at a general distance z. The disk lies in the x–y-plane of a right-handed coordinate system. We will define the area of the disk source as A_s. A ring of radius r and thickness dr is shown on the figure and we will calculate the flux reaching the detector from this ring. The distance from a general point on the ring to the center of the detector element is defined as ρ. The angle θ is the angle between the surface normal vector and the directed line segment ρ.

One must be careful in assigning the appropriate coordinate values when specifying radiance functions. Since we are assuming that the radiance is constant

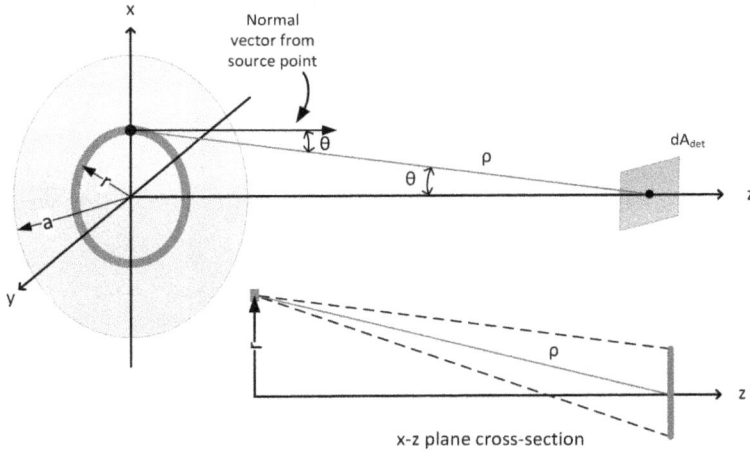

Figure 6.11. Geometry for calculating the irradiance on-axis a distance z from a disk Lambertian source. The insert in the figure shows a cross-section of the x–z-plane with the dashed lines outlining the solid angle subtended by the detector from the source point.

at all points on the disk source (Lambertian) then it is not critical. However, in general, the radiance could be a function of the position on the surface area, as well as the polar angle and the azimuthal angle at that particular surface position. For this particular situation, $L_e(r, \theta, \varphi) = L_e = $ constant.

From geometry we know that,

$$\rho^2 = z^2 + r^2 \quad \text{and} \quad z = \rho \cos \theta$$

The differential flux at the detector element from this ring of area dA_s is,

$$d^2\Phi_e = L_e \, dA_s \, d\omega \cos \theta$$

where $d\omega$ is the solid angle subtended by the detector area from a point on the ring. The dashed lines on the x–z-plane cross-section insert of figure 6.11 illustrate the solid angle boundary subtended by the detector element. From the definition of the solid angle,

$$d\omega = \frac{dA_{\text{det}} \cos \theta}{\rho^2}$$

The differential element over the surface can be written as, $dA_s = 2\pi r dr$, where the differential width of the ring on the disk surface is dr. Making substitutions of these relations into the flux equation leads to,

$$d^2\Phi_e = 2\pi L_e \, dA_{\text{det}} \frac{z^2 r}{(z^2 + r^2)^2} dr$$

Since this is a Lambertian source and the radiance is constant we only need to integrate over the distance r of the source from $r = 0$ to $r = a$,

$$d\Phi_e = 2\pi L_e dA_{\det} z^2 \int\limits_0^a \frac{r}{(z^2 + r^2)^2} dr$$

Since the irradiance at the detection area is, $E_e = \frac{d\Phi_e}{dA_{\det}}$, after integrating we obtain,

$$E_e = \pi L_e \frac{a^2}{z^2 + a^2}$$

Another alternative relation often used for the irradiance is,

$$E_e = \pi L_e \sin^2 \theta_{\max}$$

where $\theta_{\max}$ is the angle that the source subtends from the center of the detector on axis, or,

$$\sin \theta_{\max} = \frac{a}{\sqrt{z^2 + a^2}}$$

Examining the limits of the irradiance function for a disk also provides some physical insight. First, take the limit as the disk radius, a, becomes large. In this case, $E_e = \pi L_e$ which is just the radiant exitance. Second, in the limit that $z \gg a$, the irradiance function becomes,

$$E_e \approx \pi L_e \frac{a^2}{z^2} = L_e \frac{A_s}{z^2} = L_e \Omega_s$$

where Ω_s is the projected solid angle subtended by the disk source from the detection area. This relation is a fairly good assumption in many practical circumstances for optical systems where small angle approximations are used.

The implications of the approximate limits can be observed by comparing the irradiance from a disk source to that of the inverse square law. Figure 6.12 is a plot of both the exact solution (solid line) and the inverse square law (dashed line) for an example of a disk Lambertian source with radiance $L_e = 3.0$ W sr^{-1} m^{-2} and radius $a = 12.5$ mm.

The absolute difference between the curves of course depends on the magnitude of the radiance. However, as one can see in this example, the curves begin to approach one another at a distance of about four times the disk radius.

As an example of the utility of the disk Lambertian source relation, suppose that we have a large Lambertian source whose flux is collected by a lens of focal length f and diameter D. Because of conservation of radiance the radiance at the lens plane is the same as the radiance in the source. See figure 6.13(a). The lens, which is the aperture stop in this case, effectively becomes a disk Lambertian source and the irradiance in the region about the optical axis is,

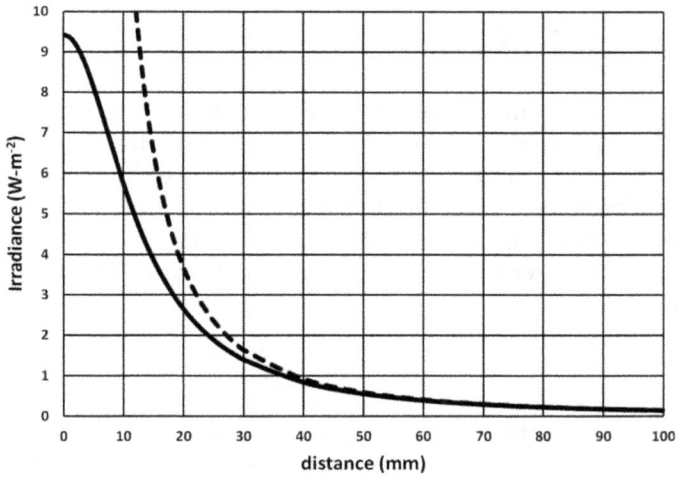

Figure 6.12. Comparison of the inverse square law irradiance approximation (dashed line) to the irradiance of a disk Lambertian source (solid line).

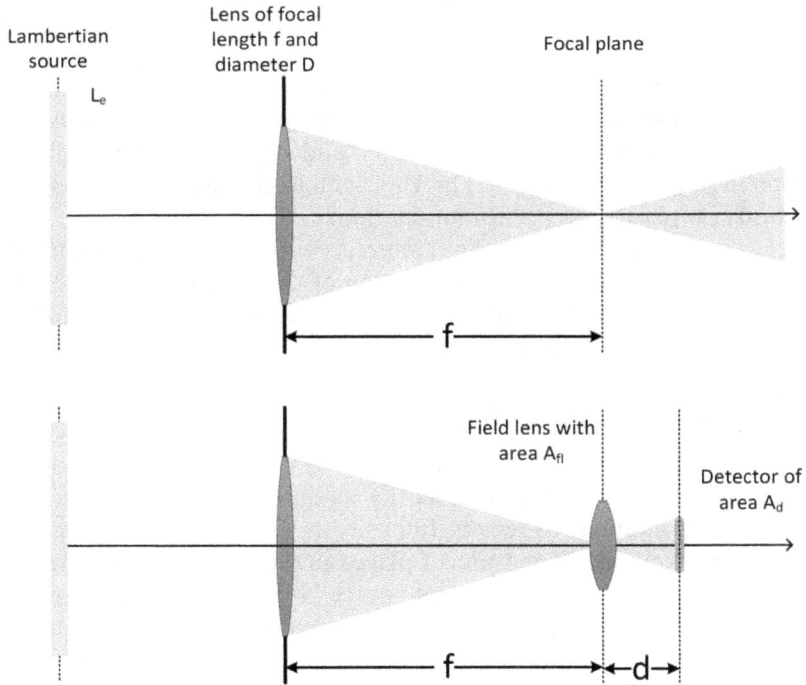

Figure 6.13. Imaging a Lambertian source (a) in back focal plane of a lens, and (b) addition of a field stop.

$$E_e = L_e \frac{A_{\text{lens}}}{f^2} = L_e \frac{\pi(D^2/4)}{f^2} = L_e \frac{\pi}{4(f/\#)^2}$$

Providing another alternative relation for the irradiance. One can also include a multiplicative lens transmittance T_{lens} to account for any lost flux within the lens.

Figure 6.13(b) shows a similar arrangement with the addition of a field lens (area A_{fl}) placed in the back focal plane becoming the field stop for this system. If the field lens images the aperture stop onto the detector a distance d from the field lens with image size equal or greater than the detector area, then the power collected by the detector of area A_d is,

$$\Phi_e = T_{\text{sys}} L_e \frac{A_d A_{fl}}{d^2}$$

where T_{sys} is the transmittance of the system accounting for losses in each lens. In practice the addition of a field lens provides a more uniform flux distribution over the detector area and also helps to minimize collection of stray energy [8].

6.5 Radiation transmission and reflection

From our earlier discussion we saw that if one knows the radiance functional dependence of each differential area and solid angle from a source then both the radiant exitance function and the intensity function can be determined. Without information about the source radiance function none of these calculations can be performed. However, a direct measure of radiance can be difficult. This is why we often resort to geometrical source models ('bottom-up' method) or measured values ('top-down' method) in applying radiometry and photometry principles for light propagation in optical systems [12]. This topic will be discussed in greater detail later in the sections on photometry and illumination.

When electromagnetic radiation light interacts with a planar surface it will reflect, transmit, scatter, and absorb. The polarization of the waves may also change. This section deals with some of the effects of beams, incident flux reflecting from surfaces, and transmission of radiation through materials.

6.5.1 Angular intensity representations

Since radiance is an angular dependent function, we need to make angular measurements of some radiometric quantity at different points across the surface area of a source or illuminated region. Intensity as a function of angle is commonly measured using a goniometer instrument placed in the far field where the source can be assumed to be essentially like a point source [15]. Once the angular intensity is measured graphical methods for displaying the angular distribution are needed. This is true for the emission from sources as well as the flux reflected from a surface or transmitted through a material.

Figure 6.14 shows one way to represent information about a source by plotting the intensity of an isotropic (point source-like) model and a Lambertian source model as a function of angle. In this plot the functions are shown on a normalized

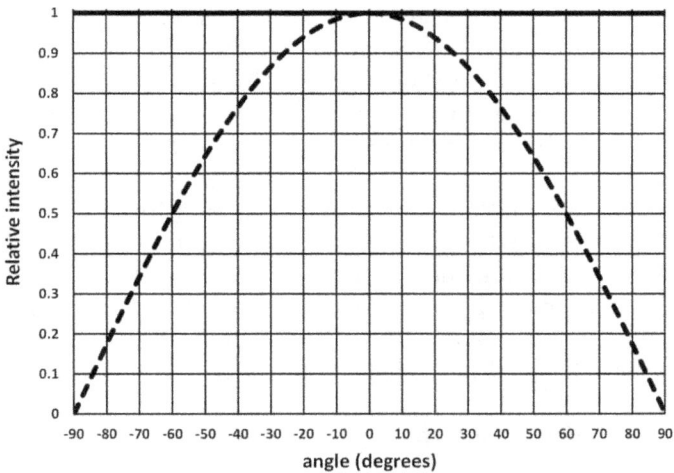

Figure 6.14. Intensity as a function of angle for a Lambertian source model (dashed line) and an isotropic/point source model (solid line).

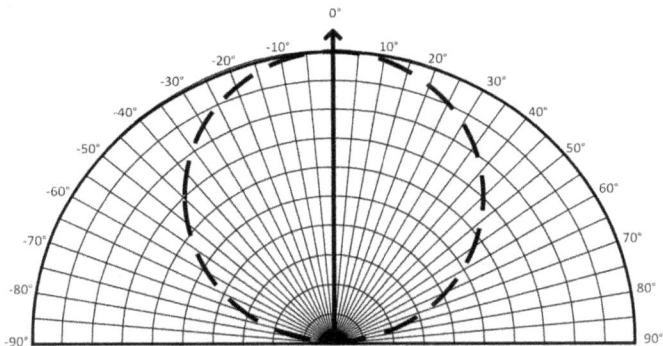

Figure 6.15. Intensity polar plots for a Lambertian source model (dashed line) and an isotropic/point source model (solid line).

intensity scale. The angular range indicates that the emitted radiation is contained within a hemisphere as one would see from a plane surface.

One of the main reasons that this graphical representation is used is because experimental data of the intensity of the output from a source is measured directly [15, 16]. This angular intensity data then allows an expected irradiance to be computed on a detector plane or other surface at any angular location from the center of the source.

Another common representation of this type of data is on a polar plot. The polar plot may show just a hemispherical range or a complete spherical radiation pattern. Figure 6.15 shows a hemispherical polar plot of the Lambertian and the isotropic source models. Polar diagrams such as these are commonly used in describing illumination from light sources and luminaires. We will use these types of diagrams extensively in later discussion about illumination systems.

The intensity plots are also consistent with our earlier descriptions of rays and ray interaction with surfaces. Not only is the ray direction provided but the amount of flux per unit solid angle is also shown. Figure 6.16 shows a typical example of a reflected ray incident on a mirror surface at an angle −30° to the surface normal. The angular intensity plot and polar plot for the ray reflected from the mirror surface are also shown.

The angular cone of radiation seen in figure 6.13 to the right of the focal plane can also be described graphically using these types of plots. Figure 6.17 shows and example of intensity plots for a uniform cone of rays between −30° and +30°.

All of the angular plots here describe radiation from a single plane. As the radiation propagates from one source plane to another detection plane then the irradiance may be calculated in the detection plane.

6.5.2 Reflected radiation from surfaces

In ray optics, the ray reflecting from a mirror surface is reflected in the same plane as the incident ray with the reflection angle equal to the incident angle as measured with respect to the surface normal. The assumption is that there is no loss of energy on reflection and the reflectance is unity. This is consistent with the definition of reflectance, R, or the ratio of the reflected power/flux to the incident power/flux.

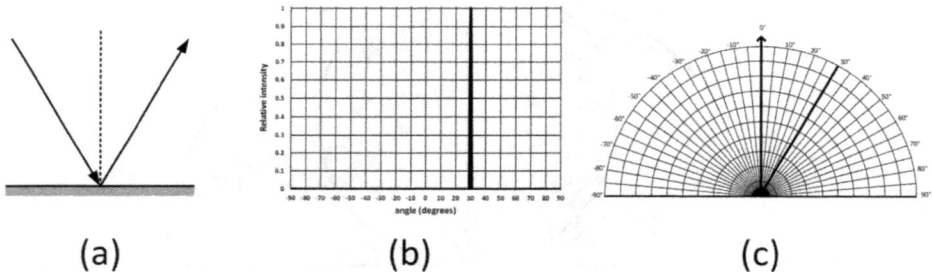

(a) (b) (c)

Figure 6.16. (a) Ray incident on a mirror surface, (b) angular intensity plot of the reflected ray, and (c) intensity polar plot.

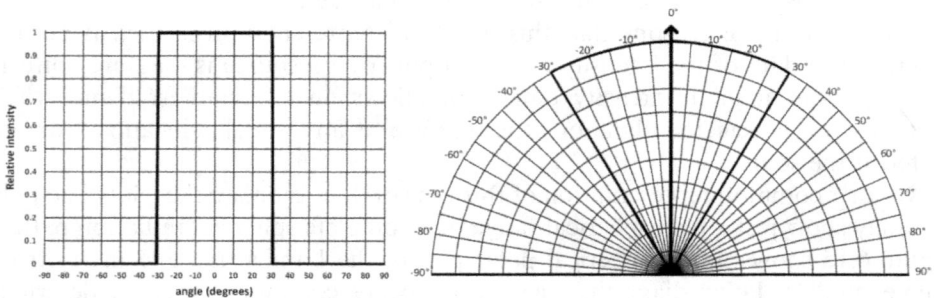

Figure 6.17. Angular intensity plot and polar plot for a cone leaving the back focal plane of the system shown in figure 6.13.

$$R = |r|^2 = \frac{\Phi_{er}}{\Phi_{ei}}$$

where r is the complex reflection coefficient at the surface, we obtained earlier in wave optics analysis. For a refractive surface, this reflectance is also a function of wavelength due to material dispersion. This also tells us that if the surface were a dielectric then the power in the reflected ray would be reduced from the incident power by the surface reflectance. This concept is shown in figures 6.16 and 6.18(a) with the reflected ray called the specular reflection.

If we examine the surface on a smaller and smaller size scale all surfaces deviate from the ideal planar surface or have a surface roughness. Roughness of surface leads to scattering also called a diffuse reflectance component. This occurs for mirror surfaces as well as surfaces of metallic and dielectric materials as described in earlier chapters. Quantifying the amount of flux lost at surfaces is important in designing systems because effects such as reduced transmitted flux, image blur, and lower image contrast degrade performance [4, 17, 18].

Radiation reflecting from a surface within the Lambertian model would result in a completely diffuse reflectance. If specular reflection is at one extreme of the reflectance from a surface and Lambertian reflectance is at the opposite extreme them most surfaces will be somewhere between. These concepts are illustrated in figure 6.18 using a geometrical polar plot representation.

The manner in which radiation reflects from surfaces depends on surface characteristics. Nicodemus defined a quantity called the Bidirectional Reflectance Distribution Function (BRDF) that gives a relationship between the incident irradiance and the resulting radiance of the scattered radiation [19, 20]. The illustrations in figure 6.18 are simple in-plane representations but the BRDF is in general a three-dimensional function. The BRDF function is defined as the ratio of the differential radiance reflected from a surface element to the differential incident irradiance on that surface element.

$$BRDF(\theta_i, \varphi_i, \theta_r, \varphi_r) = \frac{L_e(\theta_i, \varphi_i, \theta_r, \varphi_r)}{E_e(\theta_i, \varphi_i)}$$

where θ_i and φ_i are the polar and azimuthal angles of the incident radiation and θ_r and φ_r are the polar and azimuthal angles of the reflected distribution. This geometry is shown in figure 6.19. Note that the units of BRDF are per unit solid angle.

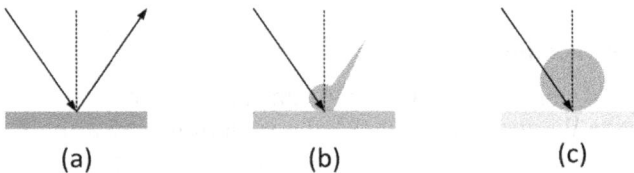

Figure 6.18. Reflectance from surfaces with different surface roughness. (a) Specular reflection at a planar surface, (b) large specular reflection and a small diffuse reflectance component, and (c) completely diffuse reflectance.

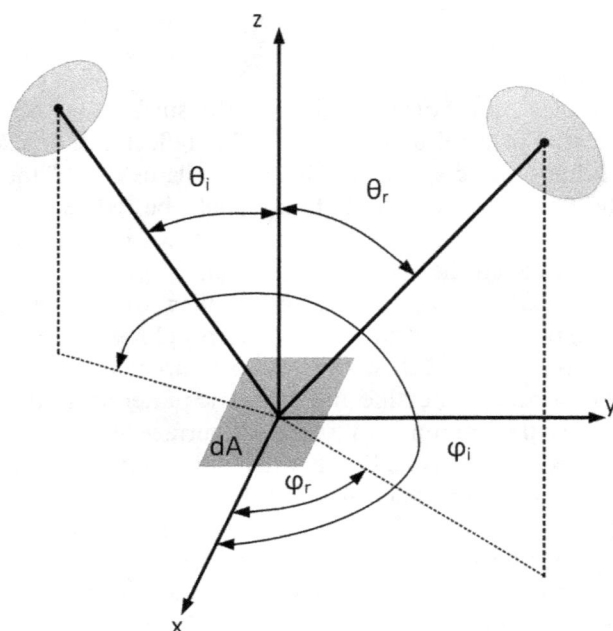

Figure 6.19. Incident irradiance falling on surface element dA and reflected radiance from dA.

For a Lambertian surface the incident irradiance results in a radiant exitance at the surface equal modified by the surface reflectance, $M_e = RE_e = \pi L_e$ thus the BRDF of a Lambertian surface is,

$$\text{BRDF}_{\text{Lambertian}} = \frac{R}{\pi}$$

For a general surface BRDF is a function of wavelength and polarization state. A number of models have been developed to provide information to designers about different types of surfaces that require only a few variables to adequately describe the scattering/reflectance effects [12, 18]. Details of these models are beyond the scope of our discussion other than to recognize their existence.

There are surfaces and surface coatings that have been developed to approximate Lambertian reflectance. For example, the thermoplastic resin material Spectralon has a reflectance of greater than 99% within the 350–1500 nm spectral range [21].

6.5.3 Transmitted radiation through materials

A ray incident on a material surface may transmit (at a refraction angle) as well as reflect. As we have discussed previously, the radiation may also absorb within the material as well as scatter. The scattering in this case is volume scattering and can be caused by numerous mechanisms including density fluctuations in the material, dislocations, and particles [17, 22].

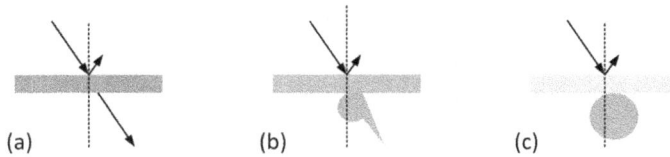

Figure 6.20. Transmittance through a material slab. (a) Refracted, reflected, and transmitted rays, (b) refracted beam spread out due to volume and surface scattering effects and a small diffuse transmittance component, and (c) complete diffuse transmittance. Details of the radiation within the material are not shown but notice the slight shift of the transmitted beams due to refraction.

Figure 6.20(a) shows an incident ray on the surface of a slab of material that follows the laws of geometrical optics. In this case there is no scattering or absorption included. The irradiance transmitted through the slab will be reduced due to the reflectance at each surface according to Fresnel reflections at each surface. At the other extreme, the material characteristics dictate that we have nothing but a complete diffuse transmittance, figure 6.20(c). As with surface reflection, figure 6.20(b) shows an intermediate case for radiation transmission where there is a transmittance distribution as well a diffuse component. Again, the material characteristics of each slab are different causing a different transmittance effect. A small, reflected ray is also shown indicating that reflection is also occurring at the incident surface as discussed in the section on Fresnel reflection.

Optical diffusers find numerous applications in systems where reducing directionality of a transmitting beam is required. Opal glass has long been employed as a near-Lambertian diffusing source and a cosine corrector for collecting radiation over a large field of view. Unfortunately, opal glass has the disadvantage of a low transmittance [23]. Figure 6.21 shows data from angular resolved scattering from an opal glass with incident light from a HeNe laser plotted along with a Lambertian model.

Ground glass diffusers, made by either sand blasting or polishing a glass substrate with a particular size grit, are a cost-effective option for situations such as reducing filament or LED 'hot spots' from a source or providing a viewing screen in imaging systems. Both opal glass and ground glass elements can be made easily into a variety of shapes and thicknesses depending on the application. An example of a relative intensity plot with angle for several different ground glass components polished with grit sizes of 120, 220, 600, and 1500 is given in figure 6.22 [24]. As one can see the larger the grit (smaller number) the lower the transmittance but the greater the diffusion.

Recently, light shaping diffusers have been developed to tailor the shape of a diffusing beam. Typical applications for homogenizing laser beams, improving backlight displays for cellphones, and shaping light exiting optical fiber bundles for illumination. The microstructure of the diffuser is designed to produce patterns such as circles, squares, lines, or other shapes. These diffusers are manufactured using techniques like lithography or holography [25, 26].

The intensity plots for some of these diffusers are provided in figure 6.23.

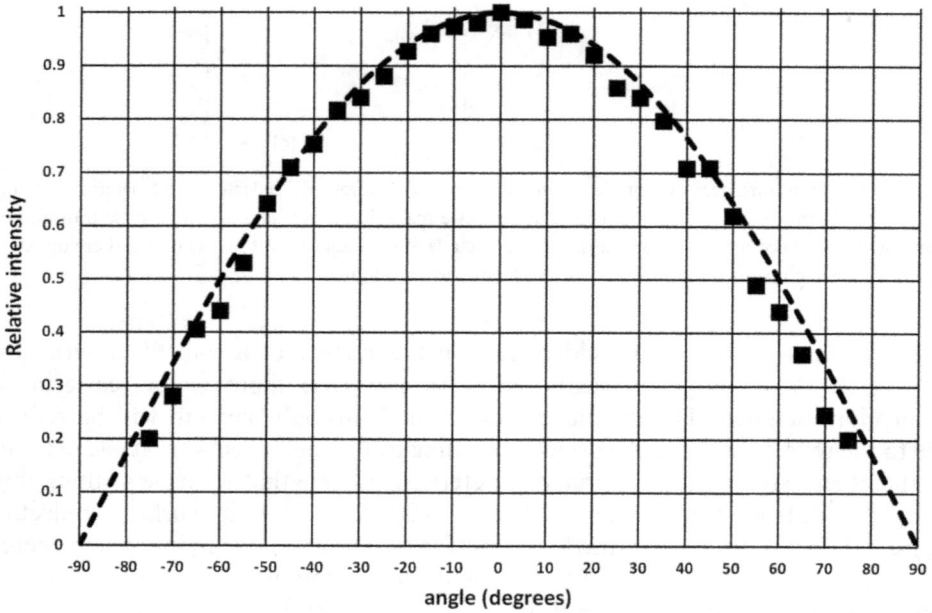

Figure 6.21. Intensity data as a function of angle from a HeNe laser incident on an opal glass sample at 0°. The dashed line is a comparison to a Lambertian model.

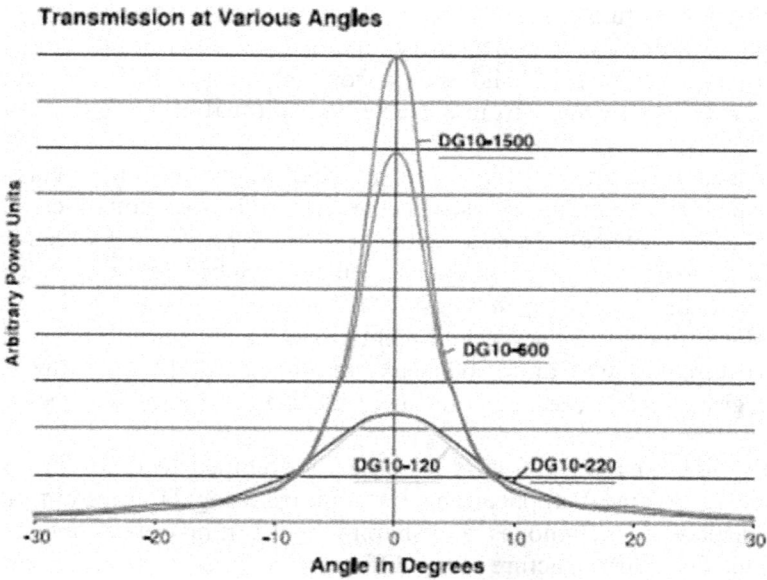

Figure 6.22. Transmission as a function of angle for different ground glass optical components [24]. (Reprinted with permission from Thorlabs, Inc. (https://www.thorlabs.com/), Product part number: DGK01 Mounted Ground Glass Diffuser Kit.)

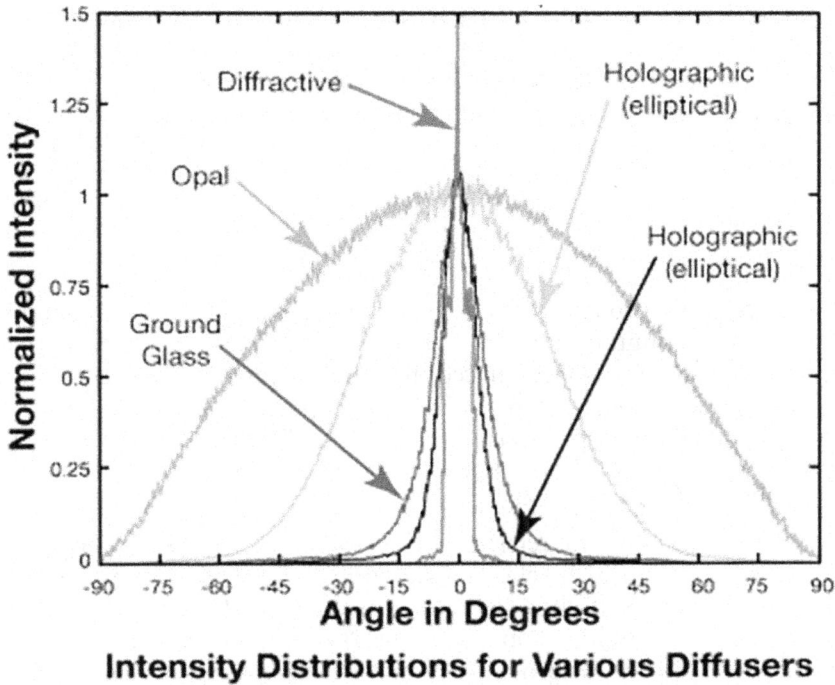

Intensity Distributions for Various Diffusers

Figure 6.23. Transmission as a function of angle for different engineered diffusers [24]. (Reprinted with permission from Thorlabs, Inc. (https://www.thorlabs.com/), Product part number: EDK01 Mounted Engineered Diffuser Kit.)

When all processes, reflectance, R, transmittance, T, and absorptance, A are included for the situation of an incident beam onto a material slab, by conservation of energy we must have,

$$T + R + A = 1$$

Scattering is sometimes included within the any of these terms depending on the situation and the details of the known parameters of a specific problem. For example, scattering may be included in reflectance if scattering is primarily known from the surface as in the case of the BRDF analysis. Scattering effects may also be included in the macroscopic absorption if there is a similar exponential decrease in power due to scattering. As in wave optics, the reduction in irradiance after propagating through distance z is then,

$$E_e(z) = E_e(0)e^{-\alpha z}$$

where $E_e(0)$ is the incident irradiance at $z = 0$. If there is absorption in the bulk of the material as well as scattering the macroscopic absorption coefficient α will have two terms, one due to the absorption (α_A) and one due to scattering (α_S) so that, $\alpha = \alpha_A + \alpha_S$.

The absorption coefficient is also a function of wavelength because of material dispersion of the slab substrate as well as any dyes or pigments doped into the slab. This results in a colored filter. When the absorption is a constant value for a large wavelength range, we call this type of filter a neutral density filter. In this situation, the transmittance through the slab is written in terms of the optical density (OD) parameter,

$$OD \equiv \log_{10}\left(\frac{1}{T}\right) = \log_{10}\left[\frac{E_e(0)}{E_e(z)}\right]$$

The optical density is also referred to as absorbance. Thus, a 2.0 OD neutral density filter will have a transmittance of $10^{-2} = 0.01$. Optical density is a useful quantity in practice. When two filters are placed together their transmittance values are multiplied together which means we add the optical density. A 2.0 OD filter followed by a 1.0 OD filter produces a total of 3.0 OD, or a transmittance of 0.001.

6.6 Spectral radiometry

One of the distinguishing characteristics of sources are their spectra. Most sources, like we saw in the case of blackbody radiation, have a wavelength dependence. However, the definitions of all radiometric quantities used so far relate back to the total radiant energy. Spectral content of the radiant sources has not really been considered, except in the case of Gaussian laser beams where a monochromatic source at a single wavelength is assumed. To account for spectral content of radiometric quantities there is a corresponding spectral radiometric quantity that describes the wavelength dependence of that quantity. Various notation is used to indicate that the radiometric quantity is spectral [3–8]. We will distinguish the spectral quantities using an additional λ subscript. For example, the spectral radiant flux would be written as,

$$\Phi_\lambda \equiv \frac{d\Phi_e(\lambda)}{d\lambda}$$

and the units of this spectral radiant flux would be specified as power per unit wavelength. So, the flux in the wavelength interval between λ and $\lambda+\Delta\lambda$ is, $\Phi_\lambda d\lambda$, giving a total radiant flux of,

$$\Phi_e = \int_0^\infty \Phi_\lambda d\lambda$$

A corresponding integral relation follows for every radiometric quantity.

Suppose an optical filter is placed directly over a source. See figure 6.24(a). The filter modifies the spectral distribution of the source resulting in a total radiant exitance of,

$$M_e = \int_0^\infty M_{\lambda-\text{source}}T(\lambda)d\lambda$$

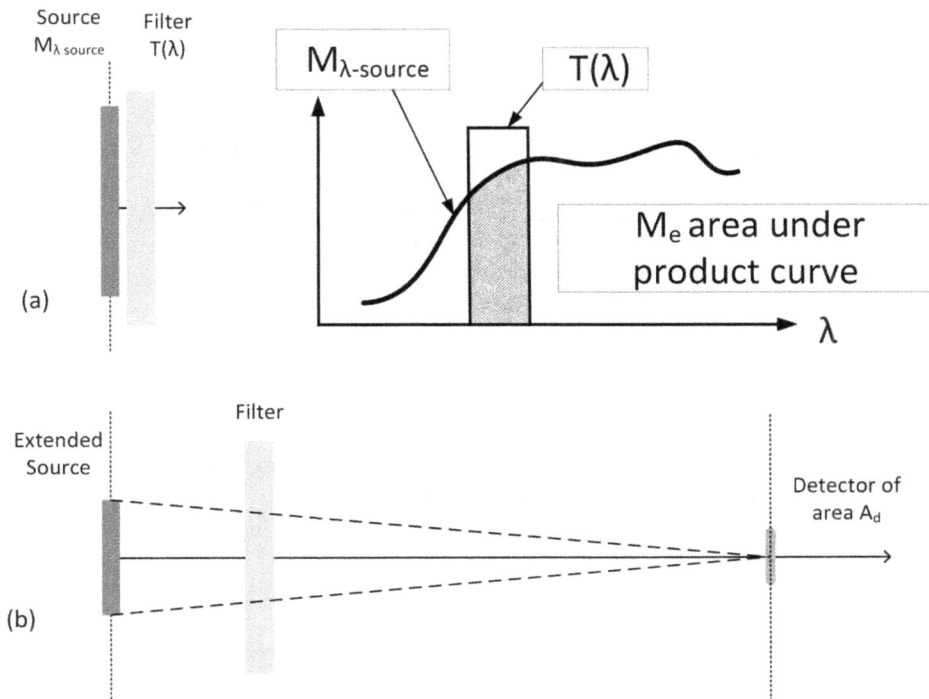

Figure 6.24. Optical filter placed behind an extended source (a) modifying the spectral radiant exitance, and (b) placed within the path between a source and detector modifying the detected spectral irradiance.

with the integrand being the new effective spectral radiant exitance. Graphically, the total radiant exitance is the area under the curve of the product of the spectral radiant exitance and the filter spectral transmittance function.

If a filter is placed somewhere in between a source and detector, as in figure 6.24(b) the spectral irradiance falling on the detector becomes,

$$E_\lambda = E_{\lambda-\text{source}}T(\lambda)$$

Detectors have their own spectral response function as discussed in the previous chapter. In this case, the total flux measured by the detectors is,

$$\Phi_e = A_d \int_0^\infty E_{\lambda-\text{source}}\mathcal{R}(\lambda)T(\lambda)d\lambda$$

Previously we found the energy density as a function of wavelength for a blackbody radiator. This was,

$$w_e(\lambda)d\lambda = \frac{8\pi hc}{\lambda^5}\frac{1}{e^{+\frac{hc}{\lambda kT}}-1}d\lambda$$

Calculating the total energy density for a blackbody we find,

$$w_e = \frac{4}{c}\sigma T^4$$

where the constant $\sigma = 5.67 \times 10^{-8}$ W m^{-2} K^{-4} is Stefan's constant. This value for energy density is a constant for a given temperature and is a statement of Stefan's law. It can be shown that the energy density is related to the radiant exitance and radiance, and since it is a constant, a blackbody source is a Lambertian source [2, 6]. From before for a Lambertian source $M_e = \pi L_e$, so,

$$M_\lambda d\lambda = \frac{c}{4}\left[\frac{8\pi hc}{\lambda^5}\frac{1}{e^{+\frac{hc}{\lambda kT}}-1}\right]d\lambda = \pi L_\lambda d\lambda$$

resulting in, $M_e = \int M_\lambda d\lambda = \sigma T^4$.

Surfaces that are not perfect blackbodies are called graybodies if a constant scaling factor can be used to match the spectrum of the radiant exitance of a blackbody. In terms of spectral radiant exitance, a graybody source can be represented as,

$$M_\lambda = \epsilon M_{\lambda-bb}$$

where $M_{\lambda-bb}$ is the blackbody spectral radiant exitance and ϵ is a constant called the emittance. In general, the emittance does not have to be a constant and can be a function of wavelength and angular variables for a radiant source. A more general description for spectral radiant emittance is,

$$\epsilon(\lambda, \theta, \varphi) = \frac{L_\lambda(\theta, \varphi)}{L_{\lambda-bb}}$$

where the spectral radiance of the source is compared to the spectral radiance of a blackbody.

6.7 Human eye

Light is the radiation contained within that region of the electromagnetic spectrum that creates a visual response in the human eye. As discussed in the previous chapter, radiometry concerns electromagnetic energy radiating at all wavelengths. In many cases the interaction of a human observer is an integral part of the system requirements. Thus, a separate terminology and set of measurement definitions have been specially formulated to use in quantifying visible light and designing optical systems in the visible portion of the spectrum.

When the eye is used to compare one source to another in an attempt to quantify an amount of visual stimulus, this field has historically been called photometry. The eye response is a function of wavelength as well as an overall sensitivity. A physical measurement standard emulating the eye's response is now used to measure photometric quantities that relate to amounts of light perceived by a human observer. The relationship between radiometry and photometry centers on the physical measurements that emulate the human eye as the detector of radiation. In this sense, photometry is a special case of spectral radiometry where the response function of the eye is known, and a photometric quantity is calculated. Before discussing photometry, we will briefly review the basic structure of the human eye and aspects of vision. Properties of the eye such as visual perception, acuity, brightness adaptation, and accommodation will also be discussed. This will naturally lead to the definitions of photometric quantities and their analogy to radiometric quantities.

6.7.1 Anatomy of the human eye

Figure 6.25 shows a cross-section of the human eye. The outer layer of the eye is comprised of the cornea, a clear portion in the front of the eye, and the sclera, the white portion forming the remainder of the roughly spherical shape. Light enters the eye and passes through a pupil whose size is controlled by the iris. The iris is a muscle that dilates and contracts to limit the amount of light accepted through the pupil. Just behind the pupil is the eye lens comprised of a membrane of clear fibers and held in place by ligaments that are attached to the ciliary muscle. Since the eye lens is not a uniform material medium, the refractive index of the eye lens is not a constant such as a piece of plastic or glass. In fact, the eye lens is a type of gradient index optical component whose value changes from the center of the lens out to its edge [27, 28]. The eye lens is somewhat pliable and is used for small focusing adjustments. The region between the cornea and the lens is filled with a fluid called the aqueous humor. The vitreous humor fills the region between the eye lens and the back of the eye. The purpose of both humors is to deliver nutrients to the eye lens

EYE ANATOMY

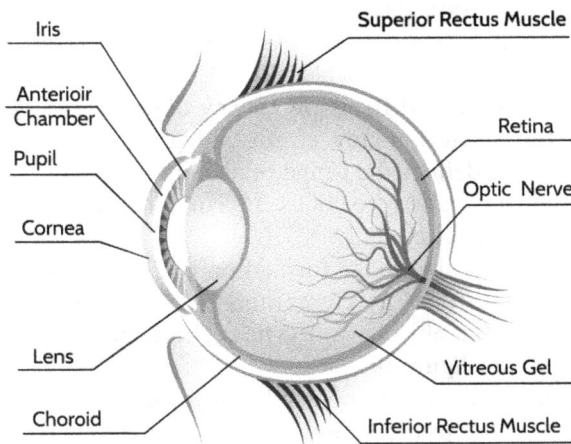

Figure 6.25. Anatomy of the human eye. (Credit: Freepik.com: Education vector created by macrovector (https://www.freepik.com/vectors/education)).

and the cornea since they do not contain blood vessels and to provide a pressure for the eye to maintain its shape.

Light detection occurs at the retinal layer opposite the cornea. The photosensitive elements are called rods and cones, a type of nerve cell. Low levels of incident light stimulate the rods. Cones are responsive to higher light levels and color and are concentrated near the center of the retina in a region known as the macula. Visual acuity is highest in this region. The choroid contains blood vessels transferring nourishment to the retina. The network of nerves within the retinal layers processes visual information and transfers this information to the brain through the optic nerve. The location of the connection of the optic nerve to the retina is called the blind spot since there can be no active rods or cones at this location.

6.7.2 Image formation by the eye

The refractive power of the eye occurs primarily at the cornea surface where the refractive index difference between the outer air and the cornea is highest. A nominal value of 1.376 is commonly used but may be as high as about 1.43 [27, 28]. There are of course differences from case-to-case for the optical properties of every portion of the eye so we must resort to using average values of refractive index and radii of curvature. Average values for the refractive index of the aqueous and vitreous humor of approximately 1.333 are often quoted. This has resulted in the development of many models for the optical properties and optical layout of the eye [29].

The eye lens is the second major refractive element. Because of the connection of the eye lens to the ciliary muscle its curvature can change which allows focusing near and far objects. This process is known as accommodation. Therefore, the cornea and eye lens form an optical system, like a compound thick, between two media, air, and the humor in front of the retina. The image formed by this type of system is real and inverted.

Because of accommodation the imaging properties of the eye are different for a relaxed eye and an accommodated eye. When the eye is completely relaxed the longest distance at which an object is clearly in focus is called the far point. As the eye begins to accommodate and focus on nearby objects there is a point where an object remains in focus without any strain called the near point. The distance between these two points defines the range of accommodation. An average value of the near point distance is often given as 25 cm but may be defined differently depending on the reference cited.

Figure 6.26(a) illustrates the image formed on the retina by a large object at the far point with a relaxed eye. For closer and smaller objects, the eye begins to accommodate and when the object is at the near point, the image is observed clearly with full accommodation, figure 6.26(b).

The simple magnifier relies on the concept of near point imaging. Suppose that we want to view a very small object of height h. The best image the eye can clearly see of this object is when it is placed at the near point. Taking the near point distance to be 25 cm, the angular extent of the small object is,

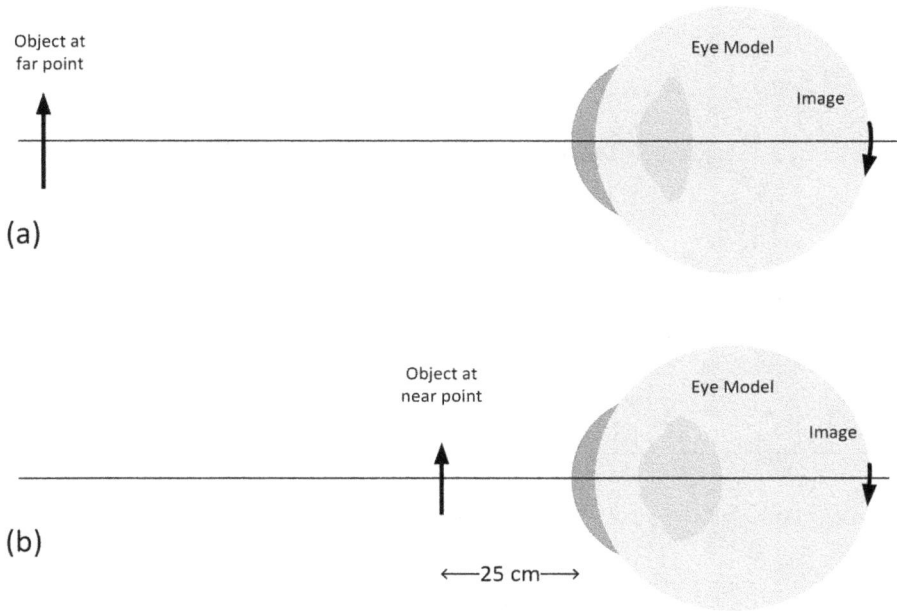

Figure 6.26. Image formed on the retina for an object at the (a) far point, and (b) the near point.

$$\theta = \frac{h}{25 \text{ cm}}$$

This angle is shown in figure 6.27(a). If the object is moved closer to the eye and a lens is inserted between the eye and the object, such that the object is at the focal length from the lens, then the eye sees a magnified view of the object. This type of magnification is angular magnification defined as,

$$M = \frac{\theta'}{\theta}$$

where θ' is the angle that the image subtends. This angle is also the ratio of the image height to the focal length, by similar triangles. See figure 6.27(b). Thus, a simple magnifier provides a magnification of,

$$M = \frac{25 \text{ cm}}{f}$$

Correcting refractive errors in the eye is common by the addition of spectacles or contact lenses. There are several conditions where corrections can be applied, and we will briefly discuss those effects. Myopia or nearsightedness occurs when the relaxed eye can only focus on nearby objects and distant objects are always out of focus. This condition may be caused by the eyeball being too long or the refractive power of the eye too high, focusing a distant object in front of the retina. The opposite condition, hyperopia or farsightedness occurs when distant objects can be seen but

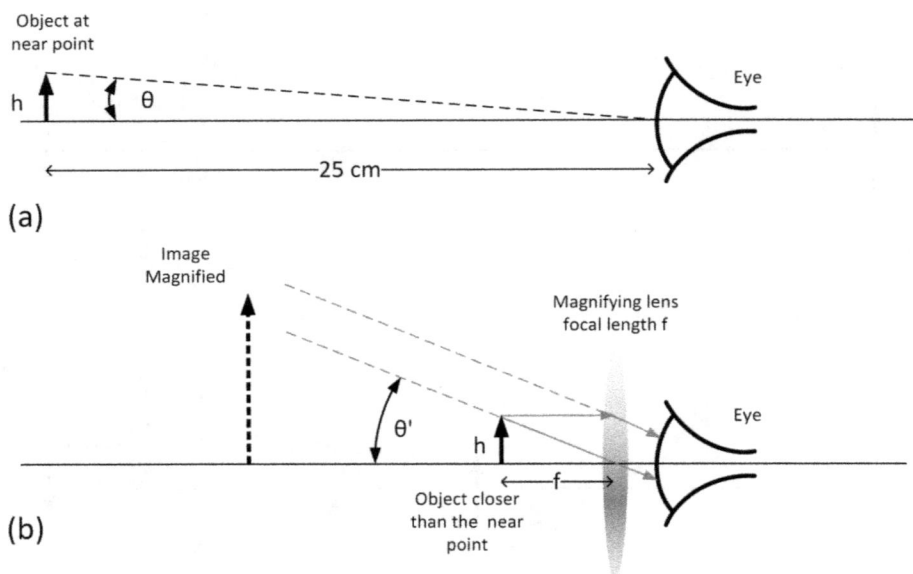

Figure 6.27. Simple magnifier.

with some accommodation. If the eyeball is too short or the refractive power of the eye is too small, then this condition is observed. Myopia is corrected with a negative focal length lens to reduce overall power of the vision system and hyperopia is corrected with a positive focal length lens. An age-related condition called presbyopia occurs because the near point becomes gradually larger with age due to hardening of the eye lens. This condition is helped with the aid of reading glasses, bifocal lenses with both near and distant correction. If the cornea is shaped such that there is a difference in refractive power along the vertical and horizontal planes of the eye, then the result is astigmatism. Astigmatism is corrected with a cylindrical type lens.

6.7.3 Detection of light by the retina

In describing the anatomy of the eye, the retina contains photosensitive elements called rods and cones. The rods are mostly responsible for low light level detection and the cones for higher illumination levels and color discrimination. There are a much greater number of rods than cones in the retina and they are not evenly distributed [30]. Cones are concentrated at the fovea whereas the rods have a higher density over the remainder of the retina. Figure 6.28 is a plot of the density of rods and cones per unit area as a function of angle centered about the fovea [31].

Rods and cones absorb light of different wavelengths. In fact, there are generally three types of cones each having a separate spectral response. Figure 6.29 shows these absorption functions for rods (Curve 498) and three types of cones (Curves 420, 534, and 564) [32]. The identifying curve number represents the wavelength in

Blind Spot

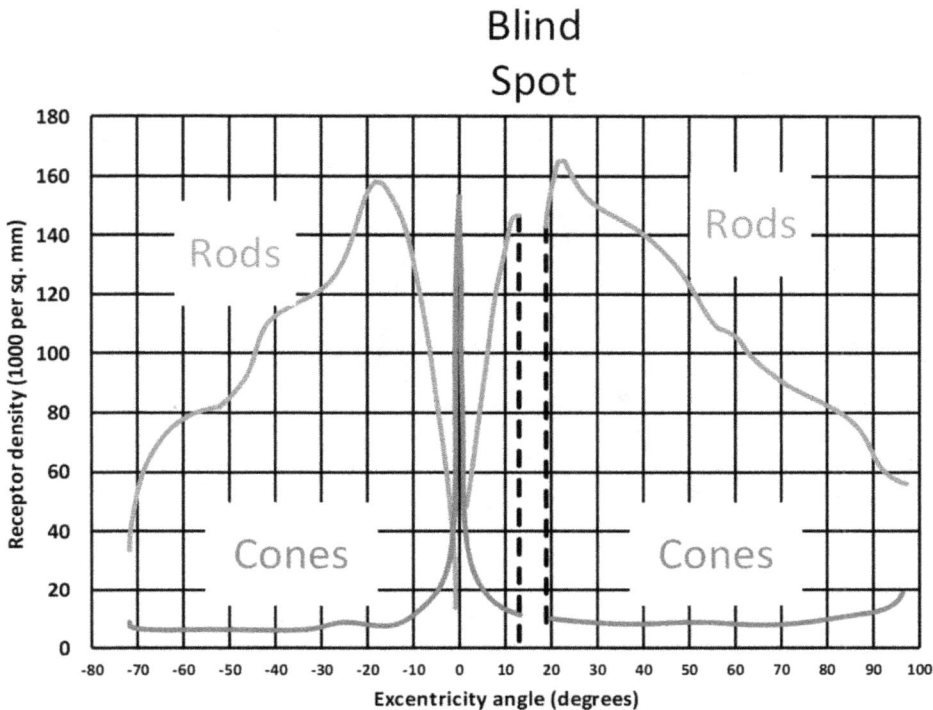

Figure 6.28. Distribution of rods and cones in the retina. (After: [31].)

nanometers of the peak of the absorption. These absorption pigments are responsible for color vision and will be discussed in more detail later.

Another effect to the visual system associated with rods and cones is adaptation. The eye has a remarkable ability to respond to a wide range of amounts of incident light. Light falling on rods or cones induces a photochemical process where particular pigment molecules decompose creating a state called bleaching [29]. In this state another incident photon cannot stimulate vision. We observe this process in going from a brightly lit area (so-called light-adapted) into the dark where the eye's sensitivity to light is low because most of his pigment is in the bleached condition. After some time in the dark the eye will again become more sensitive as the bleaching goes away and the eye becomes dark adapted. At low light levels the rods become the major receptors which is referred to as scotopic vision. When the eye has been in normal daylight conditions the cones are the major photoreceptors giving rise to photopic vision. The scotopic and photopic responses will be discussed more in the following section on photometry.

There are other temporal effects related to vision besides the relatively long-term dark adaptation. Latency is the time for cells in the retina to respond to a flash of light. Another phenomenon is called persistence of vision which is related to the finite speed of the eye to respond to fluctuations. Nominally, rates of about 30 Hz are perceived as a constant signal. This can also be detected when observing flashes of light that do not seem to stop even when the flash has stopped.

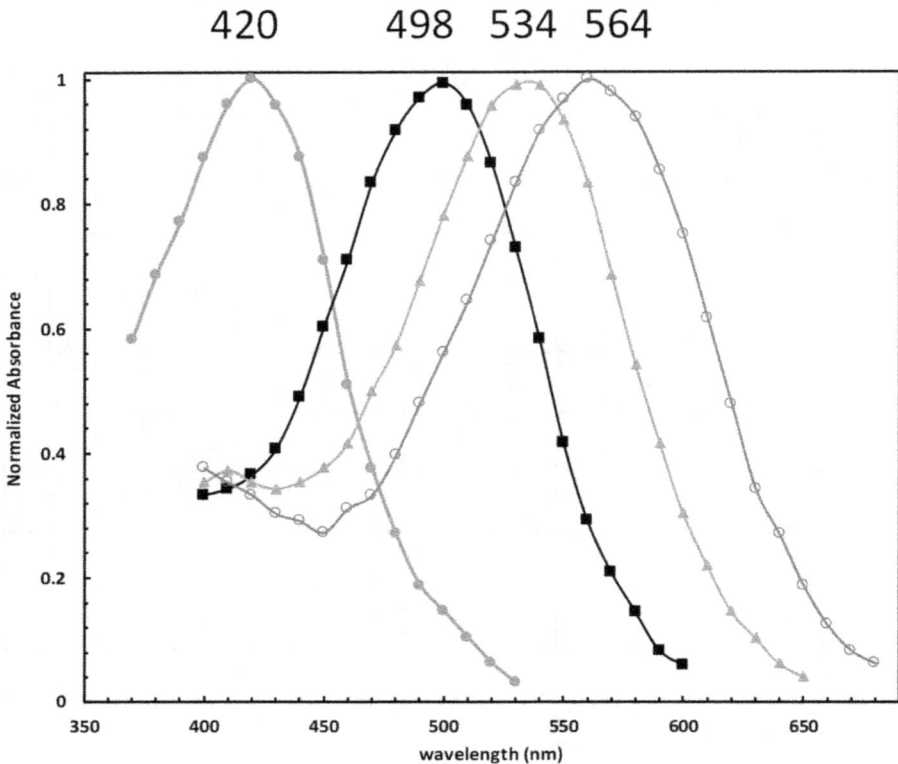

420   498 534 564

Figure 6.29. Spectral response of rods and cones in the retina. Curve 498 is the response of rods. The other curves are different types of cones and their spectral response. The number identifying each curve is the wavelength in nm of the peak response. (After: [32].)

6.7.4 Optical illusions

One of the consequences of the interaction between the eye and the brain are optical illusions. Some of our perceptions are based on experience and some are the result of physiological properties of the visual system. Sections of the retina form a complicated network of nerves that has the effect of reducing the amount of information carried to the brain. For example, image impressions are influenced by signals from neighboring regions of a scene.

Simultaneous contrast occurs when the perceived brightness of one region is influenced by a surrounding region. In the image of figure 6.30 the horizontal rectangular bar has the same gray-level. However, we perceive its brightness to change when compared to background. Another famous illusion based on simultaneous contrast is Adelson's Checkers Shadow illusion [33, 34]. The Herman grid (figure 6.31) or grid illusion is an example demonstrating perception related to high contrast images. The regions at the intersections of the black squares appear to be gray. The common explanation of this illusion is due to lateral inhibition, a neural process of the visual signals to the brain, however there are alternative descriptions [35].

Figure 6.30. Example of simultaneous contrast. The gray-level of the middle bar is the same value everywhere.

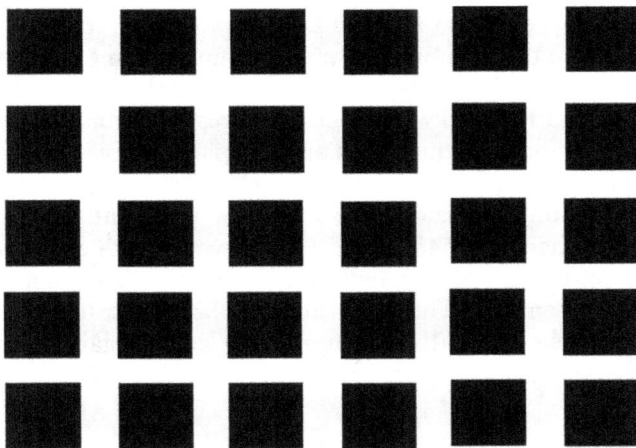

Figure 6.31. Herman grid illusion. Regions at the intersections are perceived to be gray instead of white.

There are many other examples of optical illusions. It is important to review some of these perception issues because they are often observed in optical systems where the images are viewed with the human eye or are the result of processing images [36].

6.8 Photometry

Photometry is used when we want the light measurement to relate to the response of the human eye. Optical systems designed for interactions with human observers, such as illumination systems, need to be specified using photometry since photometric units are consistent with the amount of light perceived and not the overall radiation. Figure 6.32 shows the simple situation for viewing light from a small source. The light from the source has radiant intensity I_e and is collected within the pupil of an observer's eye at a distance z. As in radiometry, the eye collects an

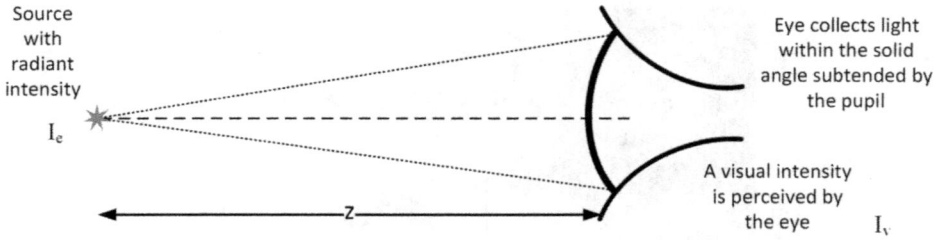

Figure 6.32. Light from a source detected within the pupil of the eye. The dotted lines indicate the solid angle subtended by the pupil.

amount of radiant flux. However, in determining the amount of flux perceived by the eye, the response function of the eye detector must be considered. Thus, the visual flux received is,

$$d\Phi_v = \mathcal{R}_{\text{eye}} d\Phi_e$$

where a subscript, 'v' is used to indicate that this quantity corresponds to a visual stimulus as opposed to the 'e' subscript for energy units from radiometry. In general, the response function can be a function of many variables such as time, frequency, the amount of illumination, and wavelength. This response function also rationalizes units between the radiometric quantities and the photometric quantities.

As discussed in chapter 1, one of the seven constants that defines SI units of measure is the luminous efficacy of 683 lumen per Watt for monochromatic radiation with a frequency of 540×10^{12} Hz (or wavelength of 555 nm). So, if the source in figure 6.32 is a monochromatic source with wavelength 555 nm, then $\mathcal{R}_{\text{eye}} = K_{\text{cd}} \equiv 683$ lumen W^{-1}. The visual unit of flux is the lumen. Now, dividing through the above relation by the common solid angle, again assuming a monochromatic source, we find,

$$I_v \equiv \frac{d\Phi_v}{d\omega} = K_{\text{cd}} I_e(555 \text{ nm})$$

which defines a luminous intensity with the standard unit candela (cd = lumen sr^{-1}). Another way of stating this is that one candela is the luminous intensity, in a given direction, of a source that emits monochromatic radiation of frequency 555 nm (540×10^{12} Hz) and has a radiant intensity in that direction of (1/683) W sr^{-1}.

Returning to figure 6.32, suppose that the radiant intensity of the source in the direction of the eye is a function of wavelength with some defined spectral distribution. To find the luminous intensity we now need to know the wavelength response function of the eye. This analysis is similar to spectral radiometry from the previous chapter. For the case in which the parameters are only a function of wavelength, we write the response function of the eye as,

$$\mathcal{R}_{\text{eye}}(\lambda) = K_m V(\lambda)$$

where the function $V(\lambda)$ is the normalized visual response function of the eye and K_m is the efficacy, maximum value of curve relating lumens to Watts. For photopic vision, dominated by cones, the value of $K_m = K_{cd} = 683$ lumen W^{-1}. This maximum value may change depending on overall illumination.

The particular visual response function and efficacy value to be used has been modified over the years [37, 38]. It is obtained from industry and international standards and depends on overall illumination conditions. Once $V(\lambda)$ is known (and the associated K_m) then photometric values can be computed from radiometric quantities. We must remember that they eye collects light from all wavelengths to provide one visual sensation. Thus, the luminous intensity under these conditions is,

$$I_v = K_m \int_0^\infty V(\lambda) I_e(\lambda) d\lambda$$

In fact, any photometric quantity may be computed by specifying the $V(\lambda)$ function and knowing the analogous radiometric wavelength function. However, conversions from radiometric units to photometric units can only be done if the spectral distributions are well known. The photometric quantities in analogy to radiometric quantities are summarized in table 6.2.

The origination of the visual response curve or spectral luminous efficiency function has a long history and is related to eye and visual response investigations. The CIE-International Commission on Illumination first adopted a spectral luminous efficiency function for photopic vision based on a 2° viewing field. This function is also known as the $\bar{y}$-function that is one of the functions used in defining the CIE 1931 Colorimetric Observer [39–41].

Photopic vision arises predominantly from photosensitive cones. The current range of adaptation associated with photopic vision is at a minimum luminance of 5 cd m^{-2}. When only rods are stimulated at light levels below 0.001 cd m^{-2} this is the region of scotopic vision. The range in between these two adaptation areas is the range call mesopic vision [7].

Table 6.2. Photometric quantities, symbols, and typical units of measure.

Quantity	Symbol	Unit (MKS)
Luminous energy	Q_v	lumen s
Luminous energy density	w_v	lumen s m^{-3}
Luminous flux	$\Phi_v = \frac{dQ_v}{dt}$	lumen
Illuminance (luminous flux areal density)	$E_v = \frac{d\Phi_v}{dA}$	lux = lumen m^{-2}
Luminous exitance	$M_v = \frac{d\Phi_v}{dA}$	lumen m^{-2}
Luminous intensity	$I_v = \frac{d\Phi_v}{d\omega}$	cd = lumen sr^{-1}
Luminance	$L_v = \frac{d^2\Phi_v}{dA \, d\omega \cos\theta}$	nit = cd m^{-2}
Luminous efficacy	$\frac{\Phi_v}{\Phi_{e-total}}$	lumen W^{-1}

Where t is time, A is the surface area receiving or emitting flux, ω is the solid angle in the given direction at a point on the surface of a source, and θ is the angle between the surface normal and the direction of the solid angle. The subscripts v explicitly show that the quantities are related to visual units.

Currently, a function $V(\lambda)$ specifies the normalized photopic visual response function and another function, $V'(\lambda)$, provides the scotopic response function. Several different curves and efficacies are typically used when mesopic vision is necessary [7]. Figure 6.33 contains plotted curves of the normalized CIE photopic and scotopic luminous efficiency functions. The CIE 1931 $V(\lambda)$ function has a peak at 555 nm as defined by SI unit standards with an efficacy of $K_m = K_{cd} = 683$ lumen W^{-1}. The scotopic function $V'(\lambda)$ has a peak at the lower wavelength value of 507 nm corresponding to the low-level illumination sensitivity of the eye's response by rods with an efficacy $K_m = 1700$ lumen W^{-1} [7].

The development of standard SI units and dimensions has clarified the units used in radiometry and photometry. However, the historical literature contains references to outdated and antiquated photometric terms and units [42, 43]. Some of these units are still used today. For example, the foot-candle (fc) = 1 lumen ft^{-2} and the phot (ph) = 1 lumen cm^{-2} are sometimes specified for illuminance. In these cases, there are simple conversions to the lux (lx) = 1 lumen m^{-2} because the differences are only in the length dimensions. Likewise, luminance units, such as a stilb (sb) = 1 cd cm^{-2} can be directly related to a nit (nt) = 1 cd m^{-2}. Unfortunately, some luminance units assume that the source is Lambertian or that the surface is perfectly diffuse in defining the unit. Thus, one sees values of π as part of the definition since luminance of a perfectly diffuse surface is the illuminance falling on the surface multiplied by π.

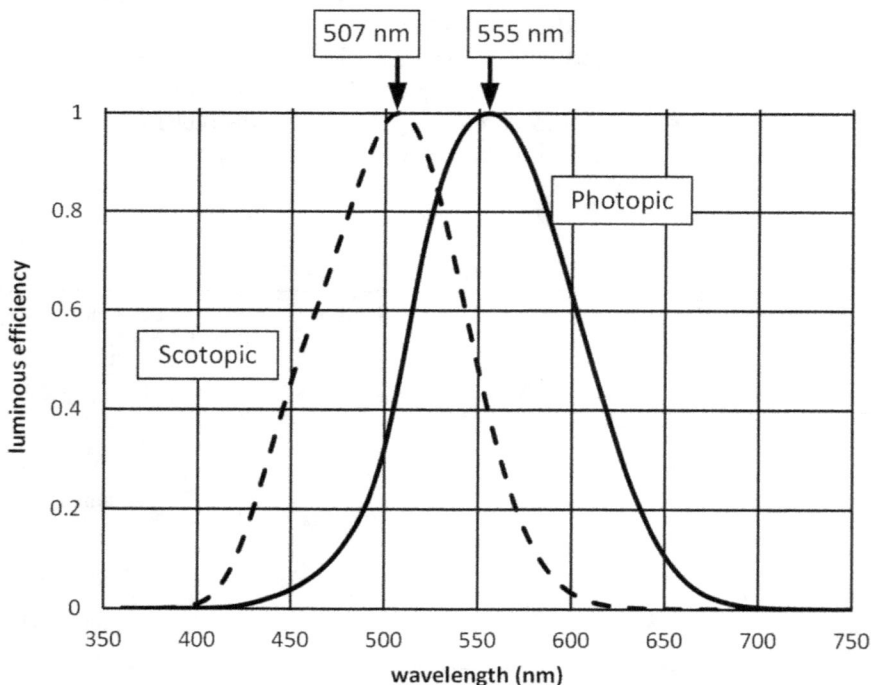

Figure 6.33. Normalized CIE photopic, $V(\lambda)$, and scotopic, $V'(\lambda)$ luminous efficiency functions. The solid curve is the photopic response and the dashed curve is the scotopic response.

Examples of this type of unit are: the Lambert (L) = 1 cd cm^{-2}/π, the footlambert = 1 cd ft^{-2}/π and the apostilb = cd m^{-2}/π. Care must be used if any of these luminance units are specified in an optical system.

It is instructive to examine the photometric output from several types of sources whose radiant spectral power/flux distribution (SPD) is known. For each case, data for the radiant flux as a function of wavelength is known from experiment. For each source, a numerical integration was performed using the simple rectangular rule to obtain the photometric flux. This calculation finds the area under the product curve of the radiant flux as a function of wavelength and the photopic response curve.

$$\Phi_v = K_m \sum V(\lambda) \Phi_e(\lambda) \Delta\lambda$$

Table 6.3 provides two plots for six different sources to allow for comparison between different characteristics that relate to visual perception. The first plot is the radiant spectral power distribution data for the source plotted along with another set of data which is simply the spectral power distribution multiplied by the normalized CIE photopic response function. This curve is the argument of the summation in the equation above. It has not been multiplied by the efficacy constant. The second graph is the same curve as in the first plot except it has been converted to a spectral luminous flux by multiplication of the efficacy. The area under this curve is the total luminous flux output from this source. In addition, for a comparison, a scaled CIE photopic response function is also shown as a dotted curve on this plot. The scaling factor was chosen to so that the value of the photopic response coincides with the value of the peak in the spectral luminous flux.

From the graphs in table 6.3, we see that the first three sources are all examples of white light sources, albeit with much different total luminous flux values. As expected, the incandescent bulb provides white light but also contains power into the infrared which is a source of heat. The fluorescent bulb spectra show strong peaks but also contains a large amount of wasted radiant flux around 450 nm. The white light LED also shows a larger fraction of the luminous flux in the blue portion of the spectrum. The blue and red LED data is given for a comparison to the white light sources as well as showing the effects arising from narrower-band sources on visual perception. In both cases, because the visual response is low in the blue and red portion of the spectrum, the radiant flux is high even though the luminous flux is low. Finally, the spectrum of an infrared LED with a peak at 750 nm is also provided. This spectrum is included to illustrate that even though the peak value of the radiant output is out of the visual response region, there is enough spectral breadth that a small fraction of the light can be detected by human vision. This explains why some infrared LEDs appear as a dull red even though their peak output is in the infrared.

6.9 Color and color measurement

Color has become an increasingly important attribute that optical engineers and designers must consider in developing systems. Advances in detector technology for digital imaging systems requires an understanding of how the system will faithfully

Table 6.3. Comparison of the spectral power distributions of six different sources. Graphs on the left are data for the radiant spectral power distribution (diamonds) and the second data set is this distribution multiplied by the normalized CIE photopic response function. The graphs on the right are the corresponding spectral luminous flux distributions compared to a scaled photopic response function.

Incandescent lamp, 60 W
The source appears white. Calculated luminous flux of 737 lumens.

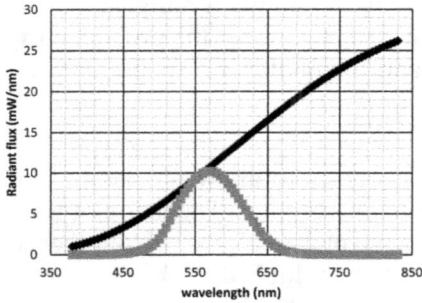

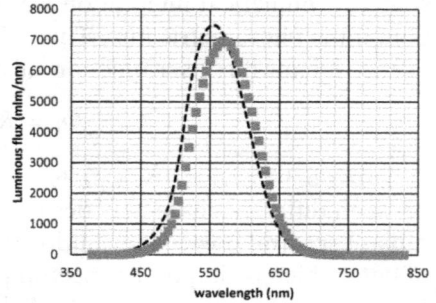

Fluorescent bulb, 40 W equivalent
The source appears white. Calculated luminous flux of 1900 lumens.

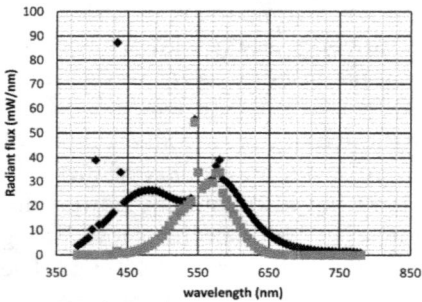

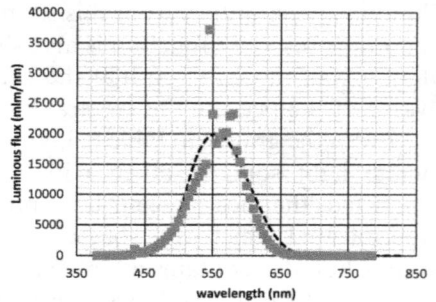

White-light LED, 5 mm indicator LED component.
The source appears white. Calculated luminous flux of 3.0 lumens

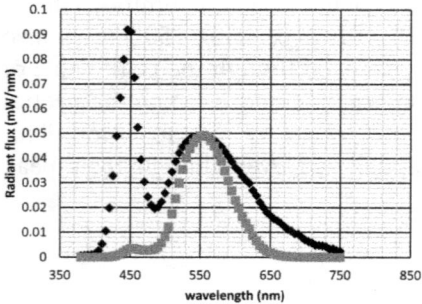

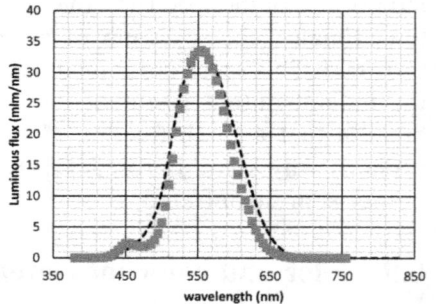

Blue LED, 5 mm indicator LED component.

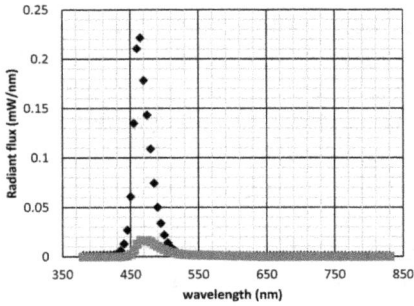

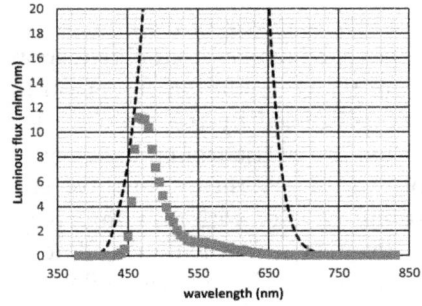

Red LED, 5 mm indicator LED component.

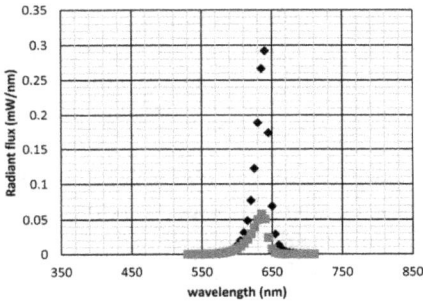

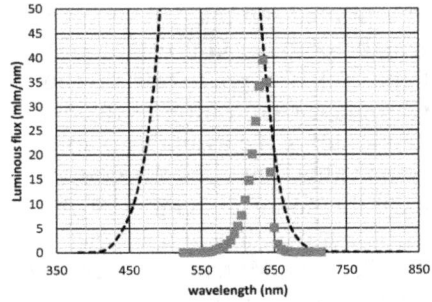

IR LED, AlGaAs LED TO-18 package 750 nm peak.

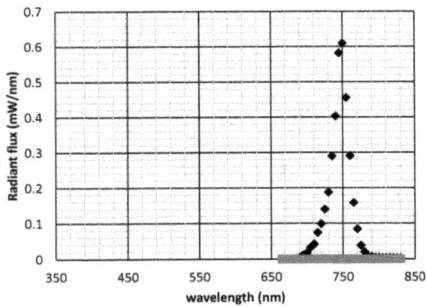

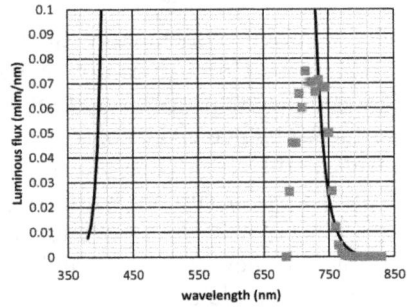

represent colors and illumination in a scene. Even if colors in the image are not affected by the optics, the color of the scene may be altered when the user wants to display or print the image. At each of these stages in the imaging process there are different color rendering systems and standards employed. The purpose of this section is to review color attributes and models important to the human perception of light sources, the appearance of objects, and measurement methods used in colorimetry.

Our discussion of the human eye indicated that there are in general three types of cones each with overlapping responses to different incident wavelengths. The signals from the cone photoreceptors in the retina are modified within the vision pathway and processed by the brain to determine the perceived color. This concept is the origin of the trichromatic theory, the first stage in color recognition [44]. The result of this theory means that we could mathematically formulate a relationship to specify a color with only three variables.

Prior to a mathematical description provided by a color measurement system, order systems have been used to represent color that could be used as a specification in communicating color. One of the earliest order systems is the three-dimensional Munsell Color System [45, 46] This system uses three common terms for color characteristics hue (H), chroma (C), and value (V) to represent the particular color of a printed sample. Hue is the parameter that gives the color its name (e.g., R for red, B for blue, GY for green-yellow) for a total of 10 different hues. Value (or lightness) is a gray-scale division ranging in scale from black (1) to white (10). Chroma (or saturation) is the purity of the color and is on a numerical scale up to a maximum of the available pigments in equal perceived saturation steps. To specify a color value, one uses the designation H V/C. So, a greenish = yellow color might have the notation 5GY 6/10. A particular light source must be used in comparing color samples. There are other color ordering systems used by different industries. The Pantone Matching System [47] and the Natural Color System [48] are examples used in printing, graphics, and design.

6.9.1 Additive and subtractive color

Color is often associated with an object as suggested by a color order system. However, an object does not really have a unique color. In an earlier chapter, we discussed mechanisms of light interactions with surfaces that alter the reflectance of a surface. The perceived color of an object is dependent on the source of illumination as well as the reflectance, absorption, and scattering from the object. If no light is reflected from a surface within a given spectral band then the object will appear dark. The color we do associate with familiar objects is from the usual way in which the object is viewed, like daylight. A self-luminous object, such as a neon sign, can also have a single-color appearance even though the light arises from the emission spectra of the gaseous mixture. In this case, the light output from the dominant spectral lines add together to form a perceived color. This is additive mixing.

The trichromatic theory is supported by color mixing experiments. If one illuminates a diffuse screen with red, green, and blue light (primary colors) in the proper proportions then the light adds, and the combination of all three primary sources appear white. This is illustrated in figure 6.34. In addition, the combination of the two primary sources produce the secondary colors, cyan (addition of blue and green), magenta (addition of blue and red), and yellow (addition of green and red).

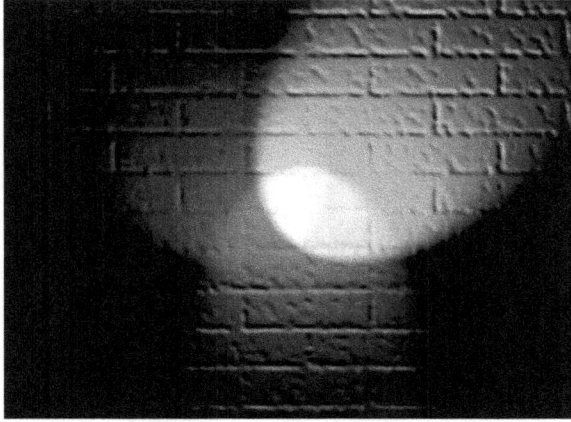

Figure 6.34. Additive primary sources of red, green, and blue producing white light. Overlap of two primary colors produces secondary colors. (Author: Bb3cxv (https://en.wikipedia.org/wiki/Image:RGB_illumination. jpg), CC BY-SA 3.0 (http://creativecommons.org/licenses/by-sa/3.0/), via Wikimedia Commons.)

The result of figure 6.34 leads to some rather simple rules for adding sources to form various colors. Suppose we define a set of variables describing the illumination associated with each of the observed colors. Take: B = blue, G = green, R = red, C = cyan, M = magenta, Y = yellow, and W = White. This simple representation requires that the amount of light for each component color be appropriately scaled. With this caveat, some simple relations result.

$$B + G + R = W$$

$$B + G = C;\ B + R = M;\ G + R = Y$$

Note that if we had a separate source of cyan, then $C + R = W$, and the addition of only two sources produces white. Two colors, such as cyan and red, when combined together to produce white are called complementary colors.

A more detailed description of the conditions and rules for algebraic additive color mixing are summarized as Grassmann's Laws [48, 49]. A set of specific conditions must be met but hold up in many common situations. According to these laws matching the color of a source should be possible by using appropriate mixtures of other color components. The experimental layout for color matching is illustrated in figure 6.35. The viewer adjusts primary sources to match the source under test.

In performing color matching, it was discovered that some test sources could not be matched without adding an additional primary to the side containing the test source. This concept is also illustrated in figure 6.35. In the experiments this amounts to subtracting (adding a negative quantity) to obtain a match. A representative set of data from a color matching experiment is shown in figure 6.36 for an average of ten observers [44]. These curves are the color matching functions for a specific set of red ($\bar{r}$), green ($\bar{g}$), and blue ($\bar{b}$) primary sources that result in a color match at a given wavelength. The bar above the variable represents an average over the number of observers. These curves would change if different primary sources and observers

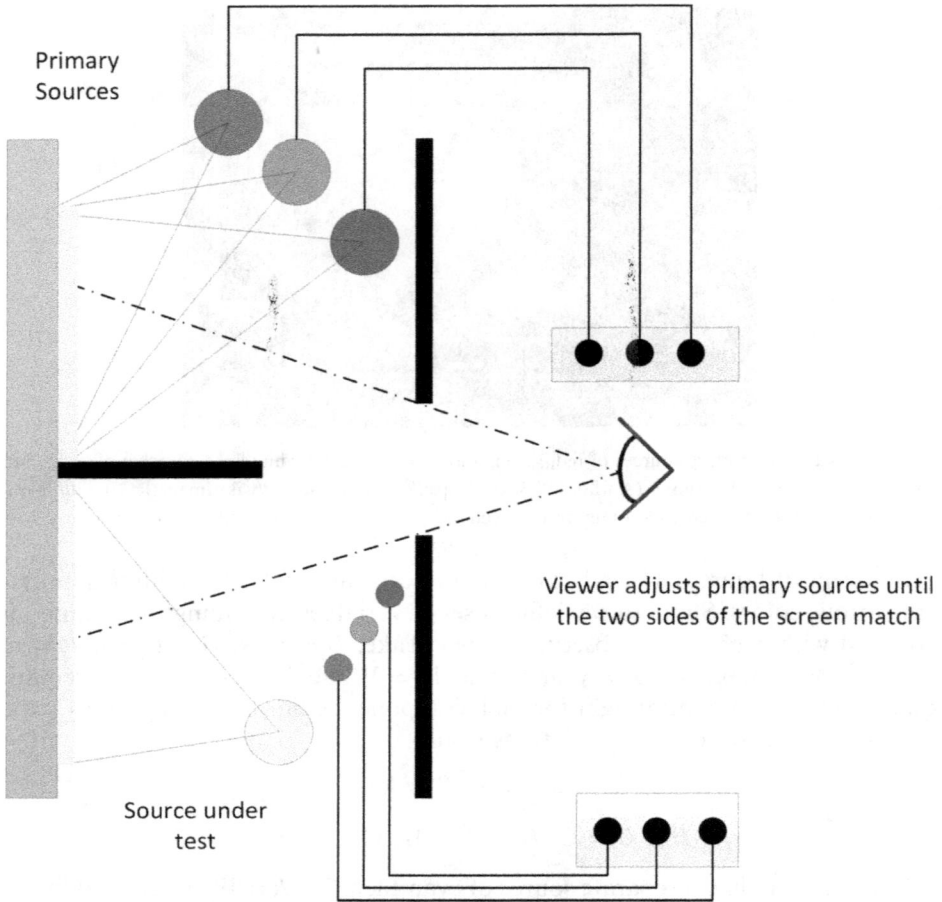

Figure 6.35. Schematic layout of color matching experiments.

were used. However, if different primary sources are used the relation between the different color matching function values is just a linear transformation [48].

Display technologies rely on additive mixing of colors to produce what we perceive as a single-color pixel. The term used for this process is partitive mixing where two colors are used together to form another color when viewed. Each effective color pixel is actually a combination of closely spaced sub-pixels. When the eye views the display,the sub-pixel colors are spatially blended together to provide the perceived color.

When pigments,inks,or dyes are mixed the combination becomes darker unlike the mixture of two sources that appear brighter. A good example is an optical color filter that is created by mixing organic dyes in a gel material that forms a substrate when dried. The dye mixture determines the spectral transmittance of the filter. When light is passed through such a filter some of the light illuminating the mixture is absorbed according to the transmittance of the filter. In general, the transmitted

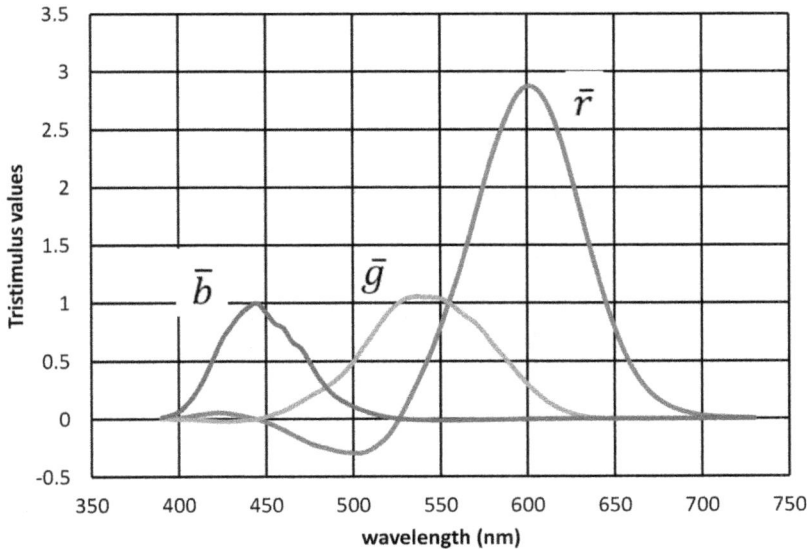

Figure 6.36. Color matching function data from a color matching experiment [44].

spectral power distribution of a filter will follow is a product of the incident spectral power distribution and the transmittance of the filter. It is a multiplicative effect at every wavelength, so that,

$$\Phi_T(\lambda) = T_f(\lambda)\Phi_I(\lambda)$$

where Φ_T is the transmitted spectral flux, Φ_I is the incident spectral flux, and T_f is the spectral transmittance of the filter. When two filters (T_1 and T_2) are placed together we still have a multiplicative effect so that the overall transmittance function of the filter is the product, $T_f = T_1 T_2$.

As with additive mixing of sources, let us extract some of the rules associated with combining ideal color filters through spectral absorption. This time we will define a set of variables associated with the transmittance of each filter. Let B = blue, G = green, R = red, C = cyan, M = magenta, and Y = yellow. In this case the incident white light illumination is ideal and has the same flux at all wavelengths. The ideal white light illumination spectra and transmittance spectra for each filter type is shown in figure 6.37. So, for an ideal white light source incident on any of these filters, the transmitted spectral flux will have the same shape as the filter transmittance.

Now let us examine what happens for two filters in tandem for incident white light. Figure 6.38 illustrates the steps as light is absorbed by a cyan filter followed by a yellow filter. After passing through the cyan filter the transmitted light is cyan and has two components, green (G) and blue (B) as predicted by additive mixtures. The effect on the incident white light beam is to remove red ($-R$) from the light. In a similar manner, when this light passes through the yellow filter the blue component

6-45

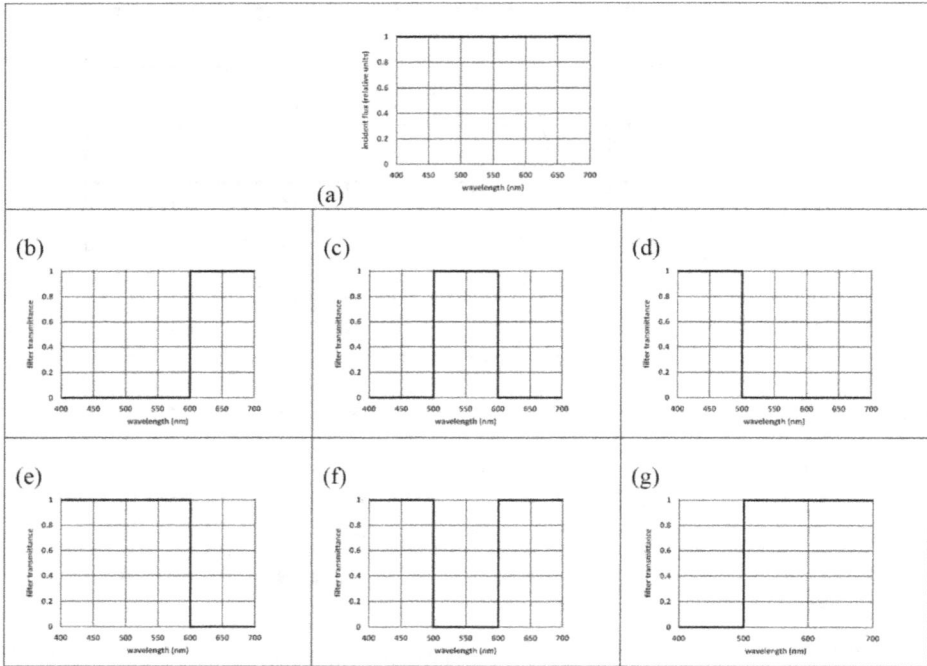

Figure 6.37. (a) White light spectral power distribution (ideal) and filter transmittance curves for ideal (b) red, (c) green, (d) blue, (e) cyan, (f) magenta, and (g) yellow.

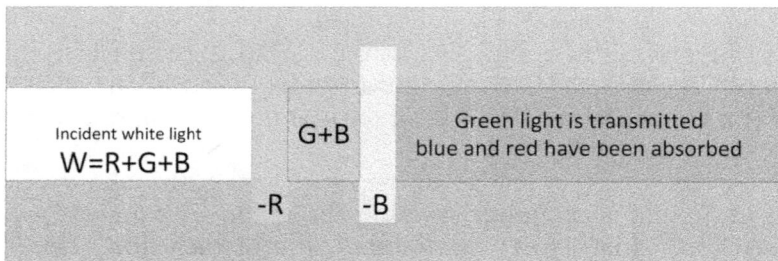

Figure 6.38. Beam of ideal white light incident on a cyan filter,followed by a yellow filter,and the resulting green output flux due to the transmittance of the filter.

is removed $(-B)$ leaving only green light. We are really only using the same additive mixture rules arrived at earlier, but these can be written as,

$$B + G + R = W$$

$$W - R = C; \; W - G = M; \; W - B = Y$$

Again,we are not really subtracting one filter spectra from another,it is multi-plicative. But the effect on the integrated color is described as subtractive mixing. Remember also that adding more pigment increases the absorption coefficient of the

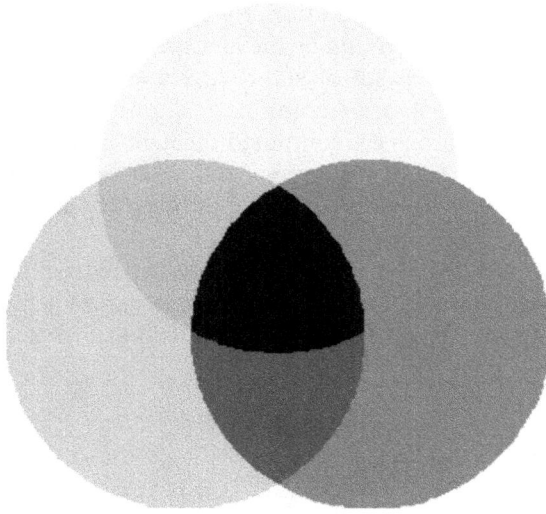

Figure 6.39. Subtractive mixing of cyan, magenta, and yellow resulting in black for all combinations. But the combination of any two results in either red, green, or blue. (Author: Jorgelrm (https://commons.wikimedia.org/wiki/File:Sintesis_sustractiva_plano.svg), CC BY-SA 3.0 (https://creativecommons.org/licenses/by-sa/3.0), via Wikimedia Commons.)

filter material and this is a logarithmic effect [3]. Adding logarithmic absorption coefficients is the same as multiplying the two transmittances. When cyan, magenta, and yellow filters are combined together in the same way we combined red, green, and blue light sources the result is shown in figure 6.39.

In practice,when all filters are combined,we do not obtain a high contrast black because of the overlap of spectra and the simple ideal algebraic rules between CMY and RGB do not often apply. Thus, color printers use CMY ink along with black, not only to save ink for black and white printing but also to increase contrast of the printed pages and reduce the registration requirements for the overlap of three-color dots. This is the standard CMYK print designation where K is black of (key).

6.9.2 CIE 1931 color spaces and chromaticity diagram

In the early 1900s it was recognized that a physical measure was needed to specify color for both light sources and reflective objects. The results of the color matching experiments indicated that this could be accomplished with a standard set of color matching functions that could be used to compute three values to define a specific color. The range of these three values is called a color space. For sources,the physical measurement required is the spectral power distribution of the source. For reflective objects,the spectral power distribution of the incident illumination as well as the reflectance from the object as a function of wavelength must be known.

A standard was developed by the Commission Internationale de L'Eclairage (CIE) in 1931 to meet this need [38, 48, 49]. Practical computational issues required that the standard color matching functions all be positive values. This was accomplished using a simple linear transformation. In addition, for consistency with luminous efficiency standards, one of the functions should be the identical to the photopic response $V(\lambda)$. These considerations lead to a set of standard curves $\bar{x}(\lambda)$, $\bar{y}(\lambda)$, $\bar{z}(\lambda)$ for the so-called 1931 CIE 2° standard observer color matching functions [39, 40, 48, 49]. Note that since the standard functions are computed through a linear transformation, they each contain information from red, green, and blue primaries, so none of these functions should be associated with red, green, or blue. These functions are shown in figure 6.40 [50].

The 2° standard observer color matching functions are used to compute the color of sources and light reflected from objects. The three values defined by the standard are the tristimulus values (X, Y, Z). For a source with known spectral radiance, $L\lambda$, these quantities are,

$$X = K_m \sum L_\lambda \bar{x}(\lambda) \Delta\lambda$$

$$Y = K_m \sum L_\lambda \bar{y}(\lambda) \Delta\lambda$$

$$Z = K_m \sum L_\lambda \bar{z}(\lambda) \Delta\lambda$$

where $K_m = 683$ lumen W^{-1} is the luminous efficacy and $\Delta\lambda$ is the size of the incremental wavelength separation for the luminance data. The value of Y is just the luminance of the source as in photometry. For sources, the tristimulus values have units of luminance.

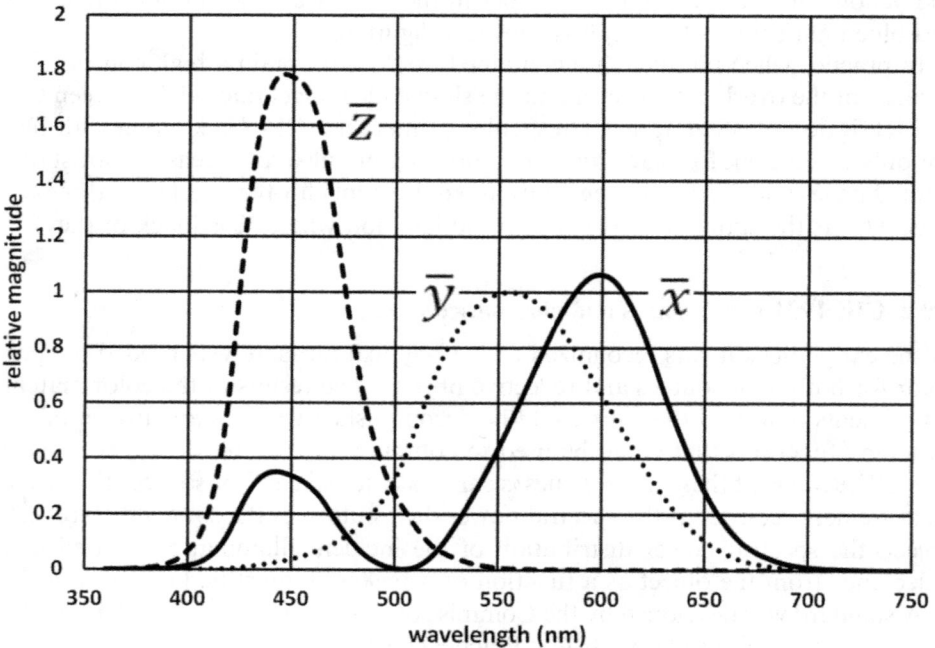

Figure 6.40. 1931 CIE color matching functions.

To compute the tristimulus values for objects the spectral distribution function of the source must be known as well as the reflectance. To assist in predictions of the color of objects when the source of illumination is not known the CIE has specified standard illuminants with known spectral power distributions [51]. For example, CIE standard illuminant A is similar to a tungsten-filament light source typical of an incandescent bulb. CIE standard illuminant D65 represents the spectrum of average daylight. If the spectral distribution of the illumination source is given as S_λ, and the spectral reflectance from the object is R_λ, then the tristimulus values are,

$$X = k \sum S_\lambda R_\lambda \bar{x}(\lambda) \Delta\lambda$$

$$Y = k \sum S_\lambda R_\lambda \bar{y}(\lambda) \Delta\lambda$$

$$Z = k \sum S_\lambda R_\lambda \bar{z}(\lambda) \Delta\lambda$$

where k is a normalization constant setting $Y = 100$ for maximum reflectance. Objects with smaller reflectance values will be scaled appropriately. Using this definition, the normalization constant is,

$$k = \frac{100}{\sum S_\lambda \bar{y}(\lambda) \Delta\lambda}$$

If the object transmits, such as a color filter, the reflectance function is simply replaced by the spectral transmittance.

Since Y represents the overall luminance, only two other parameters provide a unique color specification. This is accomplished with the following definitions involving the tristimulus values,

$$x \equiv \frac{X}{X + Y + Z}; \; y \equiv \frac{Y}{X + Y + Z}; \; z \equiv \frac{Z}{X + Y + Z} = 1 - (x + y)$$

where the values of x and y are called the chromaticity coordinates. This allows for a two-dimensional plot to be made where a unique point on the graph represents the color of a source or object. Note that x and y are dimensionless parameters. Also, these values should not be confused with the $\bar{x}(\lambda)$, $\bar{y}(\lambda)$, $\bar{y}(\lambda)$ standard observer functions. If (x,y) is given along with the overall luminance Y, then the tristimulus values may be computed.

$$X = \frac{x}{y} Y; \; Y = Y; \; Z = \frac{1 - (x + y)}{y} Y$$

Figure 6.41 is the chromaticity diagram, a plot of the bounds of possible values for an (x, y) color coordinate. The boundary is specified by a monochromatic source of a given wavelength so that any point inside the curve is a possible color coordinate, points outside are not physically realizable color values.

Figure 6.42 shows a chromaticity diagram that includes the chromaticity coordinates for some standard and well-known sources. The point labeled CIE-E is the equal energy point for a uniform spectral power distribution function.

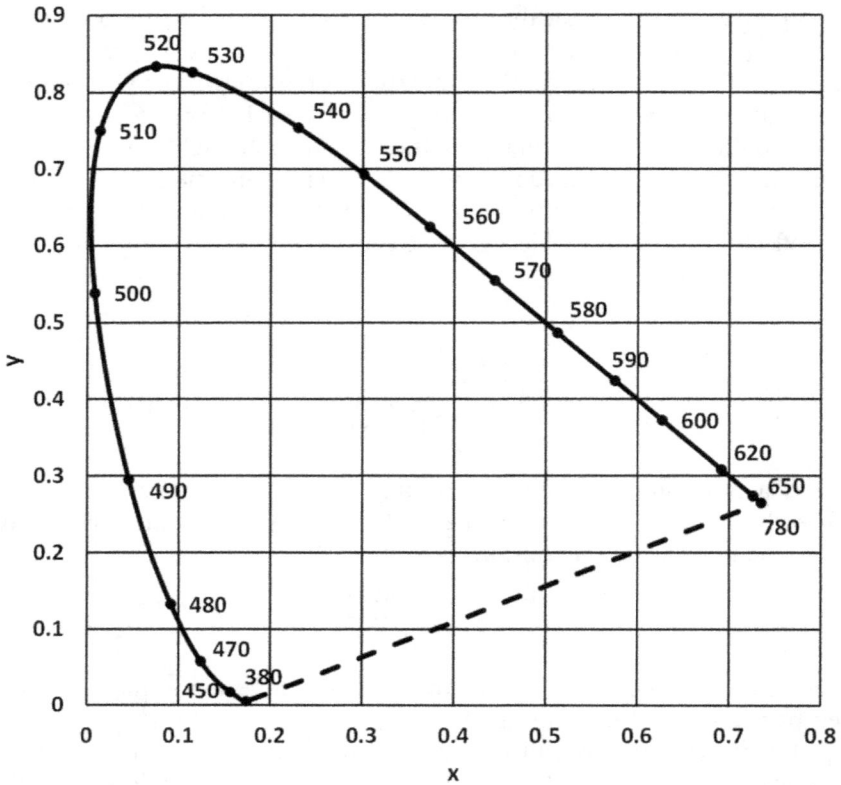

Figure 6.41. 1931 CIE Chromaticity diagram. Specific wavelengths in nanometers are labeled on the boundary representing the chromaticity coordinate of a monochromatic source with that wavelength. The dashed line is called the purple line of purple boundary. These colors are possible but there is no wavelength representing this value.

The chromaticity coordinate for the equal energy point (0.333, 0.333). Two of the standard sources CIE-A and CIE-D65 are also shown. As one can see, D65 is close to the CIE-E point indicating that this is nearly a uniform white light source as expected. Points for white, red, green, and blue light emitting diodes are also plotted. The points are near the boundary showing that these sources are not monochromatic and have some spectral width. The HeNe laser source, however, is near the boundary as expected for a nearly monochromatic source.

The chromaticity diagram is not only useful for identifying the color and the purity of sources but can also predict characteristics of color object surfaces or color filters. Three different color coordinates are plotted on the diagram of figure 6.43, point P1 for a green filter, point P2 for a red painted surface, and point P3 for a pink fabric. In addition, the D65 standard illuminant coordinates are plotted, and, in this case, this point is the white point. Drawing a line from the given D65 white point to P1 intersects a point on the boundary that is identified as the dominant wavelength. For the green filter represented by P1 this wavelength is 520 nm. When this line is

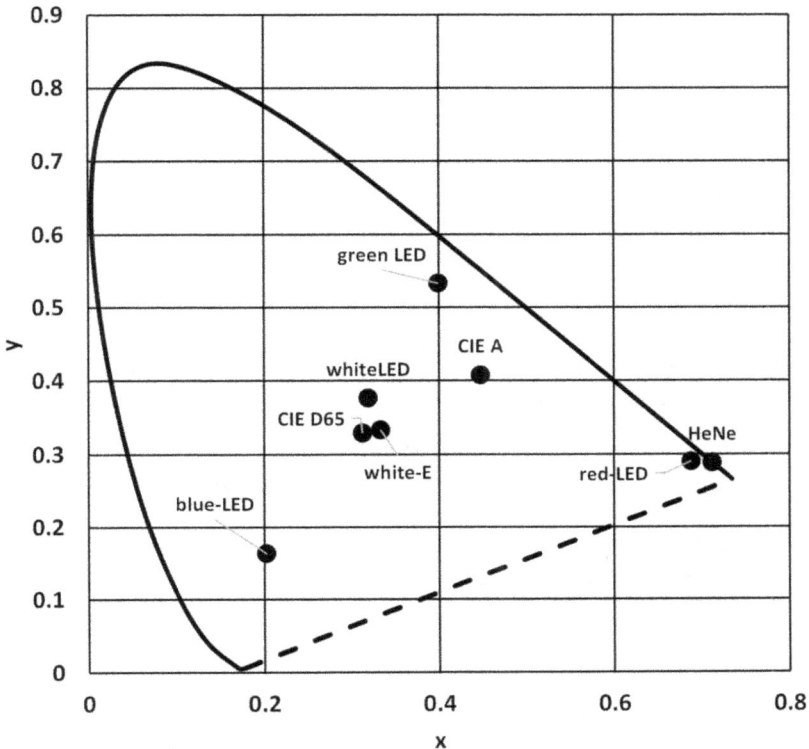

Figure 6.42. 1931 CIE Chromaticity diagram with chromaticity coordinates of several standard and well-known sources.

extended in the opposite direction it intersects the purple line. There is no complementary wavelength associated with this filter. For the red surface represented by point P2 a similar line is drawn from the source point to P2 intersecting the boundary at a dominant wavelength value of 604 nm. However, for this object it does have a complementary wavelength of 492 nm. The color represented by point P3 has no dominant wavelength since it intersects the purple line, however, it does have a complementary wavelength of 508 nm. A purity parameter is defined as the distance on the chromaticity diagram from the color point to the intersection with the boundary locus divided by the distance from the white point to the boundary.

Another class of standard sources are blackbody radiators. These sources are Lambertian emitters and are specified by their radiation temperature. The characteristic line made by connecting chromaticity coordinates of different temperature blackbody radiators is called the Planckian locus. Figure 6.44 shows the Planckian locus with several points labeled for the color coordinates of a blackbody at the given temperature in degrees Kelvin.

The temperature specifying the blackbody radiance or radiant exitance is the radiation temperature which is the same value as the thermodynamic temperature for the ideal blackbody. There are several other temperatures defined related to a

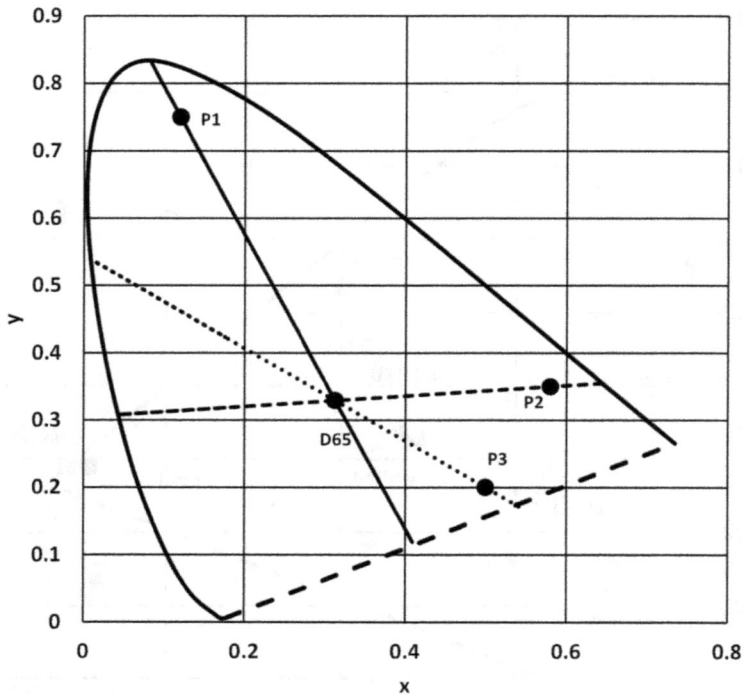

Figure 6.43. 1931 CIE Chromaticity diagram with chromaticity coordinates a green filter, a red painted surface, and a pink fabric material. The D65 point is used as the illumination source. Several other lines are drawn to illustrate the dominant and complementary wavelength of these three objects when illuminated by the D65 source.

blackbody and used in describing light sources and radiating surfaces in an attempt to use one parameter to specify the source appearance [52]. When a spectral radiance of an unknown source is measured within a fixed spectral band, the temperature of the blackbody with equal radiance within the same spectral band is called the radiance temperature [3, 8]. The color temperature of a source is the temperature of the blackbody with the same color coordinate as the source [48, 49, 52]. Thus, if the chromaticity of a source lay on the Planckian locus then the color temperature would be equal to the radiation temperature. Sources such as incandescent lamps have color coordinates on or very near the Planckian locus so determining their color temperature is straightforward.

A difficulty arises when a source that appears white has chromaticity coordinates far away from the Planckian locus. Examples of these types of sources are fluorescents and some LEDs. For these sources, the single temperature quantity used is known as the CCT [48, 49]. The CCT of an unknown source is the temperature of a blackbody source that is perceived to match the unknown. The assumption used to arrive at the CCT is to pick the blackbody source that is nearest to the chromaticity of the source. However, the 1931 CIE chromaticity diagram does not have uniformly spaced coordinates so finding the CCT is not simple in this color

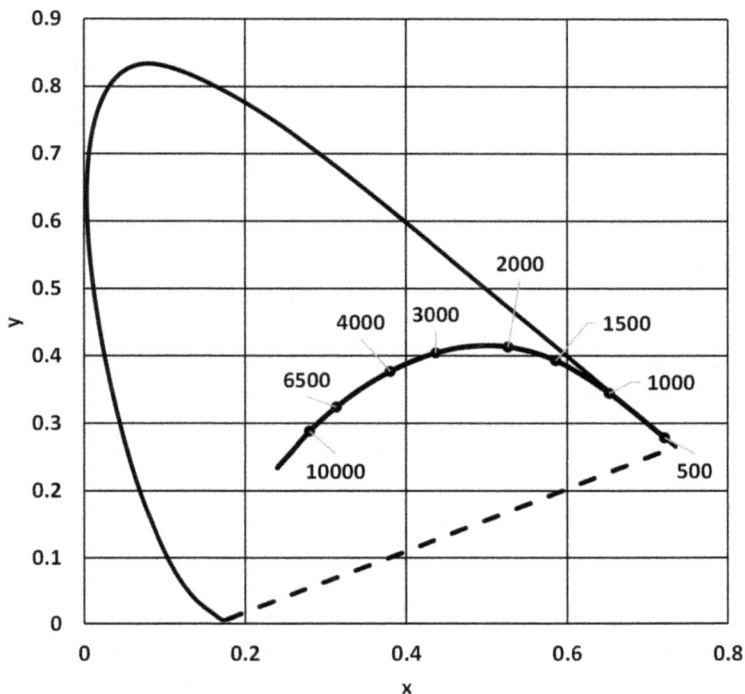

Figure 6.44. 1931 CIE Chromaticity diagram highlighting the Planckian locus, color coordinates associated with blackbody radiation of different radiation temperatures.

space. Graphical techniques have been developed using a set of isothermal lines plotted on the chromaticity diagram [48, 49, 52]. Numerous numerical and computational methods and approximations have been proposed to determine CCT from a given set of color coordinates [53]. The utility of CCT is mainly in providing a representative distinction between different sources by lamp manufacturers.

The chromaticity diagram is also useful in determining how two different color sources can be mixed to generate other colors. This concept is also apparent in finding the color coordinates of the emission from spectral calibration source lamps. Figure 6.45 shows the spectral power distribution of a mercury (Hg) and a helium (He) lamp. Both sources have large dominant characteristic spectrally narrow emission lines associated with the element's atomic transitions.

Figure 6.46 shows a chromaticity diagram with the color coordinates of the Mercury and helium sources. The color coordinate of the mercury source lies near the Planckian locus in the region of higher radiation temperatures, so this source appears nearly white to the observer even though the spectral power distribution has several large dominant monochromatic lines. Also plotted on this chromaticity diagram are the color coordinates of two fictitious monochromatic sources, S1 at 480 nm and S2 at 580 nm. If the power of these two sources is identical, the color of these two sources combined results in a color coordinate extremely near the

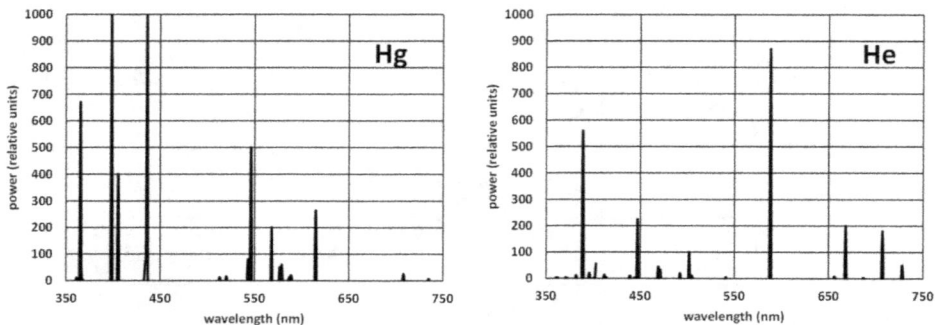

Figure 6.45. Spectral power distributions for a mercury and a helium calibration source with emission lines associated with their corresponding atomic spectra.

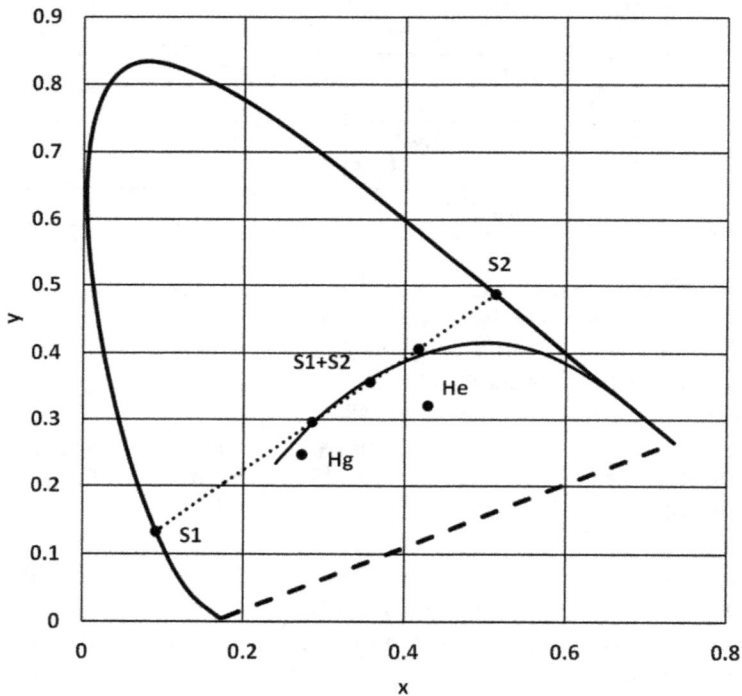

Figure 6.46. 1931 CIE Chromaticity diagram with mercury (Hg) and helium (He) sources from spectral power distributions of figure 6.45, along with the addition of two fictitious monochromatic sources S1 and S2 showing additive mixing. Any color along the line between the two monochromatic sources can be obtained by adjusting the power of the two sources.

Planckian locus with a color temperature of about 4820 K. The combination would appear as a white source. This is the point labeled, S1 + S2. When source S1 is reduced to half the power of source S2, and the sources are combined then the color coordinate shifts toward the coordinate of S2. When source S2 is reduced to half the power of source S1, and the sources are combined then the color coordinate shifts toward the coordinate of S2. The shift in the color coordinate occurs along the line between the two sources, just another example of additive mixing. Therefore, any color along the line connecting the color coordinates of these two sources can be obtained by adjusting the power of the sources.

When three sources are used, then any color within a triangle formed by the color coordinates of the three sources on a chromaticity diagram can be obtained by adjusting the power of the three sources. This concept was used in first determining the color matching functions of additive mixing. The triangle formed on a chromaticity diagram by the three source points is called a color gamut. Any optical display, whatever the mechanism employed (such as cathode ray tubes and liquid crystal devices), has a color gamut. Printed output from a color printer will also have a color gamut that depends on ink and paper used.

Although the x, y chromaticity diagram provides a unique color coordinate value from physical measurements it is not as useful in helping quantify human perception of color. Also, perceived color differences are not linear over the CIE 1931 color space. Many attempts have been made over the years to correct this disadvantage using transformations of the tristimulus values and make a uniform chromaticity scale [48, 49]. Ultimately, the particular scale will depend on the criterion that is the closest to the visual situation encountered. However, most scales still rely on a set of tristimulus values as described here. Therefore, the concepts discussed can be applied to any color space.

6.9.3 Color models based on a color gamut

Numerous standards have been developed to define expectations of color representations for use by manufacturers of color devices and products [54, 55]. Table 6.4 lists the chromaticity coordinates for the primaries specified by three different standards. The color gamut for the primaries of table 6.4 are shown in figure 6.47. The NTSC and sRGB gamut are similar with the UHDTV (ultra-high-definition television) gamut covering a much larger range of color coordinates. Most manufacturers of monitors of personal computers use the sRGB international standard that was adopted by the International Electrotechnical Commission

Table 6.4. Color gamut chromaticity coordinates and white reference for selected standard primaries. White point references for these primaries is D65 (0.3127, 0.3290).

| | Primaries | | | | | | Standard |
| | Red | | Green | | Blue | | |
Gamut name	x	y	x	y	x	y	
NTSC-PAL	0.630	0.340	0.310	0.595	0.155	0.070	SMPTE 170M-1999
sRGB	0.640	0.330	0.300	0.600	0.150	0.060	ITU-R BT.709–6
UHDTV	0.708	0.292	0.170	0.797	0.131	0.046	ITU-R BT.2020–2

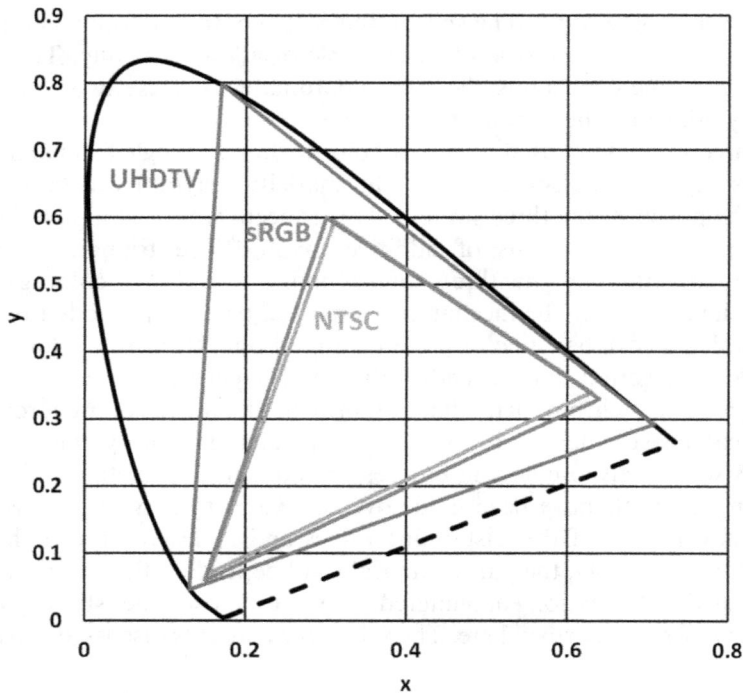

Figure 6.47. Color gamut of three selected standard primaries.

(IEC) [56]. A compatible display specified with a color gamut associated with a larger triangle can reproduce a wider range of colors on screen.

For displays and printing several color models have been developed for the limited set of color coordinates outlined by a color gamut. An RGB model is a common one used where the RGB values each form one axis of a three-dimensional coordinate system. In this model each coordinate axis has a maximum value of one, normalized to the luminance of that primary. Figure 6.48(a) illustrates the resulting RGB cube resulting from this model.

The additive RGB model is a common specification used for displays and digital photography. Another common additive color model is HSV (Hue, Saturation, and Value) or HSI (Hue, Saturation, and Intensity). Figure 6.48(b) shows the solid hexagon-cone representing the HSV space. These models were developed principally by the computer graphics community and are designed to be hardware independent since the hue identifies a specific color and the saturation and value coordinates control the amount of black and white mixed with a hue. Both models are used in specifying the color of a pixel element. When used in software settings for a display the color coordinate values must be quantized. This leads to a bit representation designation for a given color. An eight-bit value means that there are, $2^8 = 256$ quantized values ranging from 0 to 255 for each coordinate. Since there are three axes for each coordinate there are $(2^8)^3 = 16\,777\,216$ bits of data in any given color

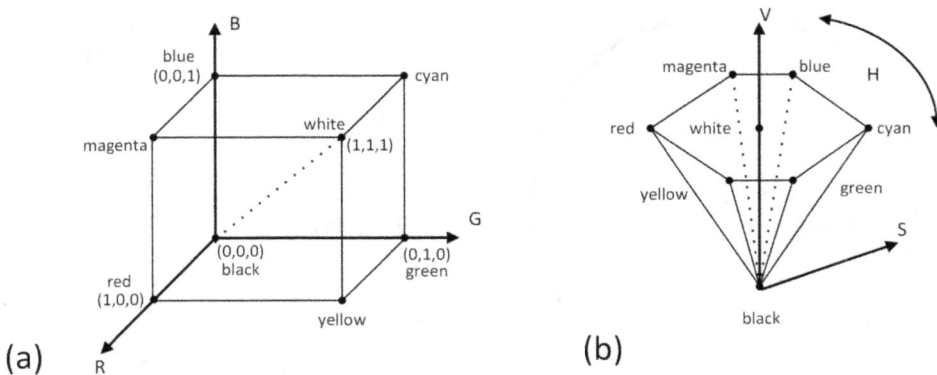

Figure 6.48. (a) RGB cube, and (b) HSV (HSI) models used for digital color representations.

space. A typical computer paint program or image processing software provides control of color settings using one or both of these models.

Since printing is a subtractive process for combining colors, the printing industry uses combinations of cyan, magenta, and yellow to create a particular color mixture. Other industries that produce paints and dyes also require a subtractive mixture of colors. Ideally, mixing all colors together should produce black but in practice the mixture is a gray color. We discussed this previously in the context of color filters. Thus, a four-color set of combinations are used, CMYK, where the K represents a black pigment. The K comes from the fact that in printing the black background is called a 'Key-plate'. There are other proprietary printing combinations used by different industries, but they all used the basic set of combinations of CMYK, just adding more compounds to increase the color gamut.

6.10 Application: measuring light flux—the integrating sphere

An integrating sphere is a hollow spherical shell with an inner surface coated with a material that approximates a Lambertian reflector. The main purpose is to provide a uniform irradiance across a detector placed at a point on the inner surface emitted by a source that has a nonuniform spatial light distribution. In addition to light flux measurements, an integrating sphere can be configured for reflectance and transmittance measurements of materials [4–6, 57]. The detection system can be calibrated for total light flux measurements or used with a spectrometer to obtain spectral radiant flux. The inner coating of the sphere is chosen for the application environment and spectral range desired. Integrating spheres may also be used as an effective Lambertian source when a general source is placed inside the sphere or when a beam enters a port opening on the sphere.

Here we will use radiometric principle already discussed to find the total flux emitted from a source placed within a sphere. Figure 6.49(a) shows the basic layout of an integrating sphere with a source placed inside the sphere and a detector placed at a port along the inner surface of the sphere. The interior is coated with a material whose reflectance approximates a Lambertian reflector but has a known reflectance

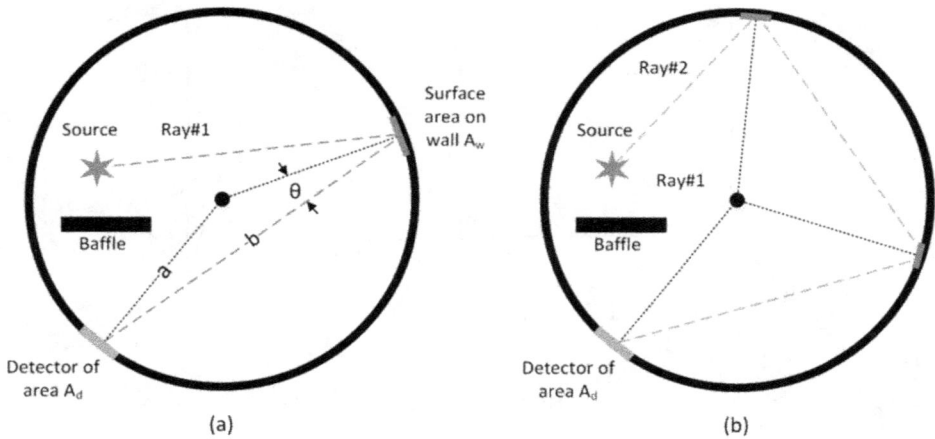

Figure 6.49. Cross-section of an integrating sphere. (a) Tracing a single ray with one reflection from the wall, and (b) tracing rays for two reflections from the wall before entering the detector. The sphere has radius, a, and dotted lines are along a normal vector at the inner surface of the sphere.

of R. A baffle is inserted in the sphere (but also coated) that blocks direct rays from the source to the detector. Ray#1 is traced from the source to a point on the wall where it reflects and illuminates the detector. The sphere has radius, a, and the chord from the wall to the detector is a length, b. The angle between the surface normal and Ray#1 and the surface normal and chord b is θ.

From the definition of radiance, the differential flux for this single Ray#1 direction is,

$$d^2\Phi_{e1} = L_e \, dA_w \, d\omega \, \cos\theta$$

where the differential solid angle is,

$$d\omega = \frac{dA_d \cos\theta}{b^2}$$

From geometry we know that the chord length, $b = 2a \cos\theta$. With these two relations,

$$d^2\Phi_{e1} = L_e dA_w \frac{dA_d \cos\theta}{(2a\cos\theta)^2}\cos\theta = \frac{dA_d}{4a^2}L_e dA_w$$

Since this is a Lambertian surface, the radiance at the wall element is,

$$L_e = \frac{M_e}{\pi} = \frac{RE_e}{\pi}$$

where E_e is the irradiance falling on the wall element from the source. So,

$$d\Phi_{e1} = R\frac{dA_d}{4\pi a^2} \int E_e dA_w = R\frac{dA_d}{4\pi a^2}\Phi_{eS}$$

where Φ_{eS} is the total flux emitted from the source.

However, the detector does not collect light from single reflections at the wall surface but multiple reflections. A drawing of two wall reflections is shown in figure 6.49(b). With two reflections the flux is a similar relation to the single reflection case except the total reflectance is now R^2 and for three reflections, R^3. Thus, the total differential radiant flux at the detector is a summation of all the multiple reflections, or,

$$d\Phi_e = \sum_i d\Phi_{ei} = R\frac{dA_d}{4\pi a^2}\Phi_{eS}[1 + R + R^2 + R^3 + ...]$$

But this infinite series can be simplified as,

$$d\Phi_e = \frac{dA_d}{4\pi a^2}\Phi_{eS}\left[\frac{R}{1 - R}\right]$$

In this simple model we have not assumed any losses due to the area of ports in the sphere (detector port or any other input or exit ports) or any coating non-uniformity, however these effects can be quantified [57]. The value of the detected flux is highly sensitive to small changes in sphere reflectance indicative of the $1 - R$ term in the denominator of the flux equation. The major advantage of using a sphere for these measurements is due to the effect of multiple reflections leading to a multiplicative effect in the detected light flux. This is important since the detected flux decreases with increasing sphere size. The larger the sphere radius the larger the source that can be accommodated since this model also assumes that the source and the stop are small compared to the area of the sphere.

6.11 Application: source-to-fiber coupling

Efficient coupling of radiant energy from fiber-to-fiber and from sources to fibers is essential for any fiber optic-based system. Radiometry principles can be employed to predict the source to fiber efficiency. Source-to-fiber coupling efficiency is defined as an optical power ratio,

$$\eta = \frac{\Phi_F}{\Phi_S}$$

where Φ_F is the total flux coupled into the fiber and Φ_S is the total output flux from the source. However, as with any radiometric problem we need the spatial distribution of the radiance of a source to compute the optical power delivered.

In fiber optic systems light emitting diodes and lasers are the common types of sources used because the sizes of their emission areas are compatible with the size of optical fibers. Light emitting diode spatial radiation distributions have been modeled using some simple analytic expressions [58, 59]. There are a wide variety of radiant distributions available from different manufacturers depending on the particular chip characteristics and whether or not encapsulation is included. One frequently used model is a cosine power law that accounts for types of sources from Lambertian to more highly directed beams [60]. The form of this type model is,

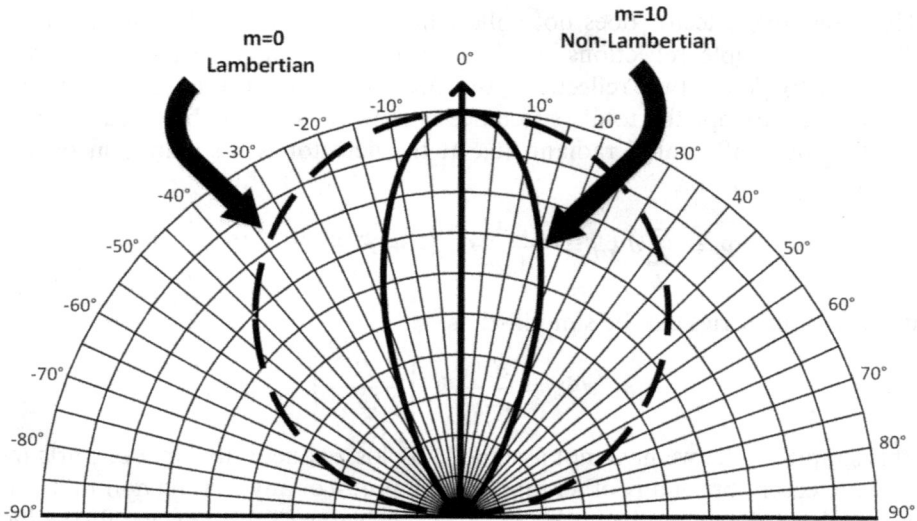

Figure 6.50. Intensity polar plot for a Lambertian source and a non-Lambertian source using the mth power cosine model.

$$L_e(\theta) = L_0\cos^m(\theta)$$

where L_0 is a constant and m is the power of the cosine. When $m = 0$ then we have a Lambertian source with radiance independent of angle, or $L_e(\theta) = L_e(0)$. Figure 6.50 shows a polar intensity plot comparing a Lambertian source and a non-Lambertian source with $m = 10$. The higher the value of m the more directed the source. This may occur due to the laser pattern itself or other optics over the source.

To compute the coupling efficiency, we need to calculate both the total power output from the source and the amount of power coupled into the fiber. Beginning with the general definition of radiance we have,

$$d^2\Phi_e = L_e(A_s, \theta, \varphi)dA_s\, d\omega\, \cos\theta$$

where the radiance is shown to depend on the position on the surface and angular variables. For simplicity, in this case we will assume that the source is circular with radius r_L and area A_s. The radiometry geometry is shown in figure 6.50. At each point on the surface of the source we must compute the output into a given solid angle $d\omega = \sin\theta\, d\theta\, d\varphi$. For the circular source we can write $dA_s = 2\pi r_s dr$. There are two different conditions to consider, (1) when the source radius is less than the fiber radius, and (2) when the source radius is greater than the fiber radius. These cases are shown in cross-section on the inserts in figure 6.51.

The total power output from this source into a hemisphere is,

$$\Phi_S = 2\pi \int_0^{r_L} r_s dr_s \int_0^{2\pi} d\varphi \int_0^{\frac{\pi}{2}} L_e(A_s, \theta, \varphi) \cos\theta \sin\theta\, d\theta$$

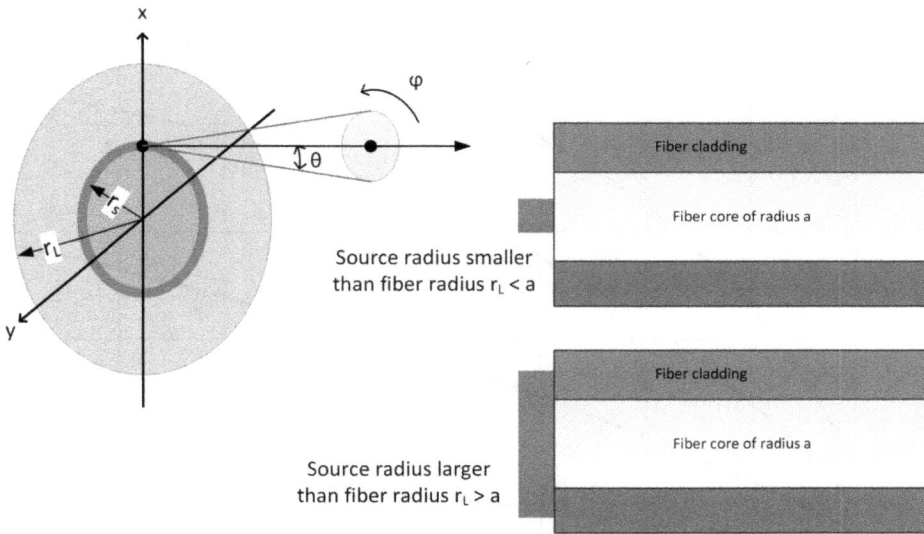

Figure 6.51. Source to fiber coupling geometry and cases for sources smaller and larger than the fiber core radius.

Using the mth power model where radiance only depends on the cosine function, the solution is,

$$\Phi_S = \pi^2 r_L^2 L_0 \frac{2}{m+2}$$

And when m = 0 for a Lambertian source this reduces to, $\Phi_S = \pi^2 r_L^2 L_0$.

Performing this calculation for step-index (constant core refractive index) fiber coupling, for $r_L < a$, the only difference is that the theta integration is now restricted to be a maximum angle related to the fiber's numerical aperture, NA, where,

$$NA \equiv \sin^2 \theta_{max}$$

For this situation, the power coupled into the fiber is,

$$\Phi_F = 2\pi \int_0^{r_L} r_s dr_s \int_0^{2\pi} d\varphi \int_0^{\theta_{max}} L_0 \cos^m \theta \cos \theta \sin \theta \, d\theta$$

Resulting in,

$$\Phi_F = \pi^2 r_L^2 L_0 \frac{2}{m+2} \left[1 - (1 - NA^2)^{\frac{m+2}{2}} \right]$$

And when $m = 0$ for a Lambertian source this reduces to, $\Phi_F = \pi^2 r_L^2 L_0 NA^2$.

For the case when $r_L > a$ the integral over the source area has an upper limit of 'a' instead of r_s. A larger source has more total output power but less becomes coupled into the fiber. In summary, the coupling efficiency for a step-index fiber becomes,

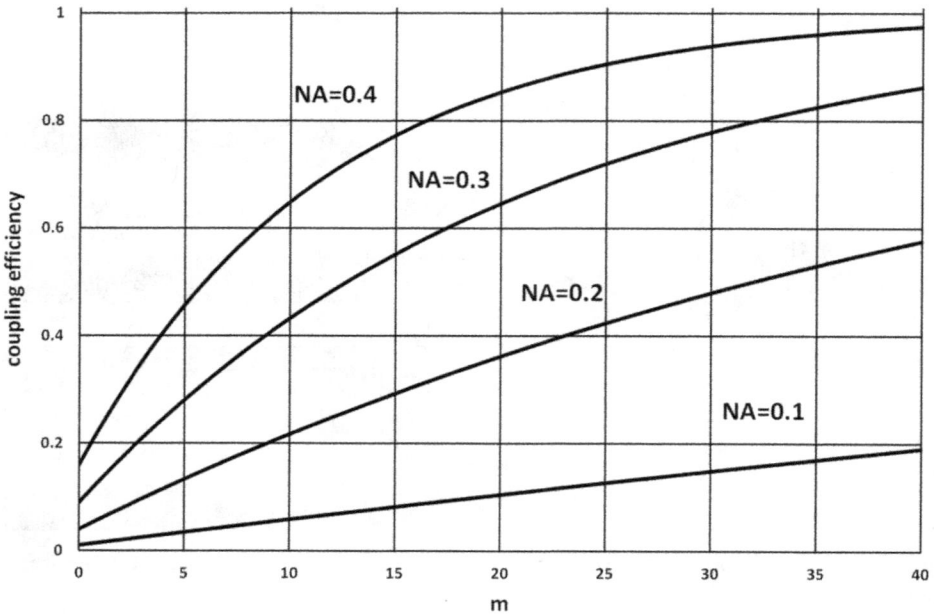

Figure 6.52. Step-index fiber coupling efficiency using an mth power cosine radiance model for numerical aperture values of 0.1, 0.2, 0.3, and 0.4.

$$\eta_{step} = \begin{cases} \left[1 - (1 - NA^2)^{\frac{m+2}{2}}\right], & r_L \leqslant a \\ \left(\dfrac{a}{r_L}\right)^2 \left[1 - (1 - NA^2)^{\frac{m+2}{2}}\right], & r_L > a \end{cases}$$

For an mth cosine power law source. The importance of increasing the directionality of the source is shown in figure 6.52, for the case of $r_L < a$. The coupling efficiency is plotted as a function of m for a family of curves with increasing numerical aperture. Typical silica fibers have NA in the 0.2 range while larger core plastic fibers may have NA values as large as 0.5.

Exercises and problems

1. A small lamp (assumed to be a point source) is located 3.00 m directly above a point P on the floor of a room and produces an irradiance at P of 1.00 W m^{-2}. (a) What is the radiant intensity of the lamp? (b) What is the irradiance produced at another point on the floor, 1.00 m distant from P?

2. The sun subtends a total angle of 0.50° at the Earth's surface, where the irradiance is 1000 W m^{-2} at normal incidence.

   (a) Compute the solid angle subtended by the Sun from the Earth.
   (b) What is the irradiance of an image of the Sun formed by a lens with a 50 mm diameter and a focal length of 500 mm?

3. A small monochromatic light source, radiating at 500 nm is rated at 500 W.
   (a) If the source radiates uniformly in all directions, determine its radiant intensity.
   (b) If the surface area of the source is 5 cm^2, determine the radiant exitance.
   (c) What is the irradiance on a screen situated 2.0 m from the source, with its surface normal to the radiant flux?
   (d) If the receiving screen contains a hole of diameter 5.0 cm, how much radiant flux gets through?

4. A parking lot is illuminated at night by identical lamps at the top of two poles 30.0 ft high and 40.0 ft apart. Assuming the lamps radiate equally in all directions, compare the irradiance at ground level for points directly under one lamp to the irradiance midway between the lamps.

5. A small size detector can rotate about its axis and is positioned so that it measures a maximum irradiance. At what rotation angle to the normal is the irradiance measured by the detector equal to 75% of the peak value?

6. An isotropic light source with a radiance of 500 W m^{-2} sr^{-1} illuminates a flat circular Lambertian diffuser (radius 5.00 cm, transmittance of 1.0) and positioned 15 cm in front of the diffuser. (a) What is the radiance leaving the diffuser? (b) Determine the irradiance at the center of a screen placed 1.732 m behind the diffuser.

7. Small night lights are required to illuminate a dark hallway when the main lighting is off. The hallway is 2.5 m high, 1.5 m wide, and 10 m long. (a) The first night light with intensity 0.625 W sr^{-1} is placed at the center of the hallway ceiling. Find the irradiance falling on the floor at a point directly below the light. (b) Compute the irradiance falling on the floor at a distance of 0.75 m from the point directly below the light. (c) A second night light with intensity 0.625 W sr^{-1} is placed at the center of the hallway ceiling and separated from the first source by 5.0 m. Now, what is the irradiance falling on the floor at a point directly below the *first* light source?

8. A blackbody radiator is used as a calibration source for an infrared detection system. The calibrator operates at a temperature of 1000 K and has a 1.0 cm diameter exit aperture. A detector whose surface area is 1.0 cm^2 is located on the axis of the calibrator's exit aperture a distance of 1.0 m away from the blackbody. (a) What is the total radiant exitance of the blackbody calibrator? (b) At what wavelength is the radiant exitance a maximum? (c) Compute the solid angle that the source subtends from the detector position 1.0 m away from the blackbody. (d) Find the optical power (radiant flux) measured by the detector.

9. Compute the total radiant exitance and peak wavelength of a blackbody source whose temperature is (a) 1000 K, (b) 2000 K, and (c) 6000 K.

10. At what temperature is the total radiant exitance of a graybody with emissivity of 0.12 equal to the total radiant exitance of a blackbody of temperature 1000 K?

11. A point source of light is buried inside a large projection optical illumination system. You need to know its radiometric characteristics, but unfortunately, you do not have the detailed plans. However, the measured irradiance from the point source at a distance $d_1 = 2.00$ m from the exit pupil of the system was 11.30 W m^{-2} and at a distance $d_2 = 3.00$ m from the exit pupil of the system the measured irradiance was 5.20 W m^{-2}. (a) How far inside the system, behind the exit pupil is the source located? (b) Compute the source intensity.

12. A darkroom safelight which has a flat face shaped like a circular disk of radius 5.00 cm is attached to the ceiling at the center of the room at a distance of 1.5 m above the darkroom workbench. The room is square of dimensions 6 m × 6 m. The surface of the safelight and the workbench top are parallel. Model the safelight as a flat disk Lambertian emitter of radiance 500 W m^{-2} sr^{-1}. (a) Determine the irradiance (at a point P) on the workbench surface directly below the safelight. (b) Find the radiant exitance of the circular disk safelight. (c) What is the irradiance at a point on the workbench, 1 m from point P? (d) What is the irradiance on the wall of the darkroom at a point level with the workbench?

13. A rectangular display room in an art museum has dimensions of 4.0 m wide by 7.0 m long. The height of the ceiling is 4.0 m. Two identical disk Lambertian source fixtures with a diameter of 0.5 m and luminance of 4.09×10^4 cd m^{-2} are used for illumination. One source is mounted flush with the ceiling in the center of the room. The other is mounted flush with the wall in the center of the 7.0 m wall. Initially the wall light is turned OFF. (a) What is the illuminance falling on the floor at the point immediately beneath the ceiling fixture? (Wall light is turned OFF). (b) Compute the illuminance falling on the floor in each corner of the room. Wall light is still turned OFF. (c) Now the wall fixture is turned ON. Find the total illuminance on the floor in the center of the room.

14. A point source whose intensity is 1.00 W sr^{-1} is located 25.0 cm in front of a lens with focal length of 10.0 cm and diameter of 2.0 cm.
    (a) Determine the solid angle subtended by the lens from the source position.
    (b) What is the intensity of the image formed by this system?
    (c) Compute the irradiance at the center of a screen placed 10.0 cm to the right of the object.
    (d) Compute the irradiance at the center of a screen placed 10.0 cm to the right of the lens.

15. The irradiance measured by a detector at 2.0 meter from a point source was 16.0 W m^{-2}.
    (a) What would be the irradiance measured by the detector at 8 m from the source?
    (b) A thin lens of focal length 1 m is placed 2 m from the source between the source and the detector (still at 8 m from the source). Now, compute the irradiance measured by the detector.

16. A new fiber optic lightpipe lighting system is being used to illuminate a room. See figure P6.16. The output radiance of the lightpipe system is 500 W m^{-2} sr^{-1}. The output from the lightpipe is directed toward a flat circular Lambertian diffuser (radius 5.00 cm) and positioned 15 cm above the diffuser. The diffuser is flush with the ceiling and located in the center of the room at a distance of 1.732 m above a desk. The surface of the ceiling and the desktop are parallel. (a) What is the radiance leaving the diffuser? (b) Determine the irradiance at a point, P, on the desktop directly below the diffuser. (c) What is the irradiance at a point, Q, on the desk, 1 m from point P?

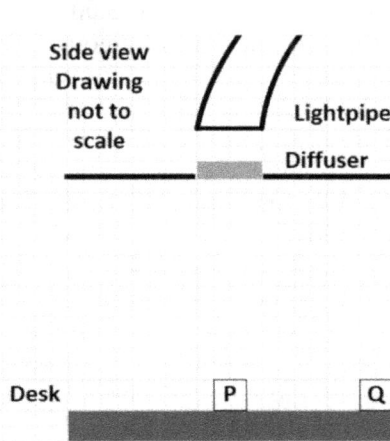

Figure P6.16. Lightpipe lighting system cross-section.

17. A Lambertian source of radiance 10 W m^{-2} sr^{-1} is made in the shape of a thick ring. The inner radius of the ring is 2.5 cm and the outer radius is 5 cm. What is the irradiance at the point P on the axis of the ring 10 cm away from the center of the ring?

18. A disk of radius R = 1.27 cm is a non-Lambertian emitter of radiance L_0 cos θ, where θ is the angle between the surface normal and the direction of observation and $L_0 = 1.0$ W m^{-2} sr^{-1}. (a) Calculate the radiant exitance of the source. (b) Find a function for the intensity of the source and plot this as a function of θ.

19. The value of the power of an incident beam with a nonuniform irradiance pattern is measured with an integrating sphere. Your sphere has a diameter of 30 cm with an inner reflectance of 98.5% and is equipped with two, 1-cm diameter ports. A detector is placed over one port and covers the entire aperture. The incident beam is directed toward the open port and is completely collected by the sphere. An irradiance of 0.5 W m^{-2} is measured by the detector. What is the incident power of the beam?

20. The strongest spectral lines in the output of an argon ion laser are at 488 nm and 514.5 nm. If a laser has an output consisting of 1.5 W at 488 nm and 2 W at 514.5 nm, what is the photometric power of the laser?

21. A red laser pointer with wavelength of 670 nm has a radiometric power of 1.00 mW. What radiometric power would be required from a green laser pointer of wavelength 510 nm if these two laser pointers are to have the same luminance?

22. An AlGaAs diode laser (1500 nm) used in a fiber optic communication system has a power level of 15 mW. What luminous flux would a human observer detect?

23. At what angle of incidence is the illuminance on a screen exactly one-half of what it is at normal incidence?

24. A small lamp (point source) is located 3.00 m from a wall producing an illuminance of 100 lux. What is the luminous intensity of the lamp?

25. Observational astronomers are often concerned about the amount of light in the sky when the Moon is full. A measurement of the illuminance produced by a full moon on a very clear night (when atmospheric losses can be neglected) gave 0.50 lm m^{-2} at the Earth surface.

    (a) Compute the solid angle that the Moon subtends from an observer on the Earth looking at the full moon. (b) What is the Moon's luminance? Assume that the Moon is a Lambertian reflector.

26. A new LED lighting system is designed to illuminate a desktop work surface in an office. The source is flush with the ceiling and located in the center of the room at a distance of 1.73 m above a desk. The surface of the ceiling and the desktop are parallel. The source is circular with radius of 0.126 m and is a Lambertian radiator with a luminance of: 30,000 cd m^{-2} = 30,000 lumens m^{-2} sr^{-1}. (a) A detector is placed just below the source on the desktop. Find the illuminance measured by the detector. (b) When the detector is translated along the desktop a distance of 1.0 m, determine the illuminance that the detector measures. (c) The detector is now moved back to the point below the source but is tilted so that its surface normal makes an angle of 45° with the axis joining the center of the source and the detector. Now what illuminance does the detector measure?

27. Compute the <u>luminance</u> of a blackbody with temperature: (a) 300 K, (b) 1000 K, and (c) 6000 K.

28. A 400.0 cd isotropic light source is suspended 80 cm above the center of a 1.0 m diameter circular table. (a) What is the illuminance at the center of the table? (b) What is the illuminance at the edge of the table? (c) The source is now replaced with a flat circular LED (radius of 0.50 cm) that has a Lambertian profile with luminance of 4.00×10^6 cd m^{-2}. The center of the source is again 80 cm above the center of the table. Now what is the illuminance at the edge of the table?

29. A small lamp used for UV curing has the spectrum shown below. The wavelength spread at each line is ±2.0 nm. The table below shows the radiometric data collected for this lamp. Assume that zero flux is contributed to the source output at all other wavelengths. (a) Determine the total <u>radiant</u> flux output from this source. (b) Determine the total <u>luminous</u> flux output from this source.

Line designation	Wavelength (nm)	Measured flux (μW nm^{-1})
A	185	60.0
B	250	80.0
C	310	50.0
D	365	60.0
E	405	20.0
F	435	60.0

30. The spectra of a mercury lamp was measured with a crude spectrometer whose wavelength spread at each line is ± 5.0 nm. The table below shows the radiometric data collected for this lamp. Assume that zero flux is contributed to the source output at all other wavelengths. (a) Determine the total <u>radiant</u> flux output from this source. (b) Determine the total <u>luminous</u> flux output from this source.

Line designation	Wavelength (nm)	Measured flux (μW nm^{-1})
A	405	33.0
B	435	65.0
C	545	76.0
D	575	80.0

31. Compute the tristimulus values (X, Y, Z) and the chromaticity coordinates (x and y) for the CIE-D65 and CIE-A standard illuminants. Plot the color coordinate on a chromaticity diagram and comment on each point's position relative to the Planckian locus.

32. The color coordinate of a green LED is (0.40, 0.50). Find the dominant wavelength of this LED for both (a) standard illuminant A and (b) standard illuminant D65.

References

[1] Miles R 2007 A light history of photometry: from Hipparchus to the Hubble Space Telescope *J. Br. Astron. Assoc.* **117** 172–86
[2] Eisberg R M and Resnick R 1985 *Quantum Physics of Atoms, Molecules, Solids, Nuclei, and Particles* 2nd edn (New York: Wiley)
[3] Palmer J M and Grant B G 2010 *The Art of Radiometry* SPIE Press Monograph vol PM184 (Bellingham, WA: SPIE Press)
[4] McCluney R 1994 *Introduction to Radiometry and Photometry* (Boston, MA: Artech House)
[5] Stimson A 1974 *Radiometry and Photometry for Engineers* (New York: Wiley)
[6] Boyd R W 1983 *Radiometry and the Detection of Optical Radiation* (New York: Wiley)
[7] *The Use of Terms and Units in Photometry—Implementation of the CIE System for Mesopic Photometry*, CIE TN 004:2016
[8] Grant B G 2011 *Field Guide to Radiometry* (Bellingham, WA: SPIE Press)

[9] ISO 80000-7:2019(en) Quantities and Units—Part 7: Light and Radiation https://iso.org/

[10] ANSI/IES RP-16–17 *Nomenclature and Definitions for Illuminating Engineering* https://ies.org/standards/definitions/

[11] Bartell F O, Dereniak E L and Wolfe W L 1981 The theory and measurement of bidirectional reflectance distribution function (BRDF) and bidirectional transmittance distribution function (BTDF) *Proc. SPIE* **257** 154–60

[12] Arecchi A V, Messadi T and Koshel R J 2007 *Field Guide to Illumination* (Bellingham, WA: SPIE Press) https://doi.org/10.1117/3.764682

[13] Young H D, Freedman R A, Ford A L and Sears F W 2004 *Sears and Zemansky's University Physics: With Modern Physics* (San Francisco, CA: Pearson Addison Wesley)

[14] Nicodemus F E 1963 Radiance *Am. J. Phys.* **31** 368

[15] CIE 70–1987 *The Measurement of Absolute Luminous Intensity Distribution* http://cie.co.at/publications/international-standards

[16] EN 13032–4:2015+A1:2019 *Light and Lighting—Measurement and Presentation of Photometric Data of Lamps and Luminaires—Part 4: LED Lamps, Modules and Luminaires*

[17] Stover J C 2012 *Optical Scattering Measurement and Analysis* 3rd edn (Bellingham, WA: SPIE Press)

[18] Harvey J E 2019 *Understanding Surface Scatter: A Linear Systems Formulation* (Bellingham, WA: SPIE Press)

[19] Nicodemus F E 1965 Directional reflectance and emissivity of an opaque surface *Appl. Opt.* **4** 767–73

[20] Nicodemus F E, Richmond J C, Hsia J J, Ginsberg I W and Limperis T 1977 *Geometrical Considerations and Nomenclature for Reflectance, NBS Monograph 160* (Washington, DC: National Bureau of Standards, US Department of Commerce)

[21] Coatings and Materials Solutions, Labsphere, Inc. https://labsphere.com/ (Accessed 22 June 2020)

[22] Bennett J M and Mattsson L 1989 *Introduction to Surface Roughness and Scattering* (Washington, DC: Optical Society of America)

[23] Hobbs P C D 2000 *Building Electro-optical Systems: Making It All Work* (New York: Wiley)

[24] Thorlabs Inc. www.thorlabs.com (Accessed 22 June 2020)

[25] *Optical Diffuser Technologies* RPC Photonics https://rpcphotonics.com/pdfs/Optical_Diffuser_Technologies_Final_030215.pdf (Accessed 22 June 2020)

[26] Sales T R M, Chakmakjian S, Morris G M and Schertler D J 2004 Engineered microlens arrays provide new control for display and lighting applications *Photonics Spectra* **38** 58–61

[27] Patel S, Marshall J and Fitzke F W 1995 Refractive index of the human corneal epithelium and stroma *J. Refract. Surg.* **11** 100–41

[28] Patel S and Tutchenko L 2019 The refractive index of the human cornea: a review *Cont. Lens Anterior Eye* **42** 575–80

[29] Pedrotti L S and Pedrotti F L 1998 *Optics and Vision* (Englewood Cliffs, NJ: Prentice Hall)

[30] Purves D, Augustine G J and Fitzpatrick D *et al* (ed) 2001 *Neuroscience* 2nd edn (Sunderland, MA: Sinauer Associates) Anatomical Distribution of Rods and Cones (available from: https://ncbi.nlm.nih.gov/books/NBK10848/) (Accessed 22 June 2020)

[31] Osterberg G 1935 Topography of the layer of rods and cones in the human retina *Acta Ophthalmol. Suppl.* **6** 1–103

[32] Bowmaker J K and Dartnall H J 1980 Visual pigments of rods and cones in a human retina *J. Physiol.* **298** 501–11

[33] Adelson E H 2000 Lightness perception and lightness illusions *The New Cognitive Neurosciences* ed M Gazzaniga 2nd edn (Cambridge, MA: MIT Press) pp 339–51

[34] Adelson E H 2005 *Checkers Shadow Illusion* http://web.mit.edu/persci/people/adelson/ Checkers Shadow_description.html

[35] Schiller P H 1982 Central connections of the retinal ON and OFF pathways *Nature* **297** 580–3

[36] Gonzalez R C and Woods R E 2018 *Digital Image Processing* 4th edn (New York: Pearson/ Prentice Hall)

[37] Sharpe L T, Stockman A, Jagla W and Jägle H 2005 A luminous efficiency function, V*(λ), for daylight adaptation *J. Vision* **5** 3

[38] ISO/CIE 11664–1:2019 [CIE LEAD, EN ISO/CIE 11664–1:2019] *Colorimetry—Part 1: CIE Standard Colorimetric Observers*

[39] CIE No. 86 1990 *CIE 1988 2° Spectral Luminous Efficiency Function for Photopic Vision* (Vienna: International Commission on Illumination)

[40] Vos J J 1978 Colorimetric and photometric properties of a 2-deg fundamental observer *Color Res. Appl.* **3** 125

[41] CIE publication 75–1988 Spectral Luminous Efficiency Functions Based Upon Brightness Matching for Monochromatic Point Sources with 2° and 10° Fields 1988

[42] Murray J J, Nicodemus F E and Wunderman I 1971 Proposed supplement to the SI nomenclature for radiometry and photometry *Appl. Opt.* **10** 1465

[43] Meyer-Arendt J R 1968 Radiometry and photometry: units and conversion factors *Appl. Opt.* **7** 2081

[44] Wyszecki G and Stiles W S 2000 Color science—concepts and methods *Quantitative Data and Formulae* 2nd edn (New York: Wiley)

[45] Munsell A H 1912 A pigment color system and notation *Am. J. Psychol.* **23** 236–44

[46] Munsell Color at https://munsell.com/ an affiliate of *X*-Rite, Incorporated (Accessed 2 July 2020)

[47] Pantone https://pantone.com/ subsidiary of *X*-Rite, Incorporated (Accessed 2 July 2020)

[48] Kruschwitz J D T 2018 *Field Guide to Colorimetry and Fundamental Color Modeling* (Bellingham, WA: SPIE Press)

[49] Malacara D 2011 *Color Vision and Colorimetry: Theory and Applications* 2nd edn (Bellingham, WA: SPIE Press)

[50] CIE data of 1931 and 1978 color matching functions is available at http://cvision.ucsd.edu and http://cvrl.org (Accessed 12 July 2020)

[51] ISO 11664–2:2007(E)/CIE S 014-2/E:2006, Colorimetry—Part 2: CIE Standard Illuminants for Colorimetry

[52] Wolfe W L 1998 *Introduction to Radiometry* (Bellingham, WA: SPIE Press)

[53] Robertson A R 1968 Computation of correlated color temperature and distribution temperature *J. Opt. Soc. Am.* **58** 1528–35

[54] Recommendation ITU-R BT.2020–2 2015 *Parameter Values for Ultra-High Definition Television Systems for Production and International Programme Exchange* (International Telecommunication Union) https://www.itu.int/rec/R-REC-BT.2020-2-201510-I/en

[55] SMPTE 170M- 1999 *Television—Composite Analog Video Signal—NTSC for Studio Applications* (Society of Motion Picture and Television Engineers) https://www.smpte.org/

[56] IEC 61966-2-1: 1999 *Multimedia Systems and Equipment—Colour Measurement and Management—Part 2-1: Colour Management* (International Electrotechnical Commission) https://www.iec.ch/

[57] *TECHNICAL GUIDE: Integrating Sphere Radiometry and Photometry* Labsphere, Inc. https://labsphere.com/ (Accessed 20 July 2020)

[58] Moreno I and Sun C-C 2008 Modeling the radiation pattern of LEDs *Opt. Express* **16** 1808–19

[59] Moreno I 2006 Spatial distribution of LED radiation *Proc. SPIE* **6342** 634216

[60] Keiser G 2011 *Optical Fiber Communications* 4th edn (New York: McGraw-Hill)

IOP Publishing

Optical Systems Design Detection Essentials
Radiometry, photometry, colorimetry, noise, and measurements
Robert M Bunch

Chapter 7

Detectors and noise

7.1 Introduction to detection

Detecting photons requires a physical mechanism to convert the energy of the photon stream into some other measurable sensation. The human eye, the first detector, collects photons and directs the light to photoreceptors called rods and cones eventually leading to a neural signal sent to the brain [1]. Traditional photographic film uses a chemical process to convert absorbed photons to a metallic grain upon development. The photoelectric effect, discussed earlier, is another example of a process where incident photons result in an electric current.

A detection system is like any other system; there is an input into the system, the system modifies this input, resulting in an output. A systems view for detecting an input optical signal power is illustrated in figure 7.1. Practical issues such as background power and noise contributions at each stage (source and background, the detector itself, and post-electronics) are also indicated. Note that the signal is only that part of the incident power carrying the information to be detected. The background may also have a power level that is detected but is not desired. It is also important to recognize that a background power is not noise. Noise is a fluctuation in a power level due to some physical mechanism. Both the signal power and the background power will in general have a separate noise contribution.

There are two major classes of detection mechanisms where the incident photon energy is converted to a measurable output. The first is a thermal detector that relies on heat generated from photon absorption to be converted into an electronic response. The second class is a photon detector that generates electrons by the absorption of photons. Photon detectors are also classified as quantum detectors and include both photoemissive detectors, and semiconductor detectors. The applications for different detector types and their electronic configurations will also be discussed. Noise mechanisms in detectors including shot noise, generation–recombination noise, Johnson noise, and $1/f$ noise are described along with the limitations

doi:10.1088/978-0-7503-2252-2ch7
7-1

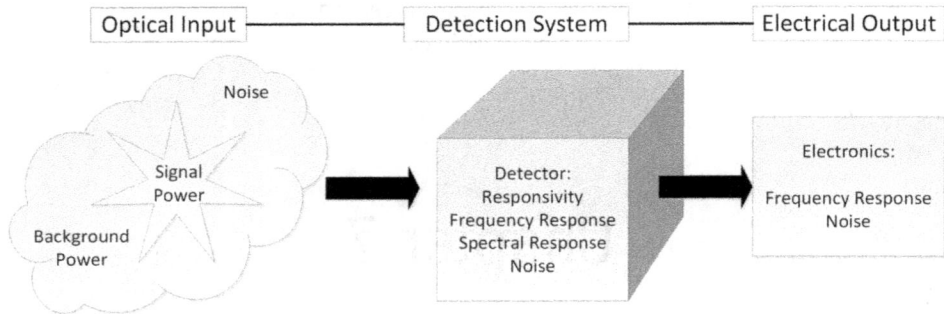

Figure 7.1. Systems view of an optical detection system.

that these noise sources impose on the detection of a signal. In addition, figures of merit used to specify detectors for use in a system, such as noise-equivalent power (NEP) and specific detectivity (D^*) will be defined.

7.2 Detector response functions

In this section we will only concern ourselves with some general principles of detection common to all types of detectors. There are two major requirements for a good optical power detection system. The first is high responsivity. Responsivity in this case is defined as the ratio of a measure of the output signal to the input optical power. Typical units of responsivity for an electronic detection system are Volts Watt^{-1} or Amps Watt^{-1}. The second detection requirement is a short response time that is a change in the output signal in time that closely follows any change in the input power with time. We will see that a system with a short response time has a high frequency response. The response of a detector to wavelength, its spectral response, is also critical in specifying detectors for a particular application. However, a spectral response specification is detector dependent based on the detector material or detection mechanism. Figure 7.2 shows a relative measure of responsivity as a function of wavelength as a comparison of a number of different detector types [2]. In this graph the responsivity also includes an area multiplier to allow for a fair comparison between detectors of different sizes.

In addition to the detection concepts above other specific terminology is used to describe detection systems. Since optical frequencies are high, many detectors are referred to as square-law detectors because they respond to optical power and not directly to the electric field where the power is proportional to the absolute square of the electric field amplitude. The detection system described here is also an example of direct detection or incoherent detection. A coherent detection system requires some type of interferometer that makes use of the coherence of the input power with respect to a local oscillator source. Quantum-limited detection relates to the special noise case where only quantum fluctuations exist and no other noise sources are present. Terminology related to other noise limiting cases will also be covered later.

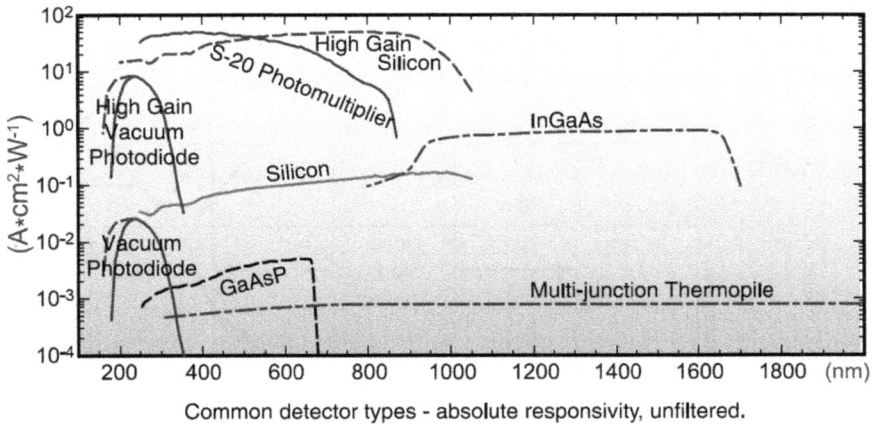

Common detector types - absolute responsivity, unfiltered.

Figure 7.2. Spectral response curves for various selected detectors. (Reprinted with permission from International Light Technologies, Light Measurement Handbook, https://www.intl-lighttech.com/light-measurement-handbook.)

7.2.1 Detector response and responsivity

Figure 7.3 illustrates many characteristics of a general detector system response relationship between output signal and input optical power. The slope of the graph is the responsivity and in this particular case the response is linear or constant slope within the limited region as shown. Linearity is rarely true over a wide range of input powers as shown in figure 7.3 as a non-linear region. Also note that most detectors will have a non-zero output for zero input signal. This value will remain constant up to some minimum detectable power. If the output measure is a current, the minimum current is referred to as the dark current when no flux is present. Detector systems often do not respond at very high input power levels and remain at a fixed output value. In most cases this is due to the detection electronics and is a condition called saturation.

To maximize detector output one can either increase the responsivity of the detector chosen or increase the detected flux. Optically this means that an increased output can be achieved by increasing detector area, increasing the solid angle of acceptance, or decreasing the $f/\#$ of the collection system. Cost and availability may limit the designer's choices of area and $f/\#$.

Let us assume that the responsivity $\mathcal{R}$ of a particular detector has units of Amperes per Watt. The output current, i, from an input monochromatic power source of flux Φ_e is,

$$i = \mathcal{R}\Phi_e + i_d$$

where i_d is the dark current for this detector. However, if we do not have a single monochromatic source then $\mathcal{R}$ is not simply a constant but is a multivariable function. For example, the responsivity will depend on wavelength, as we have already discussed. It could also be a function of incident angle, signal frequency, time, temperature, or numerous other variables.

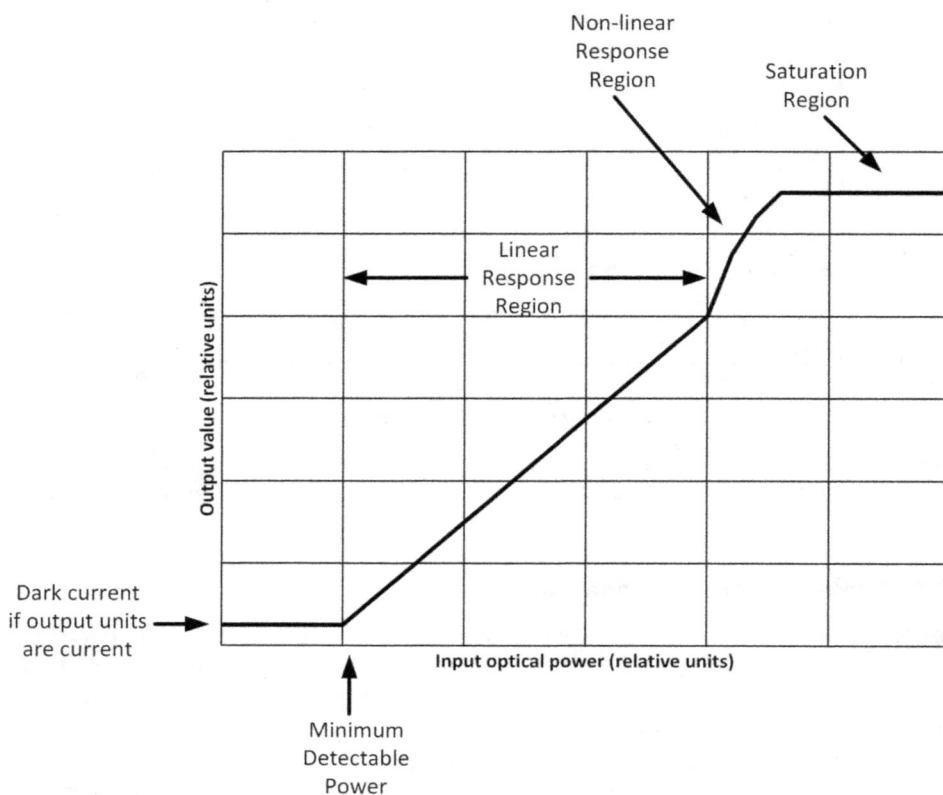

Figure 7.3. Output signal to an input optical power for a general detection system. The linear region, a non-linear region, and a saturation region are highlighted. Note the minimum detectable power value where all input power levels below this value are constant.

In many optical system applications, the important case is a responsivity that depends on wavelength. Suppose that we have two wavelength components in the source. Then each wavelength component will have its own responsivity and result in two contributions to the current. If the optical source contains many wavelength components a particular detection system responds to all of those wavelengths. So, a detector is an integration system, integrating over all components of the input weighted by the response function. If $\mathcal{R}$ only depends on wavelength, then the total output current from the detector is,

$$i = \int_0^\infty \mathcal{R}(\lambda)\Phi_e(\lambda)d\lambda + i_d$$

Another important situation is a filtered detector. In this case a transmittance filter characterized by a function $T(\lambda)$ is placed in front of the detector between the source and the detector surface area. The resulting output current is,

$$i = \int_0^\infty T(\lambda)\mathcal{R}(\lambda)\Phi_e(\lambda)d\lambda + i_d$$

7.2.2 Frequency response and bandwidth

If a detector has a short response time then any impulse in time will result in an output impulse, just as any ideal system. Typically, for an input impulse in time, the output is delayed, and the impulse is spread out in time due to detector system effects. This concept is shown in figure 7.4 with an input power impulse (delta function model) and a delayed and broadened pulse with temporal width of τ. Of course, ideally, we would want Δt and τ to be as small as possible.

Using the concept of figure 7.4 as a model, we take the input optical power ($t_0 = 0$) as,

$$\Phi_e(t) = \Phi_{e0}\delta(t)$$

In this case we choose a particular functional model for the temporal impulse response function of the detector to be,

$$i(t) = \begin{cases} 0 , & t < 0 \\ i_0 e^{-\frac{t}{\tau}} , & t \geqslant 0 \end{cases}$$

where the constant τ is a characteristic time constant. This type of model is analogous to a typical electrical system comprised of a resistor and capacitor with time constant $\tau = RC$. Instead of a temporal analysis, we will use what we know about system functions and compute the temporal frequency response function for this system. The frequency response function for our detection system must be proportional to the ratio of the Fourier transform of the impulse response, and the incident power level, or,

$$\mathcal{R}(f) \equiv \frac{\mathcal{F}\{i(t)\}}{\Phi_0} = \frac{\left(\dfrac{i_0\tau}{\Phi_{e0}}\right)}{1 + j2\pi f\tau}$$

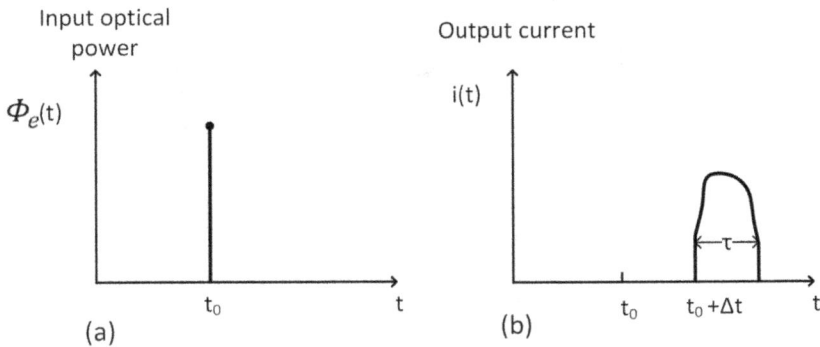

Figure 7.4. (a) Input optical power impulse, and (b) resulting output current delayed by Δt and broadened in time with a characteristic width of time τ.

The complex nature of this response function indicates that there is an inherent phase shift in the output for this system. The magnitude of this response as measured is,

$$|\mathcal{R}(f)| = \sqrt{\mathcal{R}(f)\mathcal{R}(f)^*} = \frac{\left(\dfrac{i_0\tau}{\Phi_{e0}}\right)^2}{[1 + (2\pi f\tau)^2]^{1/2}}$$

A plot of this frequency dependent response function is shown in figure 7.5. Note that when the frequency is a characteristic value $f_C = 1/2\pi\tau$, then,

$$\mathcal{R}(f_C) = \frac{\left(\dfrac{i_0\tau}{\Phi_{e0}}\right)}{\sqrt{2}}$$

which defines this characteristic frequency. This indicates that there is a band of lower frequencies where the signal is detected by this system and as the frequency increases the signal is attenuated in amplitude.

There is a specific definition used to identify the frequency bandwidth Δf for a system with a given frequency response function. This definition is,

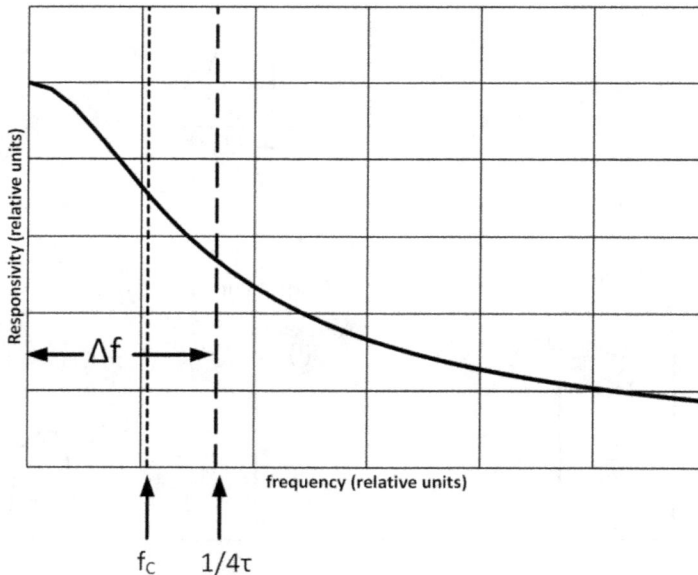

Figure 7.5. Frequency dependent response function for a model system with a temporal impulse response in the form of an exponential decay with time constant τ. Two frequencies are highlighted, the characteristic frequency f_C and the frequency $1/4\tau$ which is the maximum boundary of the frequency bandwidth.

$$\Delta f \equiv \int_0^\infty \left| \frac{\mathcal{R}(f)}{\mathcal{R}(0)} \right|^2 df$$

For the exponential decay model, the frequency bandwidth value is $1/4\tau$ and is indicated in figure 7.5.

There are other bandwidth models used commonly. One common model is for an impulse response of a simple rectangular pulse function. Following a similar analysis as above results in a value for the bandwidth to be $1/2\tau$. Both models are used in applications involving different types of detectors in systems as we shall see in later sections. In summary,

$$\Delta f_{\exp} = \frac{1}{4\tau} \quad \text{and} \quad \Delta f_{\text{rect}} = \frac{1}{2\tau}$$

where $\Delta f_{\exp}$ and Δf_{rect} are the bandwidths for an exponential decay model and a rectangular pulse, respectively.

Another temporal characteristic that is provided in specifications for many detection systems is a rise-time parameter. The concept is that the input power is a step function at time $t = 0$. Figure 7.6 shows this function along with the exponential impulse response model. For this situation, the time dependent output current becomes,

$$i(t) = i_0\left(1 - e^{-\frac{t}{\tau}}\right)$$

One common definition of rise time is the time it takes from the signal current to rise from 10% to 90% of the maximum value. Using the current from above this gives a relation for the rise time, $t_{\text{rise}} = 2.2\tau$. When the incident power suddenly drops to zero (no light) an exponential decay of the current is observed following this same model. This results in a fall time that may be modeled in a similar manner.

A chopper (segmented rotating wheel) is a common device used to externally modulate a continuous wave source. Modulating a signal to be detected allows this

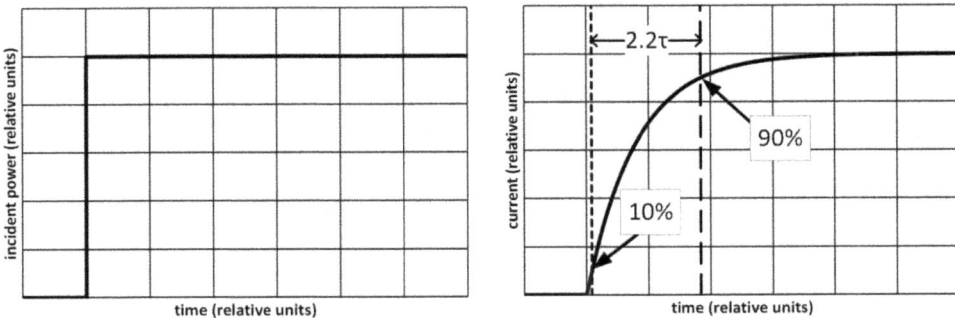

Figure 7.6. Current temporal response due to a step function increase in incident power using the exponential decay with time constant τ. Characteristic positions are also shown defining the rise-time parameter.

temporal signal to be amplified within a fixed frequency bandwidth which can increase the signal-to-noise ratio (SNR). Lock-in amplification can also be employed to measure signals in the presence of noise [3].

7.3 Thermal detectors

A thermal detector measures a temperature change related to the photon energy absorbed by the detector [4–6]. Figure 7.7 illustrates the concept of a thermal detector absorbing energy from the incident photon flux and raising the detector temperature to a value T_d above a heat sink temperature T_0. The detector and heat sink are coupled through a thermal conducting path. There are several mechanisms that can provide a measure of the temperature rise and will be discussed later for specific detectors. One of the major advantages of thermal detectors is that their spectral response is constant over a wide range of wavelengths from the visible to the far infrared. See figure 7.2. However, as this figure also shows, the responsivity is low in comparison to other detector types. Another disadvantage of thermal detectors is that their temporal response is slow because the relatively long time required to heat an object.

7.3.1 Modeling the thermal detection process

The physical concept of thermal detection is that the incident flux creates a temperature change. In heating an object at temperature T, this process can be described using a rate equation,

$$\left.\frac{dT}{dt}\right|_{\text{heating}} = \kappa_1 \Phi_e(t)$$

where the incident flux $\Phi_e(t)$ is the time dependent incident flux. The proportionality constant κ_1 accounts for issues such as the fraction of light absorbed, material properties, and detector volume. Suppose the time dependent flux is short pulse in a specific time t_1. The temperature rise from an initial temperature T_i, to a final temperature $T(t_1)$ can be obtained by simple integration, or,

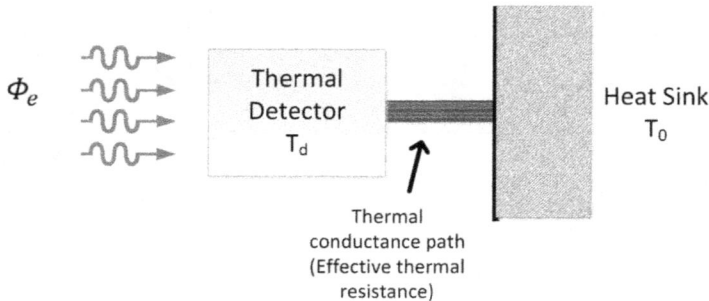

Figure 7.7. Thermal detection process. Incident flux is collected and absorbed by the thermal detector raising the temperature of the detector. The change in temperature is related to the incident photon flux.

$$\int_{T_i}^{T} dT = \int_{0}^{t_1} \kappa_1 \Phi_e(t)dt$$

Assuming that κ_1 is a constant, then performing the integrations results in,

$$\Delta T \equiv T(t_1) - T_i = \kappa_1 Q_e(t_1)$$

where Q_e is the radiant energy deposited during the duration of the pulse in time t_1. This result shows that a thermal detector operating under these conditions is an energy detector. A measure of the temperature rise ΔT is proportional to the energy in the pulse. The major assumption in this calculation is that there is no cooling taking place through conduction to a heat sink.

To consider a conduction cooling process (no radiative cooling), we can specify another rate equation for conduction to the heat sink at temperature T_0, as,

$$\left.\frac{dT}{dt}\right|_{cooling} = -\kappa_2(T - T_0)$$

where another proportionality constant κ_2 (units of inverse time) includes issues such as thermal conductivity of elements, material characteristics, and geometry. If no light enters the system after the time t_1, from above, the temperature of the detector as a function of time is,

$$T(t) = T(t_1)e^{-\frac{(t-t_1)}{\tau_{th}}}$$

where we have defined a parameter $\tau_{th} = 1/\kappa_2$, the thermal time constant for the conduction cooling process. Thus, if pulses of incident flux are of duration $t_1 \ll \tau_{th} = 1/\kappa_2$ our assumptions used above are valid and the temperature rise is proportional to the deposited energy on the detector. Detectors can be configured for pulsed sources under these conditions and are used as energy meters.

Returning to the concept shown in figure 7.7 we will examine a more general case for steady state operation of a thermal detection system. For this case, instead of a specific temperature we are interested in the difference between the detector temperature and the sink temperature or,

$$\Delta T \equiv T_d - T_0$$

The detector is thermally connected to the sink through a path with thermal conductance $G_{th} \equiv 1/R_{th}$ where R_{th} is the thermal resistance. Units of thermal conductance are [W K^{-1}]. Take the detector as having a thermal heat capacity defined as,

$$H_{th} \equiv \frac{dQ}{dT}$$

which is the heat stored when the temperature is changed by an amount dT. Units of heat capacity are [J K^{-1}]. Taking the temperature change ΔT in time interval dt, we model the energy rise as the difference between energy input and the energy transferred to the heat sink.

$$H_{th}d(\Delta T) = \varepsilon\Phi_e(t)dt - G_{th}\Delta Tdt$$

where ε is the fraction of incident power absorbed by the detector. From this model we arrive at a differential equation that can be solved for both constant input radiation and time dependent flux, such as with modulated or chopped incident radiation. The equation to be solved becomes,

$$\frac{d(\Delta T)}{dt} + \frac{1}{R_{th}H_{th}}\Delta T = \frac{\varepsilon}{H_{th}}\Phi_e(t)$$

First, take the case where a constant input flux is applied at $t = 0$ with the value of the constant incident radiation, $\Phi_e(t) = \Phi_{e0}$. After separation of variables the steady state temperature change is,

$$\Delta T = \varepsilon R_{th}\Phi_{e0}\left[1 - e^{-\frac{t}{\tau_{th}}}\right]$$

where $\tau_{th} = R_{th}H_{th}$ is again a thermal time constant. The form of this function is identical to the relations plotted in figure 7.6 except in this case the temperature change is observed instead of a current. This relation shows that after a long time, in comparison to the thermal time constant a steady state temperature difference is maintained and is proportional to the incident constant flux. There is a rise-time for all thermal detectors that is dictated by the thermal properties of the system. For these situations, the detection system provides a measure of the incident power but only after a sufficient time to reach a steady temperature.

   Next, suppose there is a modulated signal in time incident on the thermal detector. Take the incident radiant flux as a function of time oscillating at frequency f, as,

$$\Phi_e(t) = \Phi_{e0}e^{+j2\pi ft}$$

Ideally, we want the temperature change to follow the same form of this function for the detection system output to faithfully represent the incident flux. This means that we desire the temperature change to have a form,

$$\Delta T(t) = \Delta T_0 e^{+j2\pi ft}$$

where ΔT_0 is the amplitude of the temperature change observed. Substituting these functions into the differential equation above, we obtain a condition on the frequency dependence of the amplitude of the temperature change, which is,

$$\Delta T_0(f) = \frac{\varepsilon R_{th}\Phi_{e0}}{1 + j2\pi f\tau_{th}}$$

Note that the form of this equation is the same as we computed earlier for the exponential system response model. It is a direct analogy to the standard RC time constant in an electrical circuit where in this case the thermal resistance and heat capacity are analogous to the electrical resistance and capacitance, respectively.

7.3.2 Types of thermal detectors

Thermal detectors require a mechanism that senses a temperature or temperature change and provides an electrical signal (voltage or current) to be measured. The simplest thermal detector is a single junction thermocouple. When two dissimilar materials (materials with a different electrode potential such as antimony and bismuth) are joined their junction produces a potential difference that is proportional to temperature of the junction [7]. Figure 7.8 shows a diagram of a thermocouple junction optical detector concept composed of two materials, Material A and Material B. In practice, a reference voltage is required so that temperature effects of the environment can be reduced. This is accomplished using another potential or a second junction, the cold junction, as is shown in figure 7.8. The absorbing element can also be isolated from the exterior device using thin film structures to increase thermal isolation.

If we assume that there is little heating of the junction due to any current flow, the temperature change across the junction is directly proportional to the incident radiant flux absorbed by the sensing element in contact with the hot junction,

$$\Delta T_0 = \varepsilon R_{th} \Phi_e$$

where R_{th} is the thermal resistance and ε is the fraction of incident optical power absorbed by the sensing element. The open circuit voltage across the junction is then simply,

$$V_0 = \varepsilon S_{th} R_{th} \Phi_e$$

where S_{th} is the thermoelectric coefficient (or Seebeck coefficient) often expressed in units of $\mu V\ K^{-1}$. Written as a response function,

$$\mathcal{R} = \varepsilon S_{th} R_{th}$$

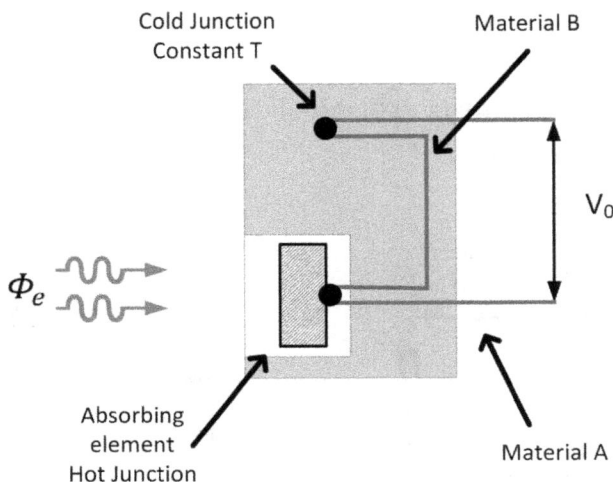

Figure 7.8. Thermocouple detector concept showing the active/hot and cold junction configuration.

Improving the response of such a detector is accomplished by increasing absorption efficiency or picking optimal materials. Another improvement can be made by adding additional thermocouples to the absorbing element so that the overall response is multiplied by the number of thermocouples. This type of device is known as a thermopile [8].

For specific applications, a narrower spectral region is often desired since thermopiles have such a large spectral response. To accommodate these specifications window materials can be used in front of the detector active region [8]. A distinct advantage of thermopile detectors is that the noise associated with the detection is only thermal noise or Johnson noise. This will be discussed later in this chapter.

Bolometers are a type of thermal detector that relies on the change in resistance of a detector element as the temperature changes. The amount of resistance change of a detector R_d with temperature T_d is governed by a materials temperature coefficient of resistance, α, where,

$$\alpha \equiv \frac{1}{R_d}\frac{dR_d}{dT_d}$$

Two common circuits are often employed, a simple voltage divider circuit or a bridge circuit. Figure 7.9(a) shows the voltage divider with an applied voltage V, provided by a battery, along with a load resistor R_L in series with the bolometer detector resistor, R_d. Light is incident on the bolometer resistor element, changing its resistance and changing the current flow in the circuit.

Analyzing the circuit to find the change in voltage Δv across the bolometer when the detector resistance changes by a small amount ΔR_d, results in,

$$\Delta v \cong \frac{V R_L \Delta R_d}{(R_L + R_d)^2}$$

The response function for this bolometer configuration becomes,

$$\mathcal{R} \equiv \frac{\Delta v}{\Delta \Phi_e} = \frac{V R_L R_d}{(R_L + R_d)^2}\frac{\varepsilon\,\alpha\,R_{th}}{[1 + (2\pi f \tau_{th})^2]^{1/2}}$$

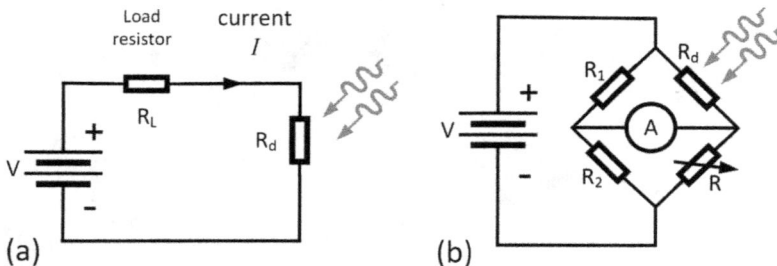

Figure 7.9. Bolometer circuits. (a) Simple voltage divider and (b) bridge circuit.

In many cases for metal bolometers the resistance $R_d \ll R_L$ and the relation above can be simplified.

With the bridge circuit of figure 7.9(b), the variable resistor R is adjusted so that no current flows through the ammeter. This condition occurs when, $R = \frac{R_2}{R_1}R_d$. When the bolometer detector resistance changes due to some incident radiation, current begins to flow in proportion to the resistance change. In any bolometer circuit, the current through the bolometer resistor should be minimized so that little heat is dissipated by the detector. One type of thermal imager uses two-dimensional arrays of microbolometers as the detecting elements [9, 10]. Advances in micro-electro-mechanical systems (MEMS) have enabled small scale thermally isolated resistive structures to be fabricated with no cooling requirements.

The pyroelectric effect is another mechanism that can be employed to design thermal detectors. This effect occurs in ferroelectric polar materials where the electric polarization changes in response to a temperature change [11]. The change in polarization produces a surface charge that can be detected. Materials such as lithium tantalate ($LiTaO_3$) and triglycerine sulfate (TGS) are common pyroelectric detector materials. Since the materials are dielectrics with surface charges, they resemble the structure of a capacitor. The equivalent electric circuit for a pyro-electric detector is shown in figure 7.10(a).

The spontaneous change in electric polarization, P, with applied temperature, T, is characterized by the pyroelectric coefficient, p,

$$p(T) = \frac{dP}{dT}$$

which has MKS units of Coulombs per square meter per degree Kelvin. Figure 7.10(b) illustrates the concept of a pyroelectric detector with surface area A. With no incident flux there are no excess charges on the surface since the two opposing polar surfaces are connected to each other through the load resistor, R_L. As soon as a time dependent radiant flux is introduced and absorbed, a change in temperature and electric polarization occurs producing a change in the surface charge, or,

$$dq = AdP = pAdT$$

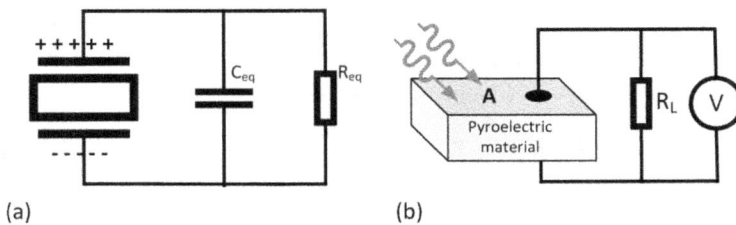

(a)   (b)

Figure 7.10. (a) Equivalent circuit for a pyroelectric detector, and (b) pyroelectric detector concept.

If the flux is applied within a time dt, there is a subsequent current flow, but only as long as the temperature is changing in time. This current is,

$$i_d = \frac{dq}{dt} = pA\frac{dT}{dt}$$

Therefore, the detected current is a time dependent quantity and only occurs as long as there is a time dependent temperature change. Because of the magnitude of the effect and the high impedance of the detector, the voltage across the detector element is normally input to a high impedance meter or interfaced to a high impedance amplifier.

Assume that the incident flux oscillates in time $\Phi_e(t)$ with frequency, f, using a similar analysis as with the bolometer. In this case, we need dT/dt, or,

$$\frac{dT}{dt} = \frac{j\,2\,\pi\,f\;\varepsilon\;R_{th}\Phi_e(t)}{[1 + (2\pi f\;\tau_{th})^2]^{1/2}}$$

The magnitude of the measured current is then from the detector is,

$$i_d = \frac{p\,A\,2\,\pi\,f\;\varepsilon\;R_{th}\Phi_e(t)}{[1 + (2\pi f\;\tau_{th})^2]^{1/2}}$$

But from the electrical equivalent circuit there is also a frequency dependent requirement placed on the system because of the equivalent resistor and capacitor of the detector passing this current. We define an equivalent electrical time constant $\tau_{eq} = R_{eq}C_{eq}$ and find the overall response function for the pyroelectric detector system as,

$$\mathcal{R} \equiv \frac{\Delta v}{\Delta \Phi_e} = \frac{p\,A\,2\,\pi\,f\;\varepsilon\;R_{th}\;R_{eq}}{[1 + (2\pi f\;\tau_{th})^2]^{1/2}\left[1 + \left(2\pi f\;\tau_{eq}\right)^2\right]^{1/2}}$$

For this detector there is both a thermal time constant and an electrical time constant limiting the frequency response of the system. The thermal time constant limits low-frequency operation and the electrical time constant limits the high frequency cutoff. A sample plot of this frequency response is shown in figure 7.11 plotted on a logarithmic frequency scale. Using a pyroelectric detector requires that the frequency range of the device be within the upper and lower frequency limits.

Pyroelectric detector systems find numerous applications including motion detection, lighting controls, intrusion sensing, temperature sensing, and pulsed laser measurement [12]. The source must provide a change in temperature or be modulated or chopped for these detectors to operate. Like bolometers, pyroelectric detectors have a broad spectral response. Devices are manufactured with windows to limit the spectral response for certain applications.

A sensor circuit used with pyroelectric detectors can be formed in either a current or voltage mode. See figure 7.12 for a schematic of typical current and voltage mode circuits. Since these detectors only operate with a modulated signal, the current-mode will have a somewhat higher frequency response than a voltage-mode circuit.

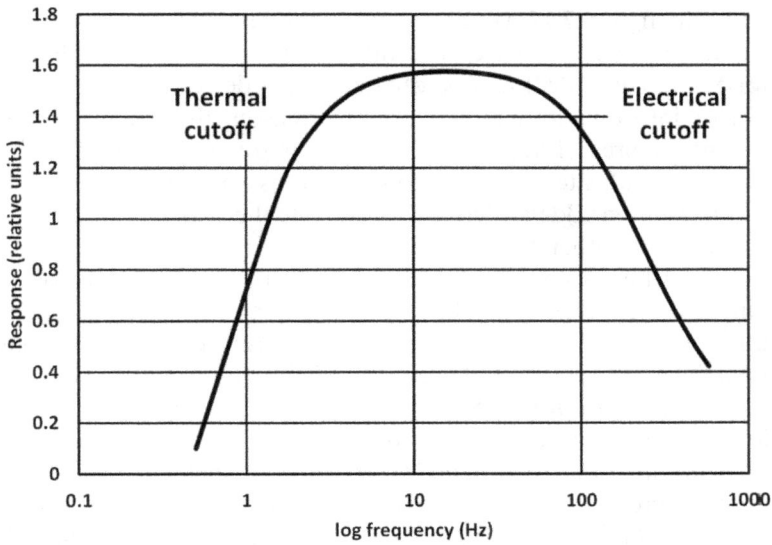

Figure 7.11. Characteristic bandpass frequency response function plot for a pyroelectric detector.

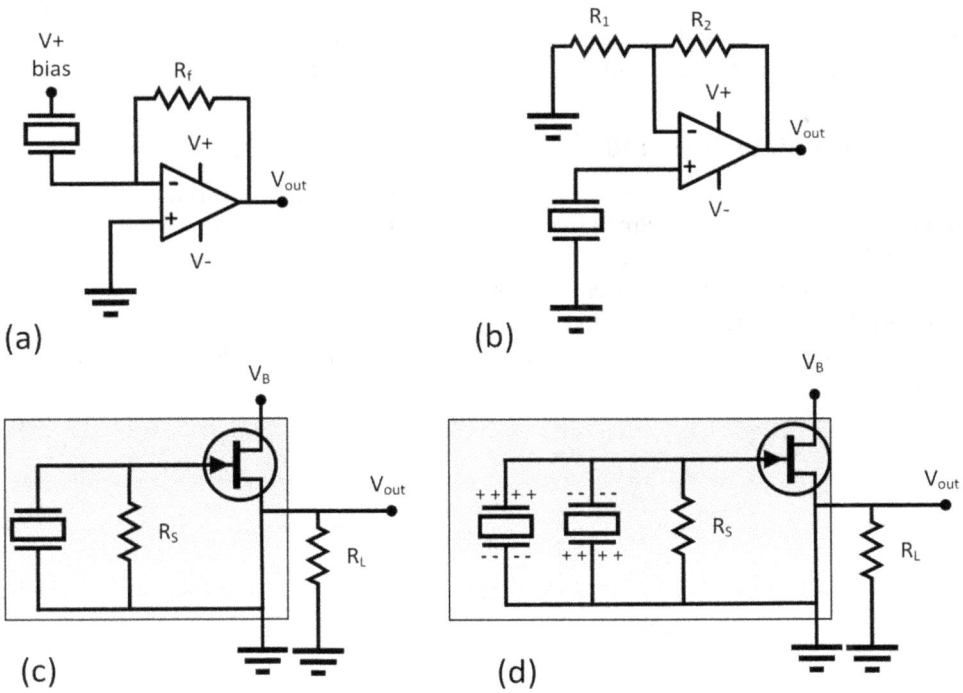

Figure 7.12. (a) Current-mode and (b) voltage mode circuits using operational amplifiers for a general pyroelectric detector. Field effect transistor circuits often combined into one device are shown in (c) and (d) the dual-opposed configuration detector component.

This is shown in figure 7.12(a). The voltage mode circuit using an operational amplifier along with a single detector element is shown in figure 7.12(b). However, reduced noise and signal drift can be obtained by a device fabricated with a field-effect transistor into one component. For this reason, the voltage mode is much more common. Figure 7.12(c) shows this circuit diagram. The box around the effective elements indicate that the component is incorporated as one device. A shunt resistor is often added internally to provide thermal stability. The schematic in figure 7.12(d) is another type of component in a dual-opposed configuration. When the incident flux is presented to the detector one element provides a positive current going pulse in time while the opposed element generates a negative going pulse in time. When the flux is removed the opposite effects occur. The output voltage signal from this device configuration provides information about the timing of the flux which is useful for some applications.

Detection systems that operate in the infrared are often chopped using a mechanical blade in order to modulate the beam of incident light. This may be necessary for a pyroelectric detector where the detector must operate within its frequency bandwidth. A chopped signal can also be more easily amplified using an AC amplifier or as a reference signal for lock-in detection [3]. When a chopped beam is utilized, another factor must be considered. This is the chopping factor which is defined as the ratio of the root-mean-square average of the periodic signal to the peak value of the signal at the chopping frequency [13, 14]. When the size of the beam is small compared to the size of the chopper blades creating a square wave signal, the chopping factor simply reduces to a value of 0.450158 [13].

7.4 Photon detectors and the photoconductor model

Photon detectors (quantum detectors) collect photon energy and release an electron in relation to each photon. Since the mechanism is related to the quantized energy of the photon, this type of detection is also called quantum detection [5]. The photoelectric effect, discussed earlier, is an example of a mechanism that can be used as a photon detector. The incident photon flux generates a current output that is typically measured by measuring the voltage across a load resistor in the detection circuit as seen in figure 7.13.

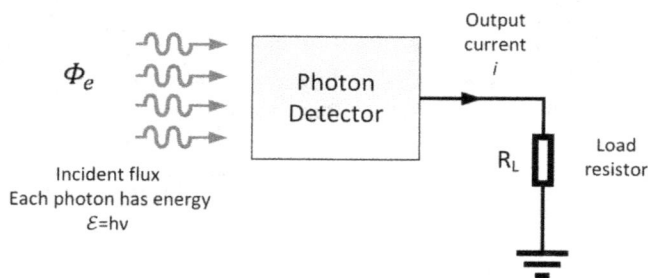

Figure 7.13. Photon detection concept where incident flux is collected by the detector causing the release of an electron in proportion to the energy of an incident photon.

For photon detectors, we can model the responsivity simply using a unit analysis and making some reasonable assumptions. This leads to the photoconductor model for quantum detection. We start with our definition of the responsivity as the ratio of the average detected current to the incident flux. We assume that this value is constant with incident flux. This assumption is valid because the average output current scales directly with the number of incident photons. This responsivity is,

$$\mathcal{R} \equiv \frac{i_{\text{avg}}}{\Phi_e}$$

The units of responsivity are A W^{-1} for this relation. Remember that current is defined as the number of electrons with charge, $e = 1.602 \times 10^{-19}$ C, flowing in the circuit or,

$$i = e \left[\frac{\text{\# electrons}}{\text{sec}} \right]$$

The value of the incident photon flux can be written in a similar manner since the optical power is the energy per unit time where $[1 \text{ W}] = [1 \text{ J s}^{-1}]$ and the energy is just the number of photons multiplied by the energy of a single photon $h\nu = hc/\lambda$. This results in a relation for the flux of,

$$\Phi_e = \frac{hc}{\lambda} \left[\frac{\text{\# photons}}{\text{sec}} \right]$$

Now, we define a new parameter called the quantum efficiency, η. Another common designation is QE. This is the ratio of the number of electrons generated in the detector to the number of photons incident on the detector. The efficiency has a maximum value of unity since at best only one electron can be freed for each incident photon. Each specific photon detector type has issues that affect its quantum efficiency, and these will be discussed later along with each specific detector. In our case, the quantum efficiency factor is,

$$\eta = \frac{\left[\frac{\text{\# electrons}}{\text{sec}} \right]}{\left[\frac{\text{\# photons}}{\text{sec}} \right]}$$

With this definition and the relations from above, the responsivity can be written as,

$$\mathcal{R} \equiv \frac{i_{\text{avg}}}{\Phi_e} = \eta \frac{e}{hc} \lambda$$

which is a constant value as expected for incident monochromatic light. If the quantum efficiency of the detector mechanism is constant the photoconductor model predicts that the spectral response will be directly proportional to the wavelength. This model also indicates that photon detectors have an inherent linear response with incident optical power.

7.5 Noise in the detection system

At every stage of the detection process issues of noise can arise and will impact the quality of the desired signal. Noise in the detection system is due to random fluctuations in optical power or from fluctuations in the electrical signal (voltage or current) detected by a detector. Sources of noise in detection include photon noise from both the signal and the background, noise present in the detector itself, and noise arising in any post-detector electronics.

In measuring the signal output from a detector as a function of time the current or voltage will fluctuate due to noise mechanisms through the system. As discussed in a previous chapter, we characterize the measurement of the signal by finding the mean value in time. For an output current the time average over a time frame from $t = 0$ to $t = T$ is defined as,

$$\langle i \rangle \equiv \frac{1}{T} \int_0^T i(t)dt$$

where $\langle i \rangle$ represents the time average value of the current or the mean value. The mean square fluctuation of this time varying signal or the variance is then,

$$\langle \Delta i^2 \rangle = \langle (i - \langle i \rangle)^2 \rangle = \frac{1}{T} \int_0^T (i(t) - \langle i \rangle)^2 dt$$

For a thermal detector, the mean value of the signal could also be from a voltage related to a temperature change. The variance is due to the noise associated with that voltage. For photon detectors that follow the photoconductor model, the mean value of the signal is usually a current related to the mean number of photon arrivals as we found earlier using Poisson statistics. The variance is due to any number of noise mechanisms in the system. The mechanisms will be discussed later. In both cases we assume that the mean value of the noise is zero and that the noise is truly a random process.

Instead of the time domain we can also describe noise characteristics in the temporal frequency domain. From Fourier analysis we define Fourier transform pairs of the current as,

$$I(f) = \mathcal{F}\{i(t)\} = \int_{-\infty}^{+\infty} i(t)e^{-j2\pi ft}dt \quad \text{and} \quad i(t) = \mathcal{F}\{I(f)\}$$

$$= \int_{-\infty}^{+\infty} I(f)e^{+j2\pi ft}df$$

But the average current signals above are actually truncated signals since the measurements are limited within the characteristic time. A truncated frequency dependent current over the integration time T is then,

$$I_T(f) = \int_0^T i(t)e^{-j2\pi ft}dt$$

Analysis of the frequency dependence of the noise contribution is specified using the power spectral density (PSD) function [5, 15, 16]. We first calculate a quantity proportional to the average electrical power, as

$$\left\langle \frac{P}{R} \right\rangle = \frac{1}{T} \int_0^T i^2(t)dt$$

where P/R has units of current-squared. Using this notation is essentially defining a power quantity divided by an effective resistance in order to preserve consistent units. Using the truncated current frequency function and the Fourier transform definition from above, this quantity can be written as,

$$\left\langle \frac{P}{R} \right\rangle = \frac{2}{T} \int_0^\infty |I_T(f)|^2 \, df$$

where we have also used the fact that the frequency dependent current must be real and positive. From these relations, the PSD function, $S_T(f)$, is defined as,

$$S_T(f) = \frac{2}{T} |I_T(f)|^2$$

This function indicates that the PSD is proportional to the average power delivered within each frequency interval. Therefore, the integral of the PSD over the measured frequency bandwidth is the average square current value,

$$\langle i^2 \rangle = \int_0^\infty S_T(f)df$$

Another Fourier analysis principle can also be used to compute the PSD. The autocorrelation function associated with the measured current within integration time T is defined as,

$$C_i(t) = \frac{1}{T} \int_0^T i(t')i(t' + t)dt'$$

The Wiener–Khintchine theorem states that the correlation function $C_i(t)$ is a Fourier transform pair with the PSD function [5, 17]. This results in,

$$S(f) = \mathfrak{F}\{C_i(t)\} = \int_{-\infty}^{+\infty} C_i(t)e^{-j2\pi ft}dt$$

While this may look complicated, in practice the autocorrelation function can be measured experimentally followed by a numerical Fourier transform on this experimental data to obtain the function [16].

Our assumption that any noise mechanism is random with a mean value of zero leads to some useful properties for detection systems. From a statistical perspective, the signal to be detected has a mean value that we want to obtain and is not zero, in other words, it is not random but has a definite average. This means that by taking several independent measurements of the mean value of the signal we can average these quantities and reduce the effect of the random noise. This is referred to a noise reduction through averaging [18, 19].

The variance of the time varying measurement will have units of current-squared or voltage-squared which is proportional to electrical power. When the noise sources are uncorrelated the electrical powers/variances of these sources add [15]. So, the standard deviation or total root-mean-square current due to several noise current fluctuations becomes,

$$i_{rms} = \sqrt{i_{1\,rms}^2 + i_{2\,rms}^2 + i_{3\,rms}^2 + \cdots}$$

7.5.1 Photon noise and Schottky's formula

As discussed previously, in a photon detector, the output current is due to electrons generated in the detector. Each electron is related to an incident photon, assuming unit quantum efficiency. The time average current measured within a characteristic detection time τ is,

$$\langle i \rangle = \frac{e \langle n \rangle}{\tau}$$

where e is the electron charge and $\langle n \rangle$ is the average number of photons collected by the detector within the characteristic time. Again, the bracket notation represents a time average.

To analyze the noise fluctuation, we use statistics to determine the mean square current deviation. Using the definition from before, and substituting the above current relation, we obtain,

$$\langle \Delta i^2 \rangle = \langle (i - \langle i \rangle)^2 \rangle = \frac{e^2}{\tau^2} \langle (n - \langle n \rangle)^2 \rangle$$

The discrete detection events for many photons follows Poisson statistics. As determined before the variance of the number of photons is equal to the average number. Thus, the mean square current deviation is,

$$\langle \Delta i^2 \rangle = \frac{e^2}{\tau^2} \langle n \rangle = \frac{e}{\tau} \langle i \rangle$$

Or the mean square deviation of the current is proportional to the mean value of the current. Since we assumed that the events take place within a characteristic time of the detection events, the bandwidth of the detection system is, $1/2\tau$. Therefore, with this substitution we arrive at Schottky's formula [5] for the mean square current deviation of photon detection, or,

$$\langle \Delta i^2 \rangle = 2\, e \langle i \rangle \, \Delta f$$

The photoconductor model provides a relationship for the average current due to an input flux photon stream. So, if only photon noise associated with the optical signal exists in the detection system, we can calculate the rms noise current contribution,

$$i_{N\,\text{rms}} = \sqrt{\langle \Delta i^2 \rangle} = \sqrt{2e\langle i \rangle\, \Delta f} = \sqrt{2\eta\, \frac{\lambda e^2}{hc} \Phi_e\, \Delta f}$$

This is referred to as quantum-limited detection since the only noise contributions are from photon energy interactions.

7.5.2 Detector noise mechanisms

Viewing the detector itself as a system, we can usually model an effective equivalent circuit that describes the photon to electron conversion process. Since the detector is a source of current, the equivalent circuit will include an effective current source. Other components of the detector include, internal resistors, internal capacitors, and an ideal diode for p–n junction type detectors. Noise contributions from all these components must be considered.

Noise produced by the detector comes from fluctuations in current or voltage that arise from the physical mechanism of the discrete photon events generating current and the flow of current within the device. Because electrical currents are a flow of discrete electrons at the microscopic level the time dependence of their motion is random. This randomness leads to noise being generated from several specific mechanisms acting on each equivalent circuit component. Here we will concentrate on a few noise sources important in detecting optical signals.

Shot noise arises because of the random emission and flow of quantized charge carriers (photoelectrons) in a device. It is observed not only in optical detectors but in numerous electronic components such as transistors and diodes. Suppose that we have an electron flow from a cathode to an anode that produces a detected signal current into an external circuit. Analyzing the physics of this process requires that the electrons flow across some potential energy barrier [16]. From an equivalent circuit perspective shot noise is an effective current generator. But this is the same argument used previously in arriving at Schottky's formula [5] for the mean square current deviation of photon detection. So, the mean square shot noise current is given by,

$$\langle i_{\text{SN}}^2 \rangle = 2e\langle i \rangle\, \Delta f$$

In this equation the average current will be from all possible current sources. Although this is shown as a single value $\langle i \rangle$ the total current may have contributions from the signal itself, the background, dark current, and could be a DC bias current depending on the detection circuit. The dark current is from thermally excited carriers and is a current that appears even though there is no incident flux. This relationship also shows that the PSD is a constant value, $2e\langle i \rangle$. When the PSD is a constant over the measurement bandwidth Δf the noise is referred to as white noise.

The detection mechanism in semiconductor detectors is through carrier generation when incident photon energy creates conduction band electrons and valence band holes producing current. Fluctuations in this process produce noise, like shot noise, except that the carriers have a lifetime (τ_c) before being recombined [16]. The carrier lifetime and thus the noise will depend on whether the semiconductor is intrinsic or extrinsic. The ratio of the carrier lifetime to the time for an average carrier to drift through the device (τ_d) is defined as the photoconductive gain, $G = \tau_c/\tau_d$. In terms of rates, the gain is the rate of drift carrier flow to the rate of generation of electron–hole pairs. Thus, this value can be larger than unity since the carrier flow rate can be increased by applying a larger potential or decreasing the device size [20]. Also, both photoexcitation and thermal excitation mechanisms for the carriers must be included. A general theory of generation–recombination noise is beyond the scope of this discussion and other references should be consulted [5, 21, 22]. Models for generation–recombination result in a mean square noise current of,

$$\langle i_{GR}^2 \rangle = \frac{4\,e\,G\langle i \rangle\,\Delta f}{[1 + (2\pi f\ \tau_c)^2]^{1/2}}$$

The average current is due to both photoexcited carriers and thermally excited carriers (like a dark current). Since this equation depends on frequency, generation–recombination noise is not in general a white noise source. However, when $f \ll 1/2\pi\tau c$, then

$$\langle i_{GR}^2 \rangle \approx 4e\,G\langle i \rangle\,\Delta f$$

which is a white noise source for this frequency region.

Johnson noise arises from the motion of charge carriers within a resistive element. Since all charge carriers will have some thermal energy at temperatures above absolute zero, they can exhibit Brownian motion and create random voltage or current fluctuations [5, 15]. This noise source is also called thermal noise and Nyquist noise. Johnson noise is present is any system which has a conductor, so it is present in all circuits.

There are several alternate approaches to arrive at the mean square Johnson noise current using thermodynamics, statistics, and circuit analysis [5, 15]. Equivalent circuit models for a current and voltage noise generator with an ideal resistor are shown in figure 7.14.

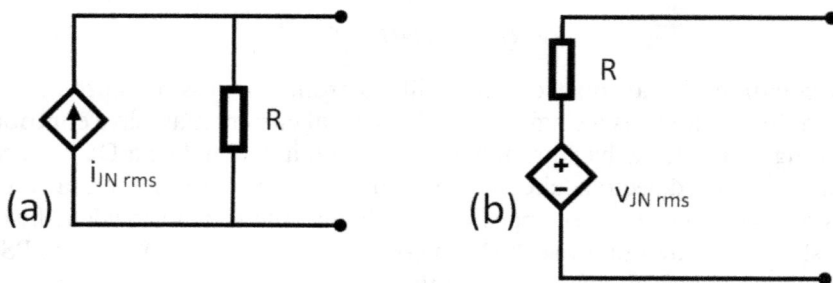

Figure 7.14. Equivalent circuits for an ideal resistor and noise generator as either (a) current noise source or (b) voltage noise source.

The result is that the mean square noise current and mean square noise voltage describing Johnson noise is,

$$\langle i_J^2 \rangle = \frac{4\,kT\,\Delta f}{R} \quad \text{and} \quad \langle v_{\mathrm{JN}}^2 \rangle = 4\,kTR\,\Delta f$$

where k is Boltzmann's constant $= 1.38 \times 10^{-23}$ J K^{-1}, T the temperature in K, R the resistance in Ω, and Δf is the noise bandwidth in Hz. Like shot noise, the PSD of Johnson noise has no frequency dependence and is a source of white noise, at least for oscillation frequencies $f \ll kT/h$. For higher frequencies in photoconductors, the generation–recombination noise decreases, and Johnson noise will dominate. As an example, Johnson noise is the main contribution to pyroelectric thermal detectors.

Another frequency dependent noise source is $1/f$ noise named for its characteristic frequency dependence. The noise effect is observed in many situations and is also known as excess noise or flicker noise depending on the device. Causes of $1/f$ noise in detectors are thought to be related to the quality of resistive contacts at electrodes and surface states [21]. We observe the effects of $1/f$ noise in detection systems as drift from calibration. An empirical relationship for root-mean-square noise current due to $1/f$ noise is,

$$\langle i_f^2 \rangle = \langle i \rangle^2 B \frac{\Delta f}{f}$$

where B is an empirically determined constant. It can be the dominant noise source for low-frequency applications but is often not included in high bandwidth systems.

Temperature noise is present in thermal detectors and in components whose parameters that are temperature sensitive. In an ideal thermal detector with no incident flux, the only noise is temperature noise [5]. The mean square temperature noise is,

$$\langle \Delta T_N^2 \rangle = \frac{4\,kR_{\mathrm{th}}T^2\,\Delta f}{[1 + (2\pi f \tau_{\mathrm{th}})^2]^{1/2}}$$

where k is Boltzmann's constant, $\tau_{\mathrm{th}} = R_{\mathrm{th}}H_{\mathrm{th}}$ is the thermal time constant, with R_{th} the thermal resistance and H_{th} the thermal heat capacity [5, 21].

7.5.3 Detection metrics and figures of merit

Figures of merit are commonly used as a simple way to compare and distinguish the performance of a device or system. In engineering, figures of merit assist with the design of systems to meet requirements. In particular, there are several metrics and figures of merit that allow designers to compare detectors and detection systems to characterize performance. We have already discussed responsivity, spectral response, and quantum efficiency as figures of merit for detection systems.

The concept of SNR as a measure of the quality of detecting a signal in the presence of noise fluctuations was introduced earlier. SNR is another figure of merit commonly used to quantify the quality of a detection system. In optical engineering, the SNR value is defined as the ratio of the optical signal power to the total noise power from all noise mechanisms. (This is different from the electrical SNR which is proportional to the square of the electrical current or voltage.) We will use SNR to describe the optical signal-to-noise unless specifically designating that the measure is an electrical signal-to-noise. Using this definition, the SNR is,

$$\text{SNR} \equiv \frac{\text{optical signal power}}{\text{noise power}} = \frac{\Phi_{es}}{\Phi_N}$$

where Φ_{es} is the radiometric flux from the signal only and Φ_N is the total noise power from all noise sources.

In a system that conforms to the photoconductive model, the signal flux is proportional to the signal current. From this, the SNR can also be written as,

$$\text{SNR} = \frac{i_s}{i_N} = \frac{\mathcal{R}\Phi_{es}}{i_N}$$

where i_s is the signal current proportional to the signal flux through the detector responsivity. The total rms noise current is specified by i_N. Contributions to the rms noise current will come from any or all of the applicable noise sources discussed earlier. Remember noise sources do not add directly but must be added in quadrature. Rearranging this equation for the total rms noise current gives,

$$i_N = \frac{\mathcal{R}\Phi_{es}}{\text{SNR}}$$

when the SNR = 1, we define another metric called the noise-equivalent power (NEP). NEP is related to the rms noise current as,

$$i_N = \mathcal{R}\,\text{NEP}$$

which is equivalent to the minimum detectable signal power. To calculate the NEP, first evaluate the SNR as a function of the signal power, set the SNR = 1, and calculate the signal power equal to NEP. The minimum detectable signal power was discussed earlier and is shown in figure 7.3. Using this definition for NEP it has units of power or Watts using the MKS system.

The specified value of the NEP quantity will depend on how the measure is made. If making a noise current measurement, the detector gain, detector characteristics, and post-detector electronics will impact the noise current measurement. One signal modulation frequency may also have a different SNR than another modulation frequency. This has led some references and manufacturers to specify a value they call NEP but normalize it to a frequency bandwidth of 1 Hz for a given incident optical wavelength. This results in specifying the NEP value in units of Watts $\text{Hz}^{-1/2}$ [3]. NEP is also often quoted at the peak of the detector spectral response curve.

To use the NEP value from a specification sheet it is necessary that an optical designer know how the value was obtained. Fortunately, most specifications include the information. If a designer needs a value at another response wavelength then,

$$i_N = \mathcal{R}(\lambda_1)\,\mathrm{NEP}(\lambda_1) = \mathcal{R}(\lambda_2)\,\mathrm{NEP}(\lambda_2)$$

can be used to obtain the desired value. If the specified NEP quantity has units of $\mathrm{W\ Hz^{-1/2}}$ then multiplying by $\sqrt{\Delta f}$ of the system under design will provide the minimum detectable optical power. In summary, the SNR may be calculated at a known NEP obtained from a specification sheet from,

$$\mathrm{SNR} = \frac{\Phi_{e\,s}}{\mathrm{NEP}}$$

But the conditions used to measure the NEP by the detector manufacturer value must be known.

While NEP is a necessary value to know in designing a detection system for a specific application it is not as useful in comparing detectors for the reasons described above. For example, the smaller the detector area the smaller the NEP. We also saw earlier that mean squared noise current for various noise mechanisms were directly proportional to the frequency bandwidth. So, measures of NEP for a given detector will be different depending on the bandwidth.

To accommodate these effects a figure of merit, specific detectivity, D^*, is defined as,

$$D^* \equiv \frac{\sqrt{A_d\,\Delta f}}{\mathrm{NEP}} = \frac{\sqrt{A_d\,\Delta f}}{\Phi_{e\,s}}\mathrm{SNR}$$

where A_d is the detector area and Δf is the bandwidth. A named unit has been defined as $(1\ \mathrm{Jones} = \mathrm{cm\ Hz^{-1/2}\ W^{-1}})$ for the units of D^*. Like NEP, the value at the peak wavelength is often quoted on a specification sheet.

The importance of the specific detectivity as a figure of merit is that it is proportional to the SNR. The larger the value the larger the SNR for a given set of detection conditions. Nevertheless, both the bandwidth and the detector area still play a role in the final SNR that can be achieved by the detection system.

The SNR is the quantity used to describe whether or not a system will detect the desired input signal flux over the noise. In some instances, one particular type of noise will dominate the total detector noise. We already discussed quantum-limited noise current. So, now we will examine some relationships for different limiting cases of the SNR.

Let us assume that for a given detector system we have possible noise contributions from the signal photon noise, $\Phi_{e\,s}$, background photon noise, $\Phi_{e\,b}$, and Johnson noise. There could of course be other types of noise contributions depending on the detection system, but this assumption shows the concept used in analyzing these systems. This would be the case for a photoconductive detector. From this assumption, we may write the SNR as,

$$\mathrm{SNR} = \frac{i_s}{i_N} = \frac{i_s}{\sqrt{\langle i_{SN}^2 \rangle_s + \langle i_{SN}^2 \rangle_b + \langle i_J^2 \rangle}}$$

where the signal and background noises arise from shot noise. Using the photo-conductive model and the definition of Johnson noise current, we obtain a general equation for the complete SNR to be,

$$\text{SNR} = \frac{\eta \frac{e}{hc}\lambda \Phi_{es}}{\sqrt{2\,\eta \frac{e^2}{hc}\lambda \Phi_{es}\,\Delta f + 2\,\eta \frac{e^2}{hc}\lambda \Phi_{eb}\,\Delta f + \frac{4\,kT\,\Delta f}{R}}}$$

This relation shows the explicit dependence on the frequency bandwidth associated with the noise power as discussed above.

First, we examine photon or quantum noise-limited detection of the signal. In this case only the shot noise of the signal flux dominates at we assume that this term is the largest of the three in the denominator. This leads to an SNR_p of,

$$\text{SNR}_p = \frac{\eta \frac{e}{hc}\lambda \Phi_{es}}{\sqrt{2\eta \frac{e^2}{hc}\lambda \Phi_{es}\,\Delta f}} = \sqrt{\frac{\eta \lambda \Phi_{es}}{2hc\,\Delta f}}$$

The photon limited NEP is then obtained by setting the $\text{SNR}_p = 1$ to find,

$$\text{NEP}_p = \frac{2hc\,\Delta f}{\eta \lambda}$$

Longer wavelengths result in smaller NEPs but only if this noise source dominates. This case is most important in that it is the maximum possible SNR.

Next, we consider the situation where the flux from the background is the major noise source. This is often referred to as BLIP (Background Limited Infrared Photodetector) [23]. From above the SNR_{BLIP} is,

$$\text{SNR}_{\text{BLIP}} = \frac{\eta \frac{e}{hc}\lambda \Phi_{es}}{\sqrt{2\eta \frac{e^2}{hc}\lambda \Phi_{eb}\,\Delta f}} = \Phi_{es}\sqrt{\frac{\eta \lambda}{2hc\,\Delta f\,\Phi_{eb}}}$$

As expected, the SNR for a detector in this case can be improved by reducing background flux. Thus, reducing radiation outside a spectral band that is not used by the signal will improve system performance. Limiting the field of view of the system will also reduce background.

The corresponding NEP is,

$$\text{NEP}_{\text{BLIP}} = \sqrt{\frac{2hc\,\Delta f\,\Phi_{eb}}{\eta \lambda}}$$

Only an increase in the detector quantum efficiency will decrease the noise once background-limited performance is achieved.

If Johnson noise dominates all shot noise, then the detection is said to be at the JOLI limit (Johnson Limit) [23]. The SNR for a photoconductor operation in this limit is,

$$\mathrm{SNR_{JOLI}} = \frac{\eta \dfrac{e}{hc}\lambda\Phi_{es}}{\sqrt{\dfrac{4\,kT\,\Delta f}{R}}}$$

Then the NEP is,

$$\mathrm{NEP_{JOLI}} = \frac{hc}{\eta e\lambda}\sqrt{\frac{4\,kT\,\Delta f}{R}}$$

With the explicit dependence on the temperature of the detector, cooling a detector reduces the noise and the $\mathrm{NEP_{JOLI}}$.

The final limiting case to discuss is shot limiting. For this situation, both the signal power and the background power contribute to shot noise resulting in a SNR, $\mathrm{SNR_{SHOT}}$ of,

$$\mathrm{SNR_{SHOT}} = \frac{\eta \dfrac{e}{hc}\lambda\Phi_{es}}{\sqrt{2\,\eta\dfrac{e^2}{hc}\lambda\,\Delta f(\Phi_{es} + \Phi_{eb})}}$$

The shot noise-limited NEP is then obtained by setting the $\mathrm{SNR_{SHOT}} = 1$, setting NEP $= \Phi_{es}$ and solving for the NEP. After some algebra, this result can be found. Since the background flux received is proportional to the detector area the background noise will often dominate and result in background-limited detection as the main source of shot noise.

7.6 Photoemissive detectors

The physical principle behind the operation of a photoemissive detector is the photoelectric effect. This effect, so important to the development of quantum mechanics, places this detector as a type know as a quantum detector. Vacuum photodiodes, photomultiplier tubes (PMTs), and microchannel plates (MCPs) are examples of photoemissive detectors that will be discussed here.

7.6.1 Vacuum photodiode or phototube

The vacuum photodiode (phototube) is a photoemissive detector consisting of an evacuated tube with a photocathode and an anode. As in the photoelectric effect, incident photons on the photocathode interact with the surface releasing electrons that can be accelerated toward the anode by an applied potential between anode and photocathode to produce a current in the external circuit. A schematic is shown in figure 7.15. These detectors follow the standard photoconductive model with a photocurrent proportional to the amount of incident radiation. The current through

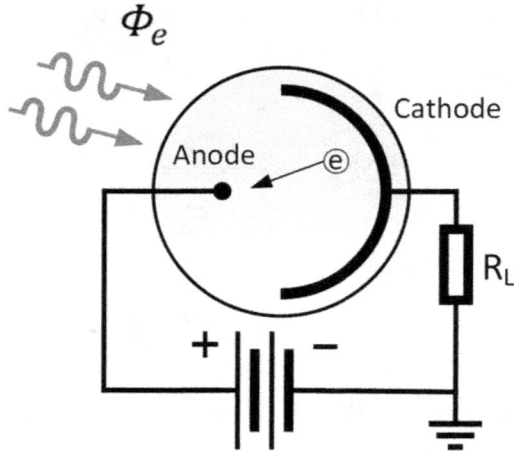

Figure 7.15. Schematic of a phototube detection.

the load resistor R_L is often monitored by measuring the voltage across the load resistor.

The photocathode material has a small work function, usually a combination of cesium with another material such as CsI, Sb_3Cs, and Na_2KSb-Cs. Different phototubes can then be used to cover a spectral response range from the ultraviolet through the visible, approximately from 120 to 600 nm. Phototubes typically have small output currents and low quantum efficiency but are linear over a wide dynamic range and have rather large detector areas compared with other detectors. These detectors are a good choice for detecting high-energy photons. They find application in chemical and medical instruments such as ultraviolet spectrometers and source monitors for ultraviolet sterilization.

The frequency response of phototubes is limited by the time of flight of the electrons released from the photocathode surface to the anode. Increasing the applied potential will decrease this characteristic time τ_c and increase frequency response. The characteristic cutoff frequency is then $f_c = 1/2\tau_c$, as with a rectangular bandwidth model.

Noise internal to the phototube is from shot noise in the signal, background, and dark current. The dark current noise is due to thermally emitted electrons and is often modeled by the Richardson–Dushman equation [11],

$$i_{d-PC} = \frac{4\pi m_e e k^2}{h^3} A_d T^2 \exp\left(-\frac{W_0}{kT}\right)$$

where m_e is the mass of electron, e is elementary charge, k is Boltzmann's constant, h is Plank's constant, A_d is the detector area, T is the temperature, and W_0 is the work function of the photocathode. Dark current can be reduced by cooling the detector. However, as mentioned before, the output current is small and typically must be amplified with post-detector electronics. Noise in the amplifier becomes the major contribution and this noise is usually limited by Johnson noise from the load resistor [5, 22].

7.6.2 Photomultiplier tube

Combining a low noise amplifier within the vacuum phototube eliminates the problem of amplifying a small signal produced from a single phototube. This detector is the PMT. Amplification is performed using the principle of secondary emission of electrons. When an electron is incident on a material the energy of the electron can collide with another electron releasing it from the surface. Within the PMT a series of dynodes are placed between the photocathode and the anode. When a photoelectron is ejected, instead of being directly swept to the anode it is accelerated to a dynode where the secondary emission process occurs. The secondary electrons are then accelerated to the next dynode in the chain releasing more secondary electrons to create a cascade effect and providing gain. This requires that a negative high voltage be applied on the photocathode with respect to the anode since negatively charged electrons are being accelerated. Figure 7.16 shows the photomultiplier concept.

A typical circuit used to provide voltage to the dynodes uses a series of resistors at each stage from the large applied negative potential. All the dynode resistors R_D shown in figure 7.16 are labeled with the same value of resistance, this is not a requirement, although it is common. As with the vacuum phototube, a large load resistor at the anode is used so that the output current from the PMT can be obtained by a voltage measurement across this resistor. The overall gain of a PMT is the result of the secondary emission process at each dynode, so if we take δ as the ratio of the number of secondary emission electrons to incident electrons then the gain for N dynode stages is $G = \delta^N$. The value of δ is approximately linear in the applied voltage over a given voltage range so there must be a precise potential difference between each dynode stage. This requires the use of a stable, high voltage

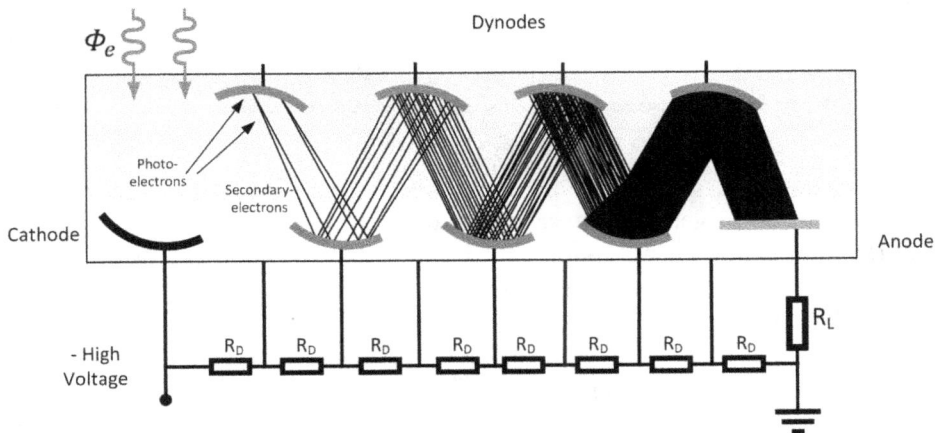

Figure 7.16. Photomultiplier tube detector schematic illustrating the cascade effect from the secondary emission of electrons.

power supply [5]. If the numbers of secondary emission electrons become too great, then they begin to repel one another causing an upper limit to the overall PMT dynamic range.

Like the phototube, the response time of a PMT is also governed by the transit time of electrons from the cathode to the anode. The location and design of the dynode structures are a critical parameter in determining the average time of flight of an electron in the tube. There are a wide variety of PMT designs and packages available on the market with different photocathode spectral responses and a range of average transit times [24]. As with our general systems model for detection shown in figure 7.4, an incident impulse of radiation will create a delay and pulse spread. Because electrons are generated by secondary emission, any given electron at the anode will have traveled over a differing path through the tube causing a pulse spread.

Advantages of the PMT are that it produces high gain (can be on the order of 10^6–10^8) with relatively low noise. The cathode current, as with a phototube, will be the sum of the photon signal current, photon background current, and the dark current of the photocathode. Like the phototube, cooling the PMT will reduce dark current. Analyzing potential noise sources, we expect to have shot noise from the incident photons (signal and background) and from the dark current. There may also be Johnson noise in the load resistor. Assume that the gain is constant. We also assume that there is noise due to fluctuations in the dynode amplification process by an amount γ which is related to the secondary electron emission factor δ [5, 22].

The output signal current at the anode will be a product of the gain and signal current produced at the photocathode. Each rms noise current will also be subject to the overall gain, G, as well as the dynode amplification γ. This means that the mean square rms shot noise current will be,

$$\langle i_{SN}^2 \rangle = 2e\gamma^2 G^2 \langle i \rangle \, \Delta f$$

Calculating the SNR for the PMT, we take the signal current at the anode and divide by the total rms noise current to obtain,

$$\text{SNR}_{\text{PMT}} = \frac{Gn\dfrac{e}{hc}\lambda \Phi_{es}}{\sqrt{2\,\eta\dfrac{e^2}{hc}\lambda\gamma^2 G^2 \Phi_{es}\,\Delta f + 2\,\eta\dfrac{e^2}{hc}\lambda\gamma^2 G^2 \Phi_{eb}\,\Delta f + 2\,\eta e\gamma^2 G^2 i_d\,\Delta f + \dfrac{4\,kT\,\Delta f}{R}}}$$

Since the gain is large, shot noise dominates the Johnson noise indicating this condition results in shot noise-limited detection. In this situation, notice that the SNR is independent of the gain. The NEP for a PMT would then be the shot noise-limited value. If the system is dark current limited then the NEP becomes,

$$\text{NEP}_{d\ \text{PMT}} = \frac{hc\gamma}{\lambda}\sqrt{\frac{2\,i_d\,\Delta f}{\eta e}}$$

7.6.3 Microchannel plate

The MCP is another type of low noise amplifier component that relies on the secondary electron emission process. Instead of a dynode chain used by a photomultiplier the concept of a channel is to replace the dynodes by a single tube. Figure 7.17(a) shows the concept of a channel electron multiplier. The interior of the tube is coated with a layer of conductive material similar to a dynode material that produces secondary electron emission. A potential is placed across the length of the tube to accelerate electrons, as in a photomultiplier. When a large number of small diameter tubes are placed in an array a MCP is formed.

Straight channels are not as effective because electrons can drift back down the tube. To mitigate this difficulty both slanted channels, shown in figure 7.17(b), and curved channels have been developed as well as the chevron-type as shown in figure 7.17(c). Channels are mostly made from glass. Development of the MCP device benefited from fiber optic technology and the ability to draw optical fibers [25].

When a photocathode is placed in front of an MCP, photoelectrons are generated when photons from in input optical image fall on the photocathode. These electrons are accelerated toward the MCP where they undergo secondary electron multiplication each time an electron strikes the channel wall. This small channel provides a region of spatial resolution for the incident photons associated with the image. The large number of electrons leaving the opposite end of the channel can then be incident onto a phosphor screen. On striking the screen, these electrons produce a cathodoluminescence that forms the output image. This entire detection, electron amplification, and display form the basis of image intensification and night vision devices. Though commonly used in image intensifiers, MCPs directly amplify charged particles and are applied to other detection areas such as x-rays and electron-beams.

7.7 Semiconductor detectors

The small band gap energy of a semiconductor makes this material a candidate for a photoconductive detector. When photons with energy greater than the band gap are

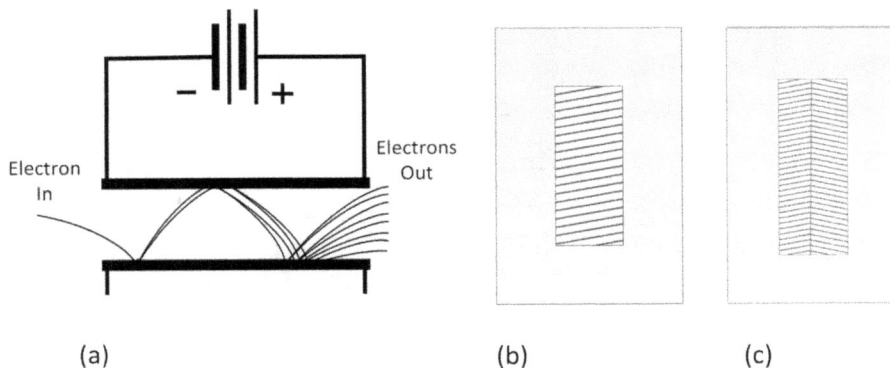

(a)　　　　　　　　　　(b)　　　　　　(c)

Figure 7.17. (a) Secondary electron emission within a single tube illustrating the concept used in microchannel plates. (b) A slanted edge plate to enhance gain. (c) Chevron concept of stacking two slanted plates together for increased gain.

incident on a semiconductor material and absorbed, electrons will be promoted from the valence band to states in the conduction and holes will be formed in the valence band. During the time that these carriers are present in the conduction and valence band the overall conductivity of the material is increased. This increased conductivity is employed to form a detector.

Figure 7.18 shows a model for a slab of a semiconductor material of length d between electrodes and cross-sectional area of A_d with light incident from above with flux Φ_e. When a potential is applied across the slab, the excess electrons and holes are then available as a current source.

Again, following the photoconductor model, the rate at which charge carriers (electrons and holes) are generated for an incident radiant flux Φ_e is,

$$r_{\text{gen}} = \eta \frac{\lambda}{hc} \Phi_e$$

where η is the quantum efficiency at the incident wavelength λ. If we take the carrier lifetime to be τ_c, the number of carriers in the device is the product of the lifetime and the generation rate. Current is produced when the carriers drift with an average velocity through the distance d of the slab. The ratio of the distance to the average drift velocity is the characteristic drift time, τ_d. Remembering from the discussion of generation–recombination noise, we defined the ratio of the carrier lifetime to the time for an average carrier to drift through the device (τ_d) as the photoconductive gain, $G = \tau_c/\tau_d$. Thus, the average carrier current is a form of the photoconductor model but explicitly includes photoconductive gain, or,

$$\langle i \rangle = \eta \frac{e}{hc} \lambda G \, \Phi_e$$

The carrier lifetime and thus the noise will depend on whether the semiconductor is intrinsic or extrinsic. Again, intrinsic photoconductors are undoped semiconductor materials while extrinsic materials are manufactured by doping materials creating states within the energy gap.

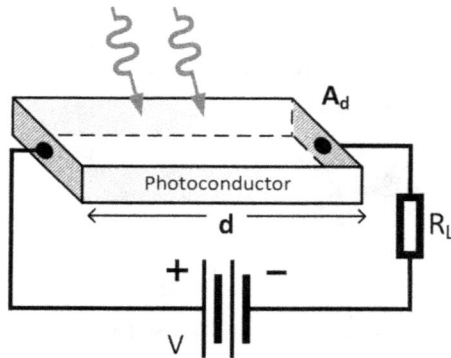

Figure 7.18. Model of a semiconductor slab photoconductor.

Many semiconductor photoconductor detector devices are responsive in the infrared. Maximum cutoff wavelength of a semiconductor photodetector is limited by the band gap energy as,

$$\lambda_c = \frac{hc}{\mathcal{E}_g}$$

Table 7.1 lists various photoconductor materials with their energy band gap and cutoff wavelength for comparison. These types of devices were originally developed as alternatives to thermal infrared detectors because of advantages in response time and sensitivity [26]. One of the most common detector material types are lead-salt compounds such as PbS, PbSe, and PbTe. The spectral response is typically between 1.0 and 6.0 μm. The alloy of mercury cadmium telluride (HgCdTe) has a spectral response that can be altered depending on the composition of the materials and ranges from about 2.0 to 15 μm. Because of the small band gap, cooling these detectors using liquid nitrogen dewars or thermoelectric coolers is necessary for most applications. Note that cooling the detector increases its dark resistance and shifts the band gap energy.

Noise contributions in semiconductor photoconductors are due to generation–recombination noise and Johnson noise, but $1/f$ noise can also be a factor for low-frequency bandwidth operation. The generation–recombination noise will occur for signal power, background power, and thermally generated carriers (similar to a dark current). Johnson noise may be in both the detector resistance as well as the load resistor employed. The NEP and D^* will depend on the noise sources present. Modulating the source or using a chopper at a frequency higher than the $1/f$ noise cutoff but less than the characteristic time constant gives the best D^* operation.

More recently, heterostructures composed of layered semiconductor materials have been developed using an epitaxial growth technique such as molecular beam epitaxy (MBE) or metal-organic chemical vapor deposition (MOCVD). For example, a thin GaAs layer deposited between AlGaAs forms such a structure as shown in figure 7.19. The AlGaAs material has a larger band gap than the GaAs layer. This creates an energy level diagram which creates potential wells and barriers as a function of position which are familiar from basic quantum physics [27]. This is referred to as a quantum well and when used as a detector material is a quantum well infrared photodetector (QWIP) [28]. Electrons can reside in states within the well

Table 7.1. Selected photoconductor materials and characteristics.

Material	Band gap energy at 300 K (eV)	Cutoff wavelength (μm)
PbS Lead sulfide	0.420	2.95
PbSe Lead selenide	0.278	4.46
CdS Cadmium sulfide	2.485	0.499
CdSe Cadmium selenide	1.80	0.690
InSb Indium antimonide	0.18	6.9

Data extracted from [31], and references therein. Cutoff wavelengths calculated from 300 K band gap energy.

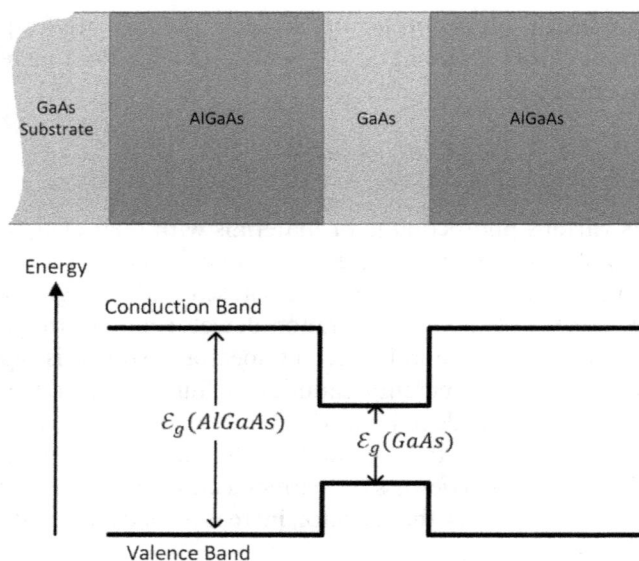

Figure 7.19. Single layer concept of a quantum well infrared photodetector.

and an incident photon can promote the electron to the conduction band increasing conductivity. Multiple layered structures and two-dimensional structures have been demonstrated forming large format focal plane infrared imaging arrays [28].

Photoconductor cells with application in the visible region of the spectrum are typically made from cadmium sulfide (CdS), cadmium selenide (CdSe), and mixtures of the two. These are low-cost devices and have many light sensing applications. They suffer from slow response times (10–100 ms) however they have reasonable photoconductive gain. They are also commonly referred to as light dependent resistors (LDR) by the electronics community [29]. Figure 7.20 shows the measured spectral responses for a CdS and a CdSe detector showing why these sensors are commonly used in the visible portion of the spectrum [30]. However, the spectral response of a specific detector will vary by manufacturer.

Components are made by depositing the semiconductor material on an insulating substrate. To maximize the area exposed to the incident illumination a serpentine pattern is used and placed between two electrodes. A representative illustration of the face of the photocell is shown in figure 7.21. The components are manufactured in a wide range of sizes and packages.

The increase in conductivity caused by carrier generation associated with an incident light flux results in a decreased resistance, however the process is not linear. The most common model used to describe the resistance as a function of illumination is a power law function,

$$R \propto E_v^{-\gamma}$$

where R is the resistance of the detector, E_v is the illumination, and γ the power law constant. Typical values of γ are 0.5–0.8 and are sometimes referred to as the

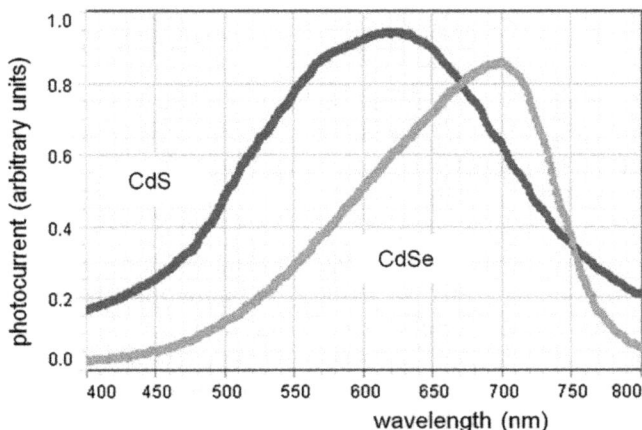

Figure 7.20. Measured spectral response curves for CdS and CdSe detectors. (Reprinted from [30] © IOP Publishing. Reproduced with permission. All rights reserved.)

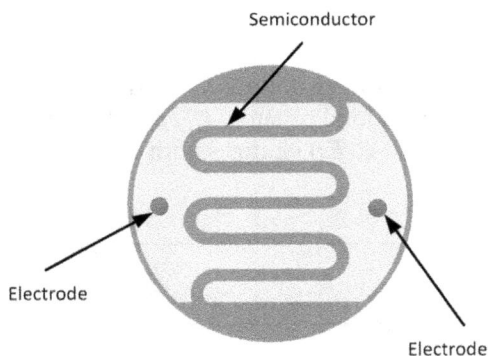

Figure 7.21. Typical light dependent resistor/photocell illustration.

sensitivity of the photocell. Devices are also characterized by a dark resistance which is the resistance value measured with no illumination falling on the detector for a specified time after exposure to light. The range of illumination and resistance covers several decades so characteristic curves for operation of a particular component are usually plotted on a Log–Log coordinate axis. The illumination value may be photometric (includes the response of the visual observer) or radiometric (optical power over a range of wavelengths). Visual/photometric measures are typically used since the applications of these detectors are usually in the visible for human observers. A standard white light source is typically used in characterizing the components. In quoting values in a specification sheet, manufacturers typically use the range of resistance change between illumination at 10 lux and 100 lux to provide the value of γ. From the model above this means that,

$$\gamma = \frac{\log_{10}(R_{100}) - \log_{10}(R_{10})}{\log_{10}(E_{100}) - \log_{10}(E_{10})} = \log_{10}\left(\frac{R_{100}}{R_{10}}\right)$$

where R_{100} is the resistance value for a 100 lux illumination and R_{10} is the resistance value for a 10 lux illumination. Specification sheets sometimes provide γ but may also provide only a characteristic curve or data. Figure 7.22 shows a family of curves from this model for different detectors with $\gamma = 0.7$ and for known resistance values at 10 lux of 1, 10, 100, and 500 kΩ.

There is often a wide variation in the value of the resistance at the 10 lux level due to manufacturing and materials, so calibration steps may be needed for a given application. This component-to-component variation has the effect of shifting the characteristic curve up or down along the resistance axis for a given value of γ. When the value of γ is known and the resistance at 10 lux has been measured or obtained from a specification sheet, the resistance of the detector for a specific illumination level E_v is,

$$R(E_v) = R(10 \text{ lux})\left(\frac{E_v}{10 \text{ lux}}\right)^{-\gamma}$$

Voltage divider circuits are usually used with these photocells so that an output voltage related to the illumination level can be measured. Figure 7.23 shows a standard voltage divider circuit incorporating a load resistor. A typical scenario is to measure the voltage across the load resistor, V_L, as a function of illumination where the resistance of the photocell $R(E_v)$ changes with incident illumination. Analyzing the circuit give us,

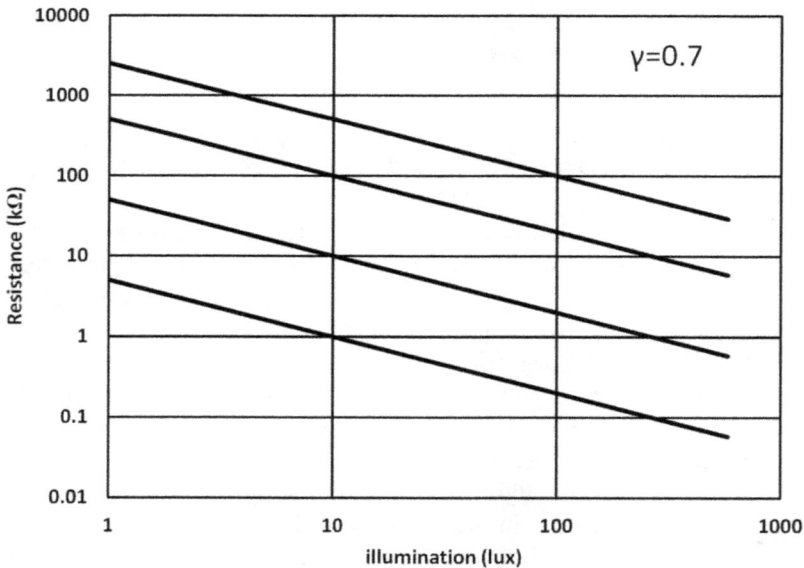

Figure 7.22. Resistance as a function of illumination according to the power law model for with $\gamma = 0.7$ and for known resistance values at 10 lux of 1, 10, 100, and 500 kΩ.

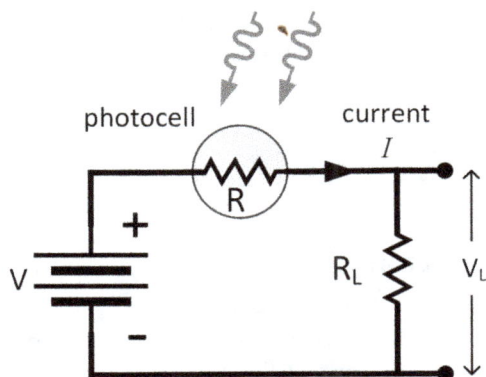

Figure 7.23. Typical voltage divider circuit for a photoconductor photocell.

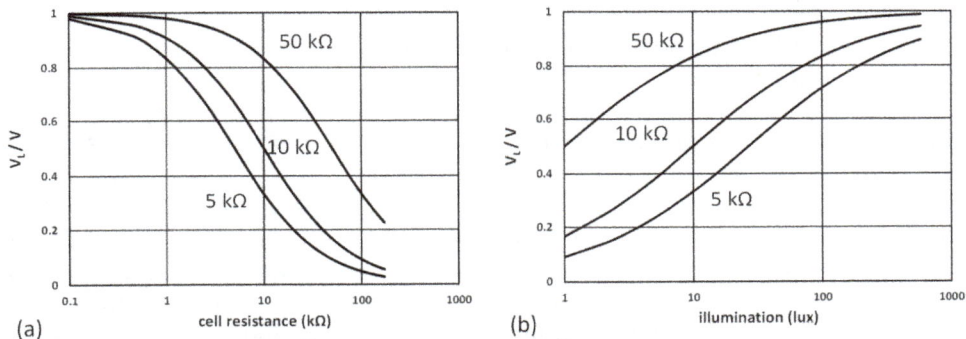

Figure 7.24. (a) Plot of curves of the voltage across the load resistor as a function of the resistance of the photocell for 5 kΩ, 10 kΩ, and 50 kΩ load resistor values. (b) Similar curves as a function of illumination with the value of the cell resistance at 10 lux equal to 10 kΩ and $\gamma = 0.7$.

$$V_L = IR_L = \frac{VR_L}{R_L + R(E_v)}$$

Figure 7.24(a) is a graph of the ratio of the output voltage to the applied bias voltage, V_L/V plotted as a function of the cell resistance for three different load resistor values ranging from 5 kΩ, 10 kΩ, and 50 kΩ. Increasing illumination results in a decreasing cell resistance. Figure 7.24(b) shows similar curves as a function of incident illumination with a fixed value of 10 kΩ cell resistance at 10 lux and $\gamma = 0.7$, as in figure 7.22. Increasing the load resistor for a given illumination level will increase the measured voltage. However, since the sensitivity of the detection circuit occurs in regions of greatest change in voltage, increasing the load resistor may come with a loss of sensitivity as shown in the figure.

As with all electronic components, photocells have a maximum power dissipation, current limits, and voltage limits. The current drawn by this circuit will range from a value where the cell resistance is equal to the dark resistance (a low current value since the dark resistance is high) to a lower value of resistance for the maximum illumination expected. In addition, the nominal resistance at a given illumination level will change based on prior illumination exhibiting a type of memory effect. This effect will necessitate a potential recalibration as well as altering the overall voltage output and sensitivity.

7.8 p–n junction detectors

Some properties of the p–n junction semiconductor diode were discussed in a previous chapter, including a description of the electronic behavior and character- istic current-voltage curve and some of the basic physics behind the operation of the device. The light emitting diode is a photonic device based on the p–n junction where a forward current allows an electron and hole within the junction to recombine producing a photon. In the reverse process, an incident photon of energy greater than the band gap energy generates an electron–hole pair allowing a current to flow through the diode. This process of photon-to-electron conversion makes the diode an efficient photodetector.

In this section, we will examine circuit models of a p–n junction detector and modes of operation for photodiode detectors. When a p–n junction device is reverse biased, the electric field within the depletion region sweeps mobile charge carriers (both electrons and holes) through the device creating current flow. The size of the electric field depends on the semiconductor material and the structure of the device. Different types of semiconductor junction photodiode detector components will be discussed including phototransistors, PIN detectors, and avalanche photodiode detectors (APD). For each type of component their optical and electronic character- istics will be provided along with typical circuit applications. Some general characteristics of different types of two-dimensional sectored and multi-element detectors will also be given.

7.8.1 Modeling the p–n junction detection process

The current produced within a p–n junction detector follows the photoconductor model as described earlier. As a system, the photodiode is then a combination of a current generator and a diode. In addition, every photodiode will also have an effective resistance and capacitance.

The current output from the current generator portion of the equivalent photodiode is provided by the photoconductor model, or

$$i_\lambda = \eta \frac{e}{hc} \lambda \Phi_e$$

where Φ_e is the incident flux with wavelength λ, and η is the quantum efficiency. Figure 7.25 shows the schematic of the equivalent circuit of a photodiode along with the schematic symbol for a single photodiode component.

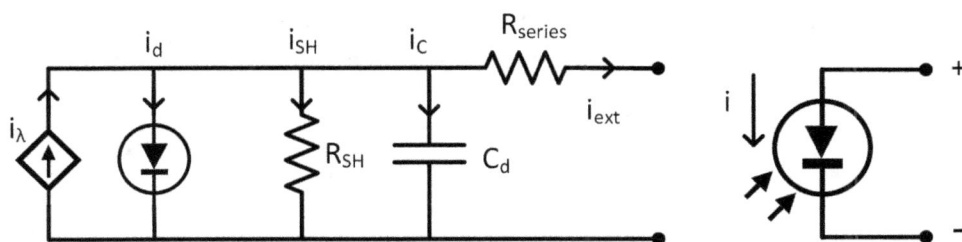

Figure 7.25. Equivalent circuit model for a photodiode and single circuit schematic symbol used for a photodiode.

The current through the ideal diode is just given by the diode characteristic equation as,

$$i_d = i_{\text{sat}}\left[\exp\left(\frac{eV_d}{kT}\right) - 1\right]$$

where i_{sat} is the reverse saturation current and V_d is the voltage across the diode. The effective capacitance of the diode, C_d, is a result of the geometry of the junction being like that of a parallel place capacitor where the capacitance is proportional to the area and inversely proportional to the thickness between regions of charge separation. Thus, this value may be controlled in part by component design. The value of the internal capacitance also depends on whether there is an applied bias voltage which alters the thickness of the depletion region. The shunt resistance is the effective resistance of the depletion region when there is no bias voltage applied. In practice, the shunt resistance is measured by applying a small voltage across the diode, measuring the current, and taking the ratio of the voltage to the current. Another way to define the shunt resistance is to take the slope of the i–V curve for a diode at the origin and equate this slope to 1/RSH [5]. In this sense, the shunt resistance is a dynamic resistance that will depend on illumination. A small series resistor is also included in this circuit model to account for the resistance of the remaining parts of the n and p regions of the diode structure. The output current into the external circuit is identified by i_{ext}. This is the current output from the physical diode when it is placed in a detection circuit and may be positive or negative depending on any bias voltage applied.

Applying Kirchhoff's law to the current junction of this equivalent circuit results in,

$$i_\lambda = i_d + i_{\text{SH}} + i_C + i_{\text{ext}}$$

The current through the capacitor is also included for completeness since the current in the circuit may be an AC current. However, this contribution is often small. In the open circuit configuration, the current into the external circuit is small. This results in,

$$\eta\frac{e\lambda}{hc}\Phi_e \cong i_{\text{sat}}\left[\exp\left(\frac{eV_d}{kT}\right) - 1\right] + \frac{V_d}{R_{\text{SH}}}$$

For large values of the shunt resistance, which is the typical case, we can solve this equation for the voltage across the diode to obtain a value called the open circuit voltage, V_{OC},

$$V_{OC} = \frac{kT}{e} ln\left(\eta \frac{e\lambda}{hci_{sat}}\Phi_e + 1\right)$$

Thus, at zero current levels the voltage across the diode is logarithmic with the incident flux.

In practice there will be some load across the photodiode. Figure 7.26 shows the simple circuit of a photodiode with a load resistor in parallel again in the photo-voltaic mode.

Placing a load resistor across the diode, allows current to flow in the completed circuit and is,

$$i \cong i_{sat}\left[\exp\left(\frac{eV_d}{kT}\right) - 1\right] - \eta \frac{e\lambda}{hc}\Phi_e$$

In this relation we are still assuming that the shunt resistance is high, and thus the shunt current contribution may be neglected. Figure 7.27 shows the characteristic i–V curve for incident flux and two values of increasing flux. The points labeled V_1 and V_2 are the values of the open circuit voltage for incident flux Φ_{e1} and Φ_{e2}, respectively. The voltage with no incident flux is zero. Also shown are two points labeled i_1 and i_2, referred to as the short circuit currents for a given amount of illumination. These values occur when the voltage across the diode is zero, meaning that both sides of the diode are at the same potential.

From the relation above, when the voltage across the diode is zero, the value of the short circuit current i_{SC} is,

$$i_{SC} = -\eta \frac{e\lambda}{hc}\Phi_e$$

Showing that this current is proportional to the incident flux on the photodiode detector.

The photovoltaic mode diode circuit is a form of a photovoltaic cell or solar cell, using the name given for its typical application. There is no bias voltage applied

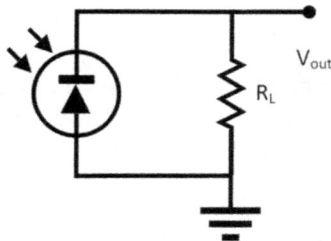

Figure 7.26. Photovoltaic mode circuit with a load resistor across the photodiode.

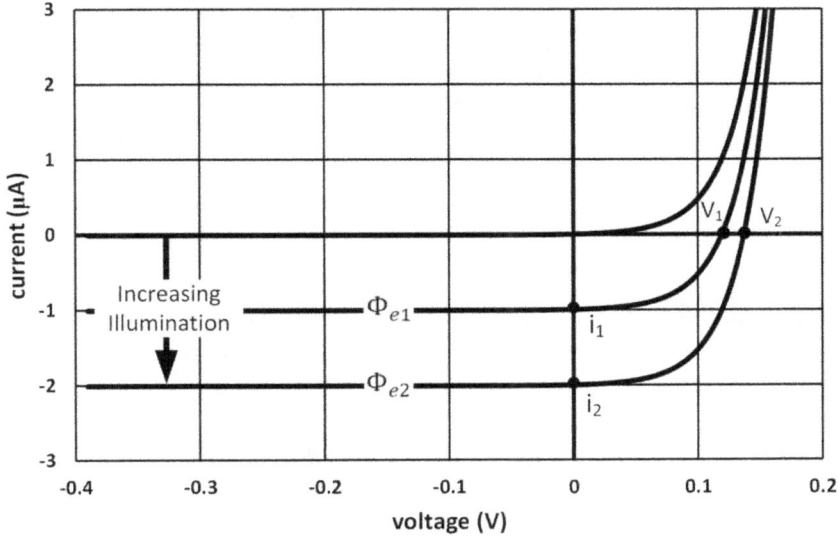

Figure 7.27. Current versus voltage characteristic curve for a photodiode in the photovoltaic mode for increasing incident flux. The open circuit voltages V_1 and V_2 and short circuit currents i_1 and i_2 for the two illumination levels are indicated. The value of the reverse saturation current for this case is 10 nA at 300 K.

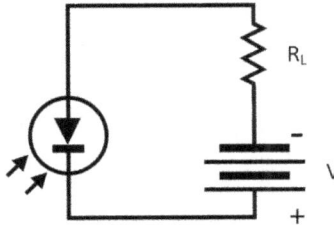

Figure 7.28. Photoconductive mode of operation for a photodiode with an externally applied bias voltage and a load resistor.

across the diode but increasing illumination provides current through the load at operating voltages less than the open circuit voltage.

The second mode of operation of a photodiode detector is the photoconductive mode. In this mode the diode has a reverse bias applied across the diode as shown in figure 7.28. In a detection application, the voltage across the load resistor is measured since it is directly related to the current through the load resistor.

To analyze this situation, we return to the current relation for the diode equivalent circuit.

$$i_\lambda = i_d + i_{SH} + i_C + i_{ext}$$

Again, we assume the current contributions from the capacitance and the shunt resistance are negligible. For a large reverse bias voltage, the voltage V_d is negative which means that $i_d \cong -i_{\text{sat}}$. But the reverse saturation current is also small in relation to the current from the incident flux. Under these conditions, we find that the current into the external circuit is equal to the photoconductor model current.

$$i_{\text{ext}} = \eta \frac{e\lambda}{hc} \Phi_e$$

Thus, in the photoconductive mode, the current into the external circuit is directly proportional to the incident flux/optical power. The voltage across the load resistor in the circuit of figure 7.28 is then,

$$V_L = i_{\text{ext}} R_L = \eta \frac{e\lambda}{hc} \Phi_e R_L$$

which is also directly proportional to the incident flux. Increasing the load resistance increases the measured voltage, however, as we shall see, this is at the cost of a reduction in the frequency response of the detector system. We will return to the photoconductive mode and photovoltaic mode later when we discuss photodiode circuit configurations.

7.8.2 Photodiode materials and p–n photodiode structure

In picking a detector for a given application a major consideration is the spectral response. The ideal photodiode response function is directly proportional to the wavelength of the incident radiation. From the photoconductor model we have,

$$\mathcal{R} \equiv \frac{i_{\text{avg}}}{\Phi_e} = \eta \frac{e}{hc} \lambda$$

In addition, an ideal detector will also have a unit quantum efficiency for all wavelengths and a slope of e/hc based on the photoconductor model. Figure 7.29 is a plot of the response as a function of wavelength for an ideal photodiode showing the linear function with wavelength.

For a p–n junction detector there are several physical mechanisms that impose limits on the spectral characteristics of a detector forcing the spectral response to deviate from the ideal curve. For realistic photodiodes, the quantum efficiency will in general be a function of wavelength. Another physical limitation on the response function is that the p–n junction photodiode will only operate at incident photon energies greater than the band gap, the minimum energy required to directly generate electron–hole pairs. As with all photoconductors, this condition imposes a limit on the upper wavelength response of the detector setting a cutoff wavelength given by,

$$\lambda_c = \frac{hc}{\mathcal{E}_g}$$

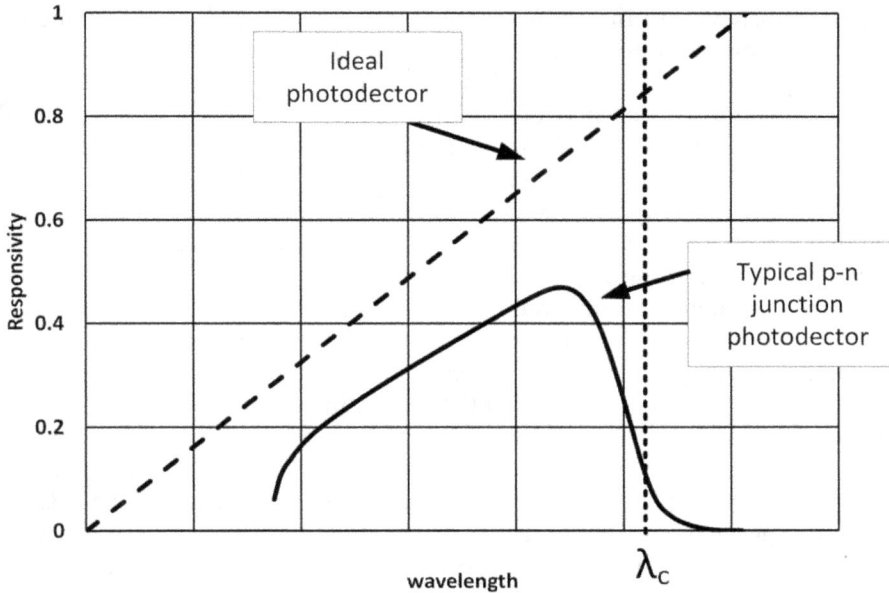

Figure 7.29. Spectral response of an ideal photodiode (dashed curve) and typical p–n junction detector response function limited by materials and other efficiency effects (solid curve). The cutoff wavelength is also identified.

At lower wavelengths, material absorption becomes high reducing the number of photons within the depletion region. The absorption is considered an external or optical contribution to the quantum efficiency along with reflectance at the surface of the diode. We may explicitly model the external portion of the quantum efficiency as,

$$\eta = (1 - R)(1 - e^{-\alpha z})\eta_{int}$$

where R is the reflectance at the diode surface, α is the wavelength dependent absorption coefficient, z is the depth from the surface into the depletion region, and η_{int} is the internal quantum efficiency [31, 32]. Internal quantum efficiency factors include issues such as surface recombination effects, and electron–hole pairs created outside depletion region not contributing to current [32]. A model sketch of the photodiode response curve including these quantum efficiency issues is also shown in figure 7.29. Note also that the reflectance is a function of wavelength, incident angle, and polarization which also affects the external quantum efficiency.

The temporal response of a photodiode detector depends on the detector mode of operation as well as material factors and device fabrication. Rise time and response of a photodiode is a complex function of several parameters including carrier drift time in the depletion region and diffusion of photocarriers when they are outside the depletion region. Material absorption is directly related to the temporal response since thinner absorption regions mean shorter carrier collection times. Typical RC time constant behavior of the photodiode and the circuit load also contribute to the overall frequency response. The photoconductive mode is preferred for high frequency response since the applied reverse bias provides a large electric field across the depletion region increasing carrier velocity and decreasing carrier drift times. Figure 7.30 illustrates this characteristic and shows the resulting electric field. Also, increasing the applied voltage increases the depletion region thickness reducing the effective capacitance. However, larger area detectors have a larger capacitance which lowers the frequency response. Considering these trade-offs is a necessary part of designing a detection system.

Silicon is the dominant material choice for many applications, especially for detection in the visible and near infrared portion of the spectrum. The devices are low cost and have relatively small dark current values. Germanium detectors have long been used due to their wider response in the near infrared. However, one of the major disadvantages of germanium as a detector is the high dark current. Other semi-conducting material combinations, such as InGaAs and InGaAsP, have been developed particularly for application in fiber optic communication systems. These detectors have replaced germanium due to their high quantum efficiency at the 1300 nm and 1550 nm wavelength bands needed by fiber optic receivers [33]. Table 7.2 lists three common photodiode materials and their energy gap and cutoff wavelength.

The lower wavelength region of a photodiode response curve is limited by material absorption. Absorption also affects the external quantum efficiency. Figure 7.31 shows data for the absorption coefficient as a function of wavelength for both silicon and germanium [34, 35]. The high absorption coefficient at lower wavelengths is apparent. This absorption decreases as the wavelength approaches the cutoff wavelength, a characteristic for all photodiode materials.

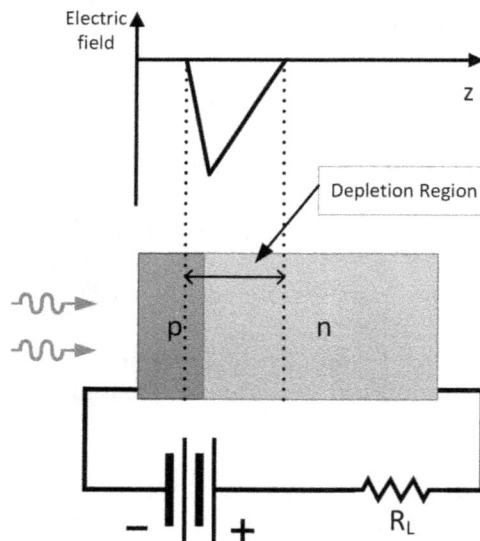

Figure 7.30. Basic operational diagram of a reverse biased p–n junction photodiode including the electric field within the depletion region.

Table 7.2. Selected photodiode materials and characteristics.

Material	Band gap energy at 300 K (eV)	Cutoff wavelength (μm)
Si Silicon	**1.1242**	**1.100**
Ge Germanium	**0.664**	**1.870**
In$_{0.53}$Ga$_{0.47}$As Indium Gallium Arsenide (lattice match to InP)	**0.752**	**1.650**

Data extracted from: (1) [31], and references therein. (2) [49]. Cutoff wavelengths calculated from 300 K band gap energy.

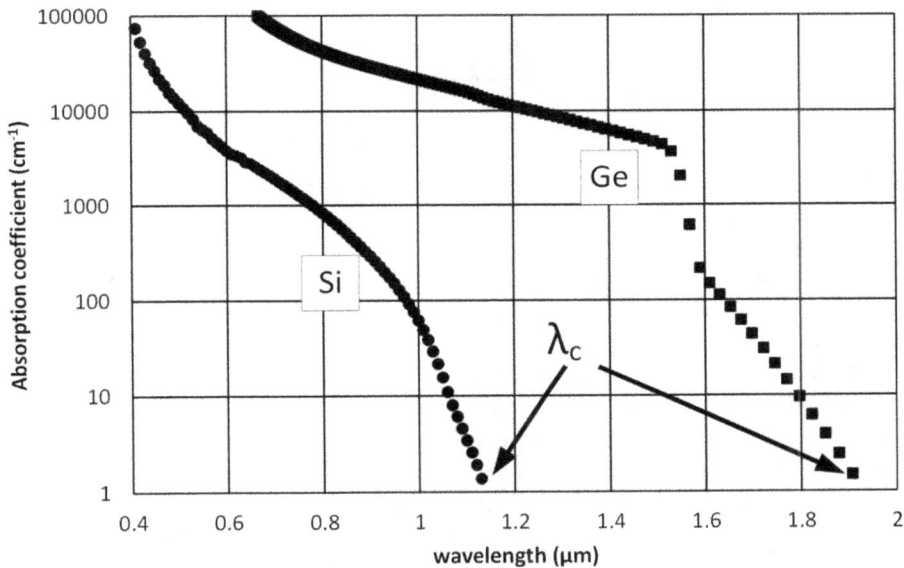

Figure 7.31. Absorption coefficient data for Si (circles) and Ge (squares). The cutoff wavelength for each material is identified.

Reflectance at the surface of the diode can be obtained using the Fresnel equations if the real and imaginary values of the refractive index are known. Figure 7.32 shows the corresponding reflectance from silicon and germanium at normal incidence as an example [34, 35]. Some detectors use antireflection coatings to reduce the reflectance and increase quantum efficiency.

7.8.3 Other photodiode structures

For high frequency detection systems an optimal junction structure would include a wide depletion region so that more photons can be absorbed and a lower effective capacitance to improve the time constant response. Unfortunately, the basic p–n

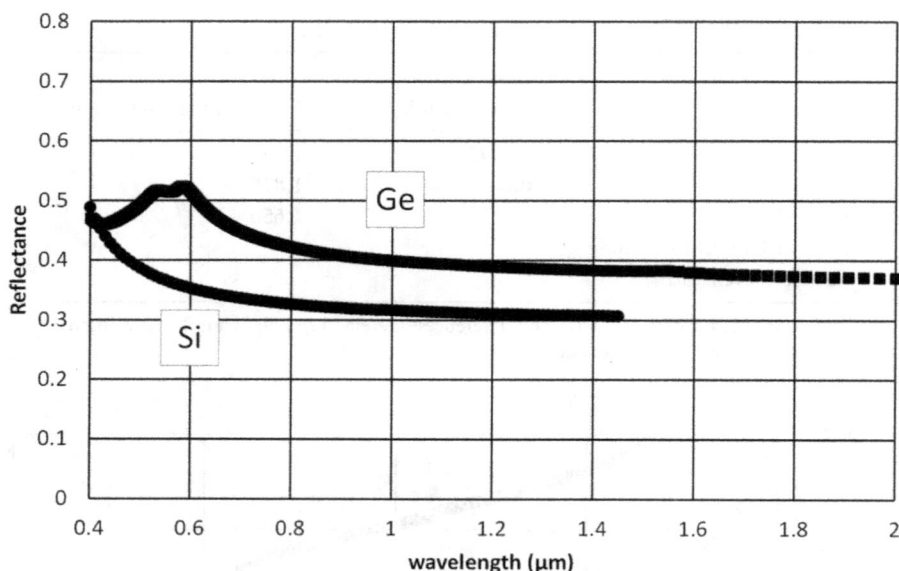

Figure 7.32. Reflectance at normal incidence for silicon and germanium whose values impact the external quantum efficiency of a photodiode.

junction structure has a relatively thin depletion region. To improve the high frequency performance of detector components, other structures have been developed and will be discussed here.

The PIN diode structure (successive layers of p-material, i-intrinsic material, and n-material) allows longer wavelength photons to be absorbed deeper into the device within the intrinsic layer. Since the intrinsic layer has little doping it has a higher resistivity than the doped layers and can sustain a higher applied voltage. Additional bias voltage can be applied to the device so that the depletion region extends all the way through the intrinsic layer and creates a relatively large electric field across the depletion region. Figure 7.33 shows the basic structure of the material layers in a PIN photodiode along with the electric field across the depletion region. The designation of n+ indicates that this is a heavily doped material which helps ohmic contact with the device electrodes [32].

A large intrinsic layer volume provides more electron–hole pairs to be created from photon absorption within the depletion region increasing the quantum efficiency. The large depletion region allows for a larger electric field to be applied so that charge carriers can be quickly swept out of the device. Extending the depletion region also reduces the effective capacitance. Therefore, the PIN photodiode structure provides a higher temporal frequency bandwidth detector than a simple p–n junction photodiode.

A phenomenon known as impact ionization (leading to avalanche breakdown of carriers) is employed by a type of detector known and an APD. When an electron–hole pair is generated from photon absorption and the energy of the carriers are

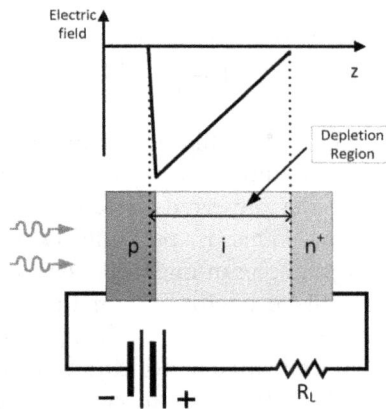

Figure 7.33. PIN photodiode structure in a reverse bias configuration showing and electric field across the depletion region.

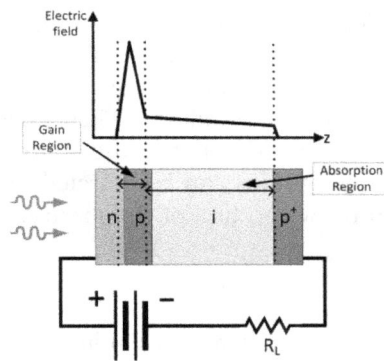

Figure 7.34. APD photodiode structure in a reverse bias configuration showing and electric field within the device layers.

large enough, an inelastic collision between this high-energy charge carrier will generate another electron–hole pair. Thus, the APD provides current gain in analogy to a PMT. The operational requirement is a large reverse bias (typically greater than 100 V) and a special junction diode structure [31].

Figure 7.34 is an illustration of the basic semiconductor layer structure of an APD. The gain region has the largest field gradient for impact ionization to occur. The intrinsic layer has a field large enough to assist with frequency response in a similar manner as the PIN photodiode, but small enough so that no further impact ionization occurs. The gain of the device is governed by impact-ionization coefficients for both electrons and holes and are semiconductor material dependent parameters. The gain also depends on the value of the large electric field in the gain region [16, 31].

In APD photodiode specifications the gain is referred to as the multiplication factor (M). It is found that an empirical relationship for the multiplicative factor is,

$$M = \frac{1}{1 - \left(\dfrac{V_J}{V_B}\right)^n}$$

where V_J is the junction voltage, V_B is the device breakdown voltage, and n is an empirical power factor between 0 and 1 for a given device [16, 31].

Avalanche photodiodes are the preferred detector when an amount of optical power is predicted to be low. This is particularly true for high bandwidth applications such as fiber optic communication systems [33]. In practice, the required higher bias voltages place conditions on the stability of power supplies and the external circuits used for detection. As the applied voltage increases up to the breakdown voltage the multiplicative factor dramatically increases. Care should be taken when using an APD so that the applied voltage does not damage the photodiode. The multiplicative factor is also dependent on temperature so that the devices may need to have some form of thermal mitigation. While the noise current in an APD is higher than with a PIN photodiode, SNRs may be higher because of the gain provided by the APD.

Another photodiode device replaces the initial p-layer with a thin metal film (like gold). The surface interface between the metal layer and the semiconductor creates a contact potential and produces a rectification. This type of diode is called a Metal-Semiconductor or Schottky barrier photodiode. Their major advantage is that they have a short time constant [32]. In addition, because absorption of the semiconductor is not involved, higher energy photons can be detected providing photodiodes that respond in the blue and ultraviolet portion of the spectrum.

7.8.4 Photodiode circuits

One of the first questions to answer when designing a circuit for the detection of radiation is what mode of operation to use, photovoltaic or photoconductive. The final decision often becomes a trade-off between the necessary speed of response and the maximum noise to be tolerated. As we have seen, a photodiode operating in the photovoltaic mode has a low-frequency response but has low noise and when operating in the photoconductive mode has a high frequency response with increased noise.

When a single load resistor is placed across the diode then it is operating in the photovoltaic mode. In this case, the voltage, V, across the diode and the voltage across the load resistor, R_L, are equal. The current through the circuit for a general incident flux Φ_e is then,

$$i = i_{\text{sat}}\left[\exp\left(\frac{eV}{kT}\right) - 1\right] - \eta\frac{e\lambda}{hc}\Phi_e = -\frac{V}{R_L}$$

There is not a simple analytic solution of this relation for the voltage across the load resistor. A graphical illustration provides the best way of understanding the situation. In figure 7.35, the current as a function of V is plotted as before showing the characteristic i–V curve for the diode. Also plotted is the current through the

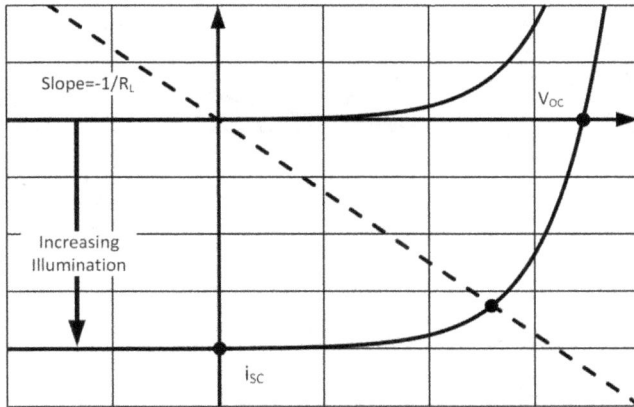

Figure 7.35. Photovoltaic mode characteristic current equation for a given incident flux (solid line) and current through the load resistor (dashed line). The intersection between these curves is the solution. The open circuit voltage and the short circuit current are also identified.

load resistor as a function of V. The intersection of these two curves provides the value of the current and voltage for a given incident flux. This is the load line analysis process. With increasing flux, the solution current becomes more negative and the voltage across the load resistor shifts to a higher value since the load line remains constant. Using this circuit to detect a signal does not provide a convenient linear relation for voltage with increasing incident flux.

This photovoltaic mode circuit is important because it is the standard operating mode used by a solar cell for converting sunlight into electricity. The solar cell is designed to optimize total power delivered to a load for a given incident flux and not necessarily give an output that relates to an input signal flux. Optimizing total power means a high current is desired at a given operating voltage. Specifications for solar cells and panels provide the open circuit voltage and the short circuit current as input parameters and these values vary widely depending on power ratings and panel designs. Typical values for open circuit voltages range from 5 to 50 V with short circuit current values up to 10 A. Matching the load resistance to cell resistance at the operating point provides optimal power. This is shown in figure 7.36 where the highlighted box in the figure is a graphical form showing the i–V power product.

Using a photovoltaic mode for optical signal detection usually requires additional electronics. Operational amplifiers with high input impedance provide a means of converting an input current to a measurable output voltage. Figure 7.37 shows two simple circuits for a photodiode operating in the photovoltaic mode. In the open circuit configuration, the current through the diode is zero and the voltage across the diode is maintained at the open circuit voltage, V_{OC}. The voltage measured at the output of the operational amplifier, V_O, is then,

$$V_O = V_{OC} \frac{R_1 + R_2}{R_1}$$

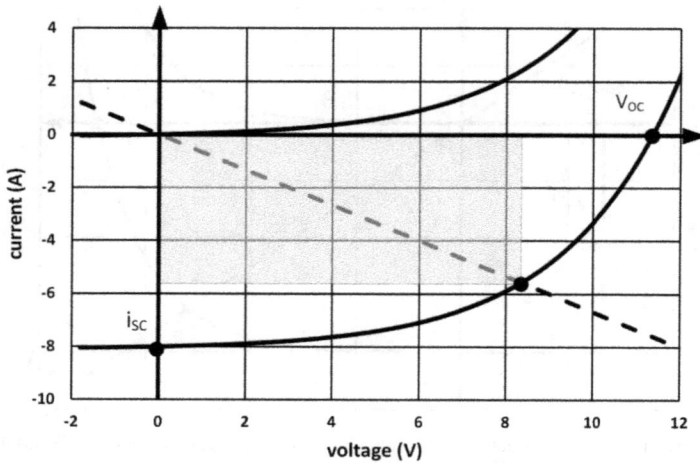

Figure 7.36. Solar cell operation illustrating the power delivered to a load at the operating voltage. For this example, the solar cell system specification sheet gives a short circuit current of 8.1 A and an open circuit voltage of 11.3 V.

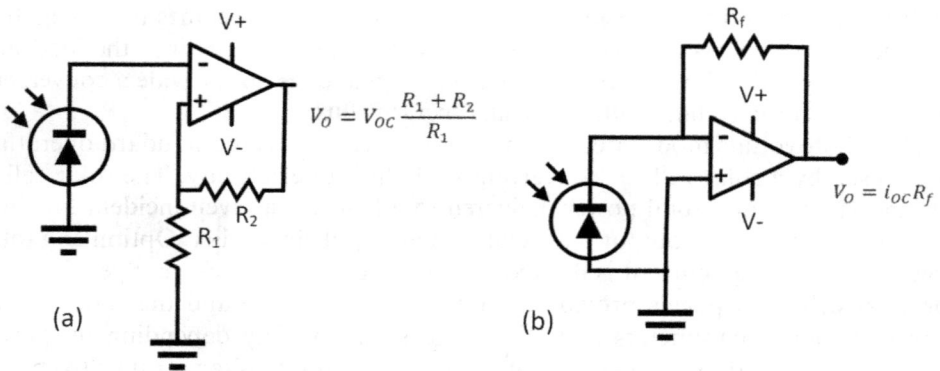

Figure 7.37. Photovoltaic mode circuits using operation amplifiers: (a) open circuit and (b) short circuit modes.

For the short circuit configuration, the current through the diode becomes the short circuit current, i_{SC}. At the output of the operational amplifier, the measured voltage is,

$$V_O = i_{OC}R_f$$

where R_f is the value of the feedback resistor. The short circuit configuration is the preferred photovoltaic mode circuit for measuring optical flux since the short circuit current is proportional to the incident flux on the detector. The voltage output is also proportional to the incident flux. Increasing the feedback resistance will increase the measured voltage but decreases frequency response.

The dominant noise mechanism for a photodiode in the photovoltaic mode is usually shot noise in the current. When a load resistor is applied then the Johnson noise in this resistor must also be considered [5].

All circuits using the photoconductive mode require a reverse bias voltage, as seen in figure 7.38. The simplest configuration is shown in figure 7.38(a). We already discussed this circuit configuration and showed that the voltage across the load resistor is proportional to the incident flux. The reverse bias voltage can have any value as long as it is less than the maximum reverse breakdown voltage of the device. In this voltage region the crystal structure of the device material is altered and can be permanently damaged due to excessive heat. There are always trade-off considerations because increasing the bias voltage decreases the junction capacitance and response time and increases the range of response to incident flux. Increasing the value of the load resistor also increases the response range but decreases the frequency response [15]. Figure 7.38(b) is a circuit amplifying the voltage across the load resistor. While the output voltage may be amplified, the noise is also amplified.

Figures 7.38(c) and (d) are examples of transimpedance circuits which convert an input current to an output voltage. The gain of the circuit is set by the feedback resistor. This value is chosen by taking the ratio of the specified output voltage range to maximum current expected from the photoconductor model. An additional capacitor (not shown on the circuit diagrams here) in parallel with the feedback resistor is often used to provide stability. This capacitance often becomes the limiting

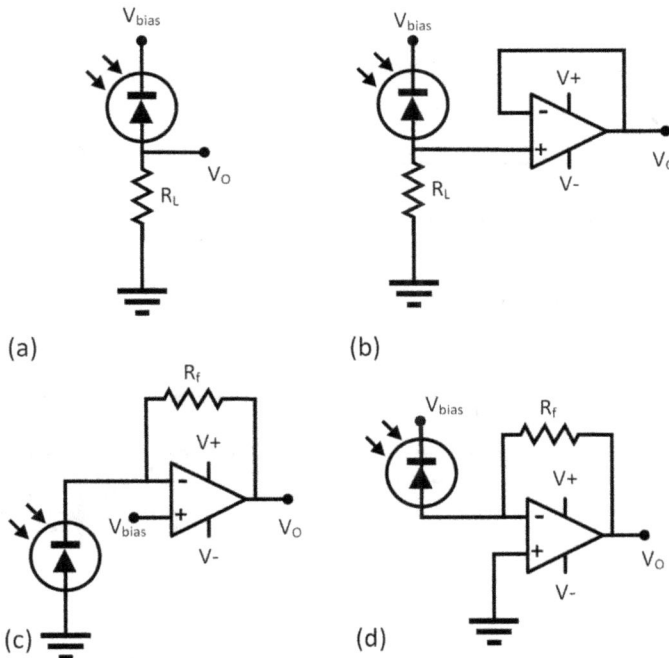

Figure 7.38. Photoconductive mode circuits. (a) Single load resistor, (b) voltage amplifier, (c) and (d) two alternative transimpedance circuits whose output voltage is proportional to the feedback resistance.

capacitance in determining the frequency cutoff for detection along with the value of the feedback resistor. The advantage of the transimpedance circuits is improved linearity and increased bandwidth for the same load/feedback resistance.

PIN photodiodes and avalanche photodiodes use photoconductive mode circuits, however, with a larger applied bias. Current limiting resistors and other electronic components are often used to protect the photodiodes from breakdown.

The choice the reverse bias voltage and the load resistor depends on the range of output voltage desired and diode current expected, as described above. Use a reverse bias high enough to detect the signal and bandwidth desired keeping it below the reverse breakdown voltage. In a similar manner to the load line analysis done for the photovoltaic mode, we can also analyze the expected performance of the photo-conductive mode circuit of figure 7.38(a). From the circuit we can see that the voltage must be limited by the reverse bias voltage $V = -V_B$ when there is zero current in the circuit. The maximum current drawn by this circuit is limited by the value of the load resistor such that,

$$i_{max} = -\frac{V_B}{R_L}$$

when the voltage across the diode is zero. These two limiting points can be plotted on the diode characteristic curve as shown in figure 7.39. In this graph, a family of characteristic curves for dark and four higher incident flux values are also plotted.

The maximum current of magnitude V_B/R_L limits the maximum flux that can be detected. This maximum flux value is,

$$\Phi_{e\ max} = \frac{V_B}{R_L} \frac{hc}{\eta e \lambda}$$

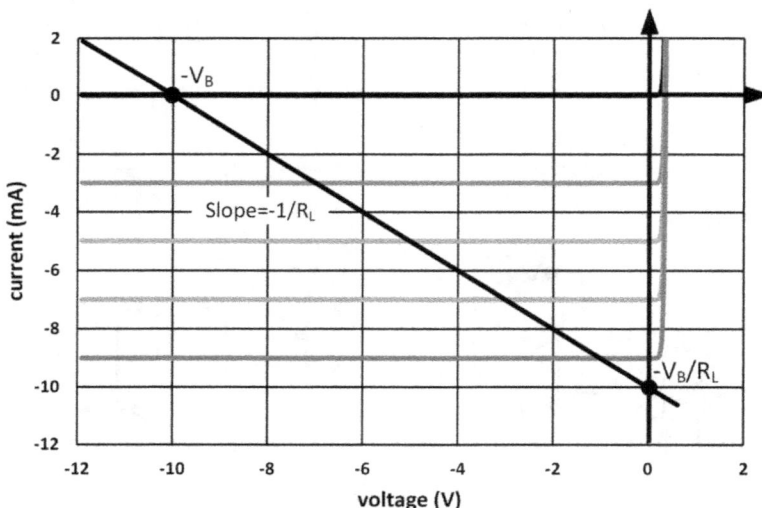

Figure 7.39. Family of curves for a detector circuit operating in the photoconductive mode showing operational limits and load line. Four curves for four different levels of illumination are shown. For this example, the bias voltage is 10 V, and the load resistance is 1000 Ω.

Any incident flux greater than this value will always produce the same maximum current. The result is a saturation condition. If a greater flux measure is needed, then the bias voltage must be increased, or the load resistor decreased. However, changing these values will also alter the frequency response of the detection system.

The load line analysis also shows how an AC input flux signal is transformed into an AC output voltage. The analysis provides limits on the characteristics of the signal to be detected. Figure 7.40 repeats the set of characteristic curves from figure 7.39 but also shows the alternating input signal current in time which is proportional to the incident flux and the resulting output voltage across the diode as a function of time. The load line acts as a transfer function. Although the voltage across the diode is shown, the voltage across the load resistor is the value most often measured in a detection system.

If the amplitude of the input flux continues to increase causing the current to becomes larger in magnitude than V_B/R_L then saturation occurs. For an AC signal, the output voltage waveform will be clipped since it cannot become greater in magnitude than V_B.

7.8.5 Phototransistors

Photomultiplier tubes and avalanche photodiodes are detector devices that provide intrinsic amplification. Phototransistor devices also provide gain of the input optical

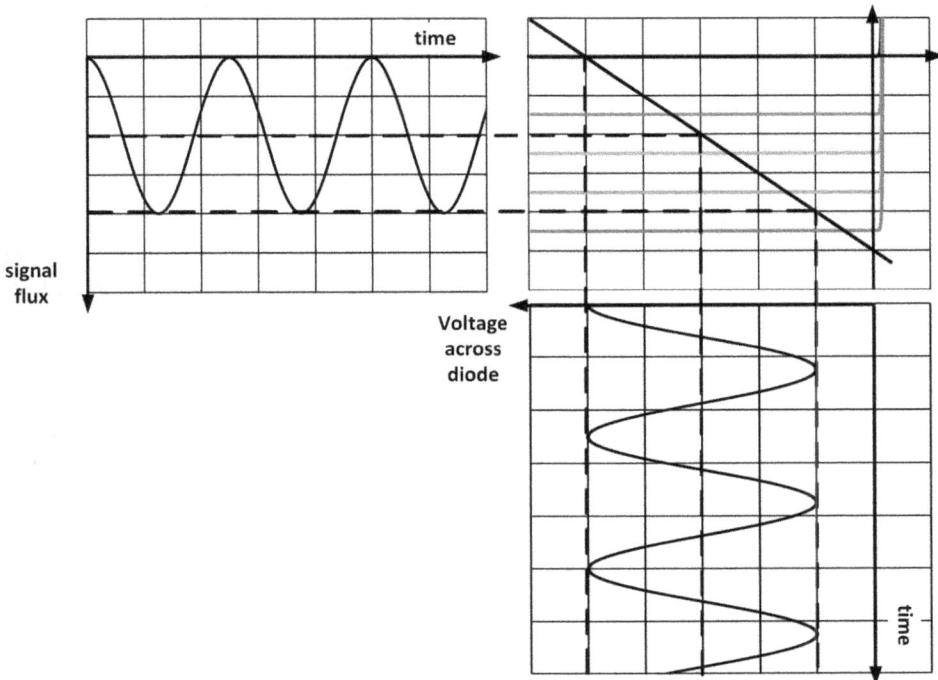

Figure 7.40. Alternating current input signal transferred to an output voltage by the photodiode detector circuit.

signal. At first glance, these devices may seem like a competing alternative to the other detectors, however, device fabrication constraints imply high effective capacitance which limits response times and leakage currents produce noise [3]. The main advantage of phototransistors is that they are inexpensive. When a detection system does not require high frequency response and noise can be tolerated then phototransistors are an appropriate design choice.

Before discussing phototransistors, we will review some of the basic characteristics of general transistors and their operation. A standard bipolar junction transistor (BJT) is a three terminal semiconductor device comprised to two junctions. A transistor acts to control current in analogy to a valve controlling flow of water in a pipe [36]. Figure 7.41 shows the schematic symbol for an NPN type transistor (by far the most common type of phototransistor) with labels for the emitter, base, and collector terminals. Also shown are DC current directions for each terminal. An effective diode equivalent is also shown in the figure.

All transistors use a small input current signal applied to the base lead to control the current through the device from the collector to the emitter. An amplification factor, β, is used to relate the current flow such that,

$$I_C = \beta I_B \text{ and } I_E = I_C + I_B = (1 + \beta)I_B$$

The amplification factor is not necessarily a constant and does depend on collector current, but these simple relations provide a useful model. The characteristic curve for transistor operation relates the collector current, I_C, to the voltage V_{CE}, the emitter to collector voltage across the device. There is a different curve for a given value of base current. Figure 7.42 shows a typical family of transistor characteristic curves. At a given value of V_{CE} the collector current increases with increasing base current as predicted.

Also shown in figure 7.42 is a load line, like the load line used with photodiodes, identifying limiting values. The voltage is limited but the supply voltage, and the collector current is limited but the load resistance within the circuit.

In a phototransistor, the base region of the transistor is the photosensitive element producing a base current proportional to the incident flux according to the photoconductor model, or,

$$i_B = \eta \frac{e}{hc} \lambda \Phi_e$$

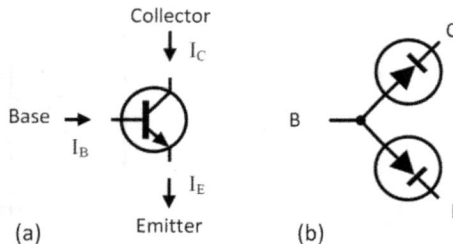

Figure 7.41. Transistor schematic symbol and simple effective diode equivalent circuit.

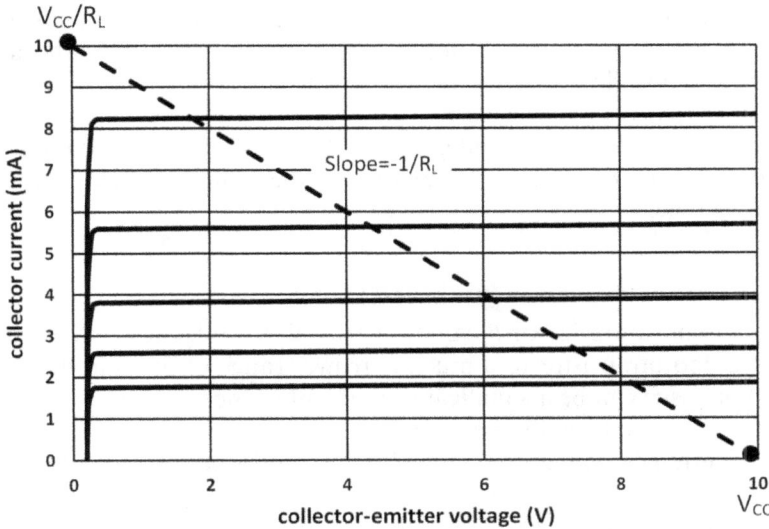

Figure 7.42. Family of characteristic curves of a transistor for different base currents along with load line used to describe the operation of the circuit. In this case $V_{CC} = 10$ V and the load resistance was 1000 Ω, dictating the limits of operation.

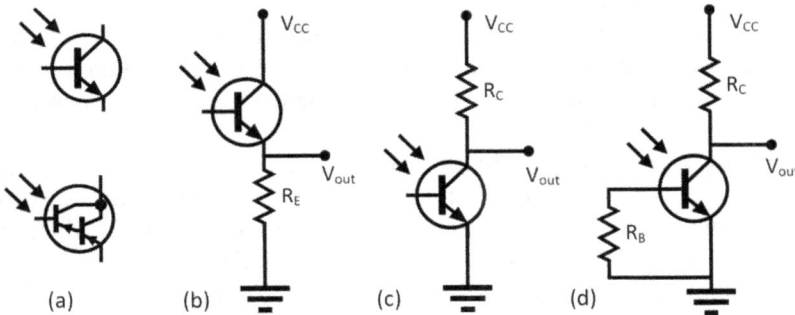

Figure 7.43. Phototransistor symbols and circuits. (a) Phototransistor and photodarlington component symbols, (b) common collector or emitter follower circuit, (c) common emitter amplifier, and (d) phototransistor amplifier with base bias resistor.

where the use of a lower-case letter for current allows for considering alternating currents related to an alternating incident flux. Figure 7.43(a) shows the schematic diagram symbol for two types of phototransistor components, a BJT phototransistor and a photodarlington transistor. With no external electrical connection at the base terminal of the phototransistor it operates similarly to a photodiode. Figure 7.43(b) is simple circuit diagram for a biased phototransistor used to measure an incident signal flux. This circuit is a standard emitter follower configuration (also called common collector) with an applied power supply voltage of V_{CC} and emitter resistance R_E to ground. There is no voltage gain in this circuit, but it is useful for impedance matching to additional detection electronics. It is called a follower because output voltage follows the input voltage on the base but with much larger current. There is a slight voltage difference due to the nominal 0.6 V drop from the base to emitter [36].

Another photodiode circuit configuration, figure 7.43(c) is the common emitter which allows for amplification. From above, the collector current through the resistor R_C, is now,

$$i_C = \beta\eta\frac{e}{hc}\lambda\Phi_e$$

Again, the amplification factor β can also depend on collector current limiting the linearity of a phototransistor to a much narrower range of incident flux than other detectors. This issue can be a significant disadvantage depending on the application and where a photodiode would be a better design choice.

The base terminal connection can also be employed to assist in biasing the transistor to a particular operating point. Figure 7.43(d) shows a circuit with the addition of a single base resistor R_B which alters the sensitivity of the output voltage. A voltage divider is another alternative.

Unlike the photodiode, the base current input can control the mode of operation of the phototransistor. Like all transistors, phototransistors can be used in switch-mode or active-mode. Switch-mode provides an on or off current flow state by operating within either saturated or cutoff regions. Saturation means that both the collector and emitter junctions are forward biased, and cutoff occurs when both junctions are reverse biased. In the active-mode, the collector junction is reverse biased, and the emitter junction is forward biased. A transistor operating in the active-mode allows signals to be amplified.

In optimizing a transistor for use as a phototransistor the near ideal characteristic curves as shown above deviate substantially. This is primarily due to a dynamic resistance internal to the device that is not linear impacting the device linearity in the active region. Figure 7.44 shows a family of more realistic characteristic curves as incident illumination is increased. Load line analysis for a particular circuit remains the same as described above.

Photodarlington devices provide even more sensitivity to incident flux than a single phototransistor. See the schematic symbol in figure 7.43(a). In the photo-darlington, two transistors are connected as a cascading pair with the output of a phototransistor coupled to the base of another transistor providing additional amplification equal to the product of the individual gain [36]. However, these devices have an even lower frequency response than a single phototransistor.

7.8.6 Sectored and multi-element detectors

In some applications, a series of closely spaced detectors or a photodiode array is required. The array is comprised of multiple detector elements, but each detector is a separate p–n photodiode that can be addressed individually. Fabricating the detectors into one package allows the detector areas to be larger and the separation

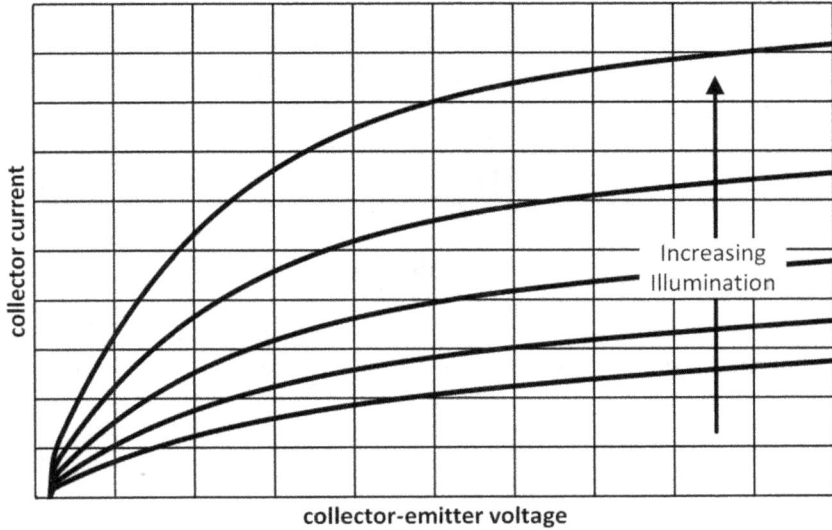

Figure 7.44. Typical phototransistor characteristic curves.

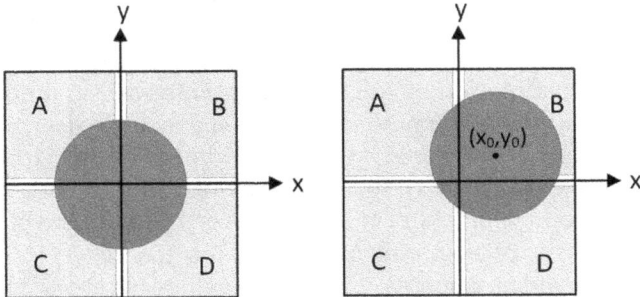

Figure 7.45. Quadrant detector used for beam alignment.

between elements to be as small as possible with the number of elements varying from two to fifty. Position sensing, scanning, beam alignment, and low-resolution imaging are examples of applications where this type of component is used. Common component packages include two-element, three-element, four-element (quadrant-cell), and linear detectors. Many three-element devices include a different color filter (red, green, and blue) covering the elements, and are used for simple color discrimination and detection.

A typical configuration using a quadrant detector for optical beam alignment is shown in figure 7.45. The current output from a photodiode depends on the incident flux but this flux is the product of the irradiance and the detector area that is illuminated. Suppose a uniform circular beam with constant irradiance (flux per unit area) is incident on the center of the quadrant detector. The four elements in each quadrant of the (x,y) coordinate system are labeled A, B, C, and D. We will use these values to represent the current (or voltage) output from each of the detectors. There is always a small gap between the individual detector elements. Because of the gaps the total current output from the entire component would be approximately the sum $A + B + C + D$, but not exactly due to these gaps.

Now suppose that the incident beam is displaced with its center at coordinate (x_0, y_0). The current output from each element is now different and proportional to the area illuminated. In general, the center coordinates of the beam are then,

$$x_0 = \frac{(B + D) - (A + C)}{A + B + C + D} \text{ and } y_0 = \frac{(A + B) - (C + D)}{A + B + C + D}$$

in terms of each current output from the detectors in each quadrant. When the beam is centered on the quadrant detector, the current output from each of the detectors would be equal and the beam's center is at the origin, (0,0), as expected. The detector outputs are often used in a feedback loop to manipulate the beam back into alignment with the center of the detector.

7.9 Array detectors and cameras

A camera is a device used to record images of objects. The form of the recorded image can vary from a physical photographic film or plate to a set of data stored in a computer file. In our current context of detectors, we wish to have a means of converting the photons from every point on an object into a two-dimensional electronic/digital representation of the image that is formed in an optical system's image plane. This may be accomplished by scanning a single detector in the image plane or scanning a linear detector array across the image plane. More conveniently, a two-dimensional array of closely spaced detectors placed in the image plane will collect data related to the amount of radiation at very point in the plane. Of course, you still need some type of electronic display to view the image.

Some of the first video cameras used a technology where a photoconductive material target was overlayed with a transparent conductive film so that a potential could be applied across the target. With no incident light the resistance of the material is high causing charge to build up uniformly across the target area, like a capacitor. The readout signal is controlled by an electron beam scanned across the back of the photoconductor probing each point in the area. These tubes are commonly called vidicons, although there are different varieties depending on target material [37]. With no illumination little additional excess charge is deposited at the point of intersection of the beam on the target. When incident photons from an image fall on a region of the photoconductor target the number of charges at that point is reduced in proportion to the incident illumination. When the scanning beam encounters this point it will recharge the region. The recharge creates an output current providing a temporal signal related to the position of the illuminating image [20, 37]. This signal is then conditioned to be sent to a display as an analog video signal conforming to standards such as NTSC or PAL [38].

Advancements in large scale integration of semiconductor and MEMS fabrication have enabled the development of two-dimensional QWIP arrays, micro-bolometer arrays, and photoconductor arrays [21, 26, 28]. The mechanisms used by these detectors for converting radiation into an electrical current or voltage were discussed earlier and generally have applicability in thermal imaging systems. Sometimes these detectors are referred to as a focal plane array (FPA) based on their typical placement in the back focal plane of the thermal detection system [39, 40].

In the visible and near infrared region of the spectrum, two different types of devices dominate the detector arrays used for still image and video imaging applications. These are the charge coupled device (CCD) and the complementary metal-oxide semiconductor (CMOS). Cameras based on array detectors can provide output signals for display in the analog standards mentioned above or in digital formats for digital television and computer monitors [38].

7.9.1 Characteristics of array detector image data and collection

Whether an FPA, thermal imager, CCD, or CMOS array, the camera devices that collect and record image plane data have a number of similar characteristics. From chapter 1, the output function of a system containing an array detector is a sampled and quantized representation of the continuous image formed by the optical system. A general description for the structure of a detector array is a rectangular grid of photosensitive elements. Figure 7.46 shows a small four-element section of a larger array with vertical and horizontal dimensions associated with each element. The separation between two elements is called the pitch. In this example there is a different pitch for the vertical (p_V) and horizontal (p_H) directions. Similarly, with the photosensitive elements, the vertical (d_V) and horizontal (d_H) dimension of each element is shown. Note that the photosensitive element does not have to be centered within the area. There is also a fixed finite gap between each area element whose size

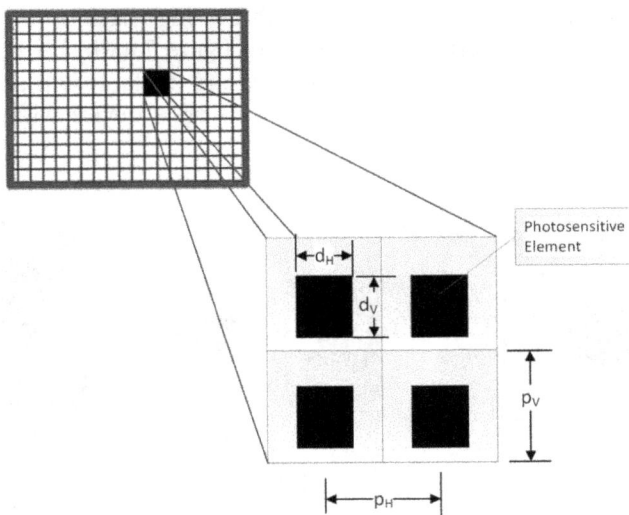

Figure 7.46. Array detector and geometry of four photosensitive elements.

is included in the pitch dimensions. The fill factor of the array is defined as the fraction of the photosensitive area to the entire area, or,

$$\text{Fill factor} = \frac{d_V d_H}{p_V p_H}$$

The area of the entire region defined by the product of the horizontal and vertical pitch dimensions is often called a pixel. However, a pixel is defined as a picture element associated with a digital image at a given coordinate and not really a physical entity. A pixel is a sample with a coordinate and a value. This may seem like semantics for a monochromatic camera, but a single pixel of a color image is comprised of several photosensitive areas all combined to provide one value [41]. We will see an example of the difference later in describing a color camera sensor array. A given pixel in a digital image file will look different depending on the display device.

All electronic imaging results in sampled and quantized image representations. The basics of this process was discussed in chapter 1. The effect of quantization on an image is illustrated in figure 7.47. The original color image is shown in the inset of figure 7.47. Successive reductions in the number of bits per pixel are shown in the remaining images displayed. First, the color image was reduced to a gray-level eight-bit image producing 256 levels (from 0 to 255) at every pixel and is shown in figure 7.47(a). Further reduction to four bits per pixel, figure 7.47(b), and two bits per pixel, figure 7.47(c), are also shown.

The effect of sampling (collecting the image with a smaller-sized array) on an image is illustrated in figure 7.48. The original eight-bit gray-scale image from figure 7.47 is shown for half the number of pixels, figure 7.48(a), one-fourth the number of pixels, figure 7.48(b), and one-eighth the number of pixels, figure 7.48(c). Several artifacts of sampled images are seen in these figures. Sharp lines appear jagged or pixelation because the sample frequency is not high enough causing aliasing [40]. A Moiré effect can also be observed in the brick background of figure 7.48(b) because the periodic pattern of the bricks and the sample spacing. Moiré artifacts will also be observed when a periodic pattern in an image is displayed on a device with an insufficient sampling frequency.

Figure 7.47. Image quantization (a) eight-bit image, (b) four-bit image, and (c) two-bit image.

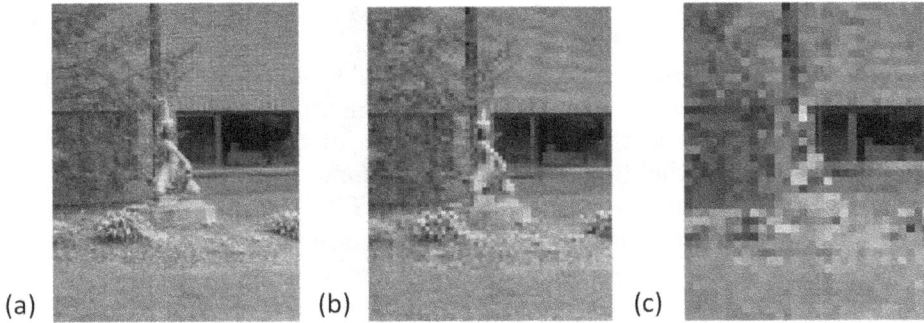

Figure 7.48. Image sampling (a) one-half the number of pixels, (b) one-fourth the number of pixels, and (c) one-eighth the number of pixels in the original gray-level image of figure 7.47.

In chapter 5 we saw that one of the system functions used to characterize image quality is the modulation transfer function. This is a functional representation of the ability of a system to transfer spatial frequency information through the system. From before, the system MTF is a cascade function and is equal to the product of the MTF of all the subsystems involved in the imaging chain. This can extend all the way through to the visual response of the human observer. Sampling by an array detector will impose an additional MTF contribution (magnitude of the detector OTF contribution) along with the MTF of the optical system/lens. Each element of the array acts essentially as an aperture with its own instantaneous field of view [42]. Thus, the system MTF is,

$$\mathrm{MTF_{sys}} = \mathrm{MTF_{opt}MTF_{sensor}}$$

where $\mathrm{MTF_{opt}}$ is the function associated with the optical subsystem and $\mathrm{MTF_{sensor}}$ is the function associated with the detector subsystem.

Reducing the lens to a single effective circular lens of diameter d and effective focal length f, gives the optical MTF spatial frequency function as,

$$\mathrm{MTF_{opt}}(\alpha) = \frac{2}{\pi}\left[\cos^{-1}\left(\frac{\alpha}{2\alpha_0}\right) - \frac{\alpha}{2\alpha_0}\sqrt{1 - \left(\frac{\alpha}{2\alpha_0}\right)^2}\right]$$

where λ is the wavelength, $\alpha^2 = f_x^2 + f_y^2$, with spatial frequencies, f_x and f_y, and $2\alpha_0 = \frac{d}{\lambda f}$. The value of $2\alpha_0$ is the cutoff spatial frequency for this system. The effect of the array, in general, has two components, one due to the sample grid or pitch spacings, the second due to the finite size of the photosensitive element. When the element geometries are rectangular, their MTF functions have a similar form but with different dimensions, or,

$$\mathrm{MTF_{sensor}}\left(f_x, f_y\right) = \left|\mathrm{sinc}\left(p_H f_x\right)\mathrm{sinc}\left(p_V f_y\right)\mathrm{sinc}(d_H f_x)\mathrm{sinc}\left(d_V f_y\right)\right|$$

The separate MTF functions are sometimes called the sample and footprint MTFs [42].

Figure 7.49 shows examples of three separate MTF curves along the horizontal axis for the sample, footprint, and sensor combination. Many plots for the MTF only include data up to the first cutoff frequency (first zero value) however, at higher spatial frequencies there is resolution even though it is spurious resolution [43]. However, for this particular example, the effects of the footprint and sampling contributions damp spatial frequencies greater than about 100 lp mm^{-1}.

Combining the sensor MTF and the optical MTF for a given lens provides the system MTF. This is shown in figure 7.50. The frequency cutoff for this circular lens is 250 lp mm^{-1}. As one can see the sensor MTF limits the overall frequency cutoff.

7.9.2 CCD and CMOS array basics

CCD and CMOS arrays both convert an input optical image formed on the plane of the array into a stream of electrical signals that can be used to reconstruct the image

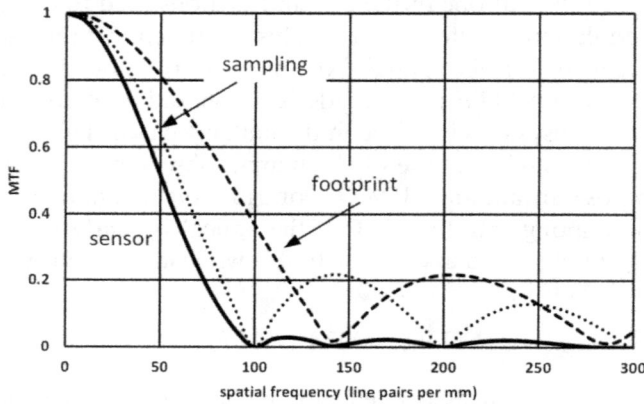

Figure 7.49. MTF curves along the horizontal direction for a sensor array detector showing the sample (dotted), footprint (dashed) contributions and their product as the sensor (solid curve). This example plot uses a sample pitch, $p_H = 10$ μm, and the sensor element dimension, $d_H = 7.0$ μm.

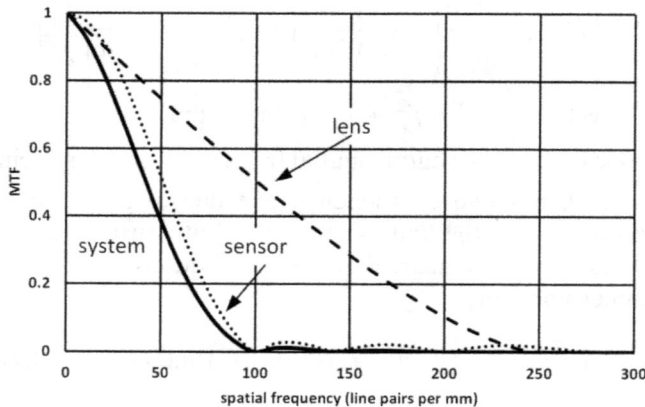

Figure 7.50. System MTF for a circular aperture lens and the sensor MTF from figure 7.49.

on a display. The main differences between the two array types are in the method used to detect incident radiation and readout the signal from each photosensitive element of the array.

The CCD is based on metal-oxide semiconductor (MOS) device structure. A simplified view of the cross-section of a single MOS detector element of a CCD array is shown in figure 7.51(a). The region under each electrode forms a photosensitive element. Again, for simplicity we assume a square structure for each array element. The layered structure is reminiscent of a capacitor and is sometimes called a MOS capacitor. An incident photon in the vicinity of a photosensitive region will create an electron–hole pair as with other photoconductors. When a positive potential is applied to an electrode the electrons will be attracted toward the positive potential and the holes will be repelled. Thus, the electrons are effectively stored in this region as in a potential well. The number of electrons in the well is then proportional to the integral of the optical power incident on the device over a specific exposure time. This is also referred to as a shutter time, like in a conventional camera.

A linear segment of a column of the CCD is shown in figure 7.51(b). There are three different instances of time showing the transfer of an amount of charge from beneath one electrode to the next electrode. This is accomplished using a series of potentials applied to the electrodes during each time sequence. In practice, the clocking cycles are not abrupt voltage changes but are varied gradually for stability of charges within a potential well. However, this charge transfer model provides insight of the readout concept used in CCDs.

A typical CCD device will have a series of horizontal and vertical square photosensitive elements. Because of their device structure CCDs can have a large fill factor and may be enhanced with micro-lenses that are fabricated over each element [44]. To better understand how data from each photosensitive element is

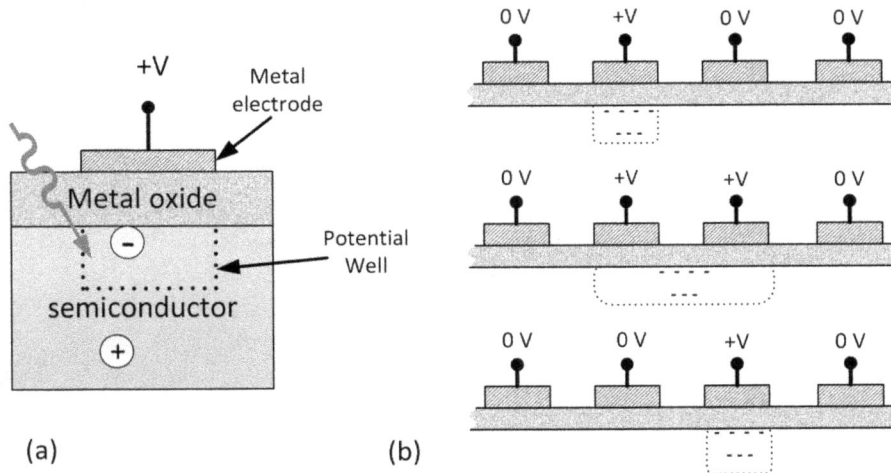

Figure 7.51. (a) Simplified cross-section of a CCD photosensitive element, and (b) illustration of the charge transfer process used to moved stored charge from on element to another showing three snapshots for the location of the stored charge for different sequences of applied potentials.

read out, figure 7.52 shows a small 3×3 element square array after exposure when the shutter is closed. The array is populated with amounts of current in each element symbolized by the letters a–i. As in most CCD readouts, data for each row is transferred downwards along every column until they reach the bottom row, which is the readout shift register, not a part of the active sensor. The transfer occurs by shifting the applied potential in time as illustrated in the clocking cycles. Charges ending up in the readout shift register are transferred to the right. An amplifier at the output of the shift register converts the current in each element to a voltage resulting in an analog output signal in time. After all cycles are completed the signal stream represents the data from each array element. Analog-to-digital conversion is typically done by additional electronic components and not incorporated on the CCD device.

There are three main types of methods used to readout CCDs, (1) interline transfer, (2) frame readout, and (3) frame transfer. The frame readout method was described above where charge is shifted through active elements. Frame transfer incorporates a replica of the entire array masked from any incident photons as a store of one shutter timeframe. Frame transfer arrays are costly since there are two effective arrays in a single device. Interline transfer methods use a separate shift register, shielded from incident photons, between each row of photosensitive elements. All data is then transferred through these shift registers and not through active elements allowing a faster readout but takes away some of the array area [45].

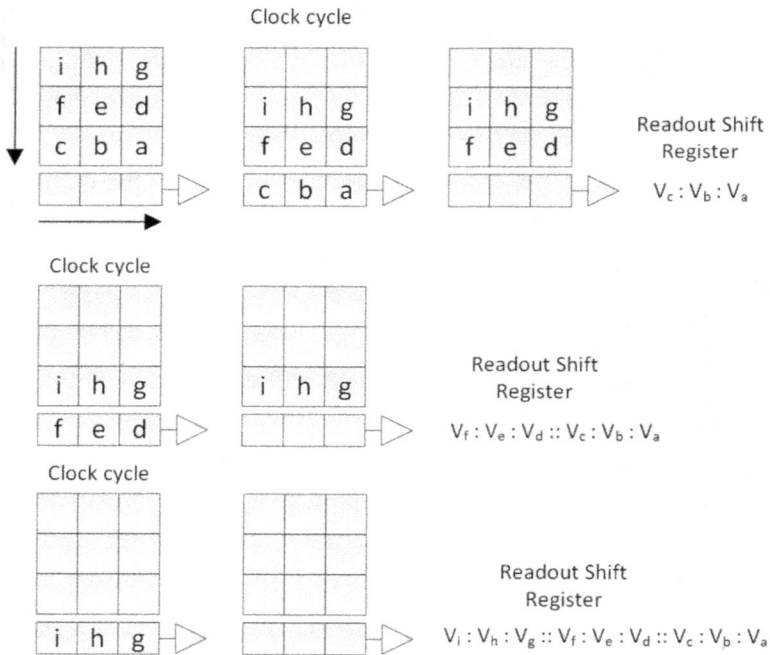

Figure 7.52. CCD full frame readout concept.

Like all photodetectors, CCDs have a quantum efficiency and responsivity which varies with wavelength dictating the spectral response characteristics of the device. Silicon-based sensors have similar spectral response to single silicon photodiodes but there are differences due to the physical structure of the material making up the CCD layered device structure.

CCDs have common sources of noise as with any photodetector, such as shot noise due to signal and dark current. Johnson noise and $1/f$ noise must also be considered as part of the noise in the on-chip readout electronics. Cooling the detector will deduce dark current. A lumped parameter called readout noise is the result of fluctuations during the current-to-voltage conversion at the array output. Since readout noise is constant with the number of electrons this is also called the noise floor [45].

In addition, there are other noise sources specific to CCD devices referred to as reset noise and pattern noise. Reset noise are fluctuations that occur due to differences in the reset level of each array element. A different reset level would translate to a different output even if the flux on two elements was constant. Fortunately, this noise can be canceled by using a correlated double sampling technique on the output signal by collecting two samples of the signal and taking the difference thus eliminating the correlated noise [44]. Pattern noise is the name given to an effect where there is a different dark current at each photosensitive element. Since this does not change frame-to-frame it appears as a source of noise to the observer. Differences in the responsivity of each element also introduces a similar variation independent of incident signal [45].

Noise calculations in CCDs are typically done by specifying the fluctuation in the number of electrons instead of a current fluctuation since there is no output current to measure from an individual element, just the number of electrons stored. If a large number of incident photons arrive during the shutter time electrons generated by the detection process will continue to be stored within a potential well up to a maximum number of electrons called the full-well capacity, n_{well}. Full-well capacity depends on the size of each element and the device design. At this point the well is saturated. As a photosensitive element nears the full-well capacity the detector response becomes nonlinear and charges will move into nearby regions of the array and create blooming. The full-well capacity is then related to the maximum voltage signal that will be measured for any array element.

At the low level with no incident photons the output voltage measured from the array element will be related to the system noise. For a single detector we used the term noise-equivalent power, NEP as a figure of merit. For CCDs, a similar term is used, noise-equivalent exposure (NEE) [45]. Like NEP this is the value of the signal when the SNR = 1, and in terms of the number of electrons it is the root-mean-square number of equivalent electrons from all noise sources in the system $\langle n_{sys} \rangle$.

The dynamic range of the CCD array is defined as the ratio of the full-well capacity to the system noise, or

$$ DR_{array} = \frac{n_{well}}{\langle n_{sys} \rangle} $$

This figure of merit basically describes the ability of the array to faithfully detect bright and dark regions in an image. This value does not account for any analog-to-digital conversion that might occur in converting the analog voltage to a set of quantized levels.

An alternative approach to the CCD uses CMOS fabrication, a common technology used in integrated circuit manufacturing. Each array element contains both a photodetector as well as electronics for reset, switching, and processing the detected signal. This is referred to as active pixel technology since each array element can be individually addressed as a separate active detector. CMOS array elements are voltage devices. Unlike the CCD, there is no need to move charge from one element to another. The output voltage signal from each element is placed on a bus and sequentially read out as a signal. Often analog-to-digital conversion is performed on the device.

Figure 7.53(a) is the equivalent circuit of a basic active element in a CMOS detector. The photodiode is the detecting element with several field-effect transistors used for controls. Figure 7.53(b) shows an overview of the array and the row and column bus control paths. During a frame capture sequence, all charge is removed from each element by initializing the reset transistor and incident radiation produces electrons in proportion to the integration time of exposure. When the exposure is complete, the charge is converted to a voltage by the readout transistor (a source follower configuration). When the row select transistor is turned on by a control signal from the row select shift register the voltage appears on the column bus that is sent to the output shift register for output. The process is then repeated for each element in the array or region of interest.

CMOS manufacturing is relatively low cost compared to CCDs and have lower power requirements. However, the noise floor is generally higher than with CCDs. Another drawback of CMOS arrays are that the circuitry required in each element reduces the fill-factor. Micro-lenses placed at every element can help alleviate some of this disadvantage.

Pattern noise in CMOS arrays is even more evident than with CCDs because of the large variation in transistor gain and offset at each element. Additional signal

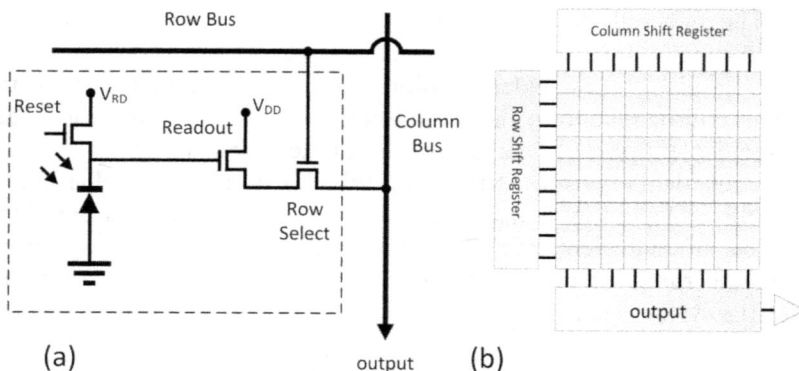

Figure 7.53. Basic CMOS element circuit and array structure with row and column select control paths.

processing can be used to reduce this effect on the CMOS chip or by requiring a calibration step in operation to subtract the pattern.

The choice of whether or not to specify a CCD or CMOS array depends on the application. Each type of device has similar attributes, such as number of elements and responsivity. Being aware of the advantages and disadvantages of both types are necessary in designing a system. Generally, when low noise is a driving requirement the CCD is the best option. CMOS is the best choice for low power applications and when cost is the major issue of the design.

7.9.3 Camera specifications and settings

The most common and standard sizes of the entire sensor array used in cameras vary from 1/3' (6 mm diagonal) up to 1' (16 mm diagonal) with other specialty sizes also available by manufacturers like the 35 mm 'full frame' sensors used in photography [46, 47]. Matching the lens used in an application to the camera sensor size is essential, the larger the sensor area the larger the required lens aperture diameter. Commercial video camera lenses are designed to form images on the sensor array in the back focal plane. To image an object closer to the camera, extension rings of various thickness can be used between the lens and camera body. These extensions are mostly useful in macro-photography, experimental systems, and in machine vision applications.

When viewing an image on a display captured by a video camera, we observe a two-dimensional mapping of the illumination on a region of the sensor array. This view is called a frame and the rate at which consecutive frames appear on the display is called the frame rate. For views by humans the frame rate must be greater than the persistence of vision so that the observed images appear stationary. Frame rate has units of frames per second and are part of video standard specifications. Higher frame rates may be needed to capture motion details without blur. But higher frame rates may not allow sufficient time for the input optical signal to be detected over the system noise. This is the exposure time or shutter time of the camera and sensor system. The higher the number of pixels to be displayed the higher the rate of data transfer needed for a fixed frame rate. Frame rates are controlled by the electronics used in cameras and displays. Thus, there is a tradeoff between numbers of pixels to be displayed, frame rate, and exposure. Picking the largest number of display pixels is not always the best choice in designing a video imaging system.

Most camera systems have a method for setting the gain of the camera. This is the gain of the amplifier prior to an analog-to-digital conversion. This may provide an image when little illumination is present, but it also amplifies the noise. Some cameras have a setting for automatic gain control (AGC) and this may be welcome in viewing a scene whose radiation level changes. However, if two images need to be compared assuming a constant illumination, AGC should be disabled. Machine vision systems often have this requirement.

Setting a proper exposure time (shutter speed, integration time) may alleviate the need for applying high gain. The frame rate may need to be reduced if there is motion in the scene. Some cameras allow a method known as binning.

This effectively combines adjacent detector elements by adding the charges and providing a greater signal. The disadvantage is the reduction in sample size and can affect the aspect ratio if the binning is not done uniformly.

The dark level (black level, brightness) is another setting provided by many cameras to allow adjustment of the output image. This value can compensate for some artifacts of the array and the environment background by subtracting a detected lower limit on the value of each output element. Like automatic gain, this value must be used with caution to properly capture an image of a dark scene.

7.9.4 Digital color cameras

For color digital display or printing a minimum of three values are required for each pixel. To capture a color still image or color video signal, each pixel in the digital data stream also requires three values. As described above, array detectors collect photons independent of their wavelength, albeit with a different response function. Thus, in every color camera some mechanism must be used to separate the three color-channels. To achieve this requirement several methods have been devised, such as: (1) three separate arrays, each covered by a different color filter, (2) one sensor with three (or more) filters moved into the optical path, and (3) using a small filter on each segment of the array detector with a specific pattern across the detector. These three concepts are shown in the illustrations of figure 7.54.

Figure 7.54(a) uses a dichroic beamsplitter or special dichroic prism arrangement to separate the incident light into red, green, and blue components. This arrangement requires registration and alignment of the array detector elements. If the detector elements are replaced by displays, the rays can be reversed, and the system becomes a color projector. The rotating color filter wheel of figure 7.54(b) is a common design where speed of image acquisition is not a major requirement. The filters can also be tailored for specific spectral bands not just RGB. This design has application in colorimetry, and hyperspectral imaging applications.

By far, the most common type of color camera array uses filters at each photo-sensitive element. Typically, the filters are red, green, and blue (although some may use cyan, magenta, and yellow complementary filters and perform subtractive operations to produce the digital image). Figure 7.54(c) shows such a filter overlay on a set of array detector elements. The particular pattern shown is called the Bayer

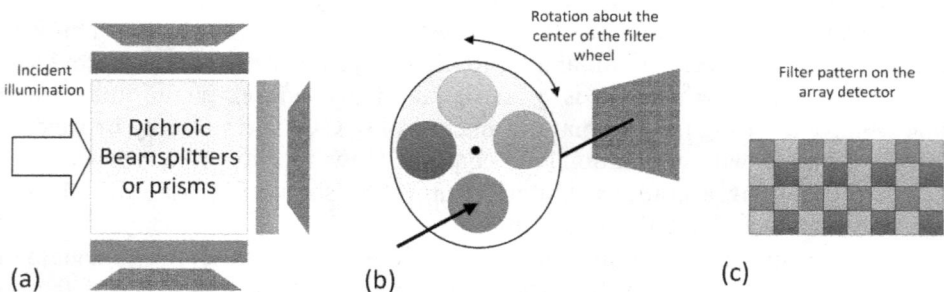

Figure 7.54. Examples of camera concepts for collecting color images.

pattern, where a square pattern of four detector elements is repeated over the array as a set of 2×2 segments [48]. In this case, there are two green filters for every red and blue filter emulating the eye response which is more sensitive in the green than in the red and blue portions of the spectrum.

Once the raw data is collected from an array it is readily available for manipulation. A simple interpretation of a filter pattern, such as the Bayer pattern, would indicate that a pixel is the simple combination of the four red, blue, and two green elements. However, this provides a lower image resolution. Since a single color in a combination of several values, algorithms are employed in most cameras to average over a wider region and increase resolution. Other types of filtering or averaging can also be performed to correct for sampling issues that create artifacts. Understanding these operations may be needed in choosing a camera for incorporation into a system.

7.10 Application: detector responsivity and linearity

Testing the linearity and measuring the responsivity of a particular or unknown detector can be straightforward with some basic optics lab equipment. Besides the detector system to be tested, the required equipment includes: a source with an output power whose maximum value is near the required maximum for a particular design, a set of neutral density filters, and an optical power meter. If the power meter is well calibrated, then this test can also act as a transfer of standards for the detector under test.

An example of this type of experimental test setup is shown in figure 7.55. For this test, an LED was used whose light was collected by a lens and directed toward the power meter filling the entire detector area. The power measurement then provides a measure of irradiance or power per unit area of the detector. One measures the power as different neutral density filters and combinations of filters are interposed between the source and detector. The specific values for the optical density of the filters does not matter just a record of the particular combination is needed as power is measured. Combinations of filter can be used to fill in a reasonable range from 0 OD up to the largest needed (smallest transmittance). For this test, the maximum optical density was 6.0 OD.

The next step is to replace the power meter with the detector under test. A silicon photodiode detector was used in reverse bias condition at 15 V and the current measured by recording the voltage across a 1000 Ω load resistor. Again, for this example the incident light beam covered the entire detector area. Output current

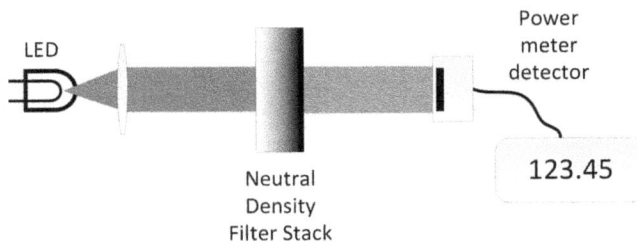

Figure 7.55. Experimental setup to test linearity of a photodiode detector.

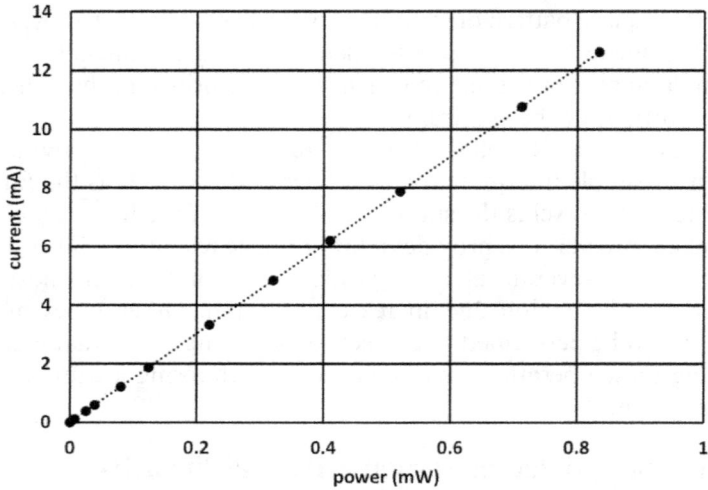

Figure 7.56. Experimental data plotted as current output verses input power showing a high degree of linearity over the test conditions. The dotted line is a trend line fit to the data.

readings were taken at every neutral density filter and filter combination used in the power meter portion of the test. The relative power falling on the detector can then be computed from the area ratio of the power meter detector area to the area of the detector under test. If a laser is used as the source and all power from the beam is collected by both detectors, then this step is not necessary. The results for this particular test are displayed in figure 7.56.

Regression analysis on this data gave a slope of $15.2 \text{ A W}^{-1} \pm 0.1 \text{ A W}^{-1}$ which is the responsivity for this detector configuration. The error in each data point value was computed by propagation of error and used to determine the error in the responsivity slope. As you can see this detector is linear over this incident power range as expected. The particular test here was done specifically to check the low light level performance for this detector, thus the relatively low input power values.

Exercises and problems

1. The temporal voltage response of a system is given by,

$$v(t) = \begin{cases} 0 & t < 0 \\ v_0 & 0 \leqslant t \leqslant \tau \\ 0 & t > \tau \end{cases}$$

where the characteristic time τ is the total observation time. Show that the noise-equivalent bandwidth (Δf) for such a system is $\Delta f = 1/2\tau$.

2. A photodetector produces on average one electron for every five incident photons from an LED with a wavelength of 780 nm. (a) Compute the quantum efficiency of this detector at this wavelength. (b) What is the responsivity? (c) How much photocurrent is generated for an incident flux of 2.0 µW?

3. A photodiode detector is setup in a photoconductive mode. The quantum efficiency of the detector is 65%. When the detector is placed in a HeNe (633 nm) laser beam with a 10 mW power output, what is the value of the current in the photoconductive circuit.

4. A blackbody calibration source (Lambertian emitter) of diameter 1.5 cm is used to calibrate infrared detectors. The radiance of the blackbody source is 100 W m^{-2} sr^{-1}. A detector under test has a 1.0 mm diameter and a NEP of 10–11 W for a 100 Hz bandwidth. At what distance from the blackbody calibrator must the detector be placed to obtain a SNR of 100?

5. You are asked to design a sensor system that detects optical radiation from the Sun in the visible spectrum between 0.400 and 0.700 micron wavelengths. The collecting lens is specified and has a focal length of 10.0 cm and diameter of 2.50 cm. The irradiance at the surface of the Earth from illumination by the Sun is 900.0 W m^{-2}. (Take the Sun radius to be 7.00×108 m and the mean Earth-Sun distance to be 1.50×1011 m). Given that the detector used in your design has a responsivity of 0.15 A W^{-1} and a rms noise current of 7.5 nA. Find: (a) the NEP for this detector, and (b) the output current when the system accepts light from the Sun. (Assume that the image of the Sun is smaller than the detector area.)

6. The spectral transmittance optical power curve for a filtered halogen source was measured with a monochromator whose wavelength resolution is ±0.5 nm (total of 1.0 nm).

   (a) Using a calibrated power meter, data was collected at sample wavelengths resulting in the spectrum shown in figure P7.6. Power units are in milliwatts and wavelength units are in nanometers. Assume zero power at all other wavelengths outside of this wavelength band and uniform power at wavelengths between each sample. Compute the total flux delivered by this source.

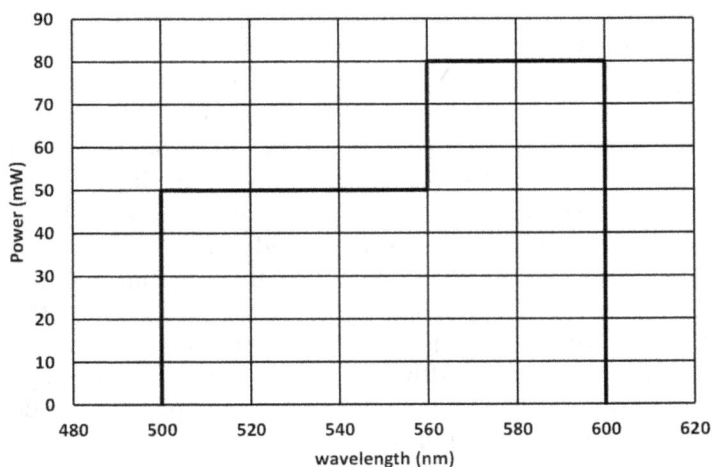

Figure P7.6. Power as a function of wavelength for a filtered source.

(b) An unknown detector was now used to record the same spectrum and the resulting voltage measurements at each wavelength are shown. Fill in the table below for the spectral response of the unknown detector. Include units for the spectral response along with the values.

Wavelength (nm)	Voltage (mV)	Response ()
500	5	
510	10	
520	15	
530	20	
540	25	
550	30	
560	55	
570	62	
580	70	
590	80	
600	5	

7. (a) Does Johnson noise depend on whether or not any photons are incident on the detector?

    (b) Determine the Johnson noise current for a receiver of noise temperature 300 K and resistance 1.00 MΩ. Assume a 1.00 MHz bandwidth.

    (c) What is the Johnson noise current if this detector was cooled to liquid nitrogen temperature of 77 K?

8. Given that $\sigma = \sqrt{\langle (v - \langle v \rangle)^2 \rangle}$. Show that $\sigma = \sqrt{\langle v^2 \rangle - \langle v \rangle^2}$ where $\langle v \rangle$ represents the average of v.

9. An InGaAs photodiode is part of a fiber optic receiver operating at room temperature (300 K) with a wavelength of 1300 nm. The detector quantum efficiency is 0.90, load resistor of 1000 Ω, and system bandwidth is 20 MHz. When a 300 nW incident optical power is detected, determine the signal-to-noise ratio. (Assume background power is negligible.)

10. A 1.30 μm wavelength diode laser has been specified as the source for a free-space optical communication system. A particular photodiode detector was chosen for this system but not all specifications were known. (a) Initial measurements showed that a continuous 6.00 μW radiant power incident on the photodiode detector produced a 5.00 μA current. What is the detector responsivity? (Assume that this is within the linear region of detector operation). (b) What is the detector's quantum efficiency at the operating wavelength of the system? (c) A modulated radiant power signal with an electrical bandwidth of 100.0 MHz and average signal power of 50.0 nW is now incident on the detector from the diode laser. The background power is significant and contributes 1.00 μW. The dominant noise is due to background power only so that we can neglect thermal/Johnson noise and post-electronic noise. Compute the signal-to-noise

ratio to determine whether or not this signal can be detected despite the large background?

11. A detector system with quantum efficiency η is only dominated by noise related to its signal flux ΦS and background flux ΦB. When detecting radiation of wavelength λ within a frequency bandwidth Δf, find a general expression for the noise-equivalent power, NEP.

12. A detector system operating at 860 nm wavelength has a Johnson noise current of 0.60 nA, a shot noise due to the signal of 0.20 nA and a shot noise due to the background flux of 0.30 nA. What is the total noise current?

13. Calculate the ratio of the current in a 0.5 V forward biased diode to the current in a -0.5 V reversed biased diode at room temperature (300 K).

14. A photodiode at zero bias voltage with resistance of 0.5 MΩ and internal capacitance of 0.5 nF is used in a room temperature (300 K) detector system. (a) Assuming an exponential decay system model, compute the frequency bandwidth expected for this detection system. (b) A direct current of 1.0 μA is measured as the output from this detector. Compute the noise current associated with shot noise. (c) Compute the Johnson noise current. (d) What is the total noise current?

15. A thermopile detector is under a constant incident flux reaching a final temperature change of 2 K at steady state. The thermal time constant of this detector is 15 ms.
    (a) Plot the detector response as a function of time.
    (b) How long does it take for the detector to reach 10% of the steady state temperature?

16. Modulated light is incident on a thermal detector whose thermal time constant is 1.0 s. At the cutoff frequency, the temperature change was determined to be 0.707 K. Plot the change in temperature of the detector as a function of frequency.

17. A photomultiplier tube (PMT) has a gain of 2.0×106 and is used to detect 1.0 μW light from an LED whose operating wavelength is 673 nm within a bandwidth of 1 Hz. The photocathode is 1% efficient and the dark current of the tube is specified to be 0.90 nA at the -1000 V operating voltage. (a) Compute the output current from the PMT. (b) Assuming that the noise-equivalent power of a PMT is only limited by shot noise in the dark current and only arises from the photocathode, compute the value of NEP.

18. A CdSe photoresistor is used to detect light within a narrow range of input flux. The detector's specification sheet gives the resistance of the detector, R_d, as a function of flux, Φ_e, to be:

$$R_d = (150 \text{ k}\Omega)\Phi_e^{\frac{1}{2}} + 1.0 \text{ k}\Omega$$

where Φ_e is in units of Watts. You have a 12.0 V battery and a 5000 Ω load resistor for use in a typical voltage divider detection circuit. (a) What is the voltage across the load resistor when the detector is in the dark? (b) Find the voltage across the load resistor when the incident flux is 100 μW.

19. A high speed InGaAs detector is made with a depletion layer thickness of 0.15 μm. What percentage of the incident photons are absorbed in this photodetector at 1310 nm if the absorption coefficient is 1.5 μm^{-1}?

20. The equivalent diode resistance can be provided in a specification sheet or found by computing the slope of the current versus voltage characteristic diode curve and solving for R = dV/di at V = 0. (a) Obtain an analytic expression for R. (b) Find the equivalent diode resistance at room temperature (300 K) for a detector with a reverse saturation current of 10^{-8} A.

21. Your supervisor asks you to measure the output from an infrared emitting diode at 830 nm with an output power level of 100 pW. An effective noise bandwidth of 1000 Hz is required. A table of candidate photovoltaic detectors is given below. Assume that the quantum efficiency of each is 0.8. Select the best detector for the job and briefly explain why you made this choice.

Detector type	i_o (A)	E_g (eV)
GaP	10^{-12}	2.25
GaAs	10^{-11}	1.4
Si	10^{-10}	1.1
InGaAs	10^{-8}	0.75
Ge	10^{-7}	0.68

22. Design a simple voltage divider circuit that will measure a sinusoidal modulated optical signal at 0.50 W peak-to-peak up to a modulation frequency of 10 MHz. The wavelength of the modulated laser source is 1300 nm. You must use one of the ThorLabs Unmounted Photodiodes (GaP, Si, InGaAs, Ge). Specify the model number you chose and briefly explain why you made this choice.

23. A 1024 × 1024 square detector array with a 6.25 μm spacing between detector elements is used to view an object on an assembly line. The 8.0 mm focal length camera lens is used and located 2.0 m above the assembly line conveyor belt. (a) What is the total field of view of the system? (b) Compute the instantaneous field of view, IFOV. (c) Find the detector footprint.

24. A single photodiode detector is setup in a photoconductive mode with a 5.0 V bias and a 1000 Ω load resistor. The quantum efficiency of the detector is 92.6% at the operating wavelength of a 670 nm high power laser source. (a) What is the value of the current through the load resistor when the input optical power is 6.0 mW? (b) Compute the maximum current allowed in this circuit. (c) What is the current through the load resistor when the optical power is increased to 12.0 mW?

References

[1] Pedrotti L S and Pedrotti F L 1998 *Optics and Vision* (Upper Saddle River, NJ: Prentice Hall)
[2] Ryer A D 1997 *Light Measurement Handbook* (Peabody, MA: International Light Technologies)

[3] Hobbs P C D 2000 *Building Electro-optical Systems: Making It All Work* (New York: Wiley)

[4] McCluney R 1994 *Introduction to Radiometry and Photometry* (Boston, MA: Artech House)

[5] Boyd R W 1983 *Radiometry and the Detection of Optical Radiation* (New York: Wiley)

[6] DiMarzio C A 2011 *Optics for Engineers* (Boca Raton, FL: CRC Press, Taylor & Francis)

[7] Wolfe W L and Zissis G J (ed) 1989 *The Infrared Handbook, Revised Edition* Prepared by the Infrared Information and Analysis (IRIA) Center, Environmental Research Institute of Michigan, for the Office of Naval Research, Department of the Navy, 3rd printing

[8] *Introduction to Thermopile Detectors* 8585 Rev E, Dexter Research Center, 7300 Huron River Drive, Dexter, MI 48130 www.DexterResearch.com (Accessed 14 September 2020)

[9] Niklaus F, Vieider C and Jakobsen H 2008 MEMS-based uncooled infrared bolometer arrays: a review *Proc. SPIE* **6836** 68360D

[10] Lohrmann D, Littleton R, Reese C, Murphy D and Vizgaitis J 2013 Uncooled long-wave infrared small pixel focal plane array and system challenges *Opt. Eng.* **52** 061305

[11] Kittel C 2004 *Introduction to Solid State Physics* 8th edn (New York: Wiley)

[12] Hossain and Rashid M H 1991 Pyroelectric detectors and their applications *IEEE Trans. Ind. Appl.* **27** 824–29

[13] Guenzer C S 1976 Chopping factors for circular and square apertures *Appl. Opt.* **15** 80–3

[14] Grant B G 2011 *Field Guide to Radiometry* (Bellingham, WA: SPIE Press)

[15] Boreman G D 2001 A users' guide to IR detectors *Proc. SPIE* **4420** 79

[16] Yariv A and Yeh P 2007 *Photonics: Optical Electronics in Modern Communications* (New York: Oxford University Press)

[17] Goodman J W 2000 *Statistical Optics* (New York: Wiley)

[18] Davenport W B Jr. and Root W L 1987 *An Introduction to the Theory of Random Signals and Noise* (New York: Wiley-IEEE Press)

[19] van Etten W 2006 *Introduction to Random Signals and Noise* (Chichester: Wiley)

[20] Wilson J and Hawkes J F B 1998 *Optoelectronics: An Introduction* (London: Prentice Hall)

[21] Dereniak E L and Crowe D G 1984 *Optical Radiation Detectors* (New York: Wiley)

[22] Kingston R H 1978 *Detection of Optical and Infrared Radiation* (Berlin: Springer)

[23] Daniels A 2010 *Field Guide to Infrared Systems, Detectors, and FPAs* 2nd edn (Bellingham, WA: SPIE Press)

[24] *Photomultiplier Tubes*, Electron Tube Division, Hamamatsu photonic www.hamamatsu.com

[25] Wiza J L 1979 Microchannel plate detectors *Nucl. Instrum. Methods* **162** 587–601

[26] Dereniak E L and Boreman G D 1996 *Infrared Detectors and Systems* (New York: Wiley)

[27] Eisberg R M and Resnick R 1985 *Quantum Physics of Atoms, Molecules, Solids, Nuclei, and Particles* 2nd edn (New York: Wiley)

[28] Gunapala S D *et al* 2014 Quantum well infrared photodetector technology and applications *IEEE J. Sel. Top. Quantum Electron.* **20** 154–65

[29] Token Electronics Industry Co., Ltd http://token.com.tw/

[30] Kraftmakher Y 2012 *Eur. J. Phys.* **33** 503

[31] Amirtharaj P M and Seiler D G 1995 Optical properties of semiconductors *Handbook of Optics, Volume II: Devices Measurements and Properties* ed M Bass, E W Van Stryland, D R Williams and W L Wolfe 2nd edn (New York: McGraw-Hill)

[32] Saito T 2011 *Spectral Properties of Semiconductor Photodiodes* https://intechopen.com/

[33] Keiser G 2010 *Optical Fiber Communications* 4th edn (New York: McGraw-Hill)

[34] Schinke C *et al* 2015 Uncertainty analysis for the coefficient of band-to-band absorption of crystalline silicon *AIP Adv.* **5** 067168

[35] Nunley T N, Fernando N S, Samarasingha N, Moya J M, Nelson C M, Medina A A and Zollner S 2016 Optical constants of germanium and thermally grown germanium dioxide from 0.5 to 6.6 eV via a multi-sample ellipsometry investigation *J. Vac. Sci. Technol.* B **34** 61205

[36] Horowitz P and Hill W 1989 *The Art of Electronics* (Cambridge: Cambridge University Press)

[37] Redington R W 1971 Introduction to the vidicon family of tubes *Photoelectronic Imaging Devices. Optical Physics and Engineering* ed L M Biberman and S Nudelman (Boston, MA: Springer) https://doi.org/10.1007/978-1-4684-2931-2_13

[38] Luther A C and Inglis A F 1999 *Video Engineering* 3rd edn (New York: McGraw-Hill)

[39] Kasunic K 2011 *Optical Systems Engineering* (New York: McGraw-Hill Education)

[40] Rogalski A 2004 Optical detectors for focal plane arrays *Opto-Electron. Rev.* **12** 221–45

[41] Gonzalez R C and Woods R E 2018 *Digital Image Processing* 4th edn (New York: Pearson/Prentice Hall)

[42] Boreman G D 2001 *Modulation Transfer Function in Optical and Electro-Optical Systems* (Bellingham, WA: SPIE Press)

[43] Goodman J W 2017 *Introduction to Fourier Optics* 4th edn (New York: WH Freeman)

[44] Tredwell T J 1995 Visible array detectors *Handbook of Optics, Volume I: Fabrication, Techniques, and Design* 2nd edn, ed M Bass, E W Van Stryland, D R Williams and W L Wolfe (New York: McGraw-Hill) ch 22

[45] Holst G C 2001 Solid state cameras *Handbook of Optics, Volume III: Classical, Vision, and X-ray Optics* 2nd edn, ed M Bass, E W Van Stryland, D R Williams and W L Wolfe (New York: McGraw-Hill) ch 4

[46] Thorlabs Inc 2020 www.thorlabs.com (Accessed 22 June 2020)

[47] Edmund Optics Inc 2020 www.edmundoptics.com (Accessed 8 June 2020)

[48] Bayer B E 1976 Color imaging array *US Patent* 3971065 issued 1976-907-20

[49] Nahory R E, Pollack M A, Johnson W D Jr and Barns R L 1978 *Appl. Phys. Lett.* **33** 659

IOP Publishing

Optical Systems Design Detection Essentials
Radiometry, photometry, colorimetry, noise, and measurements
Robert M Bunch

Chapter 8

Beam formation, modulation, and scanning

8.1 Introduction

Many optical systems are designed to produce the illumination required for another part of the system or for another optical system to detect. In particular, optical communication systems encode information by altering the flux emitted by a source to create a signal to be transmitted to a receiver. We discussed this idea earlier when we assumed that a temporal function of flux falls on a detector. Fiber optic communication links are good examples of this concept.

Creating directed beams of optical radiation provides the means of producing a maximum flux within a given target area. Beams are formed with optical components such as lenses and reflectors or are an inherent property of the source itself as with laser resonator output. Understanding some of the basic concepts of beam formation and propagation, as well as its limitations, is necessary in most optical system designs. Some of these fundamental concepts will be discussed in this chapter.

Scanning a small diameter beam over a large target area is a common task for an optical system design. Applications such as laser marking, laser printers, bar-code readers, and micromachining use various scanning methods. The basic characteristics of scanners will be described and used to compare different types of beam scanning devices.

Modulation of a beam occurs when amplitude, phase, polarization, or frequency of the propagating wave changes in time. Laser diodes and LEDs can be directly modulated by altering the drive current to the component. Other modulation methods use modulator devices external to a continuous wave source. A mechanical modulator could simply be an alternating shutter or chopper that intermittently passes and blocks the beam at some frequency. Exploiting optical properties of some materials allows modulators to be constructed that respond to electric, magnetic, and acoustic signals. Basic properties of modulators will be covered in this section.

8.2 Beams for illumination

Illumination occurs when there is a transfer of radiant flux from some source (emitter) to a target (receiver). Since the optical system does not require that any imaging occur in order to transfer radiation in these systems, the term non-imaging optics is used to describe these designs [1–5]. Flux transfer by a non-imaging optical system is illustrated in figure 8.1. The system collects radiation from a small emitter source and transfers radiation to a large target receiver, figure 8.1(a). This is the approach of a common luminaire, lamp, light fixture, or flashlight. Conversely, figure 8.1(b) shows radiation being collected from a large source emitter and transferred to a small target receiver by a system. This is a collector system or concentrator as with solar energy harvesting or focusing light from a source to a small spot [2–4].

In our earlier discussion of radiometry and photometry we defined the concept of throughput or étendue and showed that it was a conserved quantity [5]. This is particularly important in non-imaging optics because it imposes an area-solid angle tradeoff condition on any system. Transferring flux from a large area to a smaller area implies that a large area emitter will require that the system collect radiation within a small angular range for maximum flux transfer to the small area which will receive radiation within a large angular range. See figure 8.2 [2–4].

The concentration ratio, C, is defined for a three-dimensional system as, the ratio of areas

$$C = \frac{A}{A'}$$

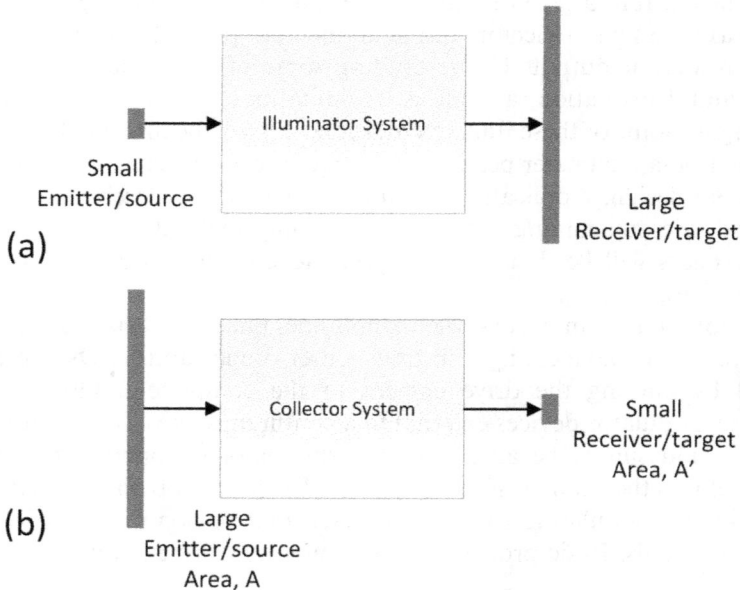

Figure 8.1. Non-imaging optical system concepts of (a) illumination, and (b) collection.

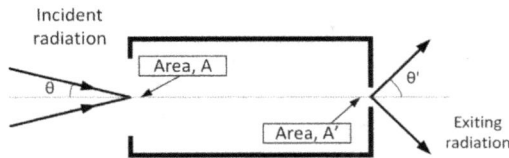

Figure 8.2. Conservation of étendue and concentration ratio of a non-imaging optical system.

And using conservation of étendue, $nA\sin^2(\theta) = n'A'\sin^2(\theta')$, the conservation ratio is,

$$C = \left(\frac{n'\sin \theta'}{n\sin \theta} \right)^2$$

A general illumination system design does not necessarily constrain the radiation to the visible region of the electromagnetic spectrum. However, many of the applications of illumination do have the human observer in mind and are restricted to visible light. When we speak of lighting we are usually talking about natural daylight or physical objects like lamps or luminaires. The importance of general lighting to the economy and energy conservation merits some discussion. Recent advances in the development of solid-state lighting has elevated the importance of lighting and luminaire design in a variety of engineering application areas such as, automotive lighting and display technology. This section reviews some basic principles of beam formation and practical specifications of luminaires used in system designs.

8.2.1 Collimated beams

The term 'beam' is commonly used to describe the directionality of a large number of light rays as they propagate from a source to a target area. We say this beam is collimated when all rays are parallel to the axis of propagation. Another way of determining a degree of collimation is whether or not the wavefronts of the beam are perpendicular to the direction of propagation. The simplest model of beam formation is a point source placed at the focal point of a lens. The image of the source is of course formed at infinity. This is shown in figure 8.3(a). If the lens is large, then this source produces a plane wave with no divergence angle. The same situation also occurs if the point source is placed at the focal point of a concave mirror. For a point source of intensity, I_e, the irradiance is,

$$E_e = \frac{I_e}{f^2}$$

a uniform irradiance at all points in a plane perpendicular to the direction of beam propagation. Of course, this is the ideal situation. When lenses have a finite pupil, the propagating wavefront is the diffracted field of the pupil function and will contain contributions from system aberrations. This results in a beam of radiation related to the pupil size but diffracting and usually diverging. In addition, when the lens cannot be considered a thin lens the focal distance is the effective focal length of the lens system.

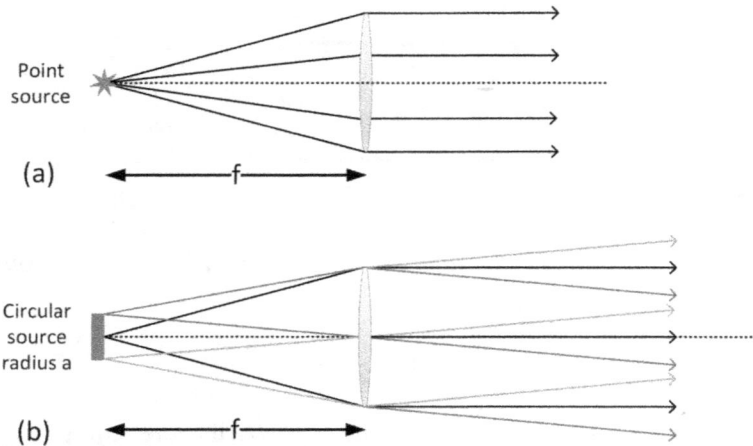

Figure 8.3. Lens collecting radiation from a point source (a) and an extended source (b) and the resulting transmitted beams.

When an extended source is placed at the front focal length of the lens the image of the source is still formed at infinity as with a point source but now the irradiance distribution is no longer a constant at all perpendicular planes. Figure 8.3(b) shows this geometry for the case of a small circular source of radius, a. Using either a lens or a mirror with an extended source is the basic approach used in searchlights [6–8]. The angle that the source subtends from the center of the lens is called the field of illumination [9]. The field of illumination can be approximated using ray optics but can also be determined from the size of the diffraction spot in diffraction-limited systems.

We can use radiometric analysis to determine the irradiance distribution at an arbitrary z-axis position to the right of the lens. Assume that the extended disk source is Lambertian for simplicity. (Actually, this assumption is fairly good for planar light emitting diode sources.) First, restricting the discussion to on-axis points, there are two regions to consider. Region 1 is defined by distances near the lens where the radius of the projected cone of the source is less than the radius of the lens. Region 2 occurs at larger distances and characterized by projections of the source which are larger than the lens radius. Each of these two regions is shown in the ray diagram of figure 8.4.

Take the diameter of the lens to be D. In Region 1, from geometry $\frac{y_1}{z} = \frac{a}{f}$ so the area of the cone at z is, πy_1^2 resulting a solid angle,

$$\Omega_1 = \frac{\pi a^2 z^2}{f^2 z^2} = \frac{\pi a^2}{f^2}$$

The irradiance in this region is then,

$$E_e(z) = L_e \Omega_1 = L_e \frac{\pi a^2}{f^2} = \text{constant}$$

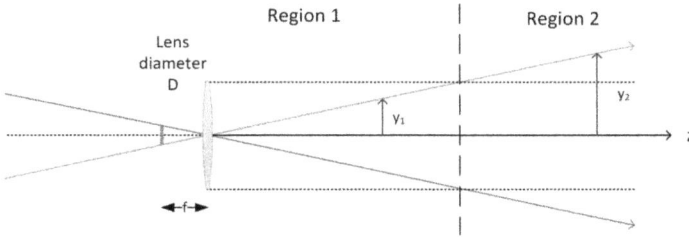

Figure 8.4. Ray diagram for the on-axis radiometric analysis of a basic searchlight.

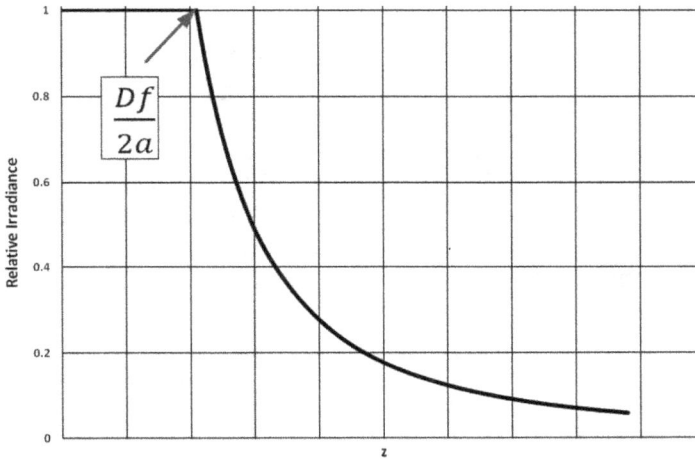

Figure 8.5. On-axis irradiance of a searchlight with a Lambertian source.

a constant for all z values in Region 1 where $z < \frac{Df}{2a}$. Repeating this calculation for Region 2, the area subtense in the solid angle definition is now just a constant value of the lens area. This results in an irradiance of,

$$E_e(z) = L_e\Omega_2 = L_e\frac{\pi D^2}{4z^2}$$

where the values of z are greater than or equal to $\frac{Df}{2a}$. If we plot the on-axis irradiance as a function of z then we obtain a graph in figure 8.5. This shows that the beam has constant irradiance out to $z = \frac{Df}{2a}$ and then falls off as an inverse square law prediction and acting like a point source. At off-axis points in Region 1 the irradiance will remain constant until the source projection is shadowed by the lens and there is no vignetting. Vignetting can also occur in Region 2 further reducing the off-axis irradiance [6, 7]. As long as the source has a finite size, the ideal planar wavefront collimated beam cannot be achieved.

A common collimation technique for laser beams is to employ a spatial filter that uses a microscope objective to focus the beam onto a pinhole creating an effective point source. The effect of the pinhole is to remove all higher order spatial frequencies

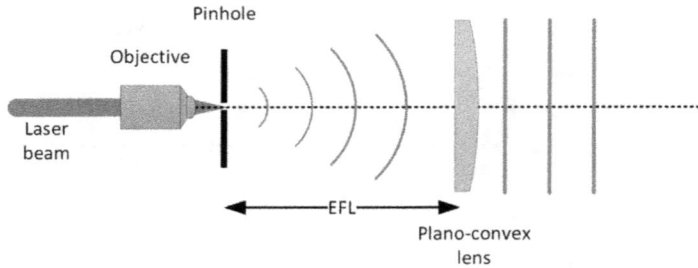

Figure 8.6. Collimating a laser beam using a spatial filter and plano-convex lens.

contained within the incident beam [7]. The radiation emerging from the pinhole can then be collected with a lens and collimated as in figure 8.3(a). The quality of the wavefronts to the right of the collimation lens depends on the collimation lens used. However, a common plano-convex lens can be effective. Aspheric lenses are also available for collimating light emitted by optical fibers and laser diodes. Figure 8.6 shows how the wavefronts propagate in such an ideal arrangement.

Lenses that are specially designed for high-efficiency illumination applications are called condenser lenses. Their purpose is to collect light from a source and form an image of the source within the entrance pupil of another optical system or collect light from a small source in its front focal plane to produce collimation. Condensers generally have high numerical apertures or a lower $f/\#$ than spherical lenses. Because aberrations can reduce illumination uniformity, condensers are often aspheric.

8.2.2 Koehler illumination

In most illumination systems the basic requirement is to uniformly illuminate an object but without observing any structure of the source. One way to do this is to use a small source but the smaller the source the lower the illumination. Suppose that an image of the source was formed on the object using a condenser. This will provide a high degree of illumination of the object, but it also means that the source is observed in the image of the object. This can work well for small fields where the source itself is somewhat uniform but is not useful for large sizes. This type of illumination is referred to as critical illumination and was used extensively in early microscopes. Another illumination method, known as diffuse illumination, uses a diffuser plate placed within the condensing part of the system, however this is often inefficient.

Koehler illumination solves many of the problems of critical and diffuse illumination. In a Koehler illumination system, the condenser lens forms an image of the source in the entrance pupil of the imaging lens. The object is placed in the exit pupil of the condenser. In this way every point on the object is illuminated by the entire source making the illumination independent of the shape of the source. Figure 8.7 shows the basic layout of a Koehler projector system. Most microscopes use some form of Koehler illumination [10, 11].

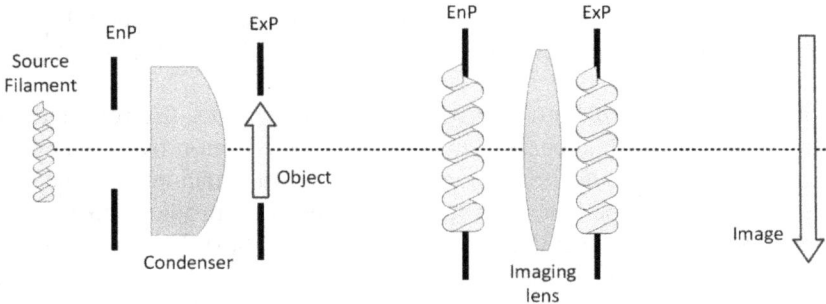

Figure 8.7. Koehler illumination used in a simple projection system.

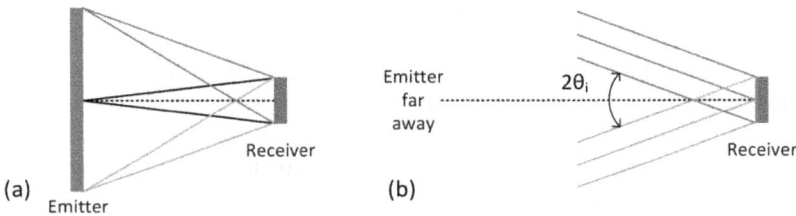

Figure 8.8. Emitter to receiver flux transfer from (a) a finite size emitter close to the receiver, and (b) a distant emitter to the receiver.

8.2.3 Illumination with reflectors

Using a reflector for illumination requires that the radiant flux from a small, finite size source emitter is transferred to a large receiver by reflections from some surface shape. This is accomplished by determining the shape of the reflector surface using some design criteria connecting the emitter output to the receiver. From a ray optics perspective, this is just a collector concentrator with the ray directions reversed thus forming a luminaire [11]. It is easier to use a concentrator concept to define parameters and examine optimal conditions needed for efficient collection and then reverse the situation replacing the collector receiver with an emitter. We will use this approach here.

To transfer the maximum amount of flux from an emitter surface to a receiver we trace rays from the edges of the emitter to the receiver area knowing that the receiver will accept flux with a solid angle subtended by the receiver. See figure 8.8(a). Now let the emitter become large and move away from the receiver as shown in figure 8.8(b). The incident rays become parallel and the flux received at each point on the receiver is confined within a complete angular range of $2\theta_i$. These limiting set of rays are called the edge rays [1, 11, 12].

By placing a small plane mirror near the receiver's edge and inclining it at angle β some additional flux can be directed toward the receiver if the angle is chosen properly for a given incident ray. We define the incident ray angle with respect to the normal of the receiver plane, which is defined as the incident angle or acceptance angle, θ_i. If the mirror angle is not large enough then the ray will be reflected away from the receiver surface. This is illustrated in figure 8.9. The dotted line is the mirror surface normal.

At a particular angle β for the given incident ray angle, the reflected ray will become an edge ray for this receiver, meaning that a reflection from the edge of the mirror intersects the edge of the receiver surface. All rays incident on the mirror at the same angle will also be collected by the receiver. This provides an optimization condition for the reflector surface as shown in figure 8.10(a). Further collection can be done by adding a second mirror but inclining it at a different optimized angle for its edge ray. The length of the second mirror can also be altered. This is shown in figure 8.10(b).

Continuing to add small mirrors shows that the shape of the reflector is changing with steeper inclination angles the further away the reflector surface moves from the receiver. Eventually the angle $\beta = 90°$ implying that there is a finite length to the reflector. Instead of segmented mirrors, we want to find a solid surface that takes all incident parallel rays and directs them to a single point. From geometry, the construction that takes all rays incident along the axis and reflects them to a single focal point is a parabola. In our case, we want parallel rays at some angle to the receiver and not along the axis of the receiver.

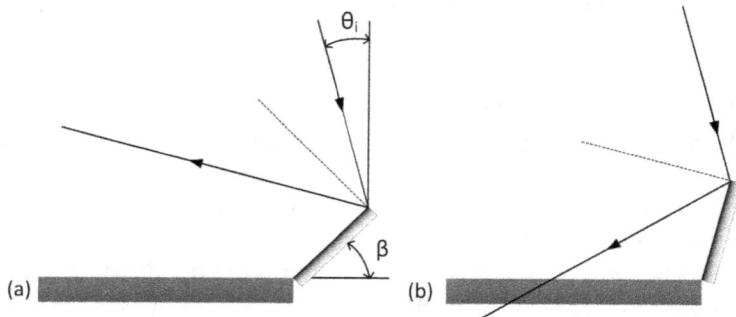

Figure 8.9. Mirror reflector placed at the edge of the receiver will direct more flux to the receiver if the mirror angle is sufficiently large. (a) Mirror angle is not large enough and (b) mirror angle is sufficient.

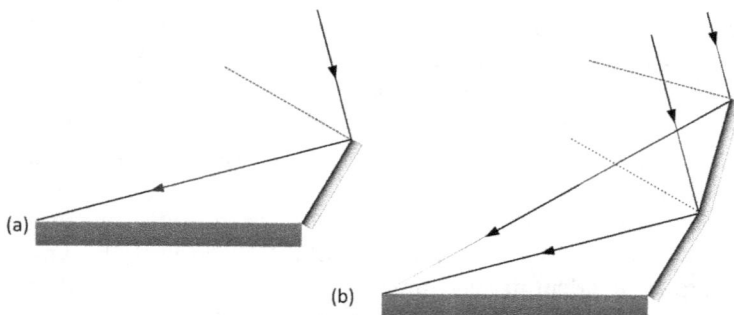

Figure 8.10. Optimized mirror angles for edge ray deflection with (a) one mirror, and (b) two mirrors.

Figure 8.11(a) shows a parabola with incident rays directed from the left to the right and being reflected toward the focal point of the parabola. If the axis of the parabola is rotated through an angle, θ_i, about the focal point, then we obtain the desired edge ray requirements. A single section of the rotated parabola is shown in figure 8.11(b).

We can now form a concentrator by truncating the rotated parabola section and placing the original parabola focal point at one end of the receiver. This defines the exit aperture size. The resulting shape is a form known as the compound parabolic concentrator (CPC) and is shown in figure 8.12 [3, 4, 11].

Figure 8.13 shows a section of a CPC with the center of the receiver placed at the origin of a coordinate system. Several points are labeled on the diagram that will be used in the discussion. The critical design parameters needed to specify a CPC are the acceptance (incident) angle, θ_i, and the diameter of the receiver, $2a'$. To find the remaining dimensions in terms of these parameters, we return to geometry and find that the radial distance, r, from the focal point of the parabola to a point on the surface of the parabolic section is given by,

$$r = \frac{2f}{1 - \cos(\theta + \theta_i)}$$

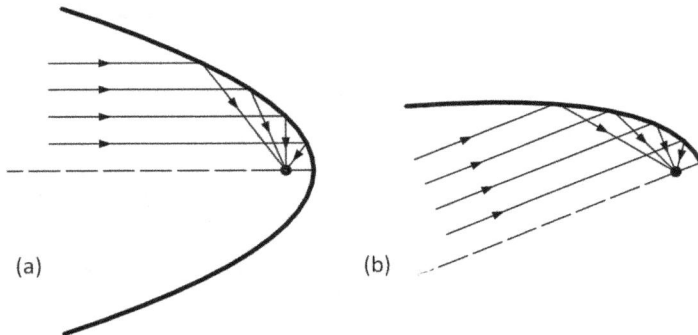

Figure 8.11. (a) Section of a parabola, and (b) a rotated parabolic section about the focal point.

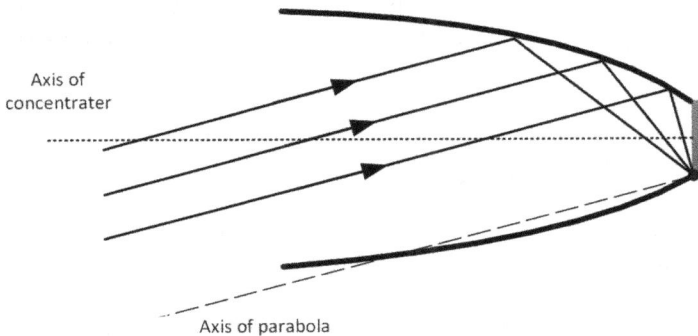

Figure 8.12. Formation of compound parabolic concentrator (CPC) from rotated and truncated parabolic sections.

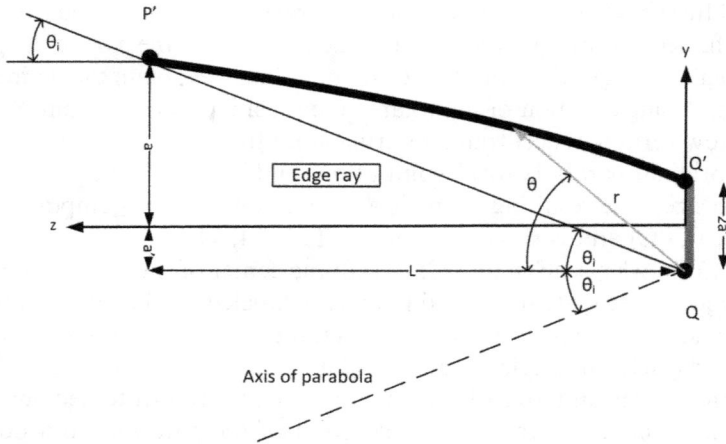

Figure 8.13. CPC section with dimensions identified to describe the characteristic shape in terms of design parameters, incident angle and receiver diameter.

where f is the focal length parameter associated with a parabola.

When $r = 2a' = QQ'$, $\theta = 90°$ providing a relation for the focal length parameter in terms of the input design variables as,

$$f = a'(1 + \sin \theta_i)$$

when $r(\theta = \theta_i) = QP'$, then,

$$QP' = \frac{a'(1 + \sin \theta_i)}{\sin^2 \theta_i}$$

From the edge ray triangle, we see that $a + a' = QP'\sin \theta_i$. Substituting the values from above results in,

$$\frac{a}{a'} = \frac{1}{\sin \theta_i}$$

indicating as expected that this is an étendue preserving system and allows the entrance aperture radius parameter, a, to be computed. In a similar manner the length L (sometimes called the height) can also be obtained and is,

$$L = a(1 + \sin \theta_i)\cot \theta_i$$

Therefore, given the receiver size and desired incident angle the CPC dimensions can be determined.

Instead of using a CPC as a concentrator, we replace the receiver with an emitter, and we have the basic form for a luminaire with an output beam radiating into a specified angular range. The output beam diameter increases as the distance from the luminaire to the detection plane increases. A plot of the normalized transmitted flux as a function of angle is shown in figure 8.14 along with a layout representation of the output beam.

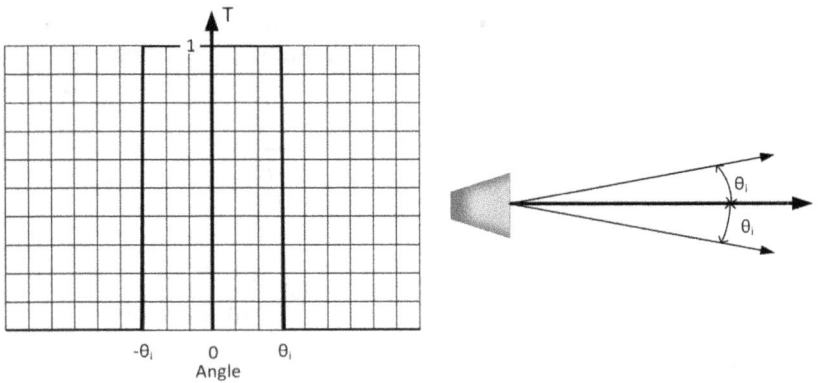

Figure 8.14. Transmittance of an ideal CPC luminaire as a function of angle about the axis of the luminaire along with a side view of the beam spread.

There are many other methods used to design reflectors that form beams with specific characteristics. As an example, the tailored edge ray method takes a known source intensity distribution and provides a reflector surface that produces a desired output intensity distribution [1]. Details of these methods are beyond the scope of this discussion.

8.3 Gaussian beams

Simple models of laser beam propagation treat the beam as a single ray. A more detailed model often used assumes that the beam has a particular size and divergence angle as a function of position along the beam and at that position the irradiance is a constant whose value is the laser power divided by the area of the beam at that point. This model is referred to as a 'top-hat' model. However, a laser beam has a particular characteristic irradiance distribution governed by the manner in which the laser is generated within the laser cavity. The form of the beam is a Gaussian irradiance distribution. In this section, the properties of a Gaussian beam are described, methods of measurement, and analysis of Gaussian beam propagation within systems.

8.3.1 Gaussian beam irradiance

Our ray optic analysis of laser cavity resonators showed that the output beam is actually comprised of low angle diverging rays. The wave analog of these rays are called transverse modes and are particular solutions to the Helmholtz equation from a wave optics field approach [13]. One approach for determining information and field solutions of the transverse modes builds on the wave optics complex field amplitudes that include quadratic phase factors. We assume a trial solution to the Helmholtz equation at any transverse plane of coordinates (x, y) as it propagates along the z-axis as,

$$u(x, y, z) = A \, e^{jp(z)} \exp\left\{ +\frac{j\pi}{\lambda q(z)}(x^2 + y^2) \right\}$$

where the functions $q(z)$ and $p(z)$ are to be determined.

Substituting the trial solution back into the Helmholtz equation we find that it satisfies the equation when,

$$\frac{dq}{dz} = 1 \quad \text{and} \quad \frac{dp}{dz} = \frac{j}{q(z)}$$

leading to,

$$q(z) = q(0) + z \quad \text{and} \quad p(z) = jln\left(\frac{q(0) + z}{q(0)}\right)$$

To find the resulting field we must substitute these expressions back into the trial solution and look for physically meaningful results related to initial conditions and beam confinement. One of the conditions for beam confinement is that $q(0)$, the initial condition on the q-parameter, must be purely imaginary [14]. We define a real distance z_0 such that $q(0) = -jz_0$. In addition, by convention, the following beam parameters are defined,

$$w(z) = w_0\sqrt{1 + \left(\frac{z}{z_o}\right)^2}$$

$$R(z) = z\left[1 + \left(\frac{z_0}{z}\right)^2\right]$$

$$\varsigma(z) = \tan^{-1}\left(\frac{z}{z_o}\right)$$

$$w_0 = \sqrt{\frac{\lambda z_0}{\pi}}$$

Resulting in the field function,

$$u(x, y, z) = A'\frac{w_0}{w(z)}\exp\left[-\frac{(x^2 + y^2)}{w^2(z)}\right]\exp\left\{j\frac{2\pi}{\lambda}\left[z + \frac{(x^2 + y^2)}{2R^2(z)} - \varsigma(z)\right]\right\}$$

The irradiance distribution is the absolute square of the field function or,

$$E_e(x, y, z) = E_{e0}\left(\frac{w_0}{w(z)}\right)^2\exp\left[-\frac{2(x^2 + y^2)}{w^2(z)}\right]$$

which is the form of a Gaussian function for all values of z. The $w(z)$ parameter is called the beam waist with $w_0 = w(0)$, the minimum beam waist when $z = 0$. The beam waist and all other beam parameters are defined in terms of the value z_0 which is the original boundary condition for a stable solution. The quantity z_0 is called the Rayleigh range. Knowing the laser wavelength and either z_0 or w_0 all Gaussian beam characteristics can be computed for any z-position.

8.3.2 Gaussian beam properties

Figure 8.15 shows three different normalized irradiance distribution representations of a Gaussian beam. These are instructive in understanding the physical features of the two-dimensional propagating beam. Because of the symmetry of a Gaussian beam, we only need to use cross-section plots in examining all the beam properties.

As mentioned, if we know the wavelength of the electromagnetic radiation and either the minimum beam radius or the Rayleigh range we can compute all parameters. To examine the properties of Gaussian beams it is useful to pick a common laser type and view graphs of each beam parameter defined earlier. In the graphs that follow we will use the example of a HeNe laser (wavelength 633 nm) with a Rayleigh range of 2.00 m resulting in a minimum beam waist radius of 0.635 mm. These are typical values for this type of beam. The irradiance distribution for this case at $z = 0$ is shown in figure 8.16. The dashed line indicates an irradiance

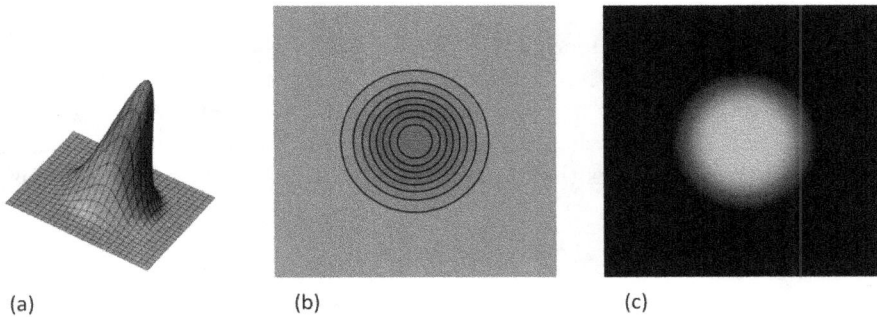

(a)   (b)   (c)

Figure 8.15. Gaussian beam representations. (a) Three-dimensional plot, (b) contour plot, and (c) density plot.

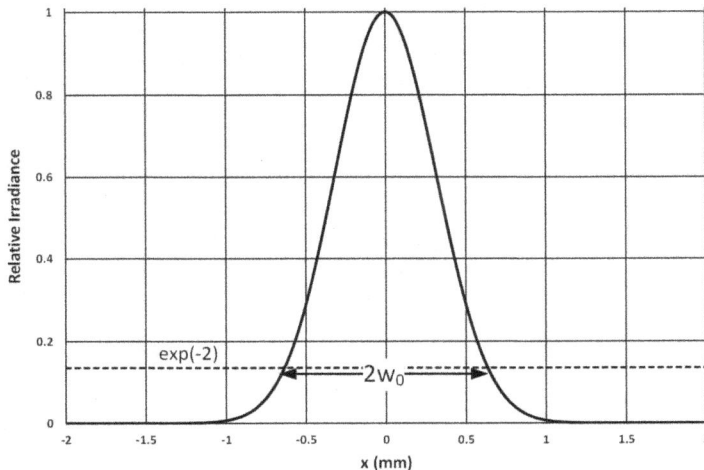

Figure 8.16. Gaussian beam irradiance distribution. The beam waist radius is 0.635 mm for a beam of wavelength 633 nm.

level of $\exp(-2)$, or where $w(z) = w_0$. This characteristic dimension along the transverse axis of the beam is called the spot size, for this example, $2w_0 = 1.27$ mm.

A plot of the beam waist as a function of the distance z along a beam is shown in figure 8.17. The origin of the z-axis is taken at a point where the waist is a minimum. This can be near the laser aperture or at some point away if optics are used to focus the beam. If we draw a line from the origin out to the value that the beam waist when $z \gg z_0$, the angle that this line makes with the z-axis defines another characteristic parameter called the beam divergence angle, θ_0. This is similar to the idea used before of a cone angle made by the radiation from sources. Under this approximation the half-angle beam divergence is,

$$\theta_0 = \frac{w(z)}{z} \approx \frac{w_0}{z_0} = \frac{\lambda}{\pi w_0}$$

So, an alternative beam parameter to use in specifying a Gaussian beam is the beam divergence angle. Any two parameters, θ_0, z_0, or w_0 specifies the beam and the other may be calculated. Note that this relation is consistent with what we found earlier in radiometric throughput or étendue. If we try to make the beam size smaller for a higher degree of focus, then the divergence angle must increase accordingly.

Figure 8.17 also indicates a range about the minimum beam waist of twice the Rayleigh range. This occurs when the beam waist value becomes, $w(z = z_0) = \sqrt{2}\,w_0$. This distance is referred to as the depth of focus of a Gaussian beam which is,

$$\text{DOF} = 2z_0 = \frac{2\pi w_0^2}{\lambda}$$

The radius of curvature identifies planes of constant phase at distances about the minimum beam waist for a Gaussian beam. By definition, the radius of curvature at

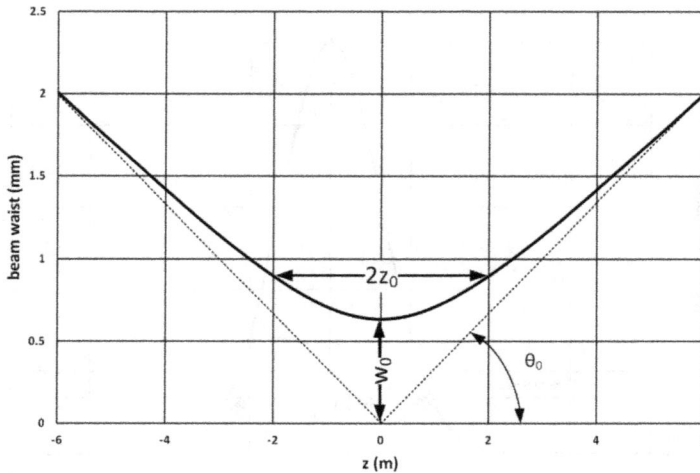

Figure 8.17. Gaussian beam waist as a function z about a minimum waist position. The beam waist radius $w_0 = 0.635$ mm with a beam of wavelength 633 nm. The dashed line is an asymptotic line to the beam waist curve from the origin. This line defines the divergence angle.

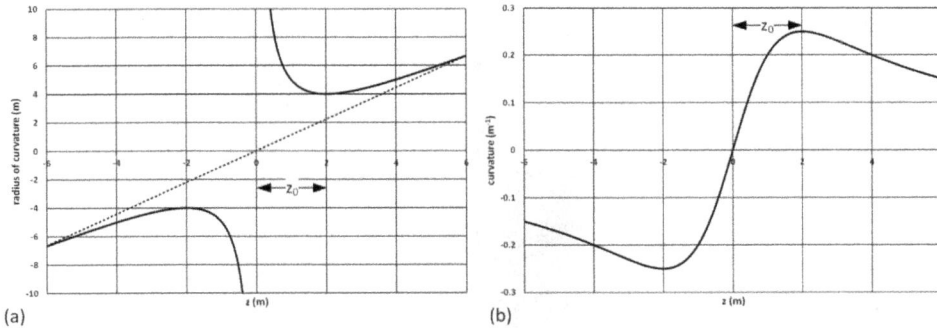

(a)   (b)

Figure 8.18. Gaussian beam radius of curvature of the wavefronts as a function of z from the minimum beam waist.

the beam waist is infinity. The radius of curvature reaches an absolute minimum at the Rayleigh range. This function is shown in figure 8.18. Lens designers often specify optical surfaces using curvature or an inverse radius of curvature of a surface. The curvature of Gaussian beam wavefronts shows that the wavefronts with the largest curvature are at the Rayleigh range distance.

Since we know the irradiance distribution that would fall on a detection plane, we can use radiometry to compute the total optical power delivered by a Gaussian beam. From the definition of irradiance, we have,

$$d\Phi_e = E_e(x, y, z)dA = E_{e0}\left(\frac{w_0}{w(z)}\right)^2 \exp\left[-\frac{2(x^2 + y^2)}{w^2(z)}\right]$$

where dA is the differential area of the detection region. By integrating over the complete detection plane at an arbitrary distance z we find the total power,

$$\Phi_{\text{etot}} = E_{e0}\left(\frac{w_0}{w(z)}\right)^2 \int\int_{-\infty}^{+\infty} \exp\left[-\frac{2(x^2 + y^2)}{w^2(z)}\right]dxdy$$

The result of this calculation is simply,

$$\Phi_{\text{etot}} = \frac{\pi}{2}E_{e0}\, w_0^2$$

which does not depend on z. This confirms simple models used with laser beams since the effective beam spot area is πw_0^2, the power is simply the product of an effective irradiance and the area of the laser spot. It is convenient to rewrite the complete irradiance distribution function in terms of the total power since the total power can be measured experimentally. Thus, the irradiance function can be completely specified, as,

$$E_e(x, y, z) = \frac{2\Phi_{\text{etot}}}{\pi w^2(z)} \exp\left[-\frac{2(x^2 + y^2)}{w^2(z)}\right]$$

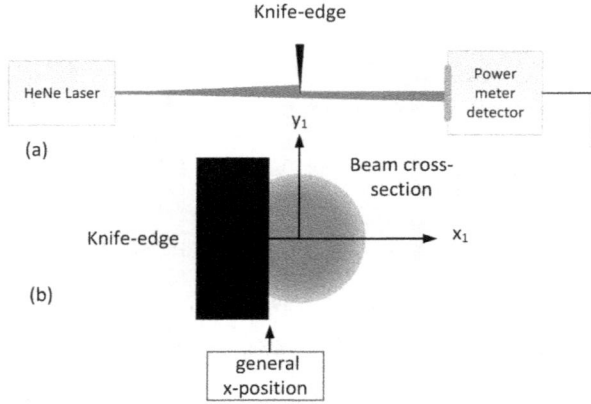

Figure 8.19. Experimental determination of the beam waist value using the scanning knife-edge method measuring optical power.

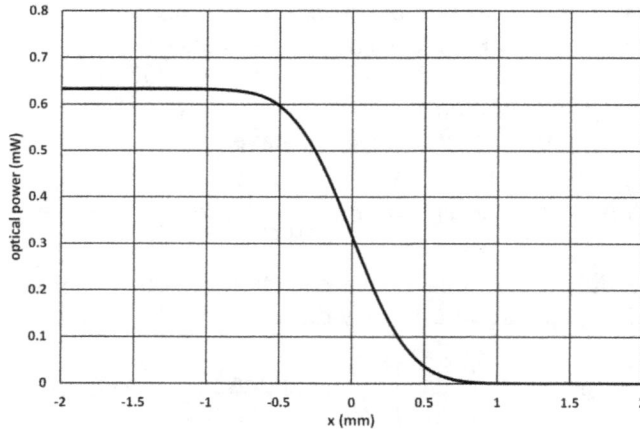

Figure 8.20. Optical power as a function of the x-position of the knife-edge.

Experimentally, a simple power measurement can be used to obtain the value of the beam waist radius [13]. This is done by scanning a knife-edge through the beam at a desired z-position in a particular (x_1, y_1) plane. Figure 8.19 shows this concept. The position of the knife-edge is specified as a general x variable.

Integrating the irradiance function of the beam along the x_1-axis from the general x to infinity as well as integrating the y_1-coordinate over all space as before,

$$\Phi_e(x) = \frac{2\Phi_{etot}}{\pi w^2(z)} \int\limits_{x}^{\infty} \int\limits_{-\infty}^{\infty} \exp\left[-\frac{2(x^2 + y^2)}{w^2(z)}\right] dx dy = \frac{\Phi_{etot}}{2}\left[1 - erf\left(\frac{\sqrt{2}x}{w(z)}\right)\right]$$

Figure 8.20 shows the power that would be predicted under the same parameters as in figure 8.18 where the total power in the beam is 0.633 mW. This function can be fit to find the beam waist $w(z)$ at this specific z-position. Another alternative is to compute the numerical derivative and find the beam waist positions from the resulting Gaussian curve representing a cut through the x_1-axis. This technique also allows for the center location of the beam to be determined at the half-power point and can be useful in alignment. Other standard methods are also available for beam waist measurements and beam quality determinations [15].

The Gaussian beams discussed so far have all had symmetric cross-sections. However, some lasers, specifically edge-emitting laser diodes, may have two different waists along the orthogonal axis directions. These types of beams are called elliptical beams. This only means that the axes must be treated separately. Optical techniques and components are available to 'circularize' elliptical beams [16].

Deviations of a measured beam waist parameter to the ideal Gaussian beam are of course common. A factor called M^2 has been developed in an attempt to provide a figure of merit for the quality of a propagating beam. This is essentially a ratio of the product of the far-field measured beam waist and the measured divergence angle to that same product for an ideal Gaussian beam. From this definition we have,

$$M^2 = \frac{w_M \theta_M}{w_0 \theta_0} = \frac{\pi}{\lambda} w_M \theta_M$$

where w_M and θ_M are the measured waist and divergence angle, respectively. It is easy to show that the Rayleigh range parameter of the beam under test is,

$$z_M = \frac{\pi w_0^2}{M^2 \lambda}$$

The closer the value M^2 is to unity the better the beam quality since the value of $M^2 = 1$ for an ideal Gaussian beam. Figure 8.21 shows a graphical illustration of the beam waist as a function of propagation distance for two different beams, the ideal Gaussian beam ($M^2 = 1$) and another beam with $M^2 > 1$.

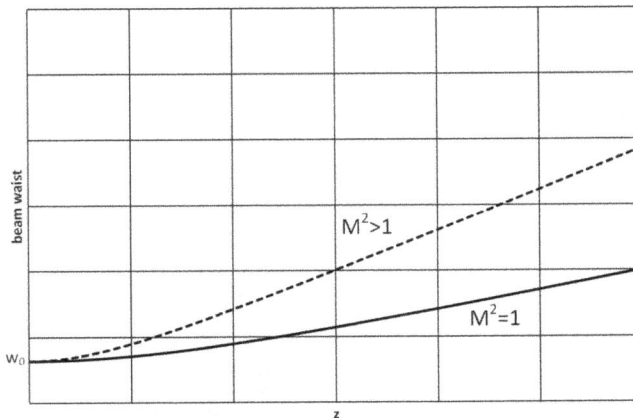

Figure 8.21. Beam waist radius as a function of z for an ideal Gaussian beam (solid curve) and a beam with $M^2 > 1$ (dashed curve).

8.3.3 Gaussian beam propagation

In arriving at the solution for the field function of a Gaussian beam we defined the complex beam parameter $q(z)$ function and also recognized the similarity of the field to a spherical wave. To obtain the final field, the following relation must be used to find the solution in terms of Gaussian beam parameters, $w(z)$ and $R(z)$. This definition is,

$$\frac{1}{q(z)} = \frac{1}{R(z)} + j\frac{\lambda}{\pi w^2(z)}$$

So, as the wave propagates from one plane to another in a system then the value of q needs to be found at each successive plane.

As a wave propagates from one plane in a system to another plane the wave diffracts. We can find the diffracted wave knowing the incident wave field using Fresnel diffraction, as was done in earlier chapters. Take an incident Gaussian wave function in Plane 1 with coordinates (x_1, y_1) using the general trial solution as,

$$u_1(x_1, y_1, z_1) = A\, e^{jp(z_1)} \exp\left\{+\frac{j\pi}{\lambda q(z_1)}(x_1^2 + y_1^2)\right\}$$

Using Fresnel diffraction, we can obtain the field in the (x_2, y_2) plane. This is shown in figure 8.22 where the characteristic spot size of the beam is larger as it propagates.

The field in Plane 2 is then,

$$u_2(x_2, y_2) = \frac{A\, e^{jp(z_1)}}{j\lambda d} e^{j\frac{2\pi}{\lambda}d} Q(x_2, y_2; d)\mathcal{F}\left\{\exp\left\{+\frac{j\pi}{\lambda q_1}(x_1^2 + y_1^2)\right\}Q(x_1, y_1; d)\right\}$$

where for simplicity, $q(z_1) = q_1$ and the quadratic phase factor notation has been employed. Using the definition of the quadratic phase factor we can combine the terms in the argument of the Fourier transform to obtain,

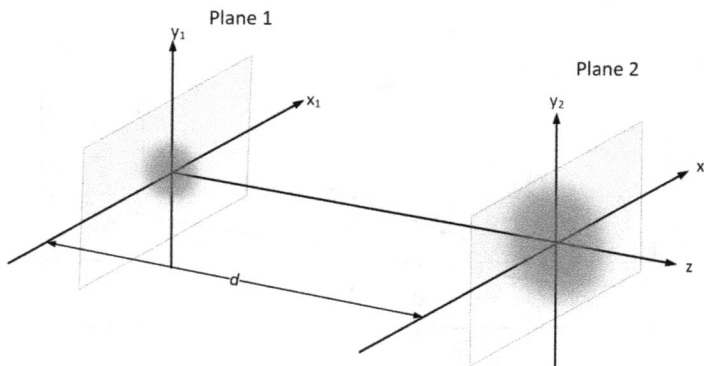

Figure 8.22. A Gaussian beam propagating from Plane 1 to Plane 2 over a distance $z = z_2 - z_1 = d$.

$$u_2(x_2, y_2) = \frac{A \, e^{jp(z_1)}}{j\lambda d} e^{j\frac{2\pi}{\lambda}d} Q(x_2, y_2; d) \mathcal{F}\left\{ \exp\left\{ +\frac{j\pi}{\lambda}(x_1^2 + y_1^2)\left[\frac{1}{q_1} + \frac{1}{d} \right] \right\} \right\}$$

The Fourier transform of a Gaussian function results in another Gaussian function. After performing this calculation, we see that the new parameter q_2 for the wave in Plane 2 is simply,

$$q_2 = q_1 + d$$

We could have arrived at this result more easily just using the definitions of q. However, the diffraction formalism shows us that the q-parameter in Gaussian beam propagation is an 'effective complex distance' and is analogous to a distance in general wave propagation.

Diffraction formalism also gives us a means of determining how a Gaussian beam can be focused by a thin lens just as was covered previously. Suppose that a thin lens is placed in Plane 1. The field leaving the plane will be modified by the phase transform of the lens of focal length f. Remember this is just another multiplicative quadratic phase factor, $Q(x_1, y_1, -f)$, that must be included in the Fourier transform argument. Using Fresnel diffraction again for propagation to the back focal plane of the lens results in the condition that,

$$\frac{1}{q_2} = \frac{1}{q_1} - \frac{1}{f}$$

For a paraxial optical system, we used ray matrices to allow us to propagate a ray at a given height at an angle with respect to the z-axis of the system. It can be shown that if we make the following definition,

$$q_2 = \frac{Aq_1 + B}{Cq_1 + D}$$

Then both the Gaussian beam translation and focusing conditions we obtained using diffraction result from the ABCD ray matrices of chapter 2. With this definition, and the ray translation matrix, T, we find,

$$T_{21} = \begin{pmatrix} 1 & z_2 - z_1 \\ 0 & 1 \end{pmatrix} = \begin{pmatrix} 1 & d \\ 0 & 1 \end{pmatrix} \quad \text{leads to} \quad q_2 = q_1 + d$$

Also, using a thin lens matrix,

$$M_{\text{thin}} = \begin{pmatrix} 1 & 0 \\ -\frac{1}{f} & 1 \end{pmatrix} \quad \text{leads to} \quad \frac{1}{q_2} = \frac{1}{q_1} - \frac{1}{f}$$

While these are specific cases, this relation can be shown to predict the beam parameters for any given system matrix [13, 14]. Given an incident Gaussian beam specified by the q_1 parameter, q_2 can be computed from known values of A, B, C, and D of the system matrix. Therefore, Gaussian beam propagation can be analyzed through any general paraxial optical system by computing the ray transfer system matrix.

8.4 Beam scanning

A general scanning system is sometimes classified as either passive or active [17]. A passive system is similar to a staring focal plane array system where the flux is collected from an object and directed to a plane from numerous points in a scene, as discussed in an earlier chapter. The scanning is done internal to the detector array. Passive scanning is also referred to as scanning for remote sensing based on its common application area [18]. An active scanning system illuminates a target object. Active scanning systems are most often referred to as spot scanners or laser scanners due to the most common source used by these systems.

This section is restricted to discussing active scanning of beams using mechanical methods. Acousto-optic and electro-optic modulators (EOMs) can also be used as scanners and will be covered in the next section on optical modulation. There are many different scanner products and devices such as piezoelectric, holographic, and flexure scanners. All types cannot be covered here, however, the basic concepts will be provided as background [18].

8.4.1 Scanning resolution

The ultimate resolution of a spot scanning system is the minimum size of the spot that can be achieved at all points on the target. For an infinitesimal point scanned along an x-axis with speed v in time t, the position x would be given by $x = vt$. If the spot originates from a focusing optical system, the size of any spot would be related to the point spread function (PSF) of the system. For a laser beam, the size of the beam waist at the scan position is usually used. The assumption used in specifying these systems is that the minimum scan resolution is the number of spots of elemental width falling within the total scan distance. This is illustrated in figure 8.23. Thus, for translational scanning, the number of spots N_T, of width, w, in scan distance, S, is,

$$N_T = \frac{S}{w}$$

Some systems scan an angular field. The resolution of an angular scan is computed similarly to the translational scan as the ratio of the total angle, Θ, to the angular width of the beam, $\Delta\theta$, or,

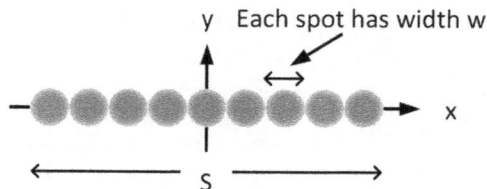

Figure 8.23. Translational scanning of beam width w over a total scan distance S.

$$N_A = \frac{\Theta}{\Delta\theta}$$

Depending on whether or not the scan beam is limited by diffraction due to the system aperture or if it is a Gaussian beam, a general relation for the angular beam width is given by,

$$\Delta\theta = a\frac{\lambda}{D}$$

where a is a parameter called the aperture shape factor [7, 18]. The value of D is the aperture diameter/beam diameter in the direction of the scan motion. Table 8.1 provides a list of various aperture shape factors for some simple uniformly illuminated, untruncated apertures. The input beam to the scanner may be truncated by another aperture and in this case, the shape factor would take on a different value.

8.4.2 Objective, pre-objective, and post-objective scanning

Objective scanning is the process where the lens, or entire system is moved either by translation or rotation. It is probably the least used type of scanner geometry because of the inertia required and motion errors introduced in moving a lens or system. However, translating or rotating a holographic element does not have these drawbacks [19].

Table 8.1. Aperture shape factors of uniformly illuminated and unobstructed scanner diameters.

Shape	Aperture diameter, D	Aperture shape factor, a
Rectangle	←—D—→	1.0
Round or elliptical	←—D—→	1.25
Keystone	←—D—→	1.5
Triangle	←—D—→	1.7
Gaussian beam at e^{-2} beam waist	←—D—→	1.27

Values after [18].

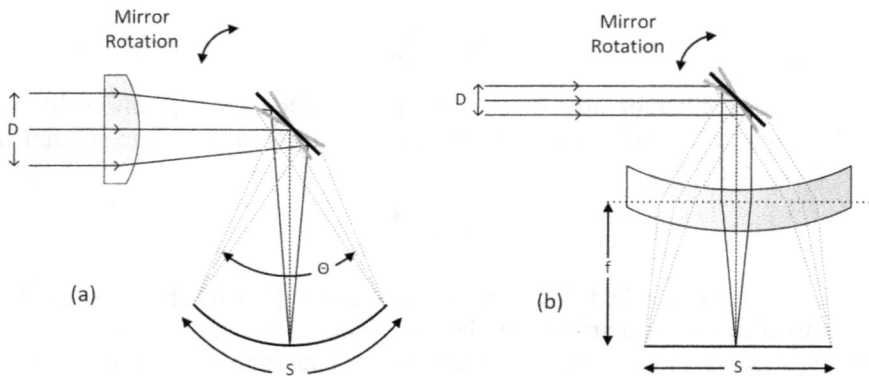

Figure 8.24. Example layouts of (a) post-objective and (b) pre-objective scanning systems.

Figure 8.24 shows the optical layouts for both post-objective and pre-objective scanners. The mirror in each case rotates about an axis in the plane of the mirror and perpendicular to the mirror normal. For high resolution requirements, well corrected lenses or more complex lens designs are necessary. The beam scans through an arc so the focal plane is curved. The curved focal plane is certainly a disadvantage, but post-objective scanning will produce high resolution over large scan angles. If the scan mirror is tilted at 45° and rotated about the optical axis, then the scan plane is perpendicular to the axis and the arc will subtend a full 360° angle. This geometry is known as internal drum scanning and has several applications [19].

Pre-objective scanning produces an incident angular field of parallel beams that are collected by a lens and focused on different positions in the back focal plane of the lens. The lens required for this system is called an F–θ lens. Its purpose is to provide a linear relationship between the spot position and the angle, which is essentially the product of the effective focal length and the angle. This lens choice is needed since a constant mirror rotation rate produces a reflected beam whose angle changes linearly in time.

8.4.3 Rotating and vibrating reflector scanning

Galvanometers were invented as a means of detecting and measuring small currents. A current carrying coil of wire sets up an induced magnetic field. If the coil is placed in a region of a fixed magnetic field, a torque on the coil proportional to the current develops due to the interaction between the two fields. The torque rotates the coil, deflecting a needle or mirror attached to the coil providing a measure of the current.

Instead of measuring a current, the galvanometer movement may be used as a scanner by supplying an appropriate current signal that deflects a mirror. Another configuration used for the galvanometric scanner is similar to a motor where the mirror is placed with its surface in the axis of rotation of a shaft. These scanners are used extensively in laser beam steering applications. Their advantages are low cost and high-speed scanning operation, but the tradeoff is limited scan precision. Introducing a servo control loop using signals from position transducers to the electronics driver for the scanner provides additional control for better positioning. They are also good choices for a two-dimensional scanning system where two mirrors, one along the x-axis direction and the other along the y-axis direction, deflects a laser beam to any point within a plane. Figure 8.25 shows this basic concept. Galvanometer scanners are sometimes referred to as vector positioners because of the beam pointing approach.

The requirement of uniform beam resolution and high-speed scanning velocities drive the system specifications to high-speed rotating polygon mirror elements. This is especially true for scanners that raster across a target. Polygon scanning system designs can be either pre-objective or post-objective configurations. The number of facets on the polygon and the raster speed is a tradeoff consideration for any design.

As an example, for determining the size and type of polygon required for a scanner system, suppose that a pre-objective system design is chosen as shown in figure 8.26. In this case, we will assume that one facet at a time scans the image plane which is called the underfilled condition. If the F–θ lens has been specified then its aperture sets the total scan angle, Θ. If not, then some assumptions must be made as to the number of resolvable points and angular beam width, $\Delta\theta$. This imposes a value for beam size of,

$$D = N_A \frac{a\lambda}{\Theta}$$

The incident beam is directed toward the polygon, falls on a facet, and reflects toward the screen. The angle between the incident beam and the line connecting the center of the output plane and the center of a facet is defined as α. This angle is commonly called the beam feed angle.

For an incident circular beam, the spot formed on the polygon surface will be elliptical. A side view of a facet and the incident beam footprint is shown in

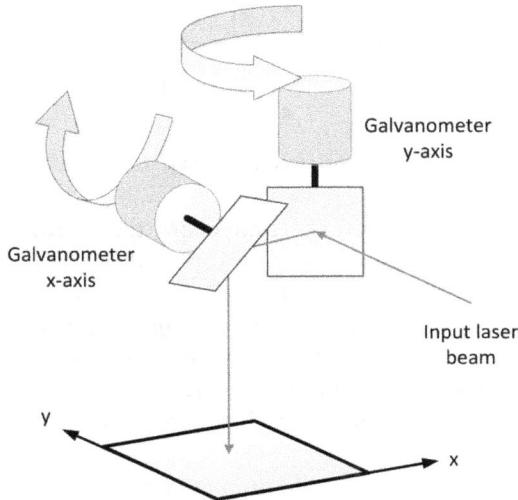

Figure 8.25. Two-dimensional scanning field of an x–y galvanometer scanner.

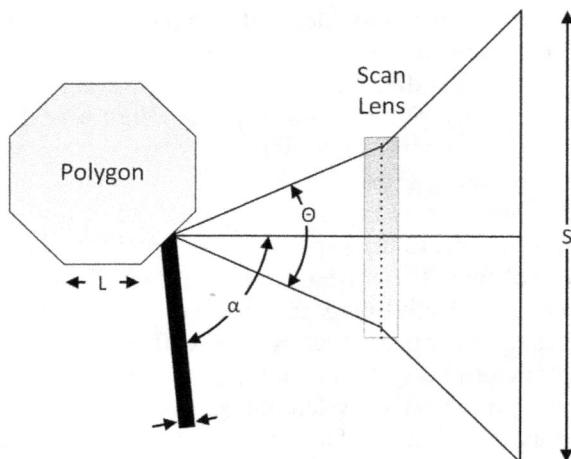

Figure 8.26. Basic layout of a pre-objective polygon scan system.

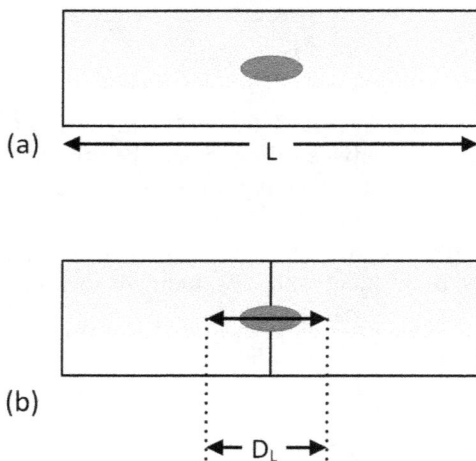

Figure 8.27. Beams intersecting a polygon facet (a) in the active region and (b) near a facet edge.

figure 8.27(a). As the polygon rotates and nears the edges of a facet the beam becomes truncated as an edge enters the beam spot ellipse. When the beam begins to be truncated a dead time begins and the active scan region stops until the beam moves past the edge and into another facet. A length D_L is used to account for the size of the elliptical beam as it is truncated by the corner and is illustrated in figure 8.27(b). Sometimes an additional safety factor is also included to ensure that sufficient dead time is accommodated which is why D_L is shown larger than the spot size [17, 18].

If the beam is not vignetted, then the angular range on the target represents the active scan time for one facet. The ratio of the active scan time to the total time it takes for one facet of length L to rotate through a point on the beam is defined as the scan duty cycle. In terms of distances, the duty cycle is,

$$\eta_d = 1 - \frac{D_L}{L} = \frac{\Theta}{4\pi} n_F$$

and can also be written in terms of the scan and the number of facets, n_F. During a design layout, if the value for the number of number of facets is not an integer then adjustments or tradeoffs must be made in other input parameters.

The required facet length can be calculated from the relation above as,

$$L = \frac{D_L}{1 - \eta_d}$$

From polygon geometry the largest diameter of the polygon in terms of the number of sides of length L, is,

$$D_P = \frac{L}{\sin\left(\dfrac{\pi}{n_F}\right)}$$

There are several issues where errors can arise in a scanner of this type. If the incident beam is not aligned perpendicular to the polygon facet plane, then the scanned line will exhibit bow. Since the polygon is rotating about its center and not the facet face then there is an inherent displacement of the beam placing requirements on the lens, especially for large apertures. Other issues such as pyramidal errors in the facets and axis wobble must also be included as errors [19].

8.5 Beam modulation

For an optical beam to carry information, one of the physical parameters of the electromagnetic wave must change in time or be modulated. Possible parameters include amplitude, phase, frequency, and direction. The carrier wave is the high frequency electromagnetic wave and is generally at a much greater frequency than the frequency of the signal. The message/signal input function changes one of the carrier parameters through some mechanism that results in a changing optical flux. Scanners are good examples of directional modulators. In earlier chapters, we discussed several cases of the time varying incident flux on a detector and the resulting analog detector signal output as either a current or voltage. A time varying incident flux originates from a source that is modulated in amplitude (sometimes called intensity) or phase.

A general form for a time dependent amplitude modulated optical flux is,

$$\Phi_e(t) = \frac{\Phi_{e0}}{2}[1 + m\, s(t)]$$

where Φ_{e0} is the maximum radiant flux, $s(t)$ is the time varying signal, and m is a constant called the modulation index or modulation depth. Since all flux values must be positive, restrictions on the parameters are,

$$|s(t)| \leqslant 1 \quad \text{and} \quad 0 \leqslant m \leqslant 1$$

Another alternative modulation scheme is to use a sub-band carrier (an oscillation at another frequency, f_S) with the signal being incorporated into the phase of the oscillation, or,

$$\Phi_e(t) = \frac{\Phi_{e0}}{2}\left\{1 + \cos\left[2\pi f_S t + ms(t)\right]\right\}$$

Figure 8.28 shows the normalized optical flux output as a function of time from a source with two different signals, using amplitude and phase modulation.

Optical modulation can be accomplished by directly controlling a device (direct or internal modulation) or using a separate device to control a transmitted beam (external modulation). Input signals may be analog or digital, but a detector/receiver will be responding to an analog flux. Additional processing, or demodulating, and digitizing the output will be required. For external modulation, the source is operated in a continuous wave mode and the signal is introduced by an external modulator using some modulation mechanism. In this section, we review different methods of modulating a source to generate the optical signal.

8.5.1 Directly modulating a source

Devices whose light output as a function of an input signal (current or voltage) can be directly modulated. LEDs and laser diodes are the predominant devices used for direct modulation since their drive current is directly related to the output radiant flux. The characteristic curves for both an LED and laser diode are shown in figure 8.29. For each device, the electrical input must be biased about a particular bias current I_B. The upper case I_B is used since this is a direct current value.

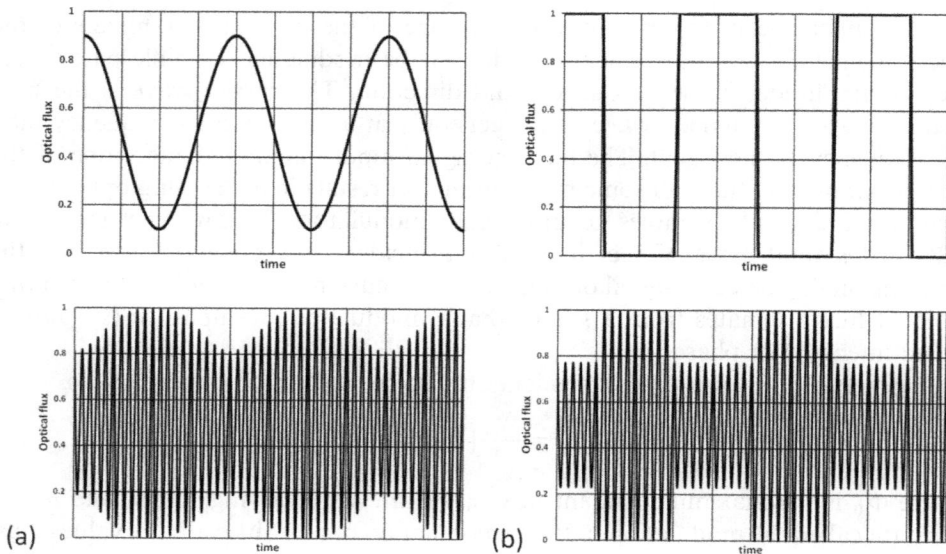

Figure 8.28. Normalized optical flux from a source with sinusoidal signal and modulation index of 0.8 in (a) amplitude and phase modulation, and (b) rectangular pulses in amplitude and phase modulation.

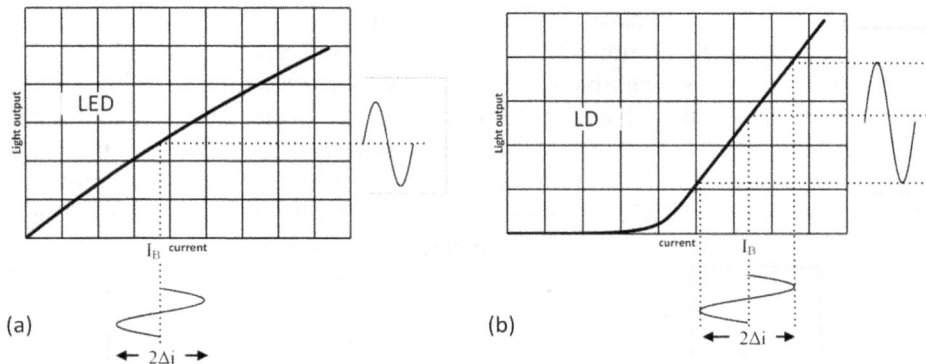

Figure 8.29. Direct modulation of (a) LED and (b) laser diode (LD) with a signal about a bias current on each of the light output versus drive current characteristic curves. The amplitude of the input current signal is Δi.

The amplitude of the input current provides radiation output that follows the transfer curve as shown by the inset oscillations in the figure.

Output from the LED is approximately linear with input current, at least within a limited current range. All LEDs suffer from a nonlinearity at high current mostly attributed to thermal effects due to self-heating within the device. Over driving the LED will distort the signal.

To modulate a laser diode the bias current must be well above the laser threshold current. If the bias current is too low the lower portion of the input current signal will be clipped causing a distorted output optical signal. The depth of modulation for the output optical signal is determined by,

$$m = \frac{\Delta i}{I_B - I_{th}}$$

where $I_{th} = 0$ for the LED. For stability, another consideration is to limit modulation depth so that the laser is never completely off.

8.5.2 Beam choppers

Beam choppers and other mechanical modulators simply transmit or block the beam from a continuous wave source, usually a laser beam. As long as the beam is passed then the modulator does not interfere or distort the character of the beam. These modulators can be used with most any wavelength source and are relatively low cost.

Beam choppers using rotating segmented wheels/blades are the most common types available. The number of segments (or holes), the size of the wheel, and speed of rotation set the modulation frequency. Many blades are coated to reduce scattering when the edge of the blade contacts the beam. The coating may also mitigate damage by high power laser beams. Many beam chopper motors are driven by a well-controlled current and incorporating feedback loops for precise rotation frequency control.

Typical application of a beam chopper is to modulate a signal so that is can be amplified within a fixed frequency bandwidth and increase the signal/noise ratio of the detected light flux. Lock-in amplification can also be employed to measure signals in the presence of noise [20]. Providing a reference signal at the beam modulation frequency is available on most devices for use with a lock-in amplifier.

There are some distinct disadvantages of beam choppers. One is that they are a fixed frequency device once a set of parameters have been chosen. They can be large and bulky and introduce vibrations within an optical system. The maximum chopping or modulation frequency is relatively low, usually less than a few kilohertz. Some diffraction can also occur about the edges of the blades.

The effect of the rotating segments on a Gaussian beam is illustrated in figure 8.30. The size of the beam and the transmitted waveform must be considered in designing a system employing a chopper modulator. The figure shows a plot of the output power as a function of time as one segment of a chopper wheel passes in front of a Gaussian beam for three different beam radii. The total beam power is held constant for each of the beams shown in this graph. In all cases there is some truncation of the beam as the edge of the blade crosses the beam. The larger the beam, the more the transmitted signal flux deviates from an ideal square-wave.

8.5.3 Electro-optic modulators

When a voltage is placed across certain materials an electrically induced birefringence is created. The amount of birefringence is dependent on material constants and the value of the applied voltage. This is known as the electro-optic effect. There are a variety of electro-optic materials available with two common crystals being lithium niobate ($LiNbO_3$) and KDP (KH_2PO_4) [14].

In general, the refractive index as a function of the applied electric field can be written in a series expansion since the induced changes are rather small [21]. That is,

$$n(E) \cong n + a_1E + \frac{1}{2}a_2E^2 + \dots$$

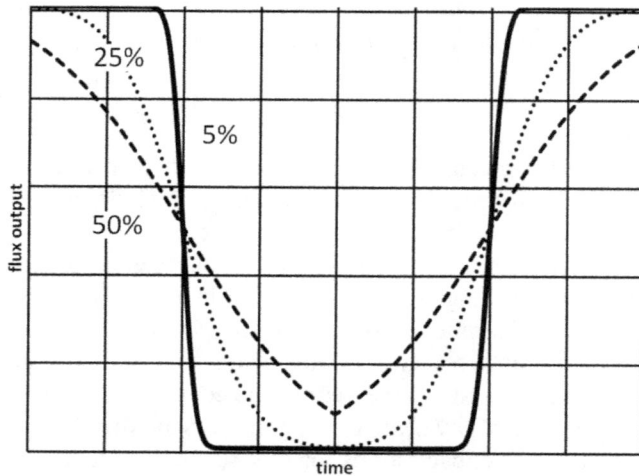

Figure 8.30. Transmitted flux from a chopped Gaussian beam. The curves shown represent a beam radius that is 5%, 25%, and 50% of the width of a segment of the chopper wheel.

where a_1 and a_2 are constants. Depending on the material, the induced birefringence will be proportional to the electric field (Pockels effect) or vary quadratically with the electric field (Kerr Effect) [21]. Since an applied voltage induces an index change the result is the creation of a voltage-controlled phase retarder which can be configured as an optical modulator.

There are numerous configurations that exploit the electro-optic effect to produce modulators and scanners [22]. Both bulk EOMs, using a single crystal, and integrated-optic waveguide modulators are available from manufacturers. The waveguide devices are used extensively as modulators in fiber optic communication systems. We will restrict our discussion to phase and amplitude external modulators to understand the basic operation of all types of electro-optic modulation devices.

Figure 8.31 is a diagram showing an input beam incident on a modulator crystal and polarized in the plane coincident with the crystal surface and perpendicular to the longitudinal direction. The longitudinally applied electric field if setup by applying a voltage on transparent electrodes on each end of the crystal. Take the amplitude of the incident electric field of the optical carrier to be E_i at the wave angular frequency ω [22].

When the voltage V is applied along the length L (longitudinal direction) a phase shift, $\Delta\varphi$, is induced on the output wave from the crystal $E_o(t)$, due to the linear electro-optic effect, or,

$$E_o(t) = E_i \cos(\omega t - \varphi)$$

where φ is the total phase shift which is the sum of the phase delay from the constant index value and the induced index. That is,

$$\varphi = \frac{2\pi}{\lambda}nL + \Delta\varphi$$

Assuming that the applied electric field is uniform in the longitudinal direction, the magnitude of the applied electric field is just $E = V/L$. The induced phase change is then given by,

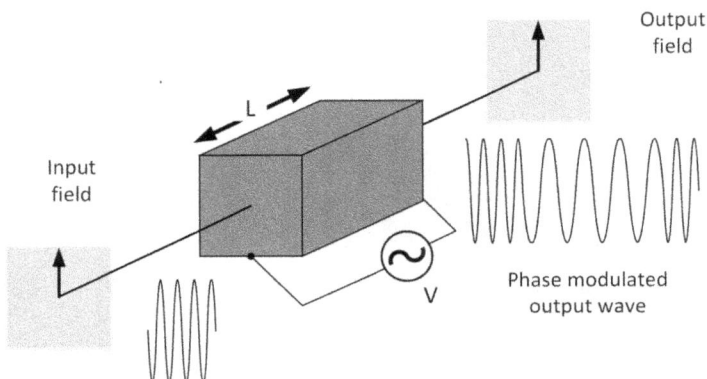

Figure 8.31. Phase modulation of an incident wave using the electro-optic effect.

$$\Delta\varphi = \frac{2\pi}{\lambda}L\Delta n = \frac{\pi}{\lambda}n^3 rLE = \frac{\pi}{\lambda}n^3 r V$$

where r is the electro-optic coefficient of the crystal. This shows that the induced phase shift is independent of L and linear in V. So, the phase of the output carrier wave can be altered with a changing or alternating applied voltage generating a phase modulated field. To generate a specific phase shift $\Delta\varphi = \pi$ for the longitudinal phase modulator, a voltage V_π is defined such that,

$$V_\pi \equiv \frac{\lambda}{r\,n^3}$$

Induced phase changes alter the polarization state of the propagating electric field. Amplitude modulation of an incident flux can be done by placing external polarizers before and after the electro-optic crystal. See figure 8.32.

To analyze the polarization at different planes as the wave propagates, we will use Jones vectors and matrices. See the appendix for a listing of these vectors and matrices. The effect of the electro-optic modulator can be modeled as a general variable phase retarder [7]. The output polarized field transmitted by the system is obtained by calculating the system matrix for a given input field. That is,

$$\vec{E}_t = M_{sys}\vec{E}_i$$

The system matrix in this case is,

$$M_{sys} = M_{-45}M_\varphi$$

Take the incident field amplitude as A. The vector just after the first polarizer is,

$\vec{E}_i = \frac{A}{\sqrt{2}}\begin{bmatrix}1\\1\end{bmatrix}$. Using the appropriate Jones matrices for these elements,

$$\vec{E}_t = \frac{Ae^{je_x}}{2\sqrt{2}}\begin{bmatrix}1 & -1\\-1 & 1\end{bmatrix}\begin{bmatrix}1 & 0\\0 & e^{-j\varphi}\end{bmatrix}\begin{bmatrix}1\\1\end{bmatrix}$$

Multiplying these matrices to compute the output polarization vector, we find that it is a polarization state oriented at $-45°$ as expected but with a new amplitude dependent on the induced phase shift or,

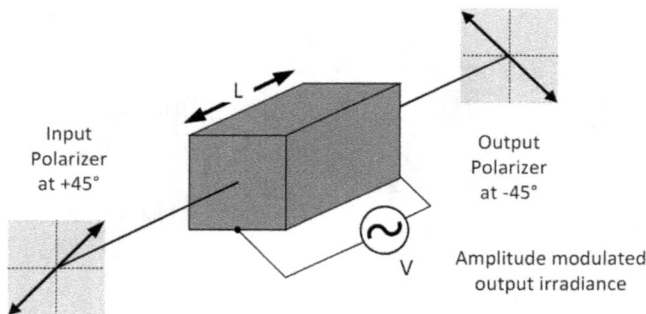

Figure 8.32. Electro-optic modulator used in an amplitude modulation configuration.

$$\vec{E}_t = \frac{Ae^{j\varepsilon_x}}{2\sqrt{2}}(1 - e^{-j\varphi})\begin{bmatrix} 1 \\ -1 \end{bmatrix}$$

The irradiance falling on a detector placed behind the output polarizer is obtained by computing the absolute square of this field or,

$$E_e = \frac{A^2}{4}(1 - \cos\varphi) = \frac{A^2}{2}\sin^2\left(\frac{\varphi}{2}\right)$$

As before, the phase is proportional to the applied voltage on the modulator and the voltage value when the phase becomes π is defined as V_π. Thus, the detected irradiance of an amplitude modulated signal can be written in a general form as,

$$E_e = E_{e0}\sin^2\left(\frac{\pi V}{2V_\pi}\right)$$

where E_{e0} is maximum transmitted irradiance when $V = V_\pi$. Figure 8.33 shows a plot of this characteristic curve for an amplitude modulated EOM where the transmittance is the ratio of the irradiance to the maximum irradiance.

Note that this is not a linear transfer curve. However, when the input voltage is biased about the midpoint voltage and is a small oscillation, the output irradiance approximately follows the input. For larger signals, the output signal waveform will be distorted [14].

8.5.4 Magneto-optic modulators

When materials are placed in magnetic fields the optical properties of the materials can change. The Zeeman effect, splitting of spectral lines when an atom is placed in a magnetic field, confirmed the concept of quantized orbital angular momentum

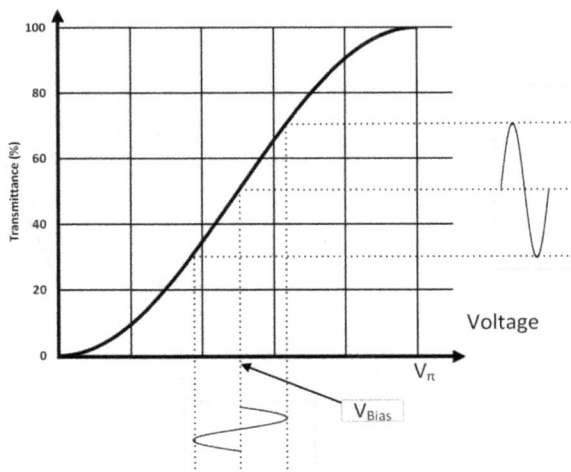

Figure 8.33. Characteristic curve for an amplitude modulated EOM.

8-31

states [23]. When polarized light is incident on some isotropic dielectric materials, the plane of polarization rotates in proportion to the strength of a magnetic field applied in the direction of propagation, a phenomenon called the Faraday effect or Faraday rotation [24]. The amount of rotation is given by the relation,

$$\beta = VBL$$

where β is the rotation angle, L the length of the material, B the magnetic induction along the direction of propagation, and V is a material constant called the Verdet constant. See figure 8.34. The Verdet constant is a function of wavelength and temperature [25]. A variety of materials possess magneto-optic properties including liquids, crystals, and glasses. Common materials such as quartz, and crown glass have known Verdet constants although their small value requires high magnetic fields, long lengths, or both. Glasses doped with rare-earth atoms have relatively large Verdet constants, such as terbium gallium garnet (TGG), and yttrium iron garnet (YIG), and are commonly used in device manufacture.

Unlike other forms of optical activity in materials the Faraday effect is independent of the direction of propagation. If the transmitted beam of figure 8.34 were reflected back through the material, the plane of polarization would rotate by an additional factor of β. This effect is exploited to make optical isolators as shown in figure 8.35. This uses two polarizers, one vertical and the other at 45° to the vertical. Flux entering from left to right will be transmitted through the system. In the process the polarized light following the vertical polarizer will rotate through 45° and be transmitted through the second polarizer. On the other hand, any flux returning back through the system will be polarized at 45°, rotated again by the rotator emerging at 90° to the vertical and be blocked by the vertical polarizer.

Faraday isolators of this form are used as components in systems using laser beams to prevent any back reflections from destabilizing the laser cavity. The main advantage is that the device is passive since the magnetic field can be produced with a simple permanent magnet.

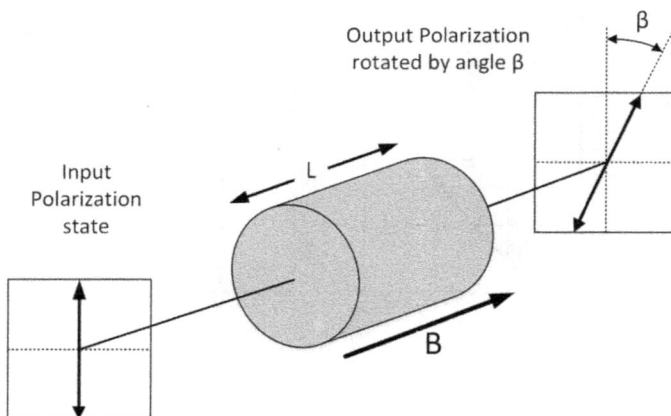

Figure 8.34. Faraday rotation.

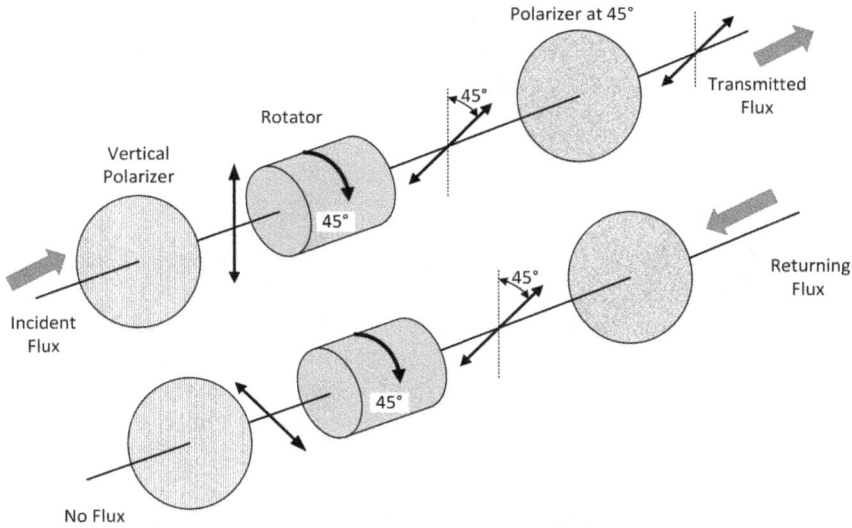

Figure 8.35. Faraday isolator concept.

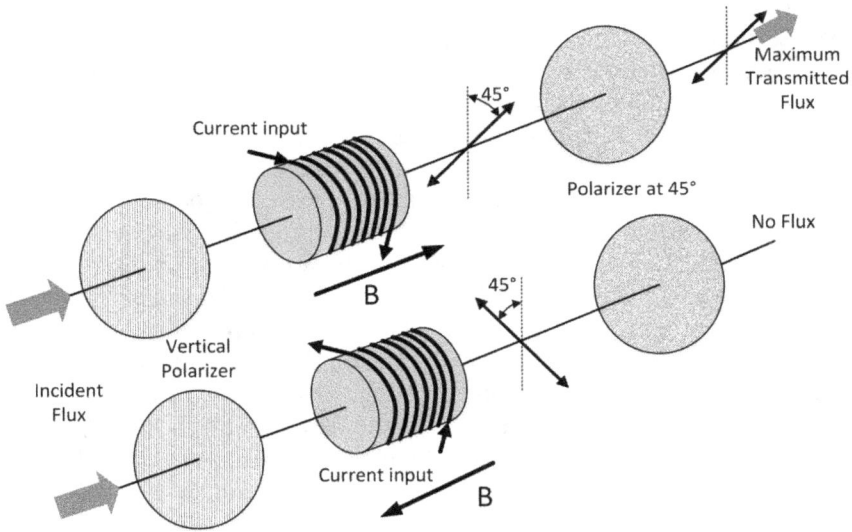

Figure 8.36. Magneto-optic switch/modulator concept.

Another application of the Faraday effect is in magneto-optic switches and modulators. Figure 8.36 illustrates this concept in a similar manner as the Faraday isolator. In this case, the magnetic induction is created with a current loop surrounding the magneto-optic active material. When current flows in one direction the light flux is passed through the system. When the current is reversed the field is reversed and the polarization rotation is in the opposite direction blocked by the polarizer at 45°. According to Malus law, the transmittance through this system with no current flow would be $T = \cos^2(45°) = 1/2$. This arrangement is used because transmittance changes are most sensitive in the 45° region and provides the most sensitive scheme for use as a modulator. For a switching application a different polaroid arrangement may be better suited so that the transmitted flux at zero current is zero. Magneto-optic switches such as this are commonly found in fiber optic applications.

8.5.5 Acousto-optic modulators

When an isotropic material is placed under stress it can become birefringent, a phenomenon known as the elasto-optic effect or photoelasticity. It is used to great effect in visualizing externally applied stress as well as internal stress within transparent solids. The induced birefringence created by the stress can be easily observed using polarized light analysis techniques [26]. Elastic stresses can also be applied in a material using an acoustic wave, called the acousto-optic effect. A wide variety of materials exhibit this effect including fused silica and flint glass [27]. There are several theoretical models used to describe the acousto-optic effect [21, 28]. However, we will concentrate here on the applications of acousto-optic modulators (AOMs).

An acousto-optic light modulator is constructed from a transparent crystal or glass slab which has an ultrasonic transducer (often a piezoelectric device) bonded onto one end. Driving the transducer with a signal at a given frequency (typically radio frequency) sets up an acoustic wave traveling at the speed of sound within the material. In regions where the acoustic pressure wave is high, the material is stressed by compression, changing in the density of the material, and leading to an increase in the refractive index in this local region. The magnitude of the relative change in refractive index is $(\Delta n/n) = 10^{-4}$. For a periodic acoustic waveform, the refractive index profile is also periodic creating a phase diffraction grating across the modulator material.

Figure 8.37 shows the top view of an AOM material of length L with the incident beam from the left. The wavelength of the acoustic wave is shown as $\Lambda = v_s/f$, where f is the frequency of the driving signal and v_s is the velocity of sound in the material. If L is small (a thin plate or a small transducer) the beam will diffract as any thin grating. However, this is rarely the case, and for most devices the grating formed within the material is a thick volume grating. Diffraction from a thick volume grating follows Bragg diffraction, leading to the name for an AOM as a Bragg cell.

An incident laser beam on the AOM front surface becomes Bragg diffracted, as illustrated in figure 8.37. The angle of diffraction (Bragg angle θ_B) depends on the diffraction grating spacing and the wavelength of the incident flux, λ, or,

$$\sin \theta_B = \frac{\lambda}{2\Lambda} = \frac{\lambda f}{2v_s}$$

One major advantage of this configuration is that because this is a continuous grating, most of the radiation is in the first order with diffraction efficiency given by,

$$\eta_B = \sin^2\left(\frac{\pi L}{\lambda \cos \theta_B} \Delta n\right)$$

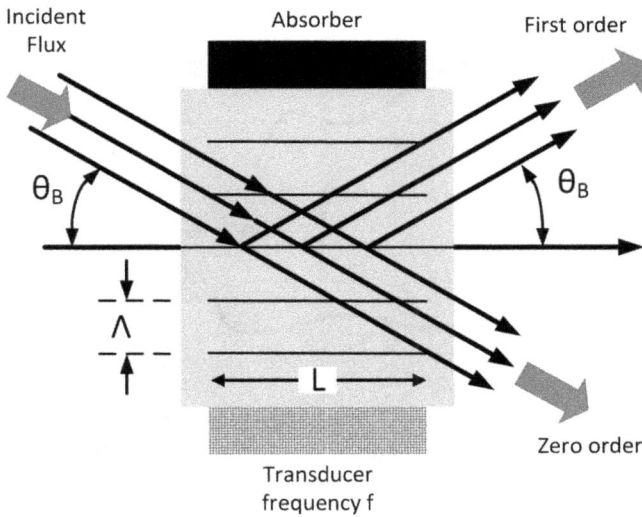

Figure 8.37. Acousto-optic modulator Bragg cell configuration.

where Δn is the amplitude of the index change introduced by the acoustic wave [14]. Since the change in index is related to the strain, which is related to the amplitude of the acoustic power, changing the acoustic power changes the diffracted flux. Thus, output flux is related to the input drive signal and modulates the diffracted beam. The rise time of a signal can be approximated by the amount of time it takes for the sound wavefront to cross the beam, or $t_{rise} = \Lambda/v_s$.

Bragg cells can also be used as scanners if the acoustic drive signal is a time varying frequency change or chirp. From the Bragg condition, the change in the Bragg angle can be written as a change in drive frequency, Δf, or

$$\Delta \theta_B = \frac{\lambda}{2v_s \cos \theta_B} \Delta f$$

So, as the frequency of the input signal changes in time the beam angle changes or deflects.

8.6 Application: optical fiber proximity sensor

Optical fibers are used as a building block in numerous sensing applications. Some of the advantages of fiber-based sensors are they can be made highly sensitive, are small size, lightweight, provide geometric versatility, have a common technology base, and are immune to electromagnetic interference (EMI) [29, 30]. The general concept of a fiber optic-based sensor is shown in figure 8.38. The sensing portion of the sensor is called an optrode that responds to some type of external measurand [30]. Incident radiation enters the optrode through a transmitting fiber and returns for measurement through a receiving fiber.

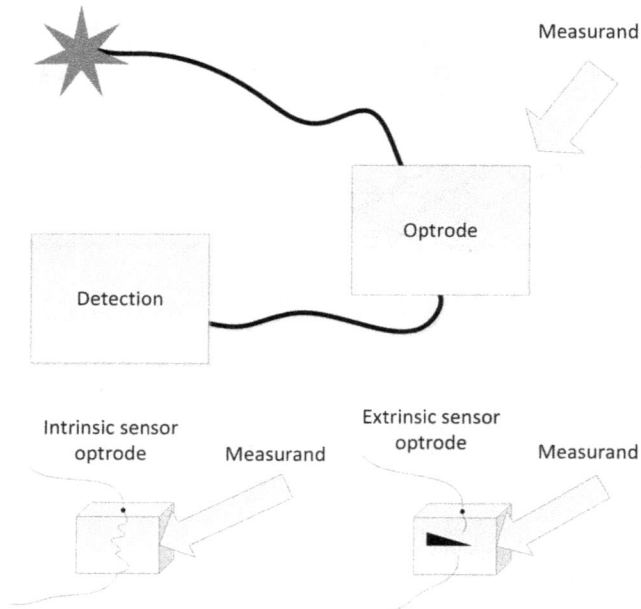

Figure 8.38. Optical fiber sensor concept showing both intrinsic and extrinsic types of sensing.

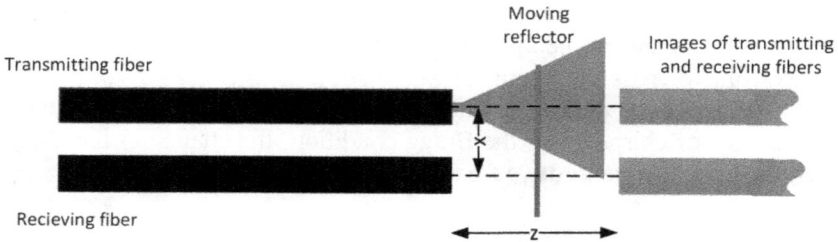

Figure 8.39. Displacement sensor using optical fiber to transmit incident radiation onto a reflector that can be translated through a different distance z.

One way to classify different types of sensors is by their sensing mechanism. An intrinsic fiber sensor uses a change in some fiber parameter such as an applied strain, altering the radiation within the fiber. In this case the radiation never exits a fiber end. On the other hand, an extrinsic fiber sensor uses a modulation mechanism outside the fiber, such as an attenuator or moving modulator, which alters the incident flux that is collected by the receiving fiber.

One particular example of a fiber sensor optrode is the moving reflector displacement/proximity sensor [30, 31]. See Figure 8.39. This sensor uses the optical fiber as both a light guide to illuminate the reflector and as a light guide to the detector. As the distance between the reflector and the transmitting/receiving fiber pair increases the amount of radiation collected by the receiving fiber is altered. Since the sensing is performed external to the fiber this type of sensor is classified as an extrinsic fiber optic sensor.

To analyze the operation of this distance sensor we want to find the amount of optical power coupled from the transmitting fiber to the receiving fiber. Because this is a reflective surface, there are effective images formed for both fibers. Optically, the radiation leaving the transmitted fiber is collected by the image of the receiving fiber. As the moving modulator is displaced in a direction perpendicular to the fiber optic axis, the detected power collected by the receiving fiber can be related to this displacement. Qualitatively, we expect that at small displacements the coupled power is low. As the displacement increases more optical power is coupled into the receiving fiber and the detected power will increase. As the displacement becomes very large the amount of coupled power will again decrease because the amount of radiation in the beam spreads out with diminishing irradiance.

One model that describes this sensor optrode uses beam propagation and approximates the irradiance profile of the radiant flux leaving the source fiber as a Gaussian beam propagating toward the detector fiber. The assumption is that the output in the far field can be approximated by a Gaussian function [31]. The irradiance at the detector fiber is then given by the irradiance profile of a Gaussian beam a distance z from the fiber,

$$E_e(x, z) = \frac{2\Phi_{etot}}{\pi w^2(z)} \exp\left[-\frac{2x^2}{w^2(z)}\right]$$

where Φ_{etot} is the total power leaving the transmitting fiber, x is the transverse separation between the receiving fiber and the transmitting fiber, $\omega(z)$ is the characteristic beam radius, and ω_0 is the initial beam waist [31]. The model used here assumes that the initial beam waist ω_0 is the input fiber radius and the divergence angle is the angle associated with the numerical aperture. One can also use ω_0 and NA as fit parameters.

An experiment was performed on this sensor configuration using 1.0-millimeter diameter plastic fiber for the transmitting and receiving fibers with separated of 2.0 mm between the fiber axes. A front surface mirror was used as the moving reflector mounted on a precision translation stage. The light source was an LED at 670 nm and the power from the receiving fiber was measured with a power meter [31]. The resulting data and model are plotted in figure 8.40 showing good agreement between the model and the experimental data. In this case only a scaling factor was applied to the model to account for the total power exiting the fiber.

8.7 Application: modulating a diode laser for detector frequency response measurement

Directly modulating a laser diode with a signal at various frequencies provides a source that can be used for determining the frequency response characteristics of detectors. The test setup is relatively simple employing a diode laser, driver, function generator, detector with reverse bias, and oscilloscope. A schematic of this arrangement is shown in figure 8.41.

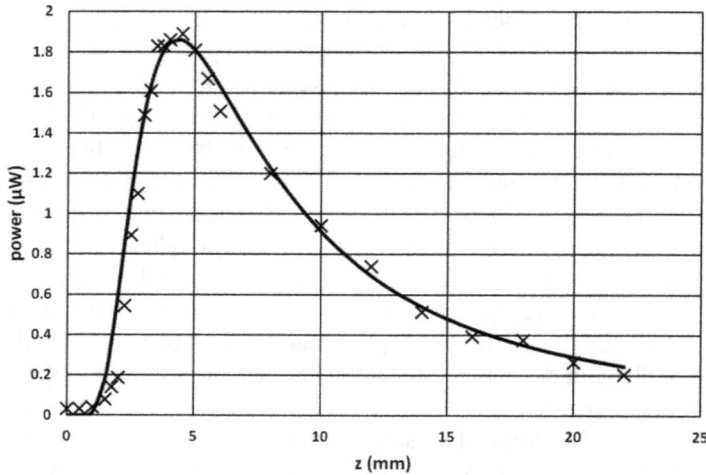

Figure 8.40. Experimental results of the optical power as a function of displacement for the fiber optic moving mirror displacement sensor configuration. The solid curve is a comparison with the Gaussian model. Only a scaling factor was used to match the model to the data.

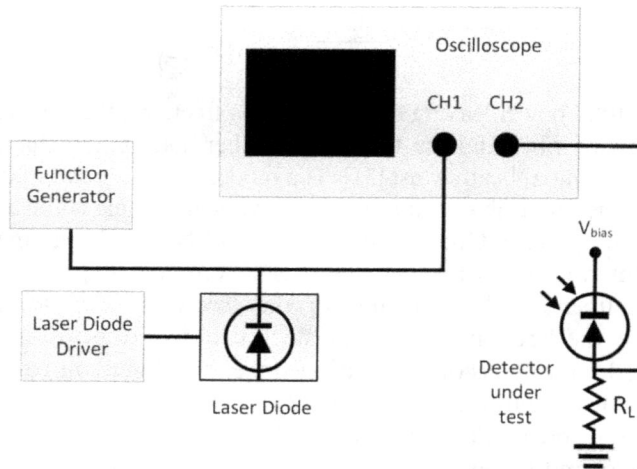

Figure 8.41. Experimental test setup for modulating a laser diode to measure the frequency response of a detector system.

The laser must be biased above the threshold current as discussed previously. The function generator provides a signal that oscillates about this current bias to produce a light output from the source that follows the same functional form chosen. The detector will detect the signal according to the photoconductive model with an output current of,

$$i(t) = \mathcal{R}\Phi_e(t)$$

where $\mathcal{R}$ is the response function of the detector. Usually, a sinusoidal signal is used to modulate the light output with a constant AC current amplitude i_0 to produce the time varying radiant flux. The voltage across the load resistor R_L as a function of time is then,

$$V_L(t) = i(t)R_L = \mathcal{R}R_L\Phi_e(t)$$

Measurements can be made of the signal amplitude or rms voltage on an oscilloscope (as shown) or with an AC voltmeter for different modulation frequencies. Keeping the AC modulation at the constant amplitude i_0, change the frequency, and monitor the voltage across the load resistor. Repeat this procedure over a wide frequency range. As the frequency increases the detected signal amplitude will remain relatively constant but begin to drop as the frequency approaches the frequency cutoff. This cutoff frequency is governed by the RC time constant of the detector system where the resistance $R = R_L$. The frequency response of the system in this case will follow the general form of,

$$|\mathcal{R}(f)| = \frac{V_L(f)}{i_0 R_L} = \frac{1}{[1 + (2\pi f\tau)^2]^{1/2}}$$

where the time constant $\tau = R_L C$ with C the effective capacitance of the system. The cutoff frequency is then defined as, $f_C = 1/2\pi R_L C$. Note that for smaller load resistance the cutoff frequency is high but the measured voltage across the load resistor is low.

Figure 8.42 shows test data of the voltage across the load resistor as a function of frequency for a PIN detector with two different values of load resistor, 5 kΩ and 50 Ω. Also shown on the plots is a fit curve to the data using the frequency response function above. Again, the tradeoff between detected signal size and frequency cutoff is apparent.

Exercises and problems

1. You are in the process of buying a CPC and you have a choice between three CPCs:

   (1) Acceptance angle = 30° and output semi-aperture (half the full small aperture size) = 5 mm,

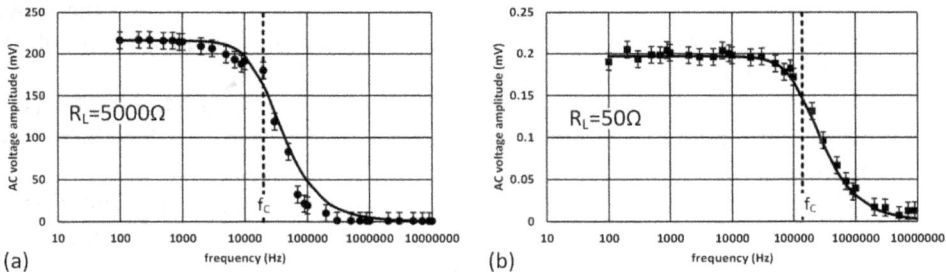

(a)   (b)

Figure 8.42. Test data taken by modulating a laser diode to determine the frequency response characteristics of a detector system. The frequency axis on both graphs is logarithmic. (a) 5 kΩ load resistor, and (b) 50 Ω load resistor.

(2) Acceptance angle = 20° and output semi-aperture (half the full small aperture size) = 5 mm, or

(3) Acceptance angle = 20° and output semi-aperture (half the full small aperture size) = 5 mm, but the input aperture size is restricted to that of case (1) above.

For each case:

(a) What is the focal length of the CPC?
(b) What is size of the input aperture?
(c) What is the length of the CPC?
(d) Make plots of each CPC.

2. An acousto-optic modulator (AOM) is made from a fused quartz block (v_s = 6000 m s^{-1}) and is driven with a 40 MHz signal in order to deflect a HeNe laser beam (633 nm). Assuming that the AOM operates in the Bragg diffraction condition, determine the angle between the incident beam and the diffracted beam.

3. The pattern of diffracted spots sketched below of HeNe laser light (633 nm) was formed in an experiment with an AOM. When an acoustic frequency of 40 MHz was applied to the AOM the spacing between dots was 4.22 mm and the distance from the AOM to the observation screen was 1 m. Determine the velocity of sound in the material.

4. A scanner is to deflect a collimated beam from a HeNe laser (633 nm) with a beam diameter of 5 mm. If we want the scanner to cover an angular range of 5°, how many resolvable spots can we achieve using the Rayleigh criterion as our definition of resolution?

5. Compute the amount of optical power in a Gaussian beam that is contained within a circle of radius equal to the beam radius.

6. A HeNe laser (633 nm) with a beam divergence of 0.8 mrad is focused with a 100 mm focal length lens. (a) Calculate the minimum beam waist diameter that can be achieved with this laser. (b) What would the Rayleigh range?

7. A 10 W argon-ion laser (wavelength 488 nm) has a minimum spot size diameter of 2.0 mm.

(a) How far will this beam travel before the spot size doubles in size? (b) How large must an aperture be made to collect only 50% of the power within the expanded beam? (c) What is the average amplitude of the electromagnetic field within this aperture?

8. A HeNe laser (633 nm wavelength) operating in a TEM$_{00}$ mode has a resonator length of 1 m and mirror radii R_1 = 3 m and R_2 = 2 m. Mirror #2 is the output mirror.

(a) Is this a stable resonator? Prove your answer using a calculation.
(b) Where is the beam waist located with respect to the output mirror?
(c) What is the radius of the minimum beam waist?
(d) What is the beam radius at the output mirror?
(e) Determine the divergence angle of the beam from the output mirror.

9. The Gaussian beam from a Nd:YAG laser at a wavelength of 1.06 μm has a total optical power of 2.0 W with a half-angle divergence of 0.50 mrad. Find the radius of the beam and depth of focus at a distance of 1.0 m from the beam waist.
10. We wish to couple a HeNe laser (0.633 μm wavelength) source into a singlemode optical fiber with a core diameter of 3.2 μm using a 20× microscope objective ($f = 8$ mm). Take the distance from the objective to the fiber tip to be, d, where we want to locate the minimum beam waist. Assume for simplicity that the incident laser is a plane wave that completely illuminates the 4 mm aperture diameter of the objective.
    (a) Compute the ABCD matrix for this situation.
    (b) Find a numerical value for d.
    (c) What is the value of the beam waist size? Will this lens be a good choice to use for this application?
11. A HeNe laser (633 nm), operating in a Gaussian beam mode, has a beam diameter of 3.5 mm at a distance of 2 m from the minimum beam waist position. At 4 m from the minimum beam waist position the beam has expanded to a diameter of 5.9 mm.
    (a) Calculate the beam divergence angle.
    (b) What is the minimum beam waist diameter?
    (c) Find the Rayleigh range parameter, z_R.
    (d) This beam is focused with a 10 cm focal length lens placed 1 m from the minimum beam waist position. What is the focused spot diameter?
    (e) Where is the focused spot relative to the lens?
12. Two laser resonator mirrors with radii of curvature $R_1 = +50$ cm and $R_2 = +100$ cm are possible candidates for a laser system. Calculate (a) the separations for which they form a marginally stable configuration, (b) the range of separations for which they would be stable, and (c) the range of separations for which they would be unstable.
13. A CO_2 laser has a Gaussian beam of radius 1.70 mm in one plane perpendicular to the direction of propagation and 3.40 mm at another plane at a distance of 0.1 m from the first. The wavelength of the laser is 10.6 μm. What is the location and size of the minimum beam waist of the laser with respect to the first plane?
14. An argon-ion laser (wavelength of 488 nm) forms a Gaussian beam waist radius of 0.25 mm. Design a single-lens optical system to focus the light to a spot size (diameter) of 100 μm. You must provide the focal length of the lens, the location of the lens from the beam waist, and the location of the focused spot with respect to the lens.
15. The beam waist radius of a laser beam from a multimode laser diode with wavelength of 1.50 μm is 3.0 mm. The half-angle beam divergence is 1.4 mrad. Compute the M^2 beam quality figure of merit for this laser.

References

[1] Koshel R J 2013 Introduction and terminology *Illumination Engineering: Design with Nonimaging Optics* ed R John Koshel (New York: Wiley-IEEE Press) ch 1

[2] Winston R, Miñano J C and Benítez P 2005 *Nonimaging Optics* (San Diego, CA: Elsevier Academic Press)

[3] Chaves J 2016 *Introduction to Nonimaging Optics* 2nd edn (Boca Raton, FL: CRC Press)

[4] Welford W T and Winston R 1978 *The Optics of Nonimaging Concentrators: Light and Solar Energy* (New York: Academic)

[5] Koshel R J 2013 Etendue *Illumination Engineering: Design with Nonimaging Optics* ed R John Koshel (New York: Wiley-IEEE Press) ch 2

[6] Klein M V and Furtak T E 1986 *Optics* 2nd edn (New York: Wiley)

[7] O'Shea D C 1985 *Elements of Modern Optical Design* (New York: Wiley)

[8] Boyd R W 1983 *Radiometry and the Detection of Optical Radiation* (New York: Wiley)

[9] Kasunic K 2011 *Optical Systems Engineering* (New York: McGraw-Hill Education)

[10] Tkaczyk T S 2010 *Field Guide to Microscopy* (Bellingham, WA: SPIE Press) https://doi.org/10.1117/3.798239

[11] Arecchi A V, Messadi T and Koshel R J 2007 *Field Guide to Illumination* (Bellingham, WA: SPIE Press) https://doi.org/10.1117/3.764682

[12] Ries H and Rabl A 1994 Edge-ray principle of nonimaging optics *J. Opt. Soc. Am.* A **11** 2627–32

[13] Sennaroglu A 2010 *Photonics and Laser Engineering: Principles, Devices, and Applications* (New York: McGraw-Hill)

[14] Yariv A and Yeh P 2007 *Photonics: Optical Electronics in Modern Communications* (New York: Oxford University Press)

[15] ISO 11146-1:2005 Lasers and laser-related equipment—Test methods for laser beam widths, divergence angles and beam propagation ratios—Part 1: Stigmatic and simple astigmatic beams https://iso.org/

[16] Thorlabs Inc 2020 www.thorlabs.com (Accessed 12 August 2020)

[17] Beiser L 2003 *Unified Optical Scanning Technology* (New York: Wiley)

[18] Beiser L and Johnson R B 1995 Scanners *Handbook of Optics, Volume II: Devices, Measurements, & Properties* 2nd edn, ed M Bass, E W Van Stryland, D R Williams and W L Wolfe (New York: McGraw-Hill) ch 19

[19] Marshall G F and Stutz G E (ed) 2012 *Handbook of Optical and Laser Scanning* 2nd edn (Boca Raton, FL: CRC Press Taylor & Francis Group)

[20] Hobbs P C D 2000 *Building Electro-optical Systems: Making It all Work* (New York: Wiley)

[21] Saleh B E A and Teich M C 2007 *Fundamentals of Photonics* 2nd edn (New York: Wiley)

[22] Maldenado T A 1995 Electro-optic modulators *Handbook of Optics, Volume II: Devices, Measurements, and Properties* 2nd edn, ed M Bass, E W Van Stryland, D R Williams and W L Wolfe (New York: McGraw-Hill) ch 13

[23] Eisberg R M and Resnick R 1985 *Quantum Physics of Atoms, Molecules, Solids, Nuclei, and Particles* 2nd edn (New York: Wiley)

[24] Schatz P N and McCaffery A J 1969 *Q. Rev. Chem. Soc.* **23** 552

[25] Zvezdin A K and Kotov V A 1997 *Modern Magnetooptics and Magnetooptical Materials* 1st edn (New York: Taylor & Francis Group)

[26] Gasvik K J 2002 *Optical Metrology* 3rd edn (New York: Wiley)

[27] Chang I C 1995 Acousto-optic devices and applications *Handbook of Optics, Volume II: Devices, Measurements, and Properties* 2nd edn, ed M Bass, E W Van Stryland, D R Williams and W L Wolfe (New York: McGraw-Hill) ch 12

[28] Guenther R 1990 *Modern Optics* (New York: Wiley)

[29] Udd E and Spillman W B Jr (ed) 2011 *Fiber Optic Sensors: An Introduction for Engineers and Scientists* 2nd edn (New York: Wiley)

[30] Spillman W B and Udd E 2014 *Field Guide to Fiber Optic Sensors* (Bellingham, WA: SPIE Press)

[31] Bunch R M 1990 Optical fiber sensor experiments for the undergraduate physics laboratory *Am. J. Phys.* **58** 74

IOP Publishing

Optical Systems Design Detection Essentials
Radiometry, photometry, colorimetry, noise, and measurements
Robert M Bunch

Appendix

Definitions of selected special functions in one dimension

Delta function	$\delta(x - x_0) = \int\limits_{-\infty}^{+\infty} e^{+j2\pi u(x-x_0)}du$	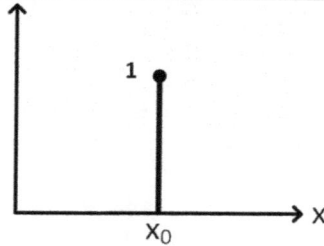
Comb function	$\text{comb}(x) = \sum\limits_{n=-\infty}^{+\infty} \delta(x - n)$ $= \sum\limits_{n=-\infty}^{+\infty} e^{j2\pi nx}$	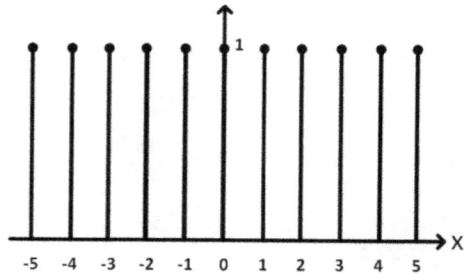
Step function	$\text{step}(x) = \begin{cases} 1, & x > 1 \\ \frac{1}{2}, & x = 0 \\ 0, & x < 0 \end{cases}$	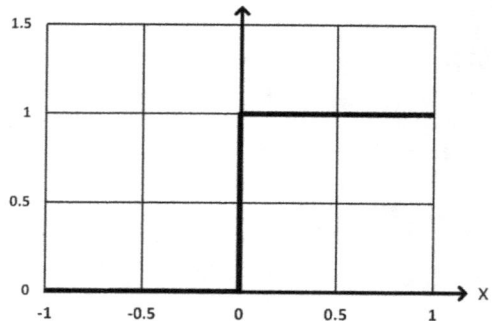

doi:10.1088/978-0-7503-2252-2ch9

(Continued)

Sign function	$\mathrm{sgn}(x) = \begin{cases} 1, & x > 1 \\ 0, & x = 0 \\ -1, & x < 0 \end{cases}$	

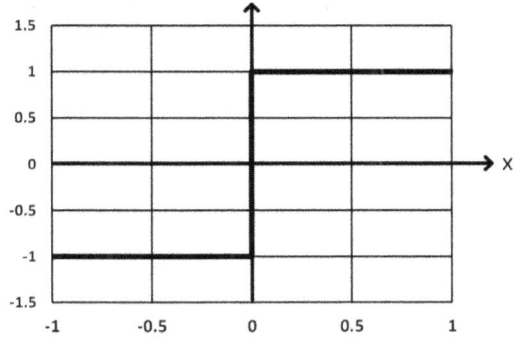

Rectangle	$\mathrm{rect}(x) = \begin{cases} 1, & |x| < \frac{1}{2} \\ \frac{1}{2}, & |x| < \frac{1}{2} \\ 0, & |x| < \frac{1}{2} \end{cases}$	

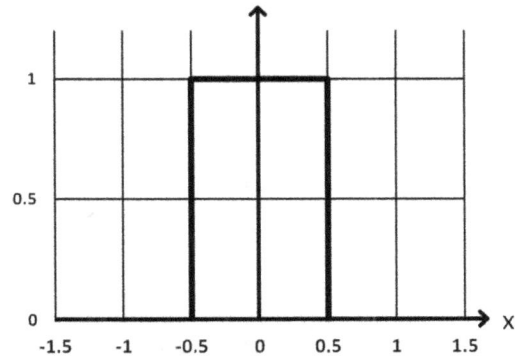

Triangle	$\mathrm{tri}(x) = \begin{cases} 1 - |x|, & |x| < 1 \\ 0, & |x| \geqslant 0 \end{cases}$	

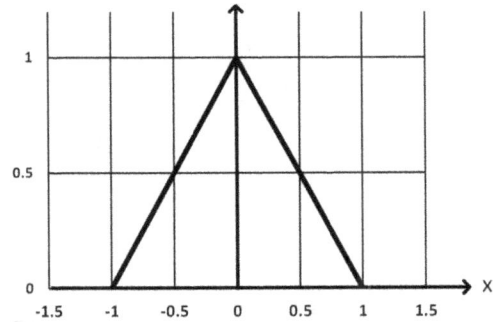

Sinc	$\mathrm{sinc}(x) = \frac{\sin(\pi x)}{\pi x}$	

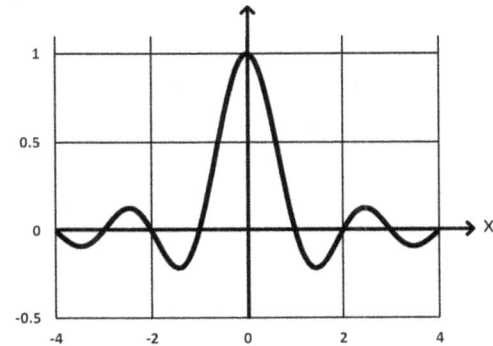

A-2

Sinc-squared

$$\text{sinc}^2(x) = \frac{\sin^2(\pi x)}{(\pi x)^2}$$

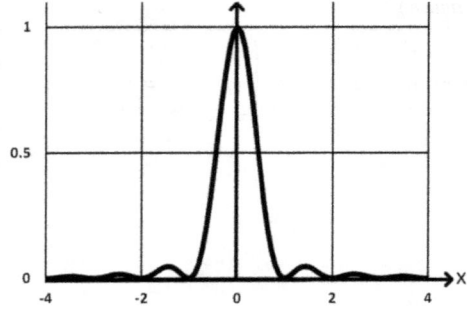

Gaussian

$$\text{Gaus}(x) = \exp(-\pi x^2)$$

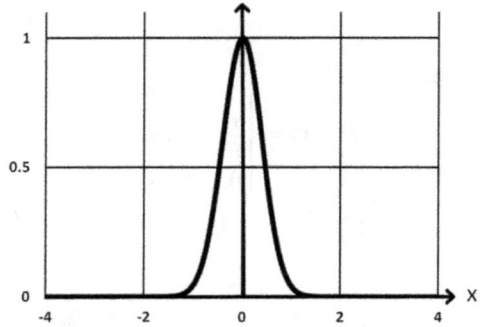

Definitions of selected two-dimensional special functions

Cylinder or circular function $r = \sqrt{x^2 + y^2}$

$$\text{circ}(r) = \begin{cases} 1, & r < 1 \\ \frac{1}{2}, & r = 1 \\ 0, & \text{otherwise} \end{cases}$$

Jinc or besinc First-order Bessel function

$$\text{jinc}(r) = 2\frac{J_1(r)}{r}$$

Somb similar to Jinc

$$\text{somb}(r) = 2\frac{J_1(\pi r)}{\pi r}$$

Gaus

$$\text{Gaus}(x, y) = \exp[-\pi(x^2 + y^2)]$$

Fourier transform properties for one-dimensional functions

Definitions

The Fourier transform of a spatial function $f(x)$ results in a function in $F(f_x)$ where the variable f_x is the spatial frequency. $\mathcal{F}\{f(x)\}$ is the Fourier transform operator notation, so that,

$$F(f_x) = \mathcal{F}\{f(x)\} = \int_{-\infty}^{+\infty} f(x)e^{-j2\pi f_x x}dx$$

The inverse Fourier transform is defined as,

$$f(x) = \mathcal{F}^{-1}\{F(f_x)\} = \int_{-\infty}^{+\infty} F(f_x)e^{+j2\pi f_x x}df_x$$

A-3

The convolution of a function $f(x)$ with another function $h(x)$ is defined as,

$$f(x) \circledast h(x) = \int_{-\infty}^{+\infty} f(w)h(x - w)dw$$

where the variable w is a dummy variable of integration.

The sifting property of the delta function is a special case of a convolution

$$f(x) = \int_{-\infty}^{+\infty} f(w)\delta(x - w)dw$$

The correlation of a function $f(x)$ with another function $h(x)$ is defined as,

$$f(x) \odot h(x) = \int_{-\infty}^{+\infty} f(w)h(x + w)dw$$

General properties of functions and their Fourier transform

The parameters used in the table: a is a constant value in space, f_0 is a constant spatial frequency value.

Property	Function	Fourier transform
Real function	$f(\pm x)$	$F(\pm f_x)$
Complex conjugate	$f^*(\pm x)$	$F^*(\pm f_x)$
Fourier integral theorem	$\mathcal{F}^{-1}\{f(x)\}$	$f(x)$
Product of two functions	$f(x)h(x)$	$F(f_x) \circledast H(f_x)$
Convolution	$f(x) \circledast h(x)$	$F(f_x)H(f_x)$
Autocorrelation	$f(x) \odot f(x)$	$|F(f_x)|^2$
Linearity	$f(x) + h(x)$	$F(f_x) + H(f_x)$
Scaling or similarity	$f\left(\frac{x}{a}\right)$	$|a|F(af_x)$
Shift	$f(x \pm a)$	$F(f_x)\exp(\pm j2\pi f_x a)$
Real function phase multiplier	$f(x)\exp(\pm j2\pi x f_0)$	$F(f_x \mp f_0)$
Derivative	$\frac{d^n f(x)}{dx^n}$	$(j2\pi f_x)^n F(f_x)$

Table of Fourier transform pairs of selected functions

The function space variable coordinate is x and the spatial frequency variable f_x.

The parameters used in the table: a and x_0 are constant values in space, f_0 is a constant spatial frequency value.

Function	Fourier transform
1	$\delta(f_x)$
$\delta(x)$	1
$\delta(ax)$	$\frac{1}{|a|}$
$\delta(x \pm x_0)$	$e^{\mp j2\pi x_0 f_x}$
$e^{\pm j2\pi f_0 x}$	$\delta(f_x \mp f_0)$

$\text{comb}\left(\frac{x}{a}\right)$ $|a|\text{comb}(af_x)$

$\text{rect}\left(\frac{x \pm x_0}{a}\right)$ $|a|\text{sinc}(af_x)e^{\pm j2\pi x_0 f_x}$

$\text{sinc}\left(\frac{x \pm x_0}{a}\right)$ $|a|\text{rect}(af_x)e^{\pm j2\pi x_0 f_x}$

$\text{tri}\left(\frac{x}{a}\right)$ $|a|\text{sinc}^2(af_x)e^{\pm j2\pi x_0 f_x}$

$\cos 2\pi f_0 x$ $\frac{1}{2}[\delta(f_x - f_0) + \delta(f_x + f_0)]$

$\sin 2\pi f_0 x$ $\frac{1}{2j}[\delta(f_x - f_0) - \delta(f_x - f_0)]$

$\text{Gaus}\left(\frac{x}{a}\right) = \exp\left(-\pi\frac{x^2}{a^2}\right)$ $|a|\text{Gaus}(af_x) = |a|\exp(-\pi a^2 f_x^2)$

$x\exp(-\pi x^2)$ $-jf_x\exp(-\pi f_x^2)$

$\exp(-|x|)$ $\dfrac{2}{[1 + (2\pi f_x)^2]}$

Properties of the quadratic phase function, $Q(x, y; d)$

The quadratic phase function appears in representations of wave propagation with wavelength λ over a general distance d. The function is purely a phase function that is quadratic in the coordinate plane perpendicular to the direction of propagation of the wave. This function is defined as:

$$Q(x, y; d) = \exp\left[j\frac{k}{2d}(x^2 + y^2)\right] = \exp\left[j\frac{\pi}{\lambda d}(x^2 + y^2)\right]$$

There are several properties of this special function that are tabulated below and are useful in computations.

Integral

$$\iint Q(x, y; d)dxdy = j\lambda d$$

Symmetry

$$Q(\pm x, \pm y; d) = Q(x, y; d)$$

Value at $d = $ infinity

$$Q(x, y; \infty) = 1$$

Scaling

$$Q(ax, ay; d) = Q\left(x, y; \frac{d}{a^2}\right)$$

Product

$$Q(x, y; d_1)Q(x, y; d_2) = Q\left(x, y; \frac{d_1 d_2}{d_1 + d_2}\right)$$

Coordinate shift

$$Q(x - u, y - v; d) = Q(x, y; d)Q(u, v; d)\exp\left(-j\frac{2\pi}{\lambda d}(xu + yv)\right)$$

Fourier transform

$$\mathcal{F}\{Q(x, y; d)\} = j\lambda d Q\left(f_x, f_y; -\frac{1}{\lambda^2 d}\right)$$

Convolution

$$Q(x, y; d_1) \circledast Q(x, y; d_2) = j\lambda \frac{d_1 d_2}{d_1 + d_2} Q(x, y; d_1 + d_2) \quad d_1 \neq d_2$$

Spectral luminous efficiency functions

$V(\lambda)$—1924 CIE Photopic response and $V'(\lambda)$—1951 Scotopic response
  From: The Colour & Vision Research Laboratory, Institute of Ophthalmology, University College London (http://www.cvrl.org/)

λ (nm)	$V(\lambda)$	$V'(\lambda)$
380	0.000 04	0.000 59
385	0.000 06	0.001 11
390	0.000 12	0.002 21
395	0.000 22	0.004 53
400	0.000 40	0.009 29
405	0.000 64	0.018 52
410	0.001 21	0.034 84
415	0.002 18	0.060 40
420	0.004 00	0.096 60
425	0.007 30	0.143 60
430	0.011 60	0.199 80
435	0.016 84	0.262 50
440	0.023 00	0.328 10
445	0.029 80	0.393 10
450	0.038 00	0.455 00
455	0.048 00	0.513 00
460	0.060 00	0.567 00
465	0.073 90	0.620 00
470	0.090 98	0.676 00
475	0.112 60	0.734 00
480	0.139 02	0.793 00
485	0.169 30	0.851 00
490	0.208 02	0.904 00
495	0.258 60	0.949 00
500	0.323 00	0.982 00

505	0.407 30	0.998 00
510	0.503 00	0.997 00
515	0.608 20	0.975 00
520	0.710 00	0.935 00
525	0.793 20	0.880 00
530	0.862 00	0.811 00
535	0.914 85	0.733 00
540	0.954 00	0.650 00
545	0.980 30	0.564 00
550	0.994 95	0.481 00
555	1.000 00	0.402 00
560	0.995 00	0.328 80
565	0.978 60	0.263 90
570	0.952 00	0.207 60
575	0.915 40	0.160 20
580	0.870 00	0.121 20
585	0.816 30	0.089 90
590	0.757 00	0.065 50
595	0.694 90	0.046 90
600	0.631 00	0.033 15
605	0.566 80	0.023 12
610	0.503 00	0.015 93
615	0.441 20	0.010 88
620	0.381 00	0.007 37
625	0.321 00	0.004 97
630	0.265 00	0.003 34
635	0.217 00	0.002 24
640	0.175 00	0.001 50
645	0.138 20	0.001 01
650	0.107 00	0.000 68
655	0.081 60	0.000 46
660	0.061 00	0.000 31
665	0.044 58	0.000 21
670	0.032 00	0.000 15
675	0.023 20	0.000 10
680	0.017 00	0.000 07
685	0.011 92	0.000 05
690	0.008 21	0.000 04
695	0.005 72	0.000 03
700	0.004 10	0.000 02
705	0.002 93	0.000 01
710	0.002 09	0.000 01
715	0.001 48	0.000 01
720	0.001 05	0.000 00

(*Continued*)

(*Continued*)

λ (nm)	V(λ)	V′(λ)
725	0.000 74	0.000 00
730	0.000 52	0.000 00
735	0.000 36	0.000 00
740	0.000 25	0.000 00
745	0.000 17	0.000 00
750	0.000 12	0.000 00
755	0.000 08	0.000 00
760	0.000 06	0.000 00
765	0.000 04	0.000 00
770	0.000 03	0.000 00
775	0.000 02	0.000 00
780	0.000 01	0.000 00

1931 2° color matching functions

From: The Colour & Vision Research Laboratory, Institute of Ophthalmology, University College London (http://www.cvrl.org/)

λ (nm)	$\bar{x}$	$\bar{y}$	$\bar{z}$
360	0.000 130	0.000 004	0.000 606
365	0.000 232	0.000 007	0.001 086
370	0.000 415	0.000 012	0.001 946
375	0.000 742	0.000 022	0.003 486
380	0.001 368	0.000 039	0.006 450
385	0.002 236	0.000 064	0.010 550
390	0.004 243	0.000 120	0.020 050
395	0.007 650	0.000 217	0.036 210
400	0.014 310	0.000 396	0.067 850
405	0.023 190	0.000 640	0.110 200
410	0.043 510	0.001 210	0.207 400
415	0.077 630	0.002 180	0.371 300
420	0.134 380	0.004 000	0.645 600
425	0.214 770	0.007 300	1.039 050
430	0.283 900	0.011 600	1.385 600
435	0.328 500	0.016 840	1.622 960
440	0.348 280	0.023 000	1.747 060
445	0.348 060	0.029 800	1.782 600
450	0.336 200	0.038 000	1.772 110
455	0.318 700	0.048 000	1.744 100
460	0.290 800	0.060 000	1.669 200
465	0.251 100	0.073 900	1.528 100
470	0.195 360	0.090 980	1.287 640

475	0.142 100	0.112 600	1.041 900
480	0.095 640	0.139 020	0.812 950
485	0.057 950	0.169 300	0.616 200
490	0.032 010	0.208 020	0.465 180
495	0.014 700	0.258 600	0.353 300
500	0.004 900	0.323 000	0.272 000
505	0.002 400	0.407 300	0.212 300
510	0.009 300	0.503 000	0.158 200
515	0.029 100	0.608 200	0.111 700
520	0.063 270	0.710 000	0.078 250
525	0.109 600	0.793 200	0.057 250
530	0.165 500	0.862 000	0.042 160
535	0.225 750	0.914 850	0.029 840
540	0.290 400	0.954 000	0.020 300
545	0.359 700	0.980 300	0.013 400
550	0.433 450	0.994 950	0.008 750
555	0.512 050	1.000 000	0.005 750
560	0.594 500	0.995 000	0.003 900
565	0.678 400	0.978 600	0.002 750
570	0.762 100	0.952 000	0.002 100
575	0.842 500	0.915 400	0.001 800
580	0.916 300	0.870 000	0.001 650
585	0.978 600	0.816 300	0.001 400
590	1.026 300	0.757 000	0.001 100
595	1.056 700	0.694 900	0.001 000
600	1.062 200	0.631 000	0.000 800
605	1.045 600	0.566 800	0.000 600
610	1.002 600	0.503 000	0.000 340
615	0.938 400	0.441 200	0.000 240
620	0.854 450	0.381 000	0.000 190
625	0.751 400	0.321 000	0.000 100
630	0.642 400	0.265 000	0.000 050
635	0.541 900	0.217 000	0.000 030
640	0.447 900	0.175 000	0.000 020
645	0.360 800	0.138 200	0.000 010
650	0.283 500	0.107 000	0.000 000
655	0.218 700	0.081 600	0.000 000
660	0.164 900	0.061 000	0.000 000
665	0.121 200	0.044 580	0.000 000
670	0.087 400	0.032 000	0.000 000
675	0.063 600	0.023 200	0.000 000
680	0.046 770	0.017 000	0.000 000
685	0.032 900	0.011 920	0.000 000
690	0.022 700	0.008 210	0.000 000

(*Continued*)

(*Continued*)

λ (nm)	$\bar{x}$	$\bar{y}$	$\bar{z}$
695	0.015 840	0.005 723	0.000 000
700	0.011 359	0.004 102	0.000 000
705	0.008 111	0.002 929	0.000 000
710	0.005 790	0.002 091	0.000 000
715	0.004 109	0.001 484	0.000 000
720	0.002 899	0.001 047	0.000 000
725	0.002 049	0.000 740	0.000 000
730	0.001 440	0.000 520	0.000 000
735	0.001 000	0.000 361	0.000 000
740	0.000 690	0.000 249	0.000 000
745	0.000 476	0.000 172	0.000 000
750	0.000 332	0.000 120	0.000 000
755	0.000 235	0.000 085	0.000 000
760	0.000 166	0.000 060	0.000 000
765	0.000 117	0.000 042	0.000 000
770	0.000 083	0.000 030	0.000 000
775	0.000 059	0.000 021	0.000 000
780	0.000 042	0.000 015	0.000 000
785	0.000 029	0.000 011	0.000 000
790	0.000 021	0.000 007	0.000 000
795	0.000 015	0.000 005	0.000 000
800	0.000 010	0.000 000	0.000 000